W9-BIF-339

Rectangular Prism

$$V = abc \qquad S = 2ac + 2ab + 2bc$$

V	volume
S	total surface area
a	width
b	length
c	height

Right Circular Cylinder

$$V = \pi r^2 h \qquad S = 2\pi r^2 + 2\pi rh$$

V	volume
S	total surface area
r	radius
h	altitude (height)

Pyramid

$$V = \frac{1}{3} Bh$$

V	volume
B	area of base
h	altitude (height)

Right Circular Cone

$$V = \frac{1}{3} \pi r^2 h \qquad S = \pi r^2 + \pi rs$$

V	volume
S	total surface area
r	radius
h	altitude (height)
s	slant height

Simple Interest

$$I = Prt \qquad A = P + Prt$$

I	simple interest
P	principal
A	amount
r	rate of interest
t	time (in years)

Temperature

$$C = \frac{5}{9}(F - 32)$$

$$F = \frac{9}{5}C + 32$$

F	Fahrenheit degrees
C	Celsius degrees

Distance-Rate-Time

$$d = rt \qquad r = \frac{d}{t} \qquad t = \frac{d}{r}$$

d	distance
r	rate
t	time

Elementary Algebra for College Students

Elementary Algebra for College Students

Third Edition

Jerome E. Kaufmann

PWS-KENT PUBLISHING COMPANY
BOSTON

PWS-KENT
Publishing Company

20 Park Plaza
Boston, Massachusetts 02116

PWS-KENT Publishing Company is a division of Wadsworth, Inc.

Library of Congress Cataloging in Publication Data

Kaufmann, Jerome E.
    Elementary algebra for college students.

    Includes index.
    1. Algebra. I. Title.
QA152.2.K38 1989   512.9   88-32922
ISBN 0-534-91573-6

Printed in the United States of America.

    90 91 92 93 — 10 9 8 7 6 5 4

Production: Susan Graham
Manufacturing Coordinator: Ellen Glisker
Interior Design: Susan Graham
Cover Design: Julie Gecha
Interior Illustration: Vantage Art
Typesetting: Polyglot Pte Ltd
Cover Printing: John Pow Company
Printing and Binding: R. R. Donnelley & Sons

Cover Photo: Donald C. Landwe/The Image Bank

Preface

When I prepared this third edition, I attempted to preserve the features that made the previous editions so successful, and at the same time I wanted to incorporate a number of improvements suggested by reviewers.

This text was written for college students who have never had an elementary algebra course, as well as for those who have had such a course but need a review before taking additional mathematics courses. The basic concepts of algebra are presented in a simple, straightforward manner. The concepts are motivated by examples and continuously reinforced by additional examples. Algebraic ideas are developed in a logical sequence, but in an easy-to-read manner without excessive technical vocabulary and formalism.

Whenever possible the algebraic concepts have been allowed to develop from their arithmetic counterparts. The following are two specific examples of this development.

1. Manipulation with simple algebraic fractions begins early (Sections 2.1 and 2.2) when reviewing operations with rational numbers.

2. Multiplying monomials—without any of the formal vocabulary—is introduced in Section 2.4 when working with exponents.

There is a common thread throughout the book, namely: "learn a skill," then "use the skill to help solve equations," and then "use equations to solve word problems." This thread influenced some other decisions.

1. Approximately 500 word problems are scattered throughout the text. Every effort has been made to start with easy ones, thereby building confidence in solving problems. Numerous problem solving suggestions are offered with special discussions in several sections. *The key is to work with various problem solving techniques and not to become overly concerned that all of the traditional types of problems are studied.*

2. Newly acquired skills are used as soon as possible to solve equations and word problems. Therefore, the concept of solving equations is introduced early and developed throughout the text. The concepts of factoring, solving equations, and solving word problems are tied together in Chapter 6.

Approximately 700 worked-out examples demonstrate a large variety of situations while still leaving some things for the students to think about in the problem sets. I have tried to use the examples to guide students in organizing their work, and also to help them decide when a shortcut may be used. (Along these lines I have used a gradual trend from showing all steps to demonstrating a suggested shortcut format.)

As recommended by the American Mathematical Association of Two-Year Colleges, some basic geometric concepts have been integrated in a problem solving setting as follows.

Section 2.5. Linear measurement concepts

Section 3.3. Complementary and supplementary angles; sum of angles of a triangle equals 180°

Section 4.1. Area and volume formulas

Section 6.3. Pythagorean Theorem

Section 10.1. More on the Pythagorean Theorem, including work with isosceles right triangles and 30°–60° right triangles.

A focal point of this revision is the problem sets. Every set of problems has been carefully analyzed to make sure that all variations of skill development exercises are contained in both the even and odd-numbered problems. Thus, a meaningful assignment can be given from either the "evens" or the "odds" and a double dosage is available by using all of them. Another consideration taken into account is that some skills need more drill than others.

In this third edition we have added a small category of problems called **Miscellaneous Problems** to some of the problem sets. These are usually either problems designed to give the students an opportunity to see what lies ahead in intermediate algebra or problems designed to broaden their background relative to a particular topic. All of them could be omitted without breaking the continuity pattern of the text; however, I feel that they do add another flexibility feature. You may wish to turn to Problem Sets 1.2, 4.6, and 6.3 to see examples of these problems.

There is a **Chapter Summary** and a **Review Problem Set** at the end of each chapter. The summaries are keyed to the section numbers and should prove helpful in pulling ideas together. The review problem sets contain problems representative of those encountered throughout the chapter. Answers to all review problem sets are included in the back of the book. In this third edition we have added some **Cumulative Review Problem Sets**. They now appear at the ends of Chapters 2, 4, 6, 7, and 10.

Section 4.2 (Problem Solving) has been divided into two sections as requested by reviewers. This provided the opportunity to offer some additional problem solving suggestions, including an attempt to get students to appreciate the role that *estimation* can play in solving problems. Other estimation ideas have also been offered in appropriate places in the text.

Finally, in this third edition we have switched to the use of solution set vocabulary and symbolism. As stated by a reviewer, this provides a better basis for the process of solving equations and inequalities. It also prepares the students better for future courses in mathematics. We have included a number of the more "unusual" equations ($2x + 3 = 2(x + 5)$) and inequalities in the new category of **Miscellaneous Problems**.

■ Acknowledgments

I would like to take this opportunity to thank the following people who served as reviewers for this manuscript. Your comments and criticisms were extremely helpful and very much appreciated: Mary K. Alter, University of Maryland; Adam Jackson, Clark College; Judy Kasabian, El Camino College; Lawrence J. Tramer, Yavapai College.

I am grateful to the staff of PWS-KENT Publishing Company for their continuous cooperation, assistance, and encouragement throughout this project. I would also like to express my gratitude to Susan Graham. She makes my life as an author so much easier by carrying out all of the details of publication in a dedicated and caring way.

Again very special thanks are due to my wife, Arlene, who continues to spend numerous hours typing and proofreading manuscripts, answer keys, solutions manuals, and test banks.

Jerome E. Kaufmann

Contents

1 Some Basic Concepts of Arithmetic and Algebra

Let's begin our study of algebra with a suggestion: keep in mind that algebra is really a generalized approach to arithmetic. Many algebraic concepts will be developed as extensions of arithmetic ideas. Use your knowledge of arithmetic to help you with your study of algebra.

1.1 Algebraic Expressions

In arithmetic, symbols such as 4, 8, 17, and π are used to represent numbers. The basic operations of addition, subtraction, multiplication, and division are commonly indicated by the symbols $+$, $-$, $\cdot$, and $\div$, respectively. Thus, specific **numerical expressions** can be formulated. For example, the indicated sum of eight and four is written as $8 + 4$.

In algebra the concept of a **variable** provides the basis for generalizing. By using x and y to represent *any* number, the expression $x + y$ can be used to represent the indicated sum of *any two* numbers. The x and y in such an expression are called variables and the phrase $x + y$ is called an **algebraic expression**. Letters of the alphabet such as x, y, z, and w are commonly used as variables. The key idea is that they represent numbers; therefore, as we review various operations and properties pertaining to numbers, we are building the foundation for our study of algebra.

Many of the notational agreements made in arithmetic are extended to algebra with a few slight modifications. The following chart summarizes these notational agreements pertaining to the four basic operations. Notice the variety of ways to write a product by including parentheses to indicate multiplication. Actually the ab form is the simplest and probably the most used form; expressions such as abc, $6x$, and $7xyz$ all indicate multiplication. Also note the various forms for indicating division; the fractional form, $\dfrac{c}{d}$, is usually used in algebra although the other forms do serve a purpose at times.

Operation	Arithmetic	Algebra	Vocabulary
addition	$4 + 6$	$x + y$	The *sum* of x and y
subtraction	$7 - 2$	$w - z$	The *difference* of w minus z
multiplication	$9 \cdot 8$	$a \cdot b$, $a(b)$, $(a)b$, $(a)(b)$, or ab	The *product* of a and b
division	$8 \div 2$, $\dfrac{8}{2}$, or $2\overline{)8}$	$c \div d$, $\dfrac{c}{d}$, or $d\overline{)c}$	The *quotient* of c divided by d

Let's now consider simplifying some numerical expressions. The following examples illustrate several important ideas pertaining to this process. Study them carefully and be sure that you agree with each result.

EXAMPLE 1   Simplify $8 + 7 - 4 + 12 - 7 + 14$.

Solution   The additions and subtractions should be performed from left to right in the order that they appear. Thus, $8 + 7 - 4 + 12 - 7 + 14$ simplifies to 30.   ■

EXAMPLE 2 Simplify $7(9 + 5)$.

Solution The parentheses indicate the product of 7 and the quantity, $9 + 5$. Perform the addition inside the parentheses first and then multiply; $7(9 + 5)$ thus simplifies to $7(14)$, which becomes 98. ■

EXAMPLE 3 Simplify $(7 + 8) \div (4 - 1)$.

Solution First, we perform the operations inside the parentheses. $(7 + 8) \div (4 - 1)$ thus becomes $15 \div 3$, which is 5. ■

A problem such as Example 3 is frequently expressed in the form $\dfrac{7 + 8}{4 - 1}$. Parentheses are not needed in this case because the fraction bar indicates that the sum of 7 and 8 is to be divided by the difference $4 - 1$. A problem may, however, contain parentheses and fraction bars, as the next example illustrates.

EXAMPLE 4 Simplify $\dfrac{(4 + 2)(7 - 1)}{9} + \dfrac{4}{7 - 3}$.

Solution First, simplify above and below the fraction bars, and then proceed to evaluate as follows.

$$\frac{(4 + 2)(7 - 1)}{9} + \frac{4}{7 - 3} = \frac{(6)(6)}{9} + \frac{4}{4}$$

$$= \frac{36}{9} + 1 = 4 + 1 = 5. \quad ■$$

EXAMPLE 5 Simplify $7 \cdot 9 + 5$.

Solution If there are no parentheses to indicate otherwise, multiplication takes precedence over addition: first perform the multiplication and then do the addition. $7 \cdot 9 + 5$, therefore, simplifies to $63 + 5$, which is 68. ■

(Compare Example 2 and Example 5 and note the difference in meanings.)

EXAMPLE 6 Simplify $8 + 4 \cdot 3 - 14 \div 2$.

Solution The multiplication and division should be done first in the order that they appear from left to right. Thus, $8 + 4 \cdot 3 - 14 \div 2$ simplifies to $8 + 12 - 7$. The addition and subtraction are now performed in the order that they appear, which simplifies $8 + 12 - 7$ to 13. ■

EXAMPLE 7 Simplify $8 \cdot 5 \div 4 + 7 \cdot 3 - 32 \div 8 + 9 \div 3 \cdot 2$.

Solution By performing the multiplications and divisions first in the order that they appear and then doing the additions and subtractions, our work could take on the following format.

$$8 \cdot 5 \div 4 + 7 \cdot 3 - 32 \div 8 + 9 \div 3 \cdot 2 = 10 + 21 - 4 + 6 = 33.$$ ■

EXAMPLE 8 Simplify $6 + 7[3(4 + 6)]$.

Solution Brackets are used for the same purpose as parentheses. In such a problem we need to simplify *from the inside out*; perform the operations in the innermost parentheses first. We would thus obtain

$$\begin{aligned} 6 + 7[3(4 + 6)] &= 6 + 7[3(10)] \\ &= 6 + 7[30] \\ &= 6 + 210 = 216. \end{aligned}$$ ■

Let us now summarize the ideas presented in the previous examples regarding **simplifying numerical expressions**. When simplifying a numerical expression the operations should be performed in the following order.

1. Perform the operations inside the symbols of inclusion (parentheses and brackets) and above and below each fraction bar. Start with the innermost inclusion symbol.

2. Perform all multiplications and divisions in the order that they appear from left to right.

3. Perform all additions and subtractions in the order that they appear from left to right.

■ Use of Variables

The concept of a variable serves as the basis for generalizing from numerical expressions to algebraic expressions. Each of the following is an example of an algebraic expression.

$$3x + 2y; \qquad 5a - 2b + c; \qquad 7(w + z);$$
$$\frac{5d + 3e}{2c - d}; \qquad 2xy + 5yz; \qquad (x + y)(x - y).$$

An algebraic expression takes on a numerical value whenever each variable in the expression is replaced by a specific number. For example, if x is replaced by 9 and z by 4, the algebraic expression $x - z$ becomes the numerical expression $9 - 4$, which simplifies to 5. We say that $x - z$ **has a value** of 5 when x equals 9 and z equals 4. The value of $x - z$ when x equals 25 and z equals 12, is 13. The general algebraic expression $x - z$ has a specific value each time x and z are replaced by numbers.

Consider the following examples that illustrate the process of finding a value of an algebraic expression. The process is often referred to as **evaluating algebraic expressions**.

EXAMPLE 9 Find the value of $3x + 2y$ when x is replaced by 5 and y by 17.

Solution The following format is convenient for such problems.

$$3x + 2y = 3(5) + 2(17) \quad \text{when } x = 5 \text{ and } y = 17$$
$$= 15 + 34$$
$$= 49.$$

In the solution to Example 9 notice that parentheses were inserted to indicate multiplication as we switched from the algebraic expression to the numerical expression. We could also use the raised dot to indicate multiplication; that is, $3(5) + 2(17)$ could be written as $3 \cdot 5 + 2 \cdot 17$. Furthermore, notice that once we have a numerical expression, our previous agreements for simplifying numerical expressions are in effect.

EXAMPLE 10 Find the value of $12a - 3b$ when $a = 5$ and $b = 9$.

Solution
$$12a - 3b = 12(5) - 3(9) \quad \text{when } a = 5, \text{ and } b = 9$$
$$= 60 - 27$$
$$= 33.$$

EXAMPLE 11 Evaluate $4xy + 2xz - 3yz$ when $x = 8$, $y = 6$, and $z = 2$.

Solution
$$4xy + 2xz - 3yz = 4(8)(6) + 2(8)(2) - 3(6)(2) \quad \text{when } x = 8, y = 6, \text{ and } z = 2$$
$$= 192 + 32 - 36$$
$$= 188.$$

EXAMPLE 12 Evaluate $\dfrac{5c + d}{3c - d}$ for $c = 12$ and $d = 4$.

Solution
$$\frac{5c + d}{3c - d} = \frac{5(12) + 4}{3(12) - 4} \quad \text{for } c = 12 \text{ and } d = 4$$
$$= \frac{60 + 4}{36 - 4}$$
$$= \frac{64}{32} = 2.$$

EXAMPLE 13   Evaluate $(2x + 5y)(3x - 2y)$ when $x = 6$ and $y = 3$.

Solution

$$
\begin{aligned}
(2x + 5y)(3x - 2y) &= (2 \cdot 6 + 5 \cdot 3)(3 \cdot 6 - 2 \cdot 3) \qquad \text{when } x = 6, \text{ and } y = 3 \\
&= (12 + 15)(18 - 6) \\
&= (27)(12) \\
&= 324.
\end{aligned}
$$ ■

One final point needs to be made at this time. Throughout this section we have been using only **whole numbers**, that is, the set $\{0, 1, 2, 3, 4, 5, \ldots\}$. (Remember that braces are used to enclose the members (elements) of a set and the three dots denote etc.)

Problem Set 1.1

For Problems 1–30, simplify each numerical expression.

1. $9 + 14 - 7$

2. $32 - 14 + 6$

3. $7(14 - 9)$

4. $8(6 + 12)$

5. $16 + 5 \cdot 7$

6. $18 - 3(5)$

7. $4(12 + 9) - 3(8 - 4)$

8. $7(13 - 4) - 2(19 - 11)$

9. $4(7) + 6(9)$

10. $8(7) - 4(8)$

11. $6 \cdot 7 + 5 \cdot 8 - 3 \cdot 9$

12. $8(13) - 4(9) + 2(7)$

13. $(6 + 9)(8 - 4)$

14. $(15 - 6)(13 - 4)$

15. $6 + 4[3(9 - 4)]$

16. $92 - 3[2(5 - 2)]$

17. $16 \div 8 \cdot 4 + 36 \div 4 \cdot 2$

18. $7 \cdot 8 \div 4 - 72 \div 12$

19. $\dfrac{8 + 12}{4} - \dfrac{9 + 15}{8}$

20. $\dfrac{19 - 7}{6} + \dfrac{38 - 14}{3}$

21. $56 - [3(9 - 6)]$

22. $17 + 2[3(4 - 2)]$

23. $7 \cdot 4 \cdot 2 \div 8 + 14$

24. $14 \div 7 \cdot 8 - 35 \div 7 \cdot 2$

25. $32 \div 8 \cdot 2 + 24 \div 6 - 1$

26. $48 \div 12 + 7 \cdot 2 \div 2 - 1$

27. $4 \cdot 9 \div 12 + 18 \div 2 + 3$

28. $5 \cdot 8 \div 4 - 8 \div 4 \cdot 3 + 6$

29. $\dfrac{6(8 - 3)}{3} + \dfrac{12(7 - 4)}{9}$

30. $\dfrac{3(17 - 9)}{4} + \dfrac{9(16 - 7)}{3}$

31. $83 - \dfrac{4(12 - 7)}{5}$

32. $78 - \dfrac{6(21 - 9)}{4}$

33. $\dfrac{4 \cdot 6 + 5 \cdot 3}{7 + 2 \cdot 3} + \dfrac{7 \cdot 9 + 6 \cdot 5}{3 \cdot 5 + 8 \cdot 2}$

34. $\dfrac{7 \cdot 8 + 4}{5 \cdot 8 - 10} + \dfrac{9 \cdot 6 - 4}{6 \cdot 5 - 20}$

For Problems 35–54, evaluate each algebraic expression for the given values of the variables.

35. $7x + 4y$ for $x = 6$ and $y = 8$

36. $8x + 6y$ for $x = 9$ and $y = 5$

37. $16a - 9b$ for $a = 3$ and $b = 4$

38. $14a - 5b$ for $a = 7$ and $b = 9$

39. $4x + 7y + 3xy$ for $x = 4$ and $y = 9$

40. $x + 8y + 5xy$ for $x = 12$ and $y = 3$

41. $14xz + 6xy - 4yz$ for $x = 8$, $y = 5$, and $z = 7$

42. $9xy - 4xz + 3yz$ for $x = 7$, $y = 3$, and $z = 2$

43. $\dfrac{54}{n} + \dfrac{n}{3}$ for $n = 9$

44. $\dfrac{n}{4} + \dfrac{60}{n} - \dfrac{n}{6}$ for $n = 12$

45. $\dfrac{y + 16}{6} + \dfrac{50 - y}{3}$ for $y = 8$

46. $\dfrac{w + 57}{9} + \dfrac{90 - w}{7}$ for $w = 6$

47. $(x + y)(x - y)$ for $x = 8$ and $y = 3$

48. $(x + 2y)(2x - y)$ for $x = 7$ and $y = 4$

49. $(5x - 2y)(3x + 4y)$ for $x = 3$ and $y = 6$

50. $(3a + b)(7a - 2b)$ for $a = 5$ and $b = 7$

51. $6 + 3[2(x + 4)]$ for $x = 7$

52. $9 + 4[3(x + 3)]$ for $x = 6$

53. $81 - 2[5(n + 4)]$ for $n = 3$

54. $78 - 3[4(n - 2)]$ for $n = 4$

For Problems 55–60, find the value of $\dfrac{bh}{2}$ for each set of values for the variables b and h.

55. $b = 8$ and $h = 12$

56. $b = 6$ and $h = 14$

57. $b = 7$ and $h = 6$

58. $b = 9$ and $h = 4$

59. $b = 16$ and $h = 5$

60. $b = 18$ and $h = 13$

For Problems 61–66, find the value of $\dfrac{Bh}{3}$ for each set of values for the variables B and h.

61. $B = 27$ and $h = 9$

62. $B = 18$ and $h = 6$

63. $B = 25$ and $h = 12$

64. $B = 32$ and $h = 18$

65. $B = 36$ and $h = 7$

66. $B = 42$ and $h = 17$

For Problems 67–72, find the value of $\dfrac{h(b_1 + b_2)}{2}$ for each set of values for the variables h, b_1, and b_2. (Subscripts are used to indicate that b_1 and b_2 are different variables.)

67. $h = 17$, $b_1 = 14$, and $b_2 = 6$

68. $h = 9$, $b_1 = 12$, and $b_2 = 16$

69. $h = 8$, $b_1 = 17$, and $b_2 = 24$

70. $h = 12$, $b_1 = 14$, and $b_2 = 5$

71. $h = 18$, $b_1 = 6$, and $b_2 = 11$

72. $h = 14$, $b_1 = 9$, and $b_2 = 7$

1.2 Prime and Composite Numbers

We say that 6 *divides* 18 because 6 times the whole number 3 produces 18; but 6 *does not divide* 19 because there is no whole number such that 6 times the number produces 19. Likewise, 5 *divides* 35 because 5 times the whole number 7 produces 35; 5 *does not divide* 42 because there is no whole number such that 5 times the number produces 42. The following general definition is used.

Definition 1.1

Given that a and b are whole numbers, with a not equal to zero, a divides b if and only if there exists a whole number k such that $a \cdot k = b$.

Remark 1   Notice the use of variables, a, b, and k, in the statement of a *general* definition. Also note that the definition merely generalizes the concept of *divides* that was introduced in the specific examples prior to the definition.

Remark 2   This is a special use of the word *divides*. Occasionally, terms in mathematics are given a special meaning in the discussion of a particular topic.

The following statements further illustrate Definition 1.1. Pay special attention to the italicized words for they indicate some of the various terminology used for this topic.

1.   8 *divides* 56 because $8 \cdot 7 = 56$.
2.   7 *does not divide* 38 because there is no whole number, k, such that $7 \cdot k = 38$.
3.   3 is a *factor* of 27 because $3 \cdot 9 = 27$.
4.   4 is *not a factor* of 38 because there is no whole number, k, such that $4 \cdot k = 38$.
5.   35 is a *multiple* of 5 because $5 \cdot 7 = 35$.
6.   29 is *not a multiple* of 7 because there is no whole number, k, such that $7 \cdot k = 29$.

The *factor* terminology is used extensively. We say that 7 and 8 are factors of 56 because $7 \cdot 8 = 56$; 4 and 14 are also factors of 56 because $4 \cdot 14 = 56$. **Factors** are sometimes called **divisors**.

Now consider two special kinds of whole numbers called **prime numbers** and **composite numbers** according to the following definition.

Definition 1.2

A **prime number** is a whole number, greater than 1, that has no factors (divisors) other than itself and 1. Whole numbers, greater than 1, that are not prime numbers are called **composite numbers**.

The prime numbers less than 50 are 2, 3, 5, 7, 11, 13, 17, 19, 23, 29, 31, 37, 41, 43, and 47. Notice that each of these has no factors other than itself and 1. An interesting point is that the set of prime numbers is an infinite set; that is, the prime numbers go on forever; there is no *largest* prime number.

Every composite number can be expressed as the product of prime numbers. Consider the following examples:

$$4 = 2 \cdot 2; \qquad 6 = 2 \cdot 3; \qquad 8 = 2 \cdot 2 \cdot 2; \qquad 10 = 2 \cdot 5; \qquad 12 = 2 \cdot 2 \cdot 3.$$

In each case a composite number has been expressed as the product of prime numbers. The indicated product form is sometimes called the **prime factorization** form of the number.

Various procedures may be used to find the prime factors of a given composite number. For our purposes, the simplest technique is to factor the given composite number into any two easily recognized factors and then to continue to factor each of these until only prime factors are obtained. Consider these examples.

$$18 = 2 \cdot 9 = 2 \cdot 3 \cdot 3; \qquad 27 = 3 \cdot 9 = 3 \cdot 3 \cdot 3;$$
$$24 = 4 \cdot 6 = 2 \cdot 2 \cdot 2 \cdot 3; \qquad 150 = 10 \cdot 15 = 2 \cdot 5 \cdot 3 \cdot 5.$$

It does not matter which two factors are chosen first. For example, one might start by expressing 18 as $3 \cdot 6$ and then factor 6 into $2 \cdot 3$, which would produce a final result of $18 = 3 \cdot 2 \cdot 3$. Either way, 18 contains two prime factors of 3 and one prime factor of 2. The order in which the prime factors are written is not important.

■ Greatest Common Factor

The prime factorization form of two composite numbers can be used to conveniently find their **greatest common factor**. Consider the following example.

$$42 = 2 \cdot 3 \cdot 7,$$
$$70 = 2 \cdot 5 \cdot 7.$$

Notice that 2 is a factor of both, as is 7. Therefore, 14 (the product of 2 and 7) is the greatest common factor of 42 and 70. In other words, 14 is the largest whole number that divides both 42 and 70. The following examples should further clarify the process of finding the greatest common factor of two or more numbers.

EXAMPLE 1   Find the greatest common factor of 48 and 60.

Solution
$$48 = 2 \cdot 2 \cdot 2 \cdot 2 \cdot 3;$$
$$60 = 2 \cdot 2 \cdot 3 \cdot 5.$$

Since two 2s and a 3 are common to both, the greatest common factor of 48 and 60 is $2 \cdot 2 \cdot 3 = 12$.   ■

EXAMPLE 2 Find the greatest common factor of 21 and 75.

Solution $21 = 3 \cdot 7$;
               $75 = 3 \cdot 5 \cdot 5$.

Since only a 3 is common to both, the greatest common factor is 3.   ■

EXAMPLE 3 Find the greatest common factor of 24 and 35.

Solution $24 = 2 \cdot 2 \cdot 2 \cdot 3$;
               $35 = 5 \cdot 7$.

Since there are no common prime factors, the greatest common factor is 1.   ■

The concept of greatest common factor can be extended to more than two numbers as the next example illustrates.

EXAMPLE 4 Find the greatest common factor of 24, 56, and 120.

Solution $24 = 2 \cdot 2 \cdot 2 \cdot 3$;
               $56 = 2 \cdot 2 \cdot 2 \cdot 7$;
               $120 = 2 \cdot 2 \cdot 2 \cdot 3 \cdot 5$.

Since three 2s are common to the numbers, the greatest common factor of 24, 56, and 120 is $2 \cdot 2 \cdot 2 = 8$.   ■

■ Least Common Multiple

This section stated earlier that 35 is a *multiple of* 5 because $5 \cdot 7 = 35$. The set of all whole numbers that are multiples of 5 consist of 0, 5, 10, 15, 20, 25, etc. In other words, 5 times each successive whole number ($5 \cdot 0 = \mathbf{0}, 5 \cdot 1 = \mathbf{5}, 5 \cdot 2 = \mathbf{10}, 5 \cdot 3 = \mathbf{15}$, etc.) produces the multiples of 5. In a like manner, the set of multiples of 4 consists of 0, 4, 8, 12, 16, etc..

It is sometimes necessary to determine the smallest common *nonzero* multiple of two or more whole numbers. The phrase **least common multiple** is used to designate this nonzero number. For example, the least common multiple of 3 and 4 is 12, which means that 12 is the smallest nonzero multiple of both 3 and 4. Stated another way, 12 is the smallest nonzero whole number that is divisible by both 3 and 4. Likewise, we would say that the least common multiple of 6 and 8 is 24.

If we cannot determine the least common multiple by inspection, then the prime factorization form of composite numbers is helpful. Study the solutions to the following examples very carefully as we develop a systematic technique for finding the least common multiple of two or more numbers.

EXAMPLE 5 Find the least common multiple of 24 and 36.

Solution Let's first express each number as a product of prime factors.

$$24 = 2 \cdot 2 \cdot 2 \cdot 3;$$
$$36 = 2 \cdot 2 \cdot 3 \cdot 3.$$

Since 24 contains three 2s, the least common multiple must have three 2s. Also, since 36 contains two 3s, we need to put two 3s in the least common multiple. The least common multiple of 24 and 36 is therefore $2 \cdot 2 \cdot 2 \cdot 3 \cdot 3 = 72$. ■

If the least common multiple is not obvious by inspection, then we can proceed as follows.

Step 1. Express each number as a product of prime factors.

Step 2. The least common multiple contains each different prime factor as many times as the *most* times it appears in any one of the factorizations from Step 1.

EXAMPLE 6 Find the least common multiple of 48 and 84.

Solution
$$48 = 2 \cdot 2 \cdot 2 \cdot 2 \cdot 3;$$
$$84 = 2 \cdot 2 \cdot 3 \cdot 7.$$

Four 2s are needed in the least common multiple because of the four 2s in 48. One 3 is needed because of the 3 in each of the numbers and one 7 is needed because of the 7 in 84. The least common multiple of 48 and 84 is $2 \cdot 2 \cdot 2 \cdot 2 \cdot 3 \cdot 7 = 336$. ■

EXAMPLE 7 Find the least common multiple of 12, 18, and 28.

Solution
$$12 = 2 \cdot 2 \cdot 3;$$
$$18 = 2 \cdot 3 \cdot 3;$$
$$28 = 2 \cdot 2 \cdot 7.$$

The least common multiple is $2 \cdot 2 \cdot 3 \cdot 3 \cdot 7 = 252$. ■

EXAMPLE 8 Find the least common multiple of 8 and 9.

Solution
$$8 = 2 \cdot 2 \cdot 2;$$
$$9 = 3 \cdot 3.$$

The least common multiple is $2 \cdot 2 \cdot 2 \cdot 3 \cdot 3 = 72$. ■

Problem Set 1.2

For Problems 1–20, classify each statement as true or false.

1. 8 divides 56

2. 9 divides 54

3. 6 does not divide 54

4. 7 does not divide 42

5. 96 is a multiple of 8

6. 78 is a multiple of 6

7. 54 is not a multiple of 4

8. 64 is not a multiple of 6

9. 144 is divisible by 4

10. 261 is divisible by 9

11. 173 is divisible by 3

12. 149 is divisible by 7

13. 11 is a factor of 143

14. 11 is a factor of 187

15. 9 is a factor of 119

16. 8 is a factor of 98

17. 3 is a prime factor of 57

18. 7 is a prime factor of 91

19. 4 is a prime factor of 48

20. 6 is a prime factor of 72

For Problems 21–30, classify each number as prime or composite.

21. 53   **22.** 57   **23.** 59   **24.** 61   **25.** 91

26. 81   **27.** 89   **28.** 97   **29.** 111   **30.** 101

For Problems 31–42, factor each composite number into a product of prime numbers. For example, $18 = 2 \cdot 3 \cdot 3$.

31. 26   **32.** 16   **33.** 36   **34.** 80

35. 49   **36.** 92   **37.** 56   **38.** 144

39. 120   **40.** 84   **41.** 135   **42.** 98

For Problems 43–54, find the greatest common factor of the given numbers.

43. 12 and 16   **44.** 30 and 36   **45.** 56 and 64

46. 72 and 96   **47.** 63 and 81   **48.** 60 and 72

49. 84 and 96   **50.** 48 and 52   **51.** 36, 72, and 90

52. 27, 54, and 63   **53.** 48, 60, and 84   **54.** 32, 80, and 96

For Problems 55–66, find the least common multiple of the given numbers.

55. 6 and 8   **56.** 8 and 12   **57.** 12 and 16

58. 9 and 12   **59.** 28 and 35   **60.** 42 and 66

61. 49 and 56   **62.** 18 and 24   **63.** 8, 12, and 28

64. 6, 10, and 12   **65.** 9, 15, and 18   **66.** 8, 14, and 24

■ Miscellaneous Problems

67. The numbers 0, 2, 4, 6, 8, etc. are multiples of 2. They are also called *even* numbers. Why is 2 the only even prime number?

68. Find the smallest nonzero whole number that is divisible by 2, 3, 4, 5, 6, 7, and 8.

69. Find the smallest whole number, greater than 1, that produces a remainder of 1 when divided by 2, 3, 4, 5, or 6.

70. What is the greatest common factor of x and y if x and y are both prime numbers and x does not equal y? Explain your answer.

71. What is the greatest common factor of x and y if x and y are nonzero whole numbers and y is a multiple of x? Explain your answer.

72. What is the least common multiple of x and y if they are both prime numbers and x does not equal y? Explain your answer.

73. What is the least common multiple of x and y if the greatest common factor of x and y is 1? Explain your answer.

Problems 74–83. Familiarity with a few basic divisibility rules can be helpful for determining the prime factors of some numbers. For example, if you can quickly recognize that 51 is divisible by 3, then you can divide 51 by 3 to find another factor of 17. Since 3 and 17 are both prime numbers, we have $51 = 3 \cdot 17$. The divisibility rules for 2, 3, 5, and 9 are as follows.

Rule for 2: A whole number is divisible by 2 if and only if the units digit of its base ten numeral is divisible by 2. (In other words, the units digit must be 0, 2, 4, 6, or 8.)

Examples: 68 is divisible by 2 because 8 is divisible by 2.

57 is not divisible by 2 because 7 is not divisible by 2.

Rule for 3: A whole number is divisible by 3 if and only if the sum of the digits of its base ten numeral is divisible by 3.

Examples: 51 is divisible by 3 because $5 + 1 = 6$ and 6 is divisible by 3.

144 is divisible by 3 because $1 + 4 + 4 = 9$ and 9 is divisible by 3.

133 is not divisible by 3 because $1 + 3 + 3 = 7$ and 7 is not divisible by 3.

Rule for 5: A whole number is divisible by 5 if and only if the units digit of its base ten numeral is divisible by 5. (In other words, the units digit must be 0 or 5.)

Examples: 115 is divisible by 5 because 5 is divisible by 5.

172 is not divisible by 5 because 2 is not divisible by 5.

Rule for 9: A whole number is divisible by 9 if and only if the sum of the digits of its base ten numeral is divisible by 9.

Examples: 765 is divisible by 9 because $7 + 6 + 5 = 18$ and 18 is divisible by 9.

147 is not divisible by 9 because $1 + 4 + 7 = 12$ and 12 is not divisible by 9.

Use the previous divisibility rules to help determine the prime factorization of each of the following numbers.

74. 118	**75.** 76	**76.** 201	**77.** 123	**78.** 85
79. 115	**80.** 117	**81.** 441	**82.** 129	**83.** 153

1.3 Integers: Addition and Subtraction

"A record temperature of 35° *below* zero was recorded on this date in 1904." "The PO stock closed *down* 3 points yesterday." "On a first down sweep around left end, Moser *lost* 7 yards." "The Widget Manufacturing Company reported *assets* of 50 million dollars and *liabilities* of 53 million dollars for 1981." Such examples simply illustrate our need for negative numbers.

The number line becomes a helpful pictorial device for our work at this time. The set of whole numbers can be associated with evenly spaced points on a line as indicated in Figure 1.1. For each nonzero whole number we can associate its *negative* to the left of zero; with 1 we associate −1, with 2 we associate −2, etc. as indicated in Figure 1.2. The set of whole numbers along with −1, −2, −3, and so on, is called the set of **integers**. The following terminology is used with reference to the integers.

$$\ldots, -3, -2, -1, 0, 1, 2, 3, \ldots \qquad \text{integers}$$
$$1, 2, 3, 4, \ldots \qquad \text{positive integers}$$
$$0, 1, 2, 3, 4, \ldots \qquad \text{nonnegative integers}$$
$$\ldots, -3, -2, -1 \qquad \text{negative integers}$$
$$\ldots, -3, -2, -1, 0 \qquad \text{nonpositive integers}$$

Figure 1.1

Figure 1.2

The symbol, −1, can be read as *negative one*, *opposite of one*, or *additive inverse of one*. The *opposite of* and *additive inverse of* terminology is very helpful when working with variables. For example, the symbol, −x, read as *opposite of x* or *additive inverse of x* emphasizes an important issue. Since x can be any integer, −x (the opposite of x) can be zero, positive, or negative. If x is a positive integer, then −x is negative. If x is a negative integer, then −x is positive. If x is zero, then −x is zero. These statements could be written as follows.

If $x = 3$, then $-x = -(3) = -3$.
If $x = -3$, then $-x = -(-3) = 3$.
If $x = 0$, then $-x = -(0) = 0$.

From this discussion we also need to recognize the following general property.

> **Property 1.1**
>
> If a is any integer, then
>
> $$-(-a) = a.$$
>
> (The opposite of the opposite of any integer is the integer itself.)

■ Addition of Integers

The number line is also a convenient pictorial aid for interpreting **addition of integers**. Consider the following examples.

Problem	Number line interpretation	Sum
$3 + 2$		$3 + 2 = 5$
$3 + (-2)$		$3 + (-2) = 1$
$-3 + 2$		$-3 + 2 = -1$
$-3 + (-2)$		$-3 + (-2) = -5$

Once a feeling of movement on the number line is acquired, then simply forming a mental image of this movement is sufficient. Consider the following addition problems and mentally picture the number line interpretation. Be sure that you agree with all of our answers.

$$5 + (-2) = 3; \qquad -6 + 4 = -2; \qquad -8 + 11 = 3;$$
$$-7 + (-4) = -11; \qquad -5 + 9 = 4; \qquad 9 + (-2) = 7;$$
$$14 + (-17) = -3; \qquad 0 + (-4) = -4; \qquad 6 + (-6) = 0.$$

The last example illustrates a general property that should be noted: Any integer plus its opposite produces zero.

Even though all problems involving addition of integers could be done by using the number line interpretation, it is sometimes convenient to be able to give a more precise description of the addition process. For this purpose we need to briefly consider the concept of absolute value. The **absolute value** of a number is the distance between the number and zero on the number line. For example, the absolute value of 6 is 6. The absolute value of −6 is also 6. The absolute value of 0 is 0. Symbolically, absolute value is denoted with vertical bars. Thus, we write

$$|6| = 6; \qquad |-6| = 6; \qquad |0| = 0.$$

Notice that the absolute value of a positive number is the number itself, but the absolute value of a negative number is its opposite. Thus, the absolute value of any number, except 0, is positive, and the absolute value of 0 is 0.

The process of **adding integers** can be precisely described using the concept of absolute value as follows.

Two Positive Integers　　The sum of two positive integers is the sum of their absolute values. (The sum of two positive integers is a positive integer.)

$$43 + 54 = |43| + |54| = 43 + 54 = 97.$$

Two Negative Integers　　The sum of two negative integers is the opposite of the sum of their absolute values. (The sum of two negative integers is a negative integer.)

$$(-67) + (-93) = -(|-67| + |-93|)$$
$$= -(67 + 93)$$
$$= -160.$$

One Positive and One Negative Integer　　The sum of a positive and a negative integer can be found by subtracting the smaller absolute value from the larger absolute value and giving the result the sign of the original number having the larger absolute value. If the integers have the same absolute value, then their sum is 0.

$$82 + (-40) = |82| - |-40|$$
$$= 82 - 40$$
$$= 42.$$

$$74 + (-90) = -(|-90| - |74|)$$
$$= -(90 - 74)$$
$$= -16.$$

$$(-17) + 17 = |-17| - |17|$$
$$= 17 - 17$$
$$= 0.$$

Zero and Another Integer   The sum of 0 and any integer is the integer itself.

$$0 + (-46) = -46$$
$$72 + 0 = 72.$$

The following examples further illustrate the previous description of adding integers.

$$-18 + (-56) = -(|-18| + |-56|) = -(18 + 56) = -74;$$
$$-71 + (-32) = -(|-71| + |-32|) = -(71 + 32) = -103;$$
$$64 + (-49) = |64| - |-49| = 64 - 49 = 15;$$
$$-56 + 93 = |93| - |-56| = 93 - 56 = 37;$$
$$-114 + 48 = -(|-114| - |48|) = -(114 - 48) = -66;$$
$$45 + (-73) = -(|-73| - |45|) = -(73 - 45) = -28;$$
$$46 + (-46) = 0; \qquad -48 + 0 = -48;$$
$$(-73) + 73 = 0; \qquad 0 + (-81) = -81.$$

It is true that this *absolute value approach* does precisely describe the process of adding integers, but don't forget about the number line interpretation. Also included in the next problem set are other physical models for interpreting addition of integers. Some people find these models very helpful.

■ Subtraction of Integers

The following examples illustrate a relationship between addition and subtraction of *whole numbers*.

$$7 - 2 = 5 \quad \text{because } 2 + 5 = 7;$$
$$9 - 6 = 3 \quad \text{because } 6 + 3 = 9;$$
$$5 - 1 = 4 \quad \text{because } 1 + 4 = 5.$$

This same relationship between addition and subtraction holds for *all integers.*

$$5 - 6 = -1 \qquad \text{because } 6 + (-1) = 5;$$
$$-4 - 9 = -13 \quad \text{because } 9 + (-13) = -4;$$
$$-3 - (-7) = 4 \quad \text{because } -7 + 4 = -3;$$
$$8 - (-3) = 11 \qquad \text{because } -3 + 11 = 8.$$

Now consider a further observation

$$5 - 6 = -1 \quad \text{and} \quad 5 + (-6) = -1;$$
$$-4 - 9 = -13 \quad \text{and} \quad -4 + (-9) = -13;$$
$$-3 - (-7) = 4 \quad \text{and} \quad -3 + 7 = 4;$$
$$8 - (-3) = 11 \quad \text{and} \quad 8 + 3 = 11.$$

Such a discussion helps us realize that subtraction of integers can be stated in terms of addition of integers. More precisely, the following general description for subtraction of integers can be given.

Subtraction of Integers　　If a and b are integers, then $a - b = a + (-b)$.

It may be helpful for you to read $a - b = a + (-b)$ as "a minus b is equal to a plus the opposite of b." Every subtraction problem can be changed to an equivalent addition problem as illustrated by the following examples.

$$6 - 13 = 6 + (-13) = -7;$$
$$9 - (-12) = 9 + 12 = 21;$$
$$-8 - 13 = -8 + (-13) = -21;$$
$$-7 - (-8) = -7 + 8 = 1.$$

It should be apparent that addition of integers is a key operation. Being able to effectively add integers is a necessary skill for further algebraic work.

　　Let's conclude this section by evaluating some algebraic expressions using negative and positive integers.

EXAMPLE 1　Evaluate each algebraic expression for the given values of the variables.

(a) $x - y$　for $x = -12$ and $y = 20$;

(b) $-a + b$　for $a = -8$ and $b = -6$;

(c) $-x - y$　for $x = 14$ and $y = -7$.

Solution

(a) $x - y = -12 - 20$　when $x = -12$ and $y = 20$
$$= -12 + (-20)$$
$$= -32.$$

(b) $-a + b = -(-8) + (-6)$　when $a = -8$ and $b = -6$
$$= 8 + (-6)$$
$$= 2.$$

(c) $-x - y = -14 - (-7)$　when $x = 14$ and $y = -7$
$$= -14 + 7$$
$$= -7.$$
■

Problem Set 1.3

For Problems 1–10, use the number line interpretation to find each sum.

1. $5 + (-3)$　　　**2.** $7 + (-4)$　　　**3.** $-6 + 2$　　　**4.** $-9 + 4$

5. $-3 + (-4)$　　**6.** $-5 + (-6)$　　**7.** $8 + (-2)$　　**8.** $12 + (-7)$

9. $5 + (-11)$　　**10.** $4 + (-13)$

For Problems 11–30, find each sum.

11. $17 + (-9)$	**12.** $16 + (-5)$	**13.** $8 + (-19)$
14. $9 + (-14)$	**15.** $-7 + (-8)$	**16.** $-6 + (-9)$
17. $-15 + 8$	**18.** $-22 + 14$	**19.** $-13 + (-18)$
20. $-15 + (-19)$	**21.** $-27 + 8$	**22.** $-29 + 12$
23. $32 + (-23)$	**24.** $27 + (-14)$	**25.** $-25 + (-36)$
26. $-34 + (-49)$	**27.** $54 + (-72)$	**28.** $48 + (-76)$
29. $-34 + (-58)$	**30.** $-27 + (-36)$	

For Problems 31–50, subtract as indicated.

31. $3 - 8$	**32.** $5 - 11$	**33.** $-4 - 9$
34. $-7 - 8$	**35.** $5 - (-7)$	**36.** $9 - (-4)$
37. $-6 - (-12)$	**38.** $-7 - (-15)$	**39.** $-11 - (-10)$
40. $-14 - (-19)$	**41.** $-18 - 27$	**42.** $-16 - 25$
43. $34 - 63$	**44.** $25 - 58$	**45.** $45 - 18$
46. $52 - 38$	**47.** $-21 - 44$	**48.** $-26 - 54$
49. $-53 - (-24)$	**50.** $-76 - (-39)$	

For Problems 51–66, add or subtract as indicated.

51. $6 - 8 - 9$	**52.** $5 - 9 - 4$
53. $-4 - (-6) + 5 - 8$	**54.** $-3 - 8 + 9 - (-6)$
55. $5 + 7 - 8 - 12$	**56.** $-7 + 9 - 4 - 12$
57. $-6 - 4 - (-2) + (-5)$	**58.** $-8 - 11 - (-6) + (-4)$
59. $-6 - 5 - 9 - 8 - 7$	**60.** $-4 - 3 - 7 - 8 - 6$
61. $7 - 12 + 14 - 15 - 9$	**62.** $8 - 13 + 17 - 15 - 19$
63. $-11 - (-14) + (-17) - 18$	**64.** $-15 + 20 - 14 - 18 + 9$
65. $16 - 21 + (-15) - (-22)$	**66.** $17 - 23 - 14 - (-18)$

The horizontal format is used extensively in algebra, but occasionally the vertical format shows up. Some exposure to the vertical format is therefore needed. Find the following *sums* for Problems 67–78.

67. $\begin{array}{r} 5 \\ -9 \\ \hline \end{array}$	**68.** $\begin{array}{r} 8 \\ -13 \\ \hline \end{array}$	**69.** $\begin{array}{r} -13 \\ -18 \\ \hline \end{array}$	**70.** $\begin{array}{r} -14 \\ -28 \\ \hline \end{array}$
71. $\begin{array}{r} -18 \\ 9 \\ \hline \end{array}$	**72.** $\begin{array}{r} -17 \\ 9 \\ \hline \end{array}$	**73.** $\begin{array}{r} -21 \\ 39 \\ \hline \end{array}$	**74.** $\begin{array}{r} -15 \\ 32 \\ \hline \end{array}$
75. $\begin{array}{r} 27 \\ -19 \\ \hline \end{array}$	**76.** $\begin{array}{r} 31 \\ -18 \\ \hline \end{array}$	**77.** $\begin{array}{r} -53 \\ 24 \\ \hline \end{array}$	**78.** $\begin{array}{r} 47 \\ -28 \\ \hline \end{array}$

For Problems 79–90, do the *subtraction* problems in vertical format.

79. $\begin{array}{r} 5 \\ 12 \\ \hline \end{array}$	**80.** $\begin{array}{r} 8 \\ 19 \\ \hline \end{array}$	**81.** $\begin{array}{r} 6 \\ -9 \\ \hline \end{array}$	**82.** $\begin{array}{r} 13 \\ -7 \\ \hline \end{array}$

83. $\begin{array}{r} -7 \\ \underline{-8} \end{array}$	**84.** $\begin{array}{r} -6 \\ \underline{-5} \end{array}$	**85.** $\begin{array}{r} 17 \\ \underline{-19} \end{array}$	**86.** $\begin{array}{r} 18 \\ \underline{-14} \end{array}$
87. $\begin{array}{r} -23 \\ \underline{16} \end{array}$	**88.** $\begin{array}{r} -27 \\ \underline{15} \end{array}$	**89.** $\begin{array}{r} -12 \\ \underline{12} \end{array}$	**90.** $\begin{array}{r} -13 \\ \underline{-13} \end{array}$

For Problems 91–100 evaluate each algebraic expression for the given values of the variables.

91. $x - y$ for $x = -6$ and $y = -13$

92. $-x - y$ for $x = -7$ and $y = -9$

93. $-x + y - z$ for $x = 3$, $y = -4$, and $z = -6$

94. $x - y + z$ for $x = 5$, $y = 6$, and $z = -9$

95. $-x - y - z$ for $x = -2$, $y = 3$, and $z = -11$

96. $-x - y + z$ for $x = -8$, $y = -7$, and $z = -14$

97. $-x + y + z$ for $x = -11$, $y = 7$, and $z = -9$

98. $-x - y - z$ for $x = 12$, $y = -6$, and $z = -14$

99. $x - y - z$ for $x = -15$, $y = 12$, and $z = -10$

100. $x + y - z$ for $x = -18$, $y = 13$, and $z = 8$

■ Miscellaneous Problems

A game such as football can also be used to interpret addition of integers. A gain of 3 yards on one play followed by a loss of 5 yards on the next play places the ball 2 yards *behind* the initial line of scrimmage and this could be expressed as $3 + (-5) = -2$. Use this "football interpretation" to find the following sums (Problems 101–110).

101. $4 + (-7)$	**102.** $3 + (-5)$	**103.** $-4 + (-6)$	**104.** $-2 + (-5)$
105. $-5 + 2$	**106.** $-10 + 6$	**107.** $-4 + 15$	**108.** $-3 + 22$
109. $-12 + 17$	**110.** $-9 + 21$		

Profits and losses pertaining to investments also provide a good physical model for interpreting addition of integers. A loss of \$25 on one investment along with a profit of \$60 on a second investment produces an overall profit of \$35. This could be written as $-25 + 60 = 35$. Use this *profit and loss* interpretation to help find the following sums (Problems 111–120).

111. $60 + (-125)$	**112.** $50 + (-85)$	**113.** $-55 + (-45)$
114. $-120 + (-220)$	**115.** $-70 + 45$	**116.** $-125 + 45$
117. $-120 + 250$	**118.** $-75 + 165$	**119.** $145 + (-65)$
120. $275 + (-195)$		

121. To find the sum of 65, 68, 71, 73, and 69 we might consider the deviation of each of these numbers from some other number, say 70. That is, consider $-5 + (-2) + 1 + 3 + (-1) = -4$. Thus, the sum of the original numbers is $(5 \cdot 70) + (-4) = 346$. Use this procedure to find the following sums.

(**a**) $84 + 78 + 82 + 75$

(**b**) $136 + 145 + 142 + 137 + 132$

(**c**) $56 + 63 + 61 + 58 + 52 + 67$

(**d**) $119 + 123 + 121 + 116 + 114 + 128$

1.4   Integers: Multiplication and Division

Multiplication of whole numbers may be interpreted as repeated addition. For example, $3 \cdot 4$ means the sum of three 4s; thus, $3 \cdot 4 = 4 + 4 + 4 = 12$. Consider the following examples that use the repeated addition idea to find the product of a positive integer and a negative integer.

$$3(-2) = -2 + (-2) + (-2) = -6;$$
$$2(-4) = -4 + (-4) = -8;$$
$$4(-1) = -1 + (-1) + (-1) + (-1) = -4.$$

Note the use of parentheses to indicate multiplication. Sometimes both numbers are enclosed in parentheses so that we have $(3)(-2)$.

When multiplying whole numbers we realize that the order in which we multiply two factors does not change the product; $2(3) = 6$ and $3(2) = 6$. Using this idea we can now handle a negative integer times a positive integer as follows.

$$(-2)(3) = (3)(-2) = (-2) + (-2) + (-2) = -6;$$
$$(-3)(2) = (2)(-3) = (-3) + (-3) = -6;$$
$$(-4)(3) = (3)(-4) = (-4) + (-4) + (-4) = -12.$$

Finally, let's consider the product of two negative integers. The following pattern helps us with the reasoning for this situation.

$$4(-3) = -12;$$
$$3(-3) = -9;$$
$$2(-3) = -6;$$
$$1(-3) = -3;$$
$$0(-3) = 0;   \qquad \text{The product of zero and any integer is zero.}$$
$$(-1)(-3) = ?$$

Certainly, to continue this pattern, the product of -1 and -3 has to be 3. In general, this type of reasoning would help us to realize that the product of any two negative integers is a positive integer.

Using the concept of absolute value it is now quite easy to precisely describe the **multiplication of integers**.

1. The product of two positive integers or two negative integers is the product of their absolute values.

2. The product of a positive and a negative integer (either order) is the opposite of the product of their absolute values.

3. The product of zero and any integer is zero.

The following examples illustrate this description of multiplication.

$$(-5)(-2) = |-5| \cdot |-2| = 5 \cdot 2 = 10;$$
$$(7)(-6) = -(|7| \cdot |-6|) = -(7 \cdot 6) = -42;$$
$$(-8)(9) = -(|-8| \cdot |9|) = -(8 \cdot 9) = -72;$$
$$(-14)(0) = 0;$$
$$(0)(-28) = 0.$$

The previous examples showed a step-by-step process for multiplying integers. In reality, however, the key issue is to remember whether the product is positive or negative. In other words, we need to remember that *the product of two positives or two negatives is positive and the product of a positive and a negative (either order) is negative*. Then we can avoid the step-by-step analysis and simply write the results as follows.

$$(7)(-9) = -63; \qquad (8)(7) = 56;$$
$$(-5)(-6) = 30; \qquad (-4)(12) = -48.$$

■ Division of Integers

By looking back at our knowledge of whole numbers we again get some guidance for our work with integers. We know, for example, that $\frac{8}{2} = 4$ because $2 \cdot 4 = 8$. In other words, the quotient of two whole numbers can be found by looking at a related multiplication problem. In the following examples we have used this same link between multiplication and division to determine the quotients.

$$\frac{8}{-2} = -4 \quad \text{because } (-2)(-4) = 8;$$

$$\frac{-10}{5} = -2 \quad \text{because } (5)(-2) = -10;$$

$$\frac{-12}{-4} = 3 \quad \text{because } (-4)(3) = -12;$$

$$\frac{0}{-6} = 0 \quad \text{because } (-6)(0) = 0;$$

$$\frac{-9}{0} \quad \text{is undefined because no number times zero produces } -9;$$

$$\frac{0}{0} \quad \text{is undefined because any number times zero produces zero.} \text{ Remember that}$$
division by zero is undefined!

A precise description for **division of integers** is as follows.

  1.   The quotient of two positive or two negative integers is the quotient of their absolute values.

2. The quotient of a positive integer and a negative integer (or a negative and a positive) is the opposite of the quotient of their absolute values.

3. The quotient of zero and any nonzero integer (zero divided by any nonzero integer) is zero.

The following examples illustrate this description of division.

$$\frac{-8}{-4} = \frac{|-8|}{|-4|} = \frac{8}{4} = 2;$$

$$\frac{-14}{2} = -\left(\frac{|-14|}{|2|}\right) = -\left(\frac{14}{2}\right) = -7;$$

$$\frac{15}{-3} = -\left(\frac{|15|}{|-3|}\right) = -\left(\frac{15}{3}\right) = -5;$$

$$\frac{0}{-4} = 0.$$

For practical purposes the key idea is to remember whether the quotient is positive or negative. We need to remember that *the quotient of two positives or two negatives is positive and the quotient of a positive and a negative or a negative and a positive is negative.* We then can simply write the quotients without showing all of the steps as follows.

$$\frac{-18}{-6} = 3; \qquad \frac{-24}{12} = -2; \qquad \frac{36}{-9} = -4.$$

■ Evaluating Algebraic Expressions

Evaluating algebraic expressions will often involve the use of two or more operations with integers. The final examples of this section illustrate such situations.

EXAMPLE 1 Find the value of $3x + 2y$ when $x = 5$ and $y = -9$.

Solution

$$\begin{aligned}
3x + 2y &= 3(5) + 2(-9) \quad \text{when } x = 5 \text{ and } y = -9 \\
&= 15 + (-18) \\
&= -3.
\end{aligned}$$ ■

EXAMPLE 2 Evaluate $-2a + 9b$ for $a = 4$ and $b = -3$.

Solution

$$\begin{aligned}
-2a + 9b &= -2(4) + 9(-3) \quad \text{when } a = 4 \text{ and } b = -3 \\
&= -8 + (-27) \\
&= -35.
\end{aligned}$$ ■

EXAMPLE 3 Find the value of $\dfrac{x - 2y}{4}$ when $x = -6$ and $y = 5$.

Solution

$$\frac{x - 2y}{4} = \frac{-6 - 2(5)}{4} \quad \text{when } x = -6 \text{ and } y = 5$$

$$= \frac{-6 - 10}{4}$$

$$= \frac{-16}{4}$$

$$= -4 \qquad ■$$

Problem Set 1.4

For Problems 1–40, find the product or quotient (multiply or divide) as indicated.

1. $5(-6)$ **2.** $7(-9)$ **3.** $\dfrac{-27}{3}$ **4.** $\dfrac{-35}{5}$

5. $\dfrac{-42}{-6}$ **6.** $\dfrac{-72}{-8}$ **7.** $(-7)(8)$ **8.** $(-6)(9)$

9. $(-5)(-12)$ **10.** $(-7)(-14)$ **11.** $\dfrac{96}{-8}$ **12.** $\dfrac{-91}{7}$

13. $14(-9)$ **14.** $17(-7)$ **15.** $(-11)(-14)$ **16.** $(-13)(-17)$

17. $\dfrac{135}{-15}$ **18.** $\dfrac{-144}{12}$ **19.** $\dfrac{-121}{-11}$ **20.** $\dfrac{-169}{-13}$

21. $(-15)(-15)$ **22.** $(-18)(-18)$ **23.** $\dfrac{112}{-8}$ **24.** $\dfrac{112}{-7}$

25. $\dfrac{0}{-8}$ **26.** $\dfrac{-8}{0}$ **27.** $\dfrac{-138}{-6}$ **28.** $\dfrac{-105}{-5}$

29. $\dfrac{76}{-4}$ **30.** $\dfrac{-114}{6}$ **31.** $(-6)(-15)$ **32.** $\dfrac{0}{-14}$

33. $(-56) \div (-4)$ **34.** $(-78) \div (-6)$ **35.** $(-19) \div 0$ **36.** $(-90) \div 15$

37. $(-72) \div 18$ **38.** $(-70) \div 5$ **39.** $(-36)(27)$ **40.** $(42)(-29)$

For Problems 41–60, simplify each numerical expression.

41. $3(-4) + 5(-7)$ **42.** $6(-3) + 5(-9)$

43. $7(-2) - 4(-8)$ **44.** $9(-3) - 8(-6)$

45. $(-3)(-8) + (-9)(-5)$ **46.** $(-7)(-6) + (-4)(-3)$

47. $5(-6) - 4(-7) + 3(2)$ **48.** $7(-4) - 8(-7) + 5(-8)$

49. $\dfrac{13 + (-25)}{-3}$ **50.** $\dfrac{15 + (-36)}{-7}$

51. $\dfrac{12 - 48}{6}$

52. $\dfrac{16 - 40}{8}$

53. $\dfrac{-7(10) + 6(-9)}{-4}$

54. $\dfrac{-6(8) + 4(-14)}{-8}$

55. $\dfrac{4(-7) - 8(-9)}{11}$

56. $\dfrac{5(-9) - 6(-7)}{3}$

57. $-2(3) - 3(-4) + 4(-5) - 6(-7)$

58. $2(-4) + 4(-5) - 7(-6) - 3(9)$

59. $-1(-6) - 4 + 6(-2) - 7(-3) - 18$

60. $-9(-2) + 16 - 4(-7) - 12 + 3(-8)$

For Problems 61–76, evaluate each algebraic expression for the given values of the variables.

61. $7x + 5y$ for $x = -5$ and $y = 9$

62. $4a + 6b$ for $a = -6$ and $b = -8$

63. $9a - 2b$ for $a = -5$ and $b = 7$

64. $8a - 3b$ for $a = -7$ and $b = 9$

65. $-6x - 7y$ for $x = -4$ and $y = -6$

66. $-5x - 12y$ for $x = -5$ and $y = -7$

67. $\dfrac{5x - 3y}{-6}$ for $x = -6$ and $y = 4$

68. $\dfrac{-7x + 4y}{-8}$ for $x = 8$ and $y = 6$

69. $3(2a - 5b)$ for $a = -1$ and $b = -5$

70. $4(3a - 7b)$ for $a = -2$ and $b = -4$

71. $-2x + 6y - xy$ for $x = 7$ and $y = -7$

72. $-3x + 7y - 2xy$ for $x = -6$ and $y = 4$

73. $-4ab - b$ for $a = 2$ and $b = -14$

74. $-5ab + b$ for $a = -1$ and $b = -13$

75. $(ab + c)(b - c)$ for $a = -2$, $b = -3$, and $c = 4$

76. $(ab - c)(a + c)$ for $a = -3$, $b = 2$, and $c = 5$

For Problems 77–82, find the value of $\dfrac{5(F - 32)}{9}$ **for each of the given values for F.**

77. $F = 59$ **78.** $F = 68$ **79.** $F = 14$ **80.** $F = -4$

81. $F = -13$ **82.** $F = -22$

For Problems 83–88, find the value of $\dfrac{9C}{5} + 32$ **for each of the given values for C.**

83. $C = 25$ **84.** $C = 35$ **85.** $C = 40$ **86.** $C = 0$

87. $C = -10$ **88.** $C = -30$

1.5 Use of Properties

We will begin this section by listing and briefly commenting on some of the basic properties of integers. We will then show how these properties facilitate manipulation with integers and also serve as a basis for some algebraic computation.

Commutative Property of Addition

If a and b are integers, then

$$a + b = b + a.$$

Commutative Property of Multiplication

If a and b are integers, then

$$ab = ba.$$

Addition and multiplication are said to be commutative operations. This means that the order in which you add or multiply two integers does not affect the result. For example, $3 + 5 = 5 + 3$ and $7(8) = 8(7)$. It is also important to realize that subtraction and division *are not* commutative operations; order does make a difference. For example, $8 - 7 \neq 7 - 8$ and $16 \div 4 \neq 4 \div 16$.

Associative Property of Addition

If a, b, and c are integers, then

$$(a + b) + c = a + (b + c).$$

Associative Property of Multiplication

If a, b, and c are integers, then

$$(ab)c = a(bc).$$

Addition and multiplication are associative operations. The associative properties can be thought of as grouping properties. For example $(-8 + 3) + 9 = -8 + (3 + 9)$. Changing the grouping of the numbers does not affect the final result. This is also true for multiplication as $[(-6)(5)](-4) = (-6)[(5)(-4)]$ illustrates. Subtraction and division *are not* associative operations. For example, $(8 - 4) - 7 = -3$ whereas $8 - (4 - 7) = 11$. An example showing that division is not associative is $(8 \div 4) \div 2 = 1$ whereas $8 \div (4 \div 2) = 4$.

Identity Property of Addition

If a is an integer, then

$$a + 0 = 0 + a = a.$$

Zero is referred to as the identity element for addition. This simply means that the sum of any integer and zero is exactly the same integer. For example, $-197 + 0 = 0 + (-197) = -197$.

Identity Property of Multiplication

If a is an integer, then

$$a(1) = 1(a) = a.$$

One is called the identity element for multiplication. The product of any integer and one is exactly the same integer. For example, $(-573)(1) = (1)(-573) = -573$.

Additive Inverse Property

For every integer a, there exists an integer $-a$, such that

$$a + (-a) = (-a) + a = 0.$$

The integer, $-a$, is called the additive inverse of a or the opposite of a. Thus, 6 and -6 are additive inverses and their sum is 0. The additive inverse of 0 is 0.

Multiplication Property of Zero

If a is an integer, then

$$(a)(0) = (0)(a) = 0.$$

The product of zero and any integer is zero. For example, $(-873)(0) = (0)(-873) = 0$.

Multiplicative Property of Negative One

If a is an integer, then

$$(a)(-1) = (-1)(a) = -a.$$

The product of any integer and -1 is the opposite of the integer. For example, $(-1)(48) = (48)(-1) = -48$.

Distributive Property

If a, b, and c are integers, then

$$a(b + c) = ab + ac.$$

The distributive property involves both addition and multiplication. We say that **multiplication distributes over addition**. For example, $3(4 + 7) = 3(4) + 3(7)$. Since $b - c = b + (-c)$, it follows that **multiplication also distributes over subtraction**. This could be stated as $a(b - c) = ab - ac$. For example, $7(8 - 2) = 7(8) - 7(2)$.

Let's now consider some examples that illustrate the use of the properties to help with certain types of manipulations.

EXAMPLE 1   Find the sum $(43 + (-24)) + 24$.

Solution   In such a problem it is much more advantageous to group -24 and 24. Thus,

$$(43 + (-24)) + 24 = 43 + ((-24) + 24)$$ 　　Associative property
$$= 43 + 0.$$ 　　for addition.
$$= 43.$$ ■

EXAMPLE 2   Find the product $[(-17)(25)](4)$.

Solution   In this problem it would be easier to group 25 and 4. Thus,

$$[(-17)(25)](4) = (-17)[(25)(4)]$$ 　　Associative property for multiplication.
$$= (-17)(100)$$
$$= -1700.$$ ■

EXAMPLE 3   Find the sum $17 + (-24) + (-31) + 19 + (-14) + 29 + 43$.

Solution   Certainly one could add in the order that the numbers appear. However, since addition is *commutative* and *associative* we could change the order and group any convenient way. For example, we could add all of the positive integers and add all of the negative integers and then add these two results. It might be convenient to use the vertical format as follows.

17		
19	−24	
29	−31	108
43	−14	−69
108	−69	39

■

For a problem such as Example 3 it might be advisable to first work out the problem by adding in the order that the numbers appear and then to use the rearranging and regrouping idea as a check. Don't forget the link between addition and subtraction. A problem such as $18 - 43 + 52 - 17 - 23$ can be changed to $18 + (-43) + 52 + (-17) + (-23)$.

EXAMPLE 4   Simplify $(-75)(-4 + 100)$.

Solution   For such a problem it might be convenient to apply the *distributive property* and then to simplify.

$$
\begin{aligned}
(-75)(-4 + 100) &= (-75)(-4) + (-75)(100) \\
&= 300 + (-7500) \\
&= -7200.
\end{aligned}
$$ ■

EXAMPLE 5   Simplify $19(-26 + 25)$.

Solution   For this problem we are better off *not* to apply the distributive property, but simply to add the numbers inside the parentheses first and then to find the indicated product. Thus,

$$19(-26 + 25) = 19(-1) = -19.$$ ■

EXAMPLE 6   Simplify $27(104) + 27(-4)$.

Solution   Keep in mind that the *distributive property* allows us to change from the form $a(b + c)$ to $ab + ac$ or from $ab + ac$ to $a(b + c)$. In this problem we want to use the latter change. Thus,

$$
\begin{aligned}
27(104) + 27(-4) &= 27(104 + (-4)) \\
&= 27(100) = 2700.
\end{aligned}
$$ ■

Examples 4, 5, and 6 illustrate an important issue. Sometimes the form $a(b + c)$ is most convenient, but at other times the form $ab + ac$ is better. A suggestion in regard to this issue, as well as to the use of the other properties, is *think first*, and decide whether or not the properties can be used to make the manipulations easier.

■ Combining Similar Terms

Algebraic expressions such as

$$3x \qquad 5y \qquad 7xy \qquad -4abc \qquad z$$

are called **terms**. A term is an indicated product and may have any number of factors. The variables involved in a term are called **literal factors** and the numerical factor is called the **numerical coefficient**. Thus, in $7xy$, the x and y are literal factors and 7 is the numerical coefficient. The numerical coefficient of the term, $-4abc$, is -4. Since $z = 1(z)$, the numerical coefficient of the term, z, is 1.

Terms having the same literal factors are called **like terms** or **similar terms**. Some examples of similar terms are

$$3x \quad \text{and} \quad 9x; \qquad 14abc \quad \text{and} \quad 29abc;$$
$$7xy \quad \text{and} \quad -15xy; \qquad 4z, 9z, \quad \text{and} \quad -14z.$$

A form of the distributive property serves as the basis for simplifying algebraic expressions containing similar terms. Consider the following examples.

$$3x + 5x = (3 + 5)x$$
$$= 8x.$$

$$-9xy + 7xy = (-9 + 7)xy$$
$$= -2xy.$$

$$18abc - 27abc = (18 - 27)abc = (18 + (-27))abc$$
$$= -9abc.$$

$$4x + x = (4 + 1)x \qquad \text{Don't forget that } x = 1(x).$$
$$= 5x.$$

More complicated expressions might first require some rearranging of terms by using the commutative property.

$$7x + 3y + 9x + 5y = 7x + 9x + 3y + 5y$$
$$= (7 + 9)x + (3 + 5)y$$
$$= 16x + 8y.$$

$$9a - 4 - 13a + 6 = 9a + (-4) + (-13a) + 6$$
$$= 9a + (-13a) + (-4) + 6$$
$$= (9 + (-13))a + 2$$
$$= -4a + 2.$$

As you become more adept at handling the various simplifying steps, you may want to do the steps mentally and thereby go directly from the given expression to the simplified form as follows.

$$19x - 14y + 12x + 16y = 31x + 2y;$$
$$17ab + 13c - 19ab - 30c = -2ab - 17c;$$
$$9x + 5 - 11x + 4 + x - 6 = -x + 3.$$

Simplifying some algebraic expressions requires repeated applications of the distributive property as the next examples illustrate.

$$5(x - 2) + 3(x + 4) = 5(x) - 5(2) + 3(x) + 3(4)$$
$$= 5x - 10 + 3x + 12$$
$$= 5x + 3x - 10 + 12$$
$$= 8x + 2.$$

$$-7(y + 1) - 4(y - 3) = -7(y) - 7(1) - 4(y) - 4(-3)$$
$$= -7y - 7 - 4y + 12 \qquad \text{Be careful with this sign.}$$
$$= -7y - 4y - 7 + 12$$
$$= -11y + 5.$$

$$5(x + 2) - (x + 3) = 5(x + 2) - 1(x + 3) \qquad \text{Remember } -a = -1a.$$
$$= 5(x) + 5(2) - 1(x) - 1(3)$$
$$= 5x + 10 - x - 3$$
$$= 5x - x + 10 - 3$$
$$= 4x + 7.$$

After you are sure of each step, a more simplified format may be used.

$$5(a + 4) - 7(a - 2) = 5a + 20 - 7a + 14$$
$$= -2a + 34.$$

$$9(z - 7) + 11(z + 6) = 9z - 63 + 11z + 66$$
$$= 20z + 3.$$

$$-(x - 2) + (x + 6) = -x + 2 + x + 6$$
$$= 8.$$

To simplify by combining similar terms aids in the process of evaluating some algebraic expressions. The last examples of this section illustrate this idea.

EXAMPLE 7 Evaluate $8x - 2y + 3x + 5y$ for $x = 3$ and $y = -4$.

Solution Let's first simplify the given expression.

$$8x - 2y + 3x + 5y = 11x + 3y$$

Now, we can evaluate for $x = 3$ and $y = -4$.

$$11x + 3y = 11(3) + 3(-4)$$
$$= 33 + (-12) = 21. \qquad ■$$

EXAMPLE 8   Evaluate $2ab + 5c - 6ab + 12c$ for $a = 2$, $b = -3$, and $c = 7$.

Solution

$$2ab + 5c - 6ab + 12c = -4ab + 17c$$
$$= -4(2)(-3) + 17(7) \qquad \text{when } a = 2, b = -3,$$
$$= 24 + 119 \qquad\qquad \text{and } c = 7$$
$$= 143.$$ ■

EXAMPLE 9   Evaluate $8(x - 4) + 7(x + 3)$ for $x = 6$.

Solution

$$8(x - 4) + 7(x + 3) = 8x - 32 + 7x + 21$$
$$= 15x - 11$$
$$= 15(6) - 11 \quad \text{when } x = 6$$
$$= 79.$$ ■

Problem Set 1.5

For Problems 1–12, state the property that justifies each statement. For example, $3 + (-4) = (-4) + 3$ because of the commutative property for addition.

1. $3(7 + 8) = 3(7) + 3(8)$

2. $(-9)(17) = 17(-9)$

3. $-2 + (5 + 7) = (-2 + 5) + 7$

4. $-19 + 0 = -19$

5. $143(-7) = -7(143)$

6. $5(9 + (-4)) = 5(9) + 5(-4)$

7. $-119 + 119 = 0$

8. $-4 + (6 + 9) = (-4 + 6) + 9$

9. $-56 + 0 = -56$

10. $5 + (-12) = -12 + 5$

11. $[5(-8)]4 = 5[-8(4)]$

12. $[6(-4)]8 = 6[-4(8)]$

For Problems 13–30, simplify each numerical expression. Don't forget to take advantage of the properties if they can be used to simplify the computation.

13. $(-18 + 56) + 18$

14. $-72 + [72 + (-14)]$

15. $36 - 48 - 22 + 41$

16. $-24 + 18 + 19 - 30$

17. $(25)(-18)(-4)$

18. $(2)(-71)(50)$

19. $(4)(-16)(-9)(-25)$

20. $(-2)(18)(-12)(-5)$

21. $37(-42 - 58)$

22. $-46(-73 - 27)$

23. $59(36) + 59(64)$

24. $-49(72) - 49(28)$

25. $15(-14) + 16(-8)$

26. $-9(14) - 7(-16)$

27. $17 + (-18) - 19 - 14 + 13 - 17$

28. $-16 - 14 + 18 + 21 + 14 - 17$

29. $-21 + 22 - 23 + 27 + 21 - 19$

30. $24 - 26 - 29 + 26 + 18 + 29 - 17 - 10$

For Problems 31–62, simplify each algebraic expression by combining similar terms.

31. $9x - 14x$

32. $12x - 14x + x$

33. $4m + m - 8m$

34. $-6m - m + 17m$

35. $-9y + 5y - 7y$

36. $14y - 17y - 19y$

37. $4x - 3y - 7x + y$

38. $9x + 5y - 4x - 8y$

39. $-7a - 7b - 9a + 3b$

40. $-12a + 14b - 3a - 9b$ **41.** $6xy - x - 13xy + 4x$ **42.** $-7xy - 2x - xy + x$

43. $5x - 4 + 7x - 2x + 9$ **44.** $8x + 9 + 14x - 3x - 14$

45. $-2xy + 12 + 8xy - 16$ **46.** $14xy - 7 - 19xy - 6$

47. $-2a + 3b - 7b - b + 5a - 9a$ **48.** $-9a - a + 6b - 3a - 4b - b + a$

49. $13ab + 2a - 7a - 9ab + ab - 6a$ **50.** $-ab - a + 4ab + 7ab - 3a - 11ab$

51. $3(x + 2) + 5(x + 6)$ **52.** $7(x + 8) + 9(x + 1)$

53. $5(x - 4) + 6(x + 8)$ **54.** $-3(x + 2) - 4(x - 10)$

55. $9(x + 4) - (x - 8)$ **56.** $-(x - 6) + 5(x - 9)$

57. $3(a - 1) - 2(a - 6) + 4(a + 5)$ **58.** $-4(a + 2) + 6(a + 8) - 3(a - 6)$

59. $-2(m + 3) - 3(m - 1) + 8(m + 4)$ **60.** $5(m - 10) + 6(m - 11) - 9(m - 12)$

61. $(y + 3) - (y - 2) - (y + 6) - 7(y - 1)$ **62.** $-(y - 2) - (y + 4) - (y + 7) - 2(y + 3)$

For Problems 63–80, simplify each algebraic expression and then evaluate the resulting expression for the given values for the variables.

63. $3x + 5y + 4x - 2y$ for $x = -2$ and $y = 3$

64. $5x - 7y - 9x - 3y$ for $x = -1$ and $y = -4$

65. $5(x - 2) + 8(x + 6)$ for $x = -6$

66. $4(x - 6) + 9(x + 2)$ for $x = 7$

67. $8(x + 4) - 10(x - 3)$ for $x = -5$

68. $-(n + 2) - 3(n - 6)$ for $n = 10$

69. $(x - 6) - (x + 12)$ for $x = -3$

70. $(x + 12) - (x - 14)$ for $x = -11$

71. $2(x + y) - 3(x - y)$ for $x = -2$ and $y = 7$

72. $5(x - y) - 9(x + y)$ for $x = 4$ and $y = -4$

73. $2xy + 6 + 7xy - 8$ for $x = 2$ and $y = -4$

74. $4xy - 5 - 8xy + 9$ for $x = -3$ and $y = -3$

75. $5x - 9xy + 3x + 2xy$ for $x = 12$ and $y = -1$

76. $-9x + xy - 4xy - x$ for $x = 10$ and $y = -11$

77. $(a - b) - (a + b)$ for $a = 19$ and $b = -17$

78. $(a + b) - (a - b)$ for $a = -16$ and $b = 14$

79. $-3x + 7x + 4x - 2x - x$ for $x = -13$

80. $5x - 6x + x - 7x - x - 2x$ for $x = -15$

Chapter 1 Summary

(1.1) To **simplify a numerical expression**, the operations should be performed in the following order.

1. Perform the operations inside the symbols of inclusion (parentheses and brackets) and above and below each fraction bar. Start with the innermost inclusion symbol.

2. Perform all multiplications and divisions in the order that they appear from left to right.

3. Perform all additions and subtractions in the order that they appear from left to right.

To **evaluate an algebraic expression**, substitute the given values for the variables into the algebraic expression and simplify the resulting numerical expression.

(1.2)   A **prime number** is a whole number, greater than 1, that has no factors (divisors) other than itself and 1. Whole numbers, greater than 1, that are not prime numbers are called **composite numbers**. Every composite number has one and only one prime factorization.

The **greatest common factor** of 6 and 8 is 2, which means that 2 is the largest whole number divisor of both 6 and 8.

The **least common multiple** of 6 and 8 is 24, which means that 24 is the smallest nonzero multiple of both 6 and 8.

(1.3)   The number line is a convenient pictorial aid for interpreting **addition of integers**.

Subtraction of integers is defined in terms of addition; $a - b$ means $a + (-b)$.

(1.4)   To **multiply integers** we must remember that *the product of two positives or two negatives is positive and the product of a positive and a negative (either order) is negative*.

To **divide integers** we must remember that *the quotient of two positives or two negatives is positive and the quotient of a positive and a negative (or a negative and a positive) is negative*.

(1.5)   The following basic properties help with numerical manipulations and serve as a basis for algebraic computations.

Commutative Properties $a + b = b + a$
$$ab = ba$$

Associative Properties $(a + b) + c = a + (b + c)$
$$(ab)c = a(bc)$$

Identity Properties $a + 0 = 0 + a = a$
$$a(1) = 1(a) = a$$

Additive Inverse Property $a + (-a) = (-a) + a = 0$

Multiplication Property of Zero $a(0) = 0(a) = 0$

Multiplication Property of Negative One $-1(a) = a(-1) = -a$

Distributive Properties $a(b + c) = ab + ac$

$a(b - c) = ab - ac$

Chapter 1 Review Problem Set

In Problems 1–10, do the indicated operations.

1. $7 + (-10)$

2. $(-12) + (-13)$

3. $8 - 13$

4. $-6 - 9$

5. $-12 - (-11)$

6. $-17 - (-19)$

7. $(13)(-12)$

8. $(-14)(-18)$

9. $(-72) \div (-12)$

10. $117 \div (-9)$

For Problems 11–15, classify each of the numbers as *prime* or *composite*.

11. 73

12. 87

13. 63

14. 81

15. 91

For Problems 16–20, express each of the numbers as the product of prime factors.

16. 24

17. 63

18. 57

19. 64

20. 84

21. Find the greatest common factor of 36 and 54.

22. Find the greatest common factor of 48, 60, and 84.

23. Find the least common multiple of 18 and 20.

24. Find the least common multiple of 15, 27, and 35.

For Problems 25–38, simplify each of the numerical expressions.

25. $(19 + 56) + (-9)$

26. $43 - 62 + 12$

27. $8 + (-9) + (-16) + (-14) + 17 + 12$

28. $19 - 23 - 14 + 21 + 14 - 13$

29. $3(-4) - 6$

30. $(-5)(-4) - 8$

31. $(5)(-2) + (6)(-4)$

32. $(-6)(8) + (-7)(-3)$

33. $(-6)(3) - (-4)(-5)$

34. $(-7)(9) - (6)(5)$

35. $\dfrac{4(-7) - (3)(-2)}{-11}$

36. $\dfrac{(-4)(9) + (5)(-3)}{1 - 18}$

37. $3 - 2[4(-3 - 1)]$

38. $-6 - [3(-4 - 7)]$

For Problems 39–50, simplify each algebraic expression by combining similar terms.

39. $12x + 3x - 7x$

40. $9y + 3 - 14y - 12$

41. $8x + 5y - 13x - y$

42. $9a + 11b + 4a - 17b$

43. $3ab - 4ab - 2a$

44. $5xy - 9xy + xy - y$

45. $3(x + 6) + 7(x + 8)$

46. $5(x - 4) - 3(x - 9)$

47. $-3(x - 2) - 4(x + 6)$

48. $-2x - 3(x - 4) + 2x$

49. $2(a - 1) - a - 3(a - 2)$

50. $-(a - 1) + 3(a - 2) - 4a + 1$

For Problems 51–64, evaluate each of the algebraic expressions for the given values of the variables.

51. $5x + 8y$　for $x = -7$ and $y = -3$

52. $7x - 9y$　for $x = -3$ and $y = 4$

53. $\dfrac{-5x - 2y}{-2x - 7}$　for $x = 6$ and $y = 4$

54. $\dfrac{-3x + 4y}{3x}$　for $x = -4$ and $y = -6$

55. $-2a + \dfrac{a - b}{a - 2}$　for $a = -5$ and $b = 9$

56. $\dfrac{2a + b}{b + 6} - 3b$　for $a = 3$ and $b = -4$

57. $5a + 6b - 7a - 2b$　for $a = -1$ and $b = 5$

58. $3x + 7y - 5x + y$　for $x = -4$ and $y = 3$

59. $2xy + 6 + 5xy - 8$　for $x = -1$ and $y = 1$

60. $7(x + 6) - 9(x + 1)$　for $x = -2$

61. $-3(x - 4) - 2(x + 8)$　for $x = 7$

62. $2(x - 1) - (x + 2) + 3(x - 4)$　for $x = -4$

63. $(a - b) - (a + b) - b$　for $a = -1$ and $b = -3$

64. $2ab - 3(a - b) + b + a$　for $a = 2$ and $b = -5$

2 The Real Numbers

The suggestion at the beginning of Chapter 1 is worth repeating. Use your knowledge of arithmetic to help your study of algebra. A thorough understanding of arithmetic concepts provides a solid basis for algebraic ideas.

2.1 Rational Numbers: Multiplication and Division

Any number that can be written in the form $\dfrac{a}{b}$, where a and b are integers and b is not zero, is called a **rational number**. (The form $\dfrac{a}{b}$ is called a fraction or sometimes a common fraction.) The following are examples of rational numbers.

$$\frac{1}{2} \qquad \frac{7}{9} \qquad \frac{15}{7} \qquad \frac{-3}{4} \qquad \frac{5}{-7} \qquad \frac{-11}{-13}$$

All integers are rational numbers because every integer can be expressed as the indicated quotient of two integers. Some examples are as follows.

$$6 = \frac{6}{1} = \frac{12}{2} = \frac{18}{3}, \text{ etc.;}$$

$$27 = \frac{27}{1} = \frac{54}{2} = \frac{81}{3}, \text{ etc.;}$$

$$0 = \frac{0}{1} = \frac{0}{2} = \frac{0}{3}, \text{ etc.}$$

In Chapter 1, our work with division involving negative integers helps with the next three examples.

$$-4 = \frac{-4}{1} = \frac{-8}{2} = \frac{-12}{3}, \text{ etc.;}$$

$$-6 = \frac{6}{-1} = \frac{12}{-2} = \frac{18}{-3}, \text{ etc.;}$$

$$10 = \frac{10}{1} = \frac{-10}{-1} = \frac{-20}{-2}, \text{ etc.}$$

The following general properties should be observed.

Property 2.1

$$\frac{-a}{b} = \frac{a}{-b} = -\frac{a}{b} \quad \text{and} \quad \frac{-a}{-b} = \frac{a}{b}.$$

Therefore, a rational number such as $\dfrac{-2}{3}$ can also be written as $\dfrac{2}{-3}$ or $-\dfrac{2}{3}$.

■ Multiplying Rational Numbers

Multiplication of rational numbers in common fraction form is defined as follows.

Definition 2.1

If a, b, c, and d are integers with b and d not equal to zero, then

$$\frac{a}{b} \cdot \frac{c}{d} = \frac{a \cdot c}{b \cdot d}.$$

To multiply rational numbers in common fractional form we simply *multiply numerators and multiply denominators*. Furthermore, we see from the definition that the rational numbers are commutative and associative with respect to multiplication. We are free to rearrange and regroup factors as we do with integers.

The following examples illustrate the definition for multiplying rational numbers.

$$\frac{1}{3} \cdot \frac{2}{5} = \frac{1 \cdot 2}{3 \cdot 5} = \frac{2}{15};$$

$$\frac{3}{4} \cdot \frac{5}{7} = \frac{3 \cdot 5}{4 \cdot 7} = \frac{15}{28};$$

$$\frac{-2}{3} \cdot \frac{7}{9} = \frac{-2 \cdot 7}{3 \cdot 9} = \frac{-14}{27} \quad \text{or} \quad -\frac{14}{27};$$

$$\frac{1}{5} \cdot \frac{9}{-11} = \frac{1 \cdot 9}{5(-11)} = \frac{9}{-55} \quad \text{or} \quad -\frac{9}{55};$$

$$-\frac{3}{4} \cdot \frac{7}{13} = \frac{-3}{4} \cdot \frac{7}{13} = \frac{-3 \cdot 7}{4 \cdot 13} = \frac{-21}{52} \quad \text{or} \quad -\frac{21}{52}.$$

The following property is used frequently when working with rational numbers. It is often called the **fundamental principle of fractions**.

Property 2.2

If b and k are nonzero integers and a is any integer, then

$$\frac{a \cdot k}{b \cdot k} = \frac{a}{b}.$$

$$\frac{\overset{1}{\cancel{3}} \cdot 4}{\cancel{7} \cdot 5} = \frac{4}{5}$$

This property provides the basis for what is often called *reducing fractions to lowest terms or expressing fractions in simplest or reduced form.* Let's do a few examples and illustrate various forms that are used when applying the property.

EXAMPLE 1   Reduce $\dfrac{12}{18}$ to lowest terms.

Solution $\dfrac{12}{18} = \dfrac{2 \cdot 6}{3 \cdot 6} = \dfrac{2}{3}.$ ■

EXAMPLE 2   Change $\dfrac{14}{35}$ to simplest form.

Solution $\dfrac{14}{35} = \dfrac{2 \cdot \cancel{7}}{5 \cdot \cancel{7}} = \dfrac{2}{5}.$ A common factor of 7 has been divided out of both the numerator and the demominator.   ■

EXAMPLE 3   Express $\dfrac{-24}{32}$ in reduced form.

Solution $\dfrac{-24}{32} = -\dfrac{3 \cdot 8}{4 \cdot 8} = -\dfrac{3}{4} \cdot \dfrac{8}{8} = -\dfrac{3}{4} \cdot 1 = -\dfrac{3}{4}.$ The multiplication property of 1 is being used.   ■

EXAMPLE 4   Reduce $-\dfrac{72}{90}$.

Solution $-\dfrac{72}{90} = -\dfrac{\cancel{2} \cdot 2 \cdot 2 \cdot \cancel{3} \cdot \cancel{3}}{\cancel{2} \cdot \cancel{3} \cdot \cancel{3} \cdot 5} = -\dfrac{4}{5}.$ The prime factored forms of the numerator and denominator may be used to help recognize common factors.   ■

The fractions may contain variables in the numerator or denominator (or both), but this creates no great difficulty. Our thought processes remain the same, as these next examples illustrate. Variables appearing in denominators represent **nonzero** integers.

EXAMPLE 5   Reduce $\dfrac{9x}{17x}$.

Solution $\dfrac{9x}{17x} = \dfrac{9 \cdot \cancel{x}}{17 \cdot \cancel{x}} = \dfrac{9}{17}.$ ■

EXAMPLE 6   Simplify $\dfrac{8x}{36y}$.

Solution $\dfrac{8x}{36y} = \dfrac{\cancel{2}\cdot\cancel{2}\cdot 2 \cdot x}{\cancel{2}\cdot\cancel{2}\cdot 3 \cdot 3 \cdot y} = \dfrac{2x}{9y}.$ ■

EXAMPLE 7   Express $\dfrac{-9xy}{30y}$ in reduced form.

Solution $\dfrac{-9xy}{30y} = -\dfrac{3 \cdot \cancel{3} \cdot x \cdot \cancel{y}}{2 \cdot \cancel{3} \cdot 5 \cdot \cancel{y}} = -\dfrac{3x}{10}.$ ■

EXAMPLE 8   Reduce $\dfrac{-7abc}{-9ac}$.

Solution $\dfrac{-7abc}{-9ac} = \dfrac{7ab\cancel{c}}{9a\cancel{c}} = \dfrac{7b}{9}.$ ■

We are now ready to consider multiplication problems with the agreement that *the final answer should be expressed in reduced form.* Study the following examples carefully because different formats have been illustrated for handling such problems.

EXAMPLE 9   Multiply $\dfrac{7}{9}\cdot\dfrac{5}{14}$.

Solution $\dfrac{7}{9}\cdot\dfrac{5}{14} = \dfrac{7\cdot 5}{9\cdot 14} = \dfrac{\cancel{7}\cdot 5}{3\cdot 3\cdot 2\cdot\cancel{7}} = \dfrac{5}{18}.$ ■

EXAMPLE 10   Find the product of $\dfrac{8}{9}$ and $\dfrac{18}{24}$.

Solution $\underset{1}{\overset{1}{\cancel{8}}} \cdot \underset{3}{\overset{2}{\cancel{18}}}$ $= \dfrac{2}{3}.$ A common factor of 8 has been divided out
$\underset{1}{\cancel{9}} \quad \underset{3}{\cancel{24}}$ of 8 and 24, and a common factor of 9 has
                                been divided out of 9 and 18.   ■

EXAMPLE 11   Multiply $\left(-\dfrac{6}{8}\right)\left(\dfrac{14}{32}\right)$.

Solution $\left(-\dfrac{6}{8}\right)\left(\dfrac{14}{32}\right) = -\dfrac{\overset{3}{\cancel{6}}\cdot\overset{7}{\cancel{14}}}{\underset{4}{\cancel{8}}\cdot\underset{16}{\cancel{32}}} = -\dfrac{21}{64}.$ ■

EXAMPLE 12 Multiply $\left(-\dfrac{9}{4}\right)\left(-\dfrac{14}{15}\right)$.

Solution $\left(-\dfrac{9}{4}\right)\left(-\dfrac{14}{15}\right) = \dfrac{\cancel{3}\cdot 3\cdot \cancel{2}\cdot 7}{2\cdot \cancel{2}\cdot \cancel{3}\cdot 5} = \dfrac{21}{10}.$ *Immediately we recognize that a negative times a negative is positive.* ■

EXAMPLE 13 Multiply $\dfrac{9x}{7y}\cdot \dfrac{14y}{45}$.

Solution $\dfrac{9x}{7y}\cdot \dfrac{14y}{45} = \dfrac{\cancel{9}\cdot x\cdot \overset{2}{\cancel{14}}\cdot \cancel{y}}{\cancel{7}\cdot \cancel{y}\cdot \underset{5}{\cancel{45}}} = \dfrac{2x}{5}.$ ■

EXAMPLE 14 Multiply $\dfrac{-6c}{7ab}\cdot \dfrac{14b}{5c}$.

Solution $\dfrac{-6c}{7ab}\cdot \dfrac{14b}{5c} = -\dfrac{2\cdot 3\cdot \cancel{c}\cdot 2\cdot \cancel{7}\cdot \cancel{b}}{\cancel{7}\cdot a\cdot \cancel{b}\cdot 5\cdot \cancel{c}}$

$= -\dfrac{12}{5a}.$ ■

■ Dividing Rational Numbers

A definition for division of rational numbers in fractional form can be motivated by the following example.

$$\dfrac{\frac{3}{4}}{\frac{2}{3}} = \left(\dfrac{\frac{3}{4}}{\frac{2}{3}}\right)\left(\dfrac{\frac{3}{2}}{\frac{3}{2}}\right) = \dfrac{\left(\frac{3}{4}\right)\left(\frac{3}{2}\right)}{1} = \left(\dfrac{3}{4}\right)\left(\dfrac{3}{2}\right) = \dfrac{9}{8}.$$

Notice that this is a form of 1 and $\dfrac{3}{2}$ is the reciprocal of $\dfrac{2}{3}$.

In other words, $\dfrac{3}{4}$ divided by $\dfrac{2}{3}$ is equivalent to $\dfrac{3}{4}$ times $\dfrac{3}{2}$.

The following definition for division should seem reasonable.

> ### Definition 2.2
>
> If b, c, and d are nonzero integers and a is any integer, then
>
> $$\frac{a}{b} \div \frac{c}{d} = \frac{a}{b} \cdot \frac{d}{c}.$$

This definition is frequently remembered as to *divide fractions, invert the divisor and multiply*. The following examples illustrate the important steps of a division problem.

$$\frac{2}{3} \div \frac{1}{2} = \frac{2}{3} \cdot \frac{2}{1} = \frac{4}{3};$$

$$\frac{5}{6} \div \frac{3}{4} = \frac{5}{6} \cdot \frac{4}{3} = \frac{5 \cdot 4}{6 \cdot 3} = \frac{5 \cdot \cancel{2} \cdot 2}{\cancel{2} \cdot 3 \cdot 3} = \frac{10}{9};$$

$$-\frac{9}{12} \div \frac{3}{6} = -\frac{\overset{3}{\cancel{9}} \cdot \overset{1}{\cancel{6}}}{\underset{2}{\cancel{12}} \cdot \underset{1}{\cancel{3}}} = -\frac{3}{2};$$

$$\left(-\frac{27}{56}\right) \div \left(-\frac{33}{72}\right) = \left(-\frac{27}{56}\right)\left(-\frac{72}{33}\right) = \frac{\overset{9}{\cancel{27}} \cdot \overset{9}{\cancel{72}}}{\underset{7}{\cancel{56}} \cdot \underset{11}{\cancel{33}}} = \frac{81}{77};$$

$$\frac{\frac{6}{7}}{2} = \frac{\overset{3}{\cancel{6}}}{7} \cdot \frac{1}{\cancel{2}} = \frac{3}{7};$$

$$\frac{5x}{7y} \div \frac{10}{28y} = \frac{5x}{7y} \cdot \frac{28y}{10} = \frac{\cancel{5} \cdot x \cdot \overset{2}{\cancel{28}} \cdot \cancel{y}}{\cancel{7} \cdot \cancel{y} \cdot \underset{2}{\cancel{10}}} = 2x.$$

Problem Set 2.1

1-73 odds

For Problems 1–24, reduce each fraction to lowest terms.

1. $\dfrac{8}{12}$ 2. $\dfrac{12}{16}$ 3. $\dfrac{16}{24}$ 4. $\dfrac{18}{32}$

5. $\dfrac{15}{9}$

6. $\dfrac{48}{36}$

7. $\dfrac{-8}{48}$

8. $\dfrac{-3}{15}$

9. $\dfrac{27}{-36}$

10. $\dfrac{9}{-51}$

11. $\dfrac{-54}{-56}$

12. $\dfrac{-24}{-80}$

13. $\dfrac{24x}{44x}$

14. $\dfrac{15y}{25y}$

15. $\dfrac{9x}{21y}$

16. $\dfrac{4y}{30x}$

17. $\dfrac{14xy}{35y}$

18. $\dfrac{55xy}{77x}$

19. $\dfrac{-20ab}{52bc}$

20. $\dfrac{-23ac}{41c}$

21. $\dfrac{-56yz}{-49xy}$

22. $\dfrac{-21xy}{-14ab}$

23. $\dfrac{65abc}{91ac}$

24. $\dfrac{68xyz}{85yz}$

For Problems 25–58, multiply or divide as indicated and express answers in reduced form.

25. $\dfrac{3}{4} \cdot \dfrac{5}{7}$

26. $\dfrac{4}{5} \cdot \dfrac{3}{11}$

27. $\dfrac{2}{7} \div \dfrac{3}{5}$

28. $\dfrac{5}{6} \div \dfrac{11}{13}$

29. $\dfrac{3}{8} \cdot \dfrac{12}{15}$

30. $\dfrac{4}{9} \cdot \dfrac{3}{2}$

31. $\dfrac{-6}{13} \cdot \dfrac{26}{9}$

32. $\dfrac{3}{4} \cdot \dfrac{-14}{12}$

33. $\dfrac{7}{9} \div \dfrac{5}{9}$

34. $\dfrac{3}{11} \div \dfrac{7}{11}$

35. $\dfrac{1}{4} \div \dfrac{-5}{6}$

36. $\dfrac{7}{8} \div \dfrac{14}{-16}$

37. $\left(-\dfrac{8}{10}\right)\left(-\dfrac{10}{32}\right)$

38. $\left(-\dfrac{6}{7}\right)\left(-\dfrac{21}{24}\right)$

39. $-9 \div \dfrac{1}{3}$

40. $-10 \div \dfrac{1}{4}$

41. $\dfrac{5x}{9y} \cdot \dfrac{7y}{3x}$

42. $\dfrac{4a}{11b} \cdot \dfrac{6b}{7a}$

43. $\dfrac{6a}{14b} \cdot \dfrac{16b}{18a}$

44. $\dfrac{5y}{8x} \cdot \dfrac{14z}{15y}$

45. $\dfrac{10x}{-9y} \cdot \dfrac{15}{20x}$

46. $\dfrac{3x}{4y} \cdot \dfrac{-8w}{9z}$

47. $ab \cdot \dfrac{2}{b}$

48. $3xy \cdot \dfrac{4}{x}$

49. $\left(-\dfrac{7x}{12y}\right)\left(-\dfrac{24y}{35x}\right)$

50. $\left(-\dfrac{10a}{15b}\right)\left(-\dfrac{45b}{65a}\right)$

51. $\dfrac{3}{x} \div \dfrac{6}{y}$

52. $\dfrac{6}{x} \div \dfrac{14}{y}$

53. $\dfrac{5x}{9y} \div \dfrac{13x}{36y}$

54. $\dfrac{3x}{5y} \div \dfrac{7x}{10y}$

55. $\dfrac{-7}{x} \div \dfrac{9}{x}$

56. $\dfrac{8}{y} \div \dfrac{28}{-y}$

57. $\dfrac{-4}{n} \div \dfrac{-18}{n}$

58. $\dfrac{-34}{n} \div \dfrac{-51}{n}$

For Problems 59–74, perform the operations as indicated and express answers in lowest terms.

59. $\dfrac{3}{4} \cdot \dfrac{8}{9} \cdot \dfrac{12}{20}$

60. $\dfrac{5}{6} \cdot \dfrac{9}{10} \cdot \dfrac{8}{7}$

61. $\left(-\dfrac{3}{8}\right)\left(\dfrac{13}{14}\right)\left(-\dfrac{12}{9}\right)$

62. $\left(-\dfrac{7}{9}\right)\left(\dfrac{5}{11}\right)\left(-\dfrac{18}{14}\right)$

63. $\left(\dfrac{3x}{4y}\right)\left(\dfrac{8}{9x}\right)\left(\dfrac{12y}{5}\right)$

64. $\left(\dfrac{2x}{3y}\right)\left(\dfrac{5y}{x}\right)\left(\dfrac{9}{4x}\right)$

65. $\left(-\dfrac{2}{3}\right)\left(\dfrac{3}{4}\right) \div \dfrac{1}{8}$

66. $\dfrac{3}{4} \cdot \dfrac{4}{5} \div \dfrac{1}{6}$

67. $\dfrac{5}{7} \div \left(-\dfrac{5}{6}\right)\left(-\dfrac{6}{7}\right)$

68. $\left(-\dfrac{3}{8}\right) \div \left(-\dfrac{4}{5}\right)\left(\dfrac{1}{2}\right)$

69. $\left(-\dfrac{6}{7}\right) \div \left(\dfrac{5}{7}\right)\left(-\dfrac{5}{6}\right)$

70. $\left(-\dfrac{4}{3}\right) \div \left(\dfrac{4}{5}\right)\left(\dfrac{3}{5}\right)$

71. $\left(\dfrac{4}{9}\right)\left(-\dfrac{9}{8}\right) \div \left(-\dfrac{3}{4}\right)$

72. $\left(-\dfrac{7}{8}\right)\left(\dfrac{4}{7}\right) \div \left(-\dfrac{3}{2}\right)$

73. $\left(\dfrac{5}{2}\right)\left(\dfrac{2}{3}\right) \div \left(-\dfrac{1}{4}\right) \div (-3)$

74. $\dfrac{1}{3} \div \left(\dfrac{3}{4}\right)\left(\dfrac{1}{2}\right) \div 2$

■ Miscellaneous Problems

75. The division problem $35 \div 7$ can be interpreted as "how many 7's are there in 35?" Likewise, a division problem such as $3 \div \dfrac{1}{2}$ can be interpreted as "how many one-halves in 3?" Use this "how many" interpretation to do the following division problems.

(a) $4 \div \dfrac{1}{2}$

(b) $3 \div \dfrac{1}{4}$

(c) $5 \div \dfrac{1}{8}$

(d) $6 \div \dfrac{1}{7}$

(e) $\dfrac{5}{6} \div \dfrac{1}{6}$

(f) $\dfrac{7}{8} \div \dfrac{1}{8}$

76. Estimation is important in mathematics. In each of the following, estimate whether the answer is larger than 1 or smaller than 1 by using the "how many" idea from Problem 75.

(a) $\dfrac{3}{4} \div \dfrac{1}{2}$

(b) $1 \div \dfrac{7}{8}$

(c) $\dfrac{1}{2} \div \dfrac{3}{4}$

(d) $\dfrac{8}{7} \div \dfrac{7}{8}$

(e) $\dfrac{2}{3} \div \dfrac{1}{4}$

(f) $\dfrac{3}{5} \div \dfrac{3}{4}$

2.2 Addition and Subtraction of Rational Numbers

Suppose that it is one-fifth of a mile between your dorm and the union and two-fifths of a mile between the union and the library along a straight line as indicated in Figure 2.1. The total distance between your dorm and the library is three-fifths of a mile and we write $\dfrac{1}{5} + \dfrac{2}{5} = \dfrac{3}{5}$.

Figure 2.1

$\dfrac{1}{5}$ mile $\dfrac{2}{5}$ mile

Dorm   Union   Library

A pizza is cut into seven equal pieces and you eat two of the pieces. How much of the pizza remains? We represent the whole pizza by $\dfrac{7}{7}$ and then conclude that $\dfrac{7}{7} - \dfrac{2}{7} = \dfrac{5}{7}$ of the pizza remains.

These examples motivate the following definition for addition and subtraction of rational numbers in $\dfrac{a}{b}$ form.

Definition 2.3

If a, b, and c are integers and b is not zero, then

$$\frac{a}{b} + \frac{c}{b} = \frac{a+c}{b}; \qquad \text{(Addition)}$$

$$\frac{a}{b} - \frac{c}{b} = \frac{a-c}{b}. \qquad \text{(Subtraction)}$$

We say that *rational numbers **with common denominators** can be added or subtracted by adding or subtracting the numerators and placing the results over the common denominator.* Consider the following examples.

$$\frac{3}{7} + \frac{2}{7} = \frac{3+2}{7} = \frac{5}{7};$$

$$\frac{7}{8} - \frac{2}{8} = \frac{7-2}{8} = \frac{5}{8};$$

$$\frac{2}{6} + \frac{1}{6} = \frac{2+1}{6} = \frac{3}{6} = \frac{1}{2}; \qquad \text{We agree to reduce final answer.}$$

$$\frac{3}{11} - \frac{5}{11} = \frac{3-5}{11} = \frac{-2}{11} \quad \text{or} \quad -\frac{2}{11};$$

$$\frac{5}{x} + \frac{7}{x} = \frac{5+7}{x} = \frac{12}{x};$$

$$\frac{9}{y} - \frac{3}{y} = \frac{9-3}{y} = \frac{6}{y}.$$

In the last two examples, to exclude division by zero, the variables x and y cannot be equal to zero. Restricting denominators to nonzero values is always necessary although we will not take the time or space to list such restrictions for every problem.

How do we add or subtract if the fractions do not have a common denominator? We use the fundamental principle of fractions, $\dfrac{a}{b} = \dfrac{a \cdot k}{b \cdot k}$, and obtain equivalent fractions having a common denominator. **Equivalent fractions** are fractions naming the same number. Consider the following example which shows the details.

EXAMPLE 1 Add $\dfrac{1}{2} + \dfrac{1}{3}$.

Solution

$$\frac{1}{2} = \frac{1 \cdot 3}{2 \cdot 3} = \frac{3}{6} \qquad \frac{1}{2} \text{ and } \frac{3}{6} \text{ are equivalent fractions naming the same number.}$$

$$\frac{1}{3} = \frac{1 \cdot 2}{3 \cdot 2} = \frac{2}{6} \qquad \frac{1}{3} \text{ and } \frac{2}{6} \text{ are equivalent fractions naming the same number.}$$

$$\frac{5}{6}$$

Notice that we chose 6 as our common denominator and 6 is the least common multiple of the original denominators 2 and 3. (Recall that the least common multiple is the smallest nonzero whole number divisible by the given numbers.) In general, we use the least common multiple of the denominators of the fractions to be added or subtracted as a **least common denominator** (LCD).

We recall from Section 1.2 that the least common multiple may be found either by inspection or by using the prime factorization forms of the numbers. Let's consider some examples involving these procedures.

EXAMPLE 2 Add $\dfrac{1}{4} + \dfrac{2}{5}$.

Solution By inspection we see that the LCD is 20. Thus, both fractions can be changed to equivalent fractions having a denominator of 20.

$$\frac{1}{4} + \frac{2}{5} = \frac{1 \cdot 5}{4 \cdot 5} + \frac{2 \cdot 4}{5 \cdot 4} = \frac{5}{20} + \frac{8}{20} = \frac{13}{20}.$$

use of fundamental
principle of fractions

EXAMPLE 3 Subtract $\dfrac{5}{8} - \dfrac{7}{12}$.

Solution By inspection the LCD is 24.

$$\frac{5}{8} - \frac{7}{12} = \frac{5 \cdot 3}{8 \cdot 3} - \frac{7 \cdot 2}{12 \cdot 2} = \frac{15}{24} - \frac{14}{24} = \frac{1}{24}.$$

If the LCD is not obvious by inspection, then the technique from Chapter 1 for finding the least common multiple can be used. So we proceed as follows.

Step 1.　Express each denominator as a product of prime factors.

Step 2.　The LCD contains each different prime factor as many times as the *most* times it appears in any one of the factorizations from Step 1.

EXAMPLE 4　Add $\dfrac{5}{18} + \dfrac{7}{24}$.

Solution　If we cannot find the LCD by inspection, then the prime factorization forms can be used.

$$18 = 2 \cdot 3 \cdot 3$$
$$24 = 2 \cdot 2 \cdot 2 \cdot 3$$
$$\text{LCD} = 2 \cdot 2 \cdot 2 \cdot 3 \cdot 3 = 72.$$

$$\frac{5}{18} + \frac{7}{24} = \frac{5 \cdot 4}{18 \cdot 4} + \frac{7 \cdot 3}{24 \cdot 3} = \frac{20}{72} + \frac{21}{72} = \frac{41}{72}.$$ ■

EXAMPLE 5　Subtract $\dfrac{3}{14} - \dfrac{8}{35}$.

Solution
$$14 = 2 \cdot 7$$
$$35 = 5 \cdot 7$$
$$\text{LCD} = 2 \cdot 5 \cdot 7 = 70.$$

$$\frac{3}{14} - \frac{8}{35} = \frac{3 \cdot 5}{14 \cdot 5} - \frac{8 \cdot 2}{35 \cdot 2} = \frac{15}{70} - \frac{16}{70} = \frac{-1}{70} \quad \text{or} \quad -\frac{1}{70}.$$ ■

EXAMPLE 6　Add $\dfrac{-5}{8} + \dfrac{3}{14}$.

Solution
$$8 = 2 \cdot 2 \cdot 2$$
$$14 = 2 \cdot 7$$
$$\text{LCD} = 2 \cdot 2 \cdot 2 \cdot 7 = 56.$$

$$\frac{-5}{8} + \frac{3}{14} = \frac{-5 \cdot 7}{8 \cdot 7} + \frac{3 \cdot 4}{14 \cdot 4} = \frac{-35}{56} + \frac{12}{56} = \frac{-23}{56} \quad \text{or} \quad -\frac{23}{56}.$$ ■

Having the denominators contain variables does not complicate the situation very much as the next examples illustrate.

EXAMPLE 7　Add $\dfrac{2}{x} + \dfrac{3}{y}$.

Solution   By inspection, the LCD is xy.

$$\frac{2}{x} + \frac{3}{y} = \frac{2 \cdot y}{x \cdot y} + \frac{3 \cdot x}{y \cdot x} = \frac{2y}{xy} + \frac{3x}{xy} = \frac{2y + 3x}{xy}.$$

commutative property

■

EXAMPLE 8   Subtract $\dfrac{3}{8x} - \dfrac{5}{12y}$.

Solution
$$\left. \begin{array}{l} 8x = 2 \cdot 2 \cdot 2 \cdot x \\ 12y = 2 \cdot 2 \cdot 3 \cdot y \end{array} \right) \quad \text{LCD} = 2 \cdot 2 \cdot 2 \cdot 3 \cdot x \cdot y = 24xy.$$

$$\frac{3}{8x} - \frac{5}{12y} = \frac{3 \cdot 3y}{8x \cdot 3y} - \frac{5 \cdot 2x}{12y \cdot 2x} = \frac{9y}{24xy} - \frac{10x}{24xy} = \frac{9y - 10x}{24xy}.$$

■

EXAMPLE 9   Add $\dfrac{7}{4a} + \dfrac{-5}{6bc}$.

Solution
$$\left. \begin{array}{l} 4a = 2 \cdot 2 \cdot a \\ 6bc = 2 \cdot 3 \cdot b \cdot c \end{array} \right) \quad \text{LCD} = 2 \cdot 2 \cdot 3 \cdot a \cdot b \cdot c = 12abc.$$

$$\frac{7}{4a} + \frac{-5}{6bc} = \frac{7 \cdot 3bc}{4a \cdot 3bc} + \frac{-5 \cdot 2a}{6bc \cdot 2a} = \frac{21bc}{12abc} + \frac{-10a}{12abc} = \frac{21bc - 10a}{12abc}.$$

■

■ Simplifying Numerical Expressions

Let's now consider simplifying numerical expressions containing rational numbers. As with integers, multiplications and divisions are done first and then the additions and subtractions are performed. In these next examples only the major steps are shown, so be sure that you can fill in all of the details.

EXAMPLE 10   Simplify $\dfrac{3}{4} + \dfrac{2}{3} \cdot \dfrac{3}{5} - \dfrac{1}{2} \cdot \dfrac{1}{5}$.

Solution
$$\frac{3}{4} + \frac{2}{3} \cdot \frac{3}{5} - \frac{1}{2} \cdot \frac{1}{5} = \frac{3}{4} + \frac{2}{5} - \frac{1}{10}$$

$$= \frac{15}{20} + \frac{8}{20} - \frac{2}{20}$$

$$= \frac{15 + 8 - 2}{20} = \frac{21}{20}.$$

■

EXAMPLE 11   Simplify $\dfrac{3}{5} \div \dfrac{8}{5} + \left(-\dfrac{1}{2}\right)\left(\dfrac{1}{3}\right) + \dfrac{5}{12}$.

Solution

$$\frac{3}{5} \div \frac{8}{5} + \left(-\frac{1}{2}\right)\left(\frac{1}{3}\right) + \frac{5}{12} = \frac{3}{5} \cdot \frac{5}{8} + \left(-\frac{1}{2}\right)\left(\frac{1}{3}\right) + \frac{5}{12}$$

$$= \frac{3}{8} + \frac{-1}{6} + \frac{5}{12}$$

$$= \frac{9}{24} + \frac{-4}{24} + \frac{10}{24}$$

$$= \frac{9 + (-4) + 10}{24}$$

$$= \frac{15}{24} = \frac{5}{8}. \qquad \text{Reduce!} \qquad ■$$

The distributive property, $a(b + c) = ab + ac$, holds true for rational numbers and (as with integers) can be used to facilitate manipulation.

EXAMPLE 12   Simplify $12\left(\dfrac{1}{3} + \dfrac{1}{4}\right)$.

Solution   Changing the form by applying the distributive property helps in this situation.

$$12\left(\frac{1}{3} + \frac{1}{4}\right) = 12\left(\frac{1}{3}\right) + 12\left(\frac{1}{4}\right)$$

$$= 4 + 3$$

$$= 7. \qquad\qquad ■$$

EXAMPLE 13   Simplify $\dfrac{5}{8}\left(\dfrac{1}{2} + \dfrac{1}{3}\right)$.

Solution   In this case it may be easier not to apply the distributive property, but to work with the expression in its given form.

$$\frac{5}{8}\left(\frac{1}{2} + \frac{1}{3}\right) = \frac{5}{8}\left(\frac{3}{6} + \frac{2}{6}\right)$$

$$= \frac{5}{8}\left(\frac{5}{6}\right)$$

$$= \frac{25}{48}. \qquad\qquad ■$$

Examples 12 and 13 emphasize a point made in Chapter 1. *Think first*, and decide whether or not the properties can be used to make the manipulations easier. This section concludes with an example illustrating the combining of similar terms having fractional coefficients.

EXAMPLE 14   Simplify $\dfrac{1}{2}x + \dfrac{2}{3}x - \dfrac{3}{4}x$ by combining similar terms.

Solution   The distributive property along with our knowledge of adding and subtracting rational numbers provides the basis for working this type of problem.

$$\frac{1}{2}x + \frac{2}{3}x - \frac{3}{4}x = \left(\frac{1}{2} + \frac{2}{3} - \frac{3}{4}\right)x$$

$$= \left(\frac{6}{12} + \frac{8}{12} - \frac{9}{12}\right)x$$

$$= \frac{5}{12}x.$$

■

Problem Set 2.2

multiples of 3

For Problems 1–60, add or subtract as indicated and express answers in lowest terms.

1. $\dfrac{2}{7} + \dfrac{3}{7}$

2. $\dfrac{3}{11} + \dfrac{5}{11}$

3. $\dfrac{7}{9} - \dfrac{2}{9}$

4. $\dfrac{11}{13} - \dfrac{6}{13}$

5. $\dfrac{3}{4} + \dfrac{9}{4}$

6. $\dfrac{5}{6} + \dfrac{7}{6}$

7. $\dfrac{11}{12} - \dfrac{3}{12}$

8. $\dfrac{13}{16} - \dfrac{7}{16}$

9. $\dfrac{1}{8} - \dfrac{5}{8}$

10. $\dfrac{2}{9} - \dfrac{5}{9}$

11. $\dfrac{5}{24} + \dfrac{11}{24}$

12. $\dfrac{7}{36} + \dfrac{13}{36}$

13. $\dfrac{8}{x} + \dfrac{7}{x}$

14. $\dfrac{17}{y} + \dfrac{12}{y}$

15. $\dfrac{5}{3y} + \dfrac{1}{3y}$

16. $\dfrac{3}{8x} + \dfrac{1}{8x}$

17. $\dfrac{1}{3} + \dfrac{1}{5}$

18. $\dfrac{1}{6} + \dfrac{1}{8}$

19. $\dfrac{15}{16} - \dfrac{3}{8}$

20. $\dfrac{13}{12} - \dfrac{1}{6}$

21. $\dfrac{7}{10} + \dfrac{8}{15}$

22. $\dfrac{7}{12} + \dfrac{5}{8}$

23. $\dfrac{11}{24} + \dfrac{5}{32}$

24. $\dfrac{5}{18} + \dfrac{8}{27}$

25. $\dfrac{5}{18} - \dfrac{13}{24}$

26. $\dfrac{1}{24} - \dfrac{7}{36}$

27. $\dfrac{5}{8} - \dfrac{2}{3}$

28. $\dfrac{3}{4} - \dfrac{5}{6}$

29. $-\dfrac{2}{13} - \dfrac{7}{39}$

30. $-\dfrac{3}{11} - \dfrac{13}{33}$

31. $-\dfrac{3}{14} + \dfrac{1}{21}$

32. $-\dfrac{3}{20} + \dfrac{14}{25}$

33. $\dfrac{3}{x} + \dfrac{4}{y}$

34. $\dfrac{5}{x} + \dfrac{8}{y}$

35. $\dfrac{7}{a} - \dfrac{2}{b}$

36. $\dfrac{13}{a} - \dfrac{4}{b}$

37. $\dfrac{2}{x} + \dfrac{7}{2x}$

38. $\dfrac{5}{2x} + \dfrac{7}{x}$

39. $\dfrac{10}{3x} - \dfrac{2}{x}$

40. $\dfrac{13}{4x} - \dfrac{3}{x}$

41. $\dfrac{1}{x} - \dfrac{7}{5x}$

42. $\dfrac{2}{x} - \dfrac{17}{6x}$

43. $\dfrac{3}{2y} + \dfrac{5}{3y}$

44. $\dfrac{7}{3y} + \dfrac{9}{4y}$

45. $\dfrac{5}{12y} - \dfrac{3}{8y}$

46. $\dfrac{9}{4y} - \dfrac{5}{9y}$

47. $\dfrac{1}{6n} - \dfrac{7}{8n}$

48. $\dfrac{3}{10n} - \dfrac{11}{15n}$

49. $\dfrac{5}{3x} + \dfrac{7}{3y}$

50. $\dfrac{3}{2x} + \dfrac{7}{2y}$

51. $\dfrac{8}{5x} + \dfrac{3}{4y}$

52. $\dfrac{1}{5x} + \dfrac{5}{6y}$

53. $\dfrac{7}{4x} - \dfrac{5}{9y}$

54. $\dfrac{2}{7x} - \dfrac{11}{14y}$

55. $-\dfrac{3}{2x} - \dfrac{5}{4y}$

56. $-\dfrac{13}{8a} - \dfrac{11}{10b}$

57. $3 + \dfrac{2}{x}$

58. $\dfrac{5}{x} + 4$

59. $2 - \dfrac{3}{2x}$

60. $-1 - \dfrac{1}{3x}$

For Problems 61–76, simplify each numerical expression expressing answers in reduced form.

61. $\dfrac{1}{4} - \dfrac{3}{8} + \dfrac{5}{12} - \dfrac{1}{24}$

62. $\dfrac{3}{4} + \dfrac{2}{3} - \dfrac{1}{6} + \dfrac{5}{12}$

63. $\dfrac{5}{6} + \dfrac{2}{3} \cdot \dfrac{3}{4} - \dfrac{1}{4} \cdot \dfrac{2}{5}$

64. $\dfrac{2}{3} + \dfrac{1}{2} \cdot \dfrac{2}{5} - \dfrac{1}{3} \cdot \dfrac{1}{5}$

65. $\dfrac{3}{4} \cdot \dfrac{6}{9} - \dfrac{5}{6} \cdot \dfrac{8}{10} + \dfrac{2}{3} \cdot \dfrac{6}{8}$

66. $\dfrac{3}{5} \cdot \dfrac{5}{7} + \dfrac{2}{3} \cdot \dfrac{3}{5} - \dfrac{1}{7} \cdot \dfrac{2}{5}$

67. $4 - \dfrac{2}{3} \cdot \dfrac{3}{5} - 6$

68. $3 + \dfrac{1}{2} \cdot \dfrac{1}{3} - 2$

69. $\dfrac{4}{5} - \dfrac{10}{12} - \dfrac{5}{6} \div \dfrac{14}{8} + \dfrac{10}{21}$

70. $\dfrac{3}{4} \div \dfrac{6}{5} + \dfrac{8}{12} \cdot \dfrac{6}{9} - \dfrac{5}{12}$

71. $24\left(\dfrac{3}{4} - \dfrac{1}{6}\right)$ (Don't forget the distributive property!)

72. $18\left(\dfrac{2}{3} + \dfrac{1}{9}\right)$

73. $64\left(\dfrac{3}{16} + \dfrac{5}{8} - \dfrac{1}{4} + \dfrac{1}{2}\right)$

74. $48\left(\dfrac{5}{12} - \dfrac{1}{6} + \dfrac{3}{8}\right)$

75. $\dfrac{7}{13}\left(\dfrac{2}{3} - \dfrac{1}{6}\right)$

76. $\dfrac{5}{9}\left(\dfrac{1}{2} + \dfrac{1}{4}\right)$

For Problems 77–92, simplify each algebraic expression by combining similar terms.

77. $\dfrac{1}{3}x + \dfrac{2}{5}x$

78. $\dfrac{1}{4}x + \dfrac{2}{3}x$

79. $\dfrac{1}{3}a - \dfrac{1}{8}a$

80. $\dfrac{2}{5}a - \dfrac{2}{7}a$

81. $\dfrac{1}{2}x + \dfrac{2}{3}x + \dfrac{1}{6}x$

82. $\dfrac{1}{3}x + \dfrac{2}{5}x + \dfrac{5}{6}x$

83. $\dfrac{3}{5}n - \dfrac{1}{4}n + \dfrac{3}{10}n$

84. $\dfrac{2}{5}n - \dfrac{7}{10}n + \dfrac{8}{15}n$

85. $n + \dfrac{4}{3}n - \dfrac{1}{9}n$

86. $2n - \dfrac{6}{7}n + \dfrac{5}{14}n$

87. $-n - \dfrac{7}{9}n - \dfrac{5}{12}n$

88. $-\dfrac{3}{8}n - n - \dfrac{3}{14}n$

89. $\dfrac{3}{7}x + \dfrac{1}{4}y + \dfrac{1}{2}x + \dfrac{7}{8}y$

90. $\dfrac{5}{6}x + \dfrac{3}{4}y + \dfrac{4}{9}x + \dfrac{7}{10}y$

91. $\dfrac{2}{9}x + \dfrac{5}{12}y - \dfrac{7}{15}x - \dfrac{13}{15}y$

92. $-\dfrac{9}{10}x - \dfrac{3}{14}y + \dfrac{2}{25}x + \dfrac{5}{21}y$

■ Miscellaneous Problems

93. The will of a deceased collector of antique automobiles specified that his cars were to be left to his three children. Half were to go to his elder son, $\dfrac{1}{3}$ to his daughter, and $\dfrac{1}{9}$ to his younger son. At the time of his death, 17 cars were in the collection. The administrator of his estate borrowed a car to make 18. Then he distributed the cars as follows.

elder son: $\dfrac{1}{2}(18) = 9$

daughter: $\dfrac{1}{3}(18) = 6$

younger son: $\dfrac{1}{9}(18) = 2$

This totaled 17 cars, so he then returned the borrowed car. Where is the error in this problem?

2.3 Real Numbers and Algebraic Expressions

Decimals—also called decimal fractions—are classified as **terminating**, **repeating** or **nonrepeating**. Some examples of each of these classifications are as follows.

$$\begin{bmatrix} 0.3 \\ 0.26 \\ 0.347 \\ 0.9865 \end{bmatrix} \quad \text{terminating decimals}$$

Technically, a terminating decimal can be thought of as repeating zeros after the last digit. For example, $0.3 = 0.30 = 0.300 = 0.3000$, etc..

$$\begin{bmatrix} 0.333333\ldots \\ 0.5466666\ldots \\ 0.14141414\ldots \\ 0.237237237\ldots \end{bmatrix}$$ repeating decimals

A repeating decimal has a block of digits that repeats indefinitely. This repeating block of digits may contain any number of digits and may or may not begin repeating immediately after the decimal point.

$$\begin{bmatrix} 0.5918654279\ldots \\ 0.26224222722229\ldots \\ 0.145117211193111148\ldots \end{bmatrix}$$ nonrepeating decimals

In Section 2.1 we defined a rational number to be any number that can be written in the form $\dfrac{a}{b}$, where a and b are integers and b is not zero. A rational number can also be defined as any number that has a terminating or repeating decimal representation. Thus, rational numbers can be expressed in either common fractional form or decimal fraction form as the next examples illustrate.

$$\begin{bmatrix} \dfrac{3}{4} = 0.75 \\[2mm] \dfrac{1}{8} = 0.125 \\[2mm] \dfrac{5}{16} = 0.3125 \\[2mm] \dfrac{7}{25} = 0.28 \end{bmatrix}$$ terminating decimals

$$\begin{bmatrix} \dfrac{1}{3} = 0.33333\ldots \\[2mm] \dfrac{2}{3} = 0.66666\ldots \\[2mm] \dfrac{1}{6} = 0.166666\ldots \\[2mm] \dfrac{1}{12} = 0.083333\ldots \\[2mm] \dfrac{14}{99} = 0.14141414\ldots \end{bmatrix}$$ repeating decimals

The nonrepeating decimals are called **irrational numbers** and do appear in other than decimal form. For example, $\sqrt{2}$, $\sqrt{3}$, and π are irrational numbers. A partial decimal representation for each of these is as follows.

$$\left[\begin{array}{l} \sqrt{2} = 1.414213562373\ldots \\ \sqrt{3} = 1.73205080756887\ldots \\ \quad \pi = 3.14159265358979\ldots \end{array}\right] \quad \text{nonrepeating decimals}$$

More work will be done with the irrationals in Chapter 9.

The rational numbers together with the irrationals form the set of **real numbers**. The following tree-diagram of the real number system is helpful for summarizing some basic ideas.

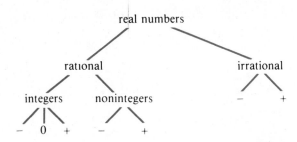

Any real number can be traced down through the diagram as follows.

5 is real, rational, an integer, and positive;

-4 is real, rational, an integer, and negative;

$\dfrac{3}{4}$ is real, rational, a noninteger, and positive;

0.23 is real, rational, a noninteger, and positive;

$-0.161616\ldots$ is real, rational, a noninteger, and negative;

$\sqrt{7}$ is real, irrational, and positive;

$-\sqrt{2}$ is real, irrational, and negative.

The properties discussed in Section 1.5 pertaining to integers are true for all real numbers and are restated here for your convenience. The multiplicative inverse property has been added to the list and is discussed following the list.

Commutative Property of Addition

If a and b are real numbers, then

$$a + b = b + a.$$

Commutative Property of Multiplication

If a and b are real numbers, then

$$ab = ba.$$

Associative Property of Addition

If a, b, and c are real numbers, then

$$(a + b) + c = a + (b + c).$$

Associative Property of Multiplication

If a, b, and c are real numbers, then

$$(ab)c = a(bc).$$

Identity Property of Addition

If a is any real number, then

$$a + 0 = 0 + a = a.$$

Identity Property of Multiplication

If a is any real number, then

$$a(1) = 1(a) = a.$$

Additive Inverse Property

For every real number a, there exists a real number $-a$, such that

$$a + (-a) = (-a) + a = 0.$$

Multiplication Property of Zero

If a is any real number, then

$$a(0) = 0(a) = 0.$$

Multiplicative Property of Negative One

If a is any real number, then

$$a(-1) = -1(a) = -a.$$

Multiplicative Inverse Property

For every nonzero real number a, there exists a real number $\dfrac{1}{a}$, such that

$$a\left(\frac{1}{a}\right) = \frac{1}{a}(a) = 1.$$

Distributive Property

If a, b, and c are real numbers, then

$$a(b + c) = ab + ac.$$

The number $\dfrac{1}{a}$ is called the **multiplicative inverse** or the **reciprocal of** a. For example, the reciprocal of 2 is $\dfrac{1}{2}$ and $2\left(\dfrac{1}{2}\right) = \dfrac{1}{2}(2) = 1$. Likewise, the reciprocal of $\dfrac{1}{2}$ is $\dfrac{1}{\frac{1}{2}} = 2$. Therefore, 2 and $\dfrac{1}{2}$ are said to be reciprocals (or multiplicative inverses) of each other. Also, $\dfrac{2}{5}$ and $\dfrac{5}{2}$ are multiplicative inverses and $\left(\dfrac{2}{5}\right)\left(\dfrac{5}{2}\right) = 1$. Since division by zero is undefined, zero does not have a reciprocal.

■ Basic Operations with Decimals

The basic operations with decimals may be related to the corresponding operation with common fractions. For example, $0.3 + 0.4 = 0.7$ because $\dfrac{3}{10} + \dfrac{4}{10} = \dfrac{7}{10}$ and $0.37 - 0.24 = 0.13$ because $\dfrac{37}{100} - \dfrac{24}{100} = \dfrac{13}{100}$. In general, to add or subtract decimals, we write the numbers in a column so that the decimal points are lined up, and then we add or subtract as we do with integers.

Addition

2.14	5.214
3.12	3.162
5.16	7.218
10.42	8.914
	24.508

Subtraction

7.6	14.231
4.9	8.785
2.7	5.446

Examples such as the following can be used to help formulate a general rule for multiplying decimals.

Since $\dfrac{7}{10} \cdot \dfrac{3}{10} = \dfrac{21}{100}$, then $(0.7)(0.3) = 0.21$.

Since $\dfrac{9}{10} \cdot \dfrac{23}{100} = \dfrac{207}{1000}$, then $(0.9)(0.23) = 0.207$.

Since $\dfrac{11}{100} \cdot \dfrac{13}{100} = \dfrac{143}{10000}$, then $(0.11)(0.13) = 0.0143$.

In general, to multiply decimals we (1) multiply the numbers ignoring the decimal points, and then (2) insert the decimal point in the product so that the number of digits to the right of the decimal point in the product is equal to the sum of the numbers of digits to the right of the decimal point in each factor.

$$(0.7) \quad (0.3) \ = \ 0.21.$$

one digit to right $+$ one digit to right $=$ two digits to right

$$(0.9) \quad (0.23) \ = \ 0.207.$$

one digit to right $+$ two digits to right $=$ three digits to right

$$(0.11) \quad (0.13) \ = \ 0.0143.$$

two digits to right $+$ two digits to right $=$ four digits to right

The vertical format is frequently used when multiplying decimals.

41.2	one digit to right		0.021	three digits to right
0.13	two digits to right		0.03	two digits to right
1236			0.00063	five digits to right
412				
5.356	three digits to right			

Notice that in the last example we actually multiplied $3 \cdot 21$ and then inserted three 0's to the left so that there would be five digits to the right of the decimal point.

Once again let's look at some links between common fractions and decimals.

$$\text{Since } \frac{6}{10} \div 2 = \frac{\overset{3}{\cancel{6}}}{10} \cdot \frac{1}{\cancel{2}} = \frac{3}{10}, \quad \text{then } 2\overline{)0.6}^{\,0.3};$$

$$\text{Since } \frac{39}{100} \div 13 = \frac{\overset{3}{\cancel{39}}}{100} \cdot \frac{1}{\cancel{13}} = \frac{3}{100}, \quad \text{then } 13\overline{)0.39}^{\,0.03};$$

$$\text{Since } \frac{85}{100} \div 5 = \frac{\overset{17}{\cancel{85}}}{100} \cdot \frac{1}{\cancel{5}} = \frac{17}{100}, \quad \text{then } 5\overline{)0.85}^{\,0.17}.$$

In general, to divide a decimal by a nonzero whole number we (1) place the decimal point in the quotient directly above the decimal point in the dividend

$$\left(\text{divisor} \overline{)\text{dividend,}}^{\,\text{quotient}} \right)$$

and then (2) divide as with whole numbers, except that in the division process, 0's are placed in the quotient immediately to the right of the decimal point in order to show the correct place value.

$$\begin{array}{r} 0.121 \\ 4\overline{)0.484} \end{array} \qquad \begin{array}{r} 0.24 \\ 32\overline{)7.68} \\ \underline{6\,4} \\ 1\,28 \\ \underline{1\,28} \end{array} \qquad \begin{array}{r} 0.0\overset{\frown}{1}9 \\ 12\overline{)0.228} \\ \underline{12} \\ 108 \\ \underline{108} \end{array} \qquad \begin{array}{l} \text{The zero is needed to show} \\ \text{the correct place value.} \end{array}$$

Don't forget that *division can be checked by multiplication.* For example, since $(12)(0.019) = 0.228$ we know that our last division example is correct.

Problems involving division by a decimal are easily handled by changing to an equivalent problem having a whole number divisor. Consider the following examples in which the original division problem has been changed to fractional form to show the reasoning involved in the procedure.

$$0.6\overline{)0.24} \rightarrow \frac{0.24}{0.6} = \left(\frac{0.24}{0.6}\right)\left(\frac{10}{10}\right) = \frac{2.4}{6} \rightarrow 6\overline{)2.4}^{\,0.4};$$

$$0.12\overline{)0.156} \rightarrow \frac{0.156}{0.12} = \left(\frac{0.156}{0.12}\right)\left(\frac{100}{100}\right) = \frac{15.6}{12} \rightarrow \begin{array}{r} 1.3 \\ 12\overline{)15.6} \\ \underline{12} \\ 36 \\ \underline{36} \end{array}$$

$$1.3\overline{)0.026} \rightarrow \frac{0.026}{1.3} = \left(\frac{0.026}{1.3}\right)\left(\frac{10}{10}\right) = \frac{0.26}{13} \rightarrow 13\overline{)0.26.} \atop \underline{26}$$

The format commonly used with such problems is as follows.

$$\begin{array}{r} 5.6 \\ 0{,}21{,}\overline{)1{,}17{,}6} \\ \underline{1\ 05} \\ 12\ 6 \\ \underline{12\ 6} \end{array}$$

The arrows indicate that the divisor and dividend have been multiplied by 100, which changes the divisor to a whole number.

$$\begin{array}{r} 0.04 \\ 3{,}7{,}\overline{)0{,}1{,}48} \\ \underline{1\ 48} \end{array}$$

The divisor and dividend have been multiplied by 10.

Our agreements for operating with positive and negative integers extend to all real numbers. For example, the product of two negative real numbers is a positive real number. Make sure that you agree with the following results. (You may need to do some work on scratch paper since the steps are not shown.)

$0.24 + (-0.18) = 0.06;$	$(-0.4)(0.8) = -0.32;$
$-7.2 + 5.1 = -2.1;$	$(-0.5)(-0.13) = 0.065;$
$-0.6 + (-0.8) = -1.4;$	$(1.4) \div (-0.2) = -7;$
$2.4 - 6.1 = -3.7;$	$(-0.18) \div (0.3) = -0.6;$
$0.31 - (-0.52) = 0.83;$	$(-0.24) \div (-4) = 0.06.$
$(0.2)(-0.3) = -0.06;$	

Numerical and algebraic expressions may contain the decimal form as well as the fractional form of rational numbers. We continue to follow the agreement that multiplications and divisions are done *first* and then the additions and subtractions unless parentheses indicate otherwise. The following examples illustrate a variety of situations involving both the decimal and fractional form of rational numbers.

EXAMPLE 1 Simplify $6.3 \div 7 + (4)(2.1) - (0.24) \div (-0.4)$.

Solution
$$\begin{aligned} 6.3 \div 7 + (4)(2.1) - (0.24) \div (-0.4) &= 0.9 + 8.4 - (-0.6) \\ &= 0.9 + 8.4 + 0.6 \\ &= 9.9. \end{aligned}$$ ■

EXAMPLE 2 Evaluate $\dfrac{3}{5}a - \dfrac{1}{7}b$ for $a = \dfrac{5}{2}$ and $b = -1$.

Solution
$$\frac{3}{5}a - \frac{1}{7}b = \frac{3}{5}\left(\frac{5}{2}\right) - \frac{1}{7}(-1) \quad \text{for } a = \frac{5}{2} \text{ and } b = -1$$

$$= \frac{3}{2} + \frac{1}{7}$$

$$= \frac{21}{14} + \frac{2}{14}$$

$$= \frac{23}{14}.$$ ■

EXAMPLE 3 Evaluate $\frac{1}{2}x + \frac{2}{3}x - \frac{1}{5}x$ for $x = -\frac{3}{4}$.

Solution First, let's *combine similar terms* by using the distributive property.

$$\frac{1}{2}x + \frac{2}{3}x - \frac{1}{5}x = \left(\frac{1}{2} + \frac{2}{3} - \frac{1}{5}\right)x$$

$$= \left(\frac{15}{30} + \frac{20}{30} - \frac{6}{30}\right)x$$

$$= \frac{29}{30}x.$$

Now we can evaluate.

$$\frac{29}{30}x = \frac{29}{30}\left(-\frac{3}{4}\right) \quad \text{when } x = -\frac{3}{4}$$

$$= \frac{29}{\underset{10}{\cancel{30}}}\left(-\frac{\overset{1}{\cancel{3}}}{4}\right) = -\frac{29}{40}.$$ ■

EXAMPLE 4 Evaluate $2x + 3y$ for $x = 1.6$ and $y = 2.7$.

Solution
$$2x + 3y = 2(1.6) + 3(2.7) \quad \text{when } x = 1.6 \text{ and } y = 2.7$$
$$= 3.2 + 8.1$$
$$= 11.3.$$ ■

EXAMPLE 5 Evaluate $0.9x + 0.7x - 0.4x + 1.3x$ for $x = 0.2$.

Solution   First, let's *combine similar terms* by using the distributive property.

$$0.9x + 0.7x - 0.4x + 1.3x = (0.9 + 0.7 - 0.4 + 1.3)x$$
$$= 2.5x.$$

Now we can evaluate.

$$2.5x = (2.5)(0.2) \quad \text{for } x = 0.2$$
$$= 0.5.$$

■

Problem Set 2.3

For Problems 1–32, perform the indicated operations.

1. $0.37 + 0.25$ **2.** $7.2 + 4.9$ **3.** $2.93 - 1.48$

4. $14.36 - 5.89$ **5.** $(7.6) + (-3.8)$ **6.** $(6.2) + (-2.4)$

7. $(-4.7) + 1.4$ **8.** $(-14.1) + 9.5$ **9.** $-3.8 + 11.3$

10. $-2.5 + 14.8$ **11.** $6.6 - (-1.2)$ **12.** $18.3 - (-7.4)$

13. $-11.5 - (-10.6)$ **14.** $-14.6 - (-8.3)$ **15.** $-17.2 - (-9.4)$

16. $-21.4 - (-14.2)$ **17.** $(0.4)(2.9)$ **18.** $(0.3)(3.6)$

19. $(-0.8)(0.34)$ **20.** $(-0.7)(0.67)$ **21.** $(9)(-2.7)$

22. $(8)(-7.6)$ **23.** $(-0.7)(-64)$ **24.** $(-0.9)(-56)$

25. $(-0.12)(-0.13)$ **26.** $(-0.11)(-0.15)$ **27.** $1.56 \div 1.3$

28. $7.14 \div 2.1$ **29.** $5.92 \div (-0.8)$ **30.** $-2.94 \div 0.6$

31. $-0.266 \div (-0.7)$ **32.** $-0.126 \div (-0.9)$

For Problems 33–46, simplify each of the numerical expressions.

33. $16.5 - 18.7 + 9.4$ **34.** $17.7 + 21.2 - 14.6$

35. $0.34 - 0.21 - 0.74 + 0.19$ **36.** $-5.2 + 6.8 - 4.7 - 3.9 + 1.3$

37. $0.76(0.2 + 0.8)$ **38.** $9.8(1.8 - 0.8)$

39. $0.6(4.1) + 0.7(3.2)$ **40.** $0.5(74) - 0.9(87)$

41. $7(0.6) + 0.9 - 3(0.4) + 0.4$ **42.** $-5(0.9) - 0.6 + 4.1(6) - 0.9$

43. $(0.96) \div (-0.8) + 6(-1.4) - 5.2$ **44.** $(-2.98) \div 0.4 - 5(-2.3) + 1.6$

45. $5(2.3) - 1.2 - 7.36 \div 0.8 + 0.2$ **46.** $0.9(12) \div 0.4 - 1.36 \div 17 + 9.2$

For Problems 47–60, simplify each algebraic expression by combining similar terms.

47. $x - 0.4x - 1.8x$ **48.** $-2x + 1.7x - 4.6x$

49. $5.4n - 0.8n - 1.6n$ **50.** $6.2n - 7.8n - 1.3n$

51. $-3t + 4.2t - 0.9t + 0.2t$ **52.** $7.4t - 3.9t - 0.6t + 4.7t$

53. $3.6x - 7.4y - 9.4x + 10.2y$

54. $5.7x + 9.4y - 6.2x - 4.4y$

55. $0.3(x - 4) + 0.4(x + 6) - 0.6x$

56. $0.7(x + 7) - 0.9(x - 2) + 0.5x$

57. $6(x - 1.1) - 5(x - 2.3) - 4(x + 1.8)$

58. $4(x + 0.7) - 9(x + 0.2) - 3(x - 0.6)$

59. $5(x - 0.5) + 0.3(x - 2) - 0.7(x + 7)$

60. $-8(x - 1.2) + 6(x - 4.6) + 4(x + 1.7)$

For Problems 61–74, evaluate each algebraic expression for the given values for the variables. Don't forget that for some problems it might be helpful to combine similar terms first and then to evaluate.

61. $x + 2y + 3z$ for $x = \dfrac{3}{4}, y = \dfrac{1}{3}$, and $z = -\dfrac{1}{6}$

62. $2x - y - 3z$ for $x = -\dfrac{2}{5}, y = -\dfrac{3}{4}$, and $z = \dfrac{1}{2}$

63. $\dfrac{3}{5}y - \dfrac{2}{3}y - \dfrac{7}{15}y$ for $y = -\dfrac{5}{2}$

64. $\dfrac{1}{2}x + \dfrac{2}{3}x - \dfrac{3}{4}x$ for $x = \dfrac{7}{8}$

65. $-x - 2y + 4z$ for $x = 1.7, y = -2.3$, and $z = 3.6$

66. $-2x + y - 5z$ for $x = -2.9, y = 7.4$, and $z = -6.7$

67. $5x - 7y$ for $x = -7.8$ and $y = 8.4$

68. $8x - 9y$ for $x = -4.3$ and $y = 5.2$

69. $0.7x + 0.6y$ for $x = -2$ and $y = 6$

70. $0.8x + 2.1y$ for $x = 5$ and $y = -9$

71. $1.2x + 2.3x - 1.4x - 7.6x$ for $x = -2.5$

72. $3.4x - 1.9x + 5.2x$ for $x = 0.3$

73. $-3a - 1 + 7a - 2$ for $a = 0.9$

74. $5x - 2 + 6x + 4$ for $x = -1.1$

■ Miscellaneous Problems

75. Without doing the actual dividing, defend the statement "$\dfrac{1}{7}$ produces a repeating decimal."

(*Hint*: Think about the possible remainders when dividing by 7.)

76. Express each of the following in repeating decimal form.

(a) $\dfrac{1}{7}$

(b) $\dfrac{2}{7}$

(c) $\dfrac{4}{9}$

(d) $\dfrac{5}{6}$

(e) $\dfrac{3}{11}$

(f) $\dfrac{1}{12}$

77. (a) How can we tell that $\dfrac{5}{16}$ will produce a terminating decimal?

(b) How can we tell that $\dfrac{7}{15}$ will not produce a terminating decimal?

2.4 Exponents

Exponents are used to indicate repeated multiplication. For example, $5 \cdot 5 \cdot 5$ can be written as 5^3 where the 3 indicates that 5 is to be used as a factor 3 times. The following general definition is helpful.

Definition 2.4

If n is a positive integer and b is any real number, then

$$b^n = \underbrace{bbb \cdots b}_{n \text{ factors of } b}.$$

The b is referred to as the **base** and n as the **exponent**. The expression b^n can be read as "b to the nth **power**." The terms **squared** and **cubed** are frequently associated with exponents of 2 and 3, respectively. For example, b^2 is read as "b squared" and b^3 as "b cubed." An exponent of 1 is usually not written, so b^1 is written as b.

The following examples further illustrate the concept of an exponent.

$$2^3 = 2 \cdot 2 \cdot 2 = 8; \qquad (0.6)^2 = (0.6)(0.6) = 0.36;$$

$$3^5 = 3 \cdot 3 \cdot 3 \cdot 3 \cdot 3 = 243; \qquad \left(\frac{1}{2}\right)^4 = \frac{1}{2} \cdot \frac{1}{2} \cdot \frac{1}{2} \cdot \frac{1}{2} = \frac{1}{16};$$

$$(-5)^2 = (-5)(-5) = 25; \qquad -5^2 = -(5 \cdot 5) = -25.$$

We especially want to call your attention to the last two examples. Notice that $(-5)^2$ means that -5 is the base and is to be used as a factor twice. However, -5^2 means that 5 is the base and that after 5 is squared we take the opposite of that result.

Exponents provide a way of writing algebraic expressions in a compact form. Sometimes we need to change from the compact form to an expanded form as these next examples illustrate.

$$x^4 = x \cdot x \cdot x \cdot x; \qquad (2x)^3 = (2x)(2x)(2x);$$
$$2y^3 = 2 \cdot y \cdot y \cdot y; \qquad (-2x)^3 = (-2x)(-2x)(-2x);$$
$$-3x^5 = -3 \cdot x \cdot x \cdot x \cdot x \cdot x; \qquad -x^2 = -(x \cdot x).$$
$$a^2 + b^2 = a \cdot a + b \cdot b;$$

At other times we need to change from an expanded form to a more compact form using the exponent notation. The next examples illustrate this idea.

$3 \cdot x \cdot x = 3x^2;$

$2 \cdot 5 \cdot x \cdot x \cdot x = 10x^3;$

$3 \cdot 4 \cdot x \cdot x \cdot y = 12x^2y;$

$7 \cdot a \cdot a \cdot a \cdot b \cdot b = 7a^3b^2;$

$(2x)(3y) = 2 \cdot x \cdot 3 \cdot y = 2 \cdot 3 \cdot x \cdot y = 6xy;$

$(3a^2)(4a) = 3 \cdot a \cdot a \cdot 4 \cdot a = 3 \cdot 4 \cdot a \cdot a \cdot a = 12a^3;$

$(-2x)(3x) = -2 \cdot x \cdot 3 \cdot x = -2 \cdot 3 \cdot x \cdot x = -6x^2.$

The commutative and associative properties for multiplication have allowed us to rearrange and regroup factors in the last three examples above.

The concept of *exponent* can be used to extend our work with a variety of concepts previously introduced. Study the following examples very carefully; it should help you to pull together many ideas.

EXAMPLE 1 Simplify $4x^2 + 7x^2 - 2x^2$ by combining similar terms.

Solution By applying the distributive property we obtain

$$4x^2 + 7x^2 - 2x^2 = (4 + 7 - 2)x^2$$
$$= 9x^2.$$ ■

EXAMPLE 2 Simplify $-8x^3 + 9y^2 + 4x^3 - 11y^2$ by combining similar terms.

Solution By rearranging terms and then applying the distributive property we obtain

$$-8x^3 + 9y^2 + 4x^3 - 11y^2 = -8x^3 + 4x^3 + 9y^2 - 11y^2$$
$$= (-8 + 4)x^3 + (9 - 11)y^2$$
$$= -4x^3 - 2y^2.$$ ■

As soon as you feel comfortable with this process of combining similar terms, you may want to do some of the steps mentally. Then your work may appear as follows.

$9a^2 + 6a^2 - 12a^2 = 3a^2;$

$6x^2 + 7y^2 - 3x^2 - 11y^2 = 3x^2 - 4y^2;$

$7x^2y + 5xy^2 - 9x^2y + 10xy^2 = -2x^2y + 15xy^2.$

The next two examples illustrate the use of exponents when reducing fractions.

EXAMPLE 3 Reduce $\dfrac{8x^2y}{12xy}$.

Solution $\dfrac{8x^2y}{12xy} = \dfrac{2 \cdot 2 \cdot 2 \cdot x \cdot x \cdot y}{2 \cdot 2 \cdot 3 \cdot x \cdot y} = \dfrac{2x}{3}.$ ■

EXAMPLE 4 Reduce $\dfrac{15a^2b^3}{25a^3b}$.

Solution $\dfrac{15a^2b^3}{25a^3b} = \dfrac{3 \cdot \cancel{5} \cdot \cancel{a} \cdot \cancel{a} \cdot \cancel{b} \cdot b \cdot b}{\cancel{5} \cdot 5 \cdot \cancel{a} \cdot \cancel{a} \cdot a \cdot \cancel{b}} = \dfrac{3b^2}{5a}.$ ■

Exponents may be used when multiplying and dividing fractions as illustrated in the next three examples.

EXAMPLE 5 Multiply $\left(\dfrac{4x}{6y}\right)\left(\dfrac{12y^2}{7x^2}\right)$ and express answer in reduced form.

Solution $\left(\dfrac{4x}{6y}\right)\left(\dfrac{12y^2}{7x^2}\right) = \dfrac{4 \cdot \overset{2}{\cancel{12}} \cdot \cancel{x} \cdot \cancel{x} \cdot y}{\cancel{6} \cdot 7 \cdot \cancel{y} \cdot x \cdot x} = \dfrac{8y}{7x}.$ ■

EXAMPLE 6 Multiply and simplify

$$\left(\dfrac{8a^3}{9b}\right)\left(\dfrac{12b^2}{16a}\right).$$

Solution $\left(\dfrac{8a^3}{9b}\right)\left(\dfrac{12b^2}{16a}\right) = \dfrac{\cancel{8} \cdot \overset{\overset{2}{\cancel{4}}}{\cancel{12}} \cdot \cancel{a} \cdot a \cdot a \cdot \cancel{b} \cdot b}{\underset{3}{\cancel{9}} \cdot \underset{2}{\cancel{16}} \cdot \cancel{b} \cdot \cancel{a}} = \dfrac{2a^2b}{3}.$ ■

EXAMPLE 7 Divide and express in reduced form

$$\dfrac{-2x^3}{3y^2} \div \dfrac{4}{9xy}.$$

Solution $\dfrac{-2x^3}{3y^2} \div \dfrac{4}{9xy} = -\dfrac{2x^3}{3y^2} \cdot \dfrac{9xy}{4} = -\dfrac{\cancel{2} \cdot \overset{3}{\cancel{9}} \cdot x \cdot x \cdot x \cdot x \cdot \cancel{y}}{\cancel{3} \cdot \underset{2}{\cancel{4}} \cdot \cancel{y} \cdot y} = -\dfrac{3x^4}{2y}.$ ■

The next two examples illustrate the use of exponents when adding and subtracting fractions.

EXAMPLE 8 Add $\dfrac{4}{x^2} + \dfrac{7}{x}$.

Solution   The LCD is x^2. Thus,

$$\frac{4}{x^2}+\frac{7}{x}=\frac{4}{x^2}+\frac{7\cdot x}{x\cdot x}=\frac{4}{x^2}+\frac{7x}{x^2}=\frac{4+7x}{x^2}. \qquad ■$$

EXAMPLE 9   Subtract $\dfrac{3}{xy}-\dfrac{4}{y^2}$.

Solution

$$\left.\begin{array}{l} xy=x\cdot y \\ y^2=y\cdot y \end{array}\right) \qquad \text{The LCD is } xy^2.$$

$$\frac{3}{xy}-\frac{4}{y^2}=\frac{3\cdot y}{xy\cdot y}-\frac{4\cdot x}{y^2\cdot x}=\frac{3y}{xy^2}-\frac{4x}{xy^2}$$

$$=\frac{3y-4x}{xy^2}. \qquad ■$$

Remember that exponents are used to indicate repeated multiplication. Therefore, to simplify numerical expressions containing exponents we proceed as follows.

1. Perform the operations inside the symbols of inclusion (parentheses and brackets) and above and below each fraction bar. Start with the innermost inclusion symbol.
2. Compute all indicated powers.
3. Perform all multiplications and divisions in the order that they appear from left to right.
4. Perform all additions and subtractions in the order that they appear from left to right.

Keep these steps in mind as we evaluate some algebraic expressions containing exponents.

EXAMPLE 10   Evaluate $3x^2-4y^2$ for $x=-2$ and $y=5$.

Solution

$$\begin{aligned} 3x^2-4y^2 &=3(-2)^2-4(5)^2 \quad \text{when } x=-2 \text{ and } y=5 \\ &=3(-2)(-2)-4(5)(5) \\ &=12-100 \\ &=-88. \end{aligned} \qquad ■$$

EXAMPLE 11   Find the value of a^2-b^2 when $a=\dfrac{1}{2}$ and $b=-\dfrac{1}{3}$.

Solution

$$a^2 - b^2 = \left(\frac{1}{2}\right)^2 - \left(-\frac{1}{3}\right)^2 \quad \text{when } a = \frac{1}{2} \text{ and } b = -\frac{1}{3}$$

$$= \frac{1}{4} - \frac{1}{9}$$

$$= \frac{9}{36} - \frac{4}{36}$$

$$= \frac{5}{36}.$$ ■

EXAMPLE 12 Evaluate $5x^2 + 4xy$ for $x = 0.4$ and $y = -0.3$.

Solution

$$5x^2 + 4xy = 5(0.4)^2 + 4(0.4)(-0.3) \quad \text{when } x = 0.4 \text{ and } y = -0.3$$

$$= 5(0.16) + 4(-0.12)$$

$$= 0.80 + (-0.48)$$

$$= 0.32.$$ ■

Problem Set 2.4

For Problems 1–20, find the value of each numerical expression. For example, $2^4 = 2 \cdot 2 \cdot 2 \cdot 2 = 16$.

1. 2^6 **2.** 2^7 **3.** 3^4 **4.** 4^3

5. $(-2)^3$ **6.** $(-2)^5$ **7.** -3^2 **8.** -3^4

9. $(-4)^2$ **10.** $(-5)^4$ **11.** $\left(\frac{2}{3}\right)^4$ **12.** $\left(\frac{3}{4}\right)^3$

13. $-\left(\frac{1}{2}\right)^3$ **14.** $-\left(\frac{3}{2}\right)^3$ **15.** $\left(-\frac{3}{2}\right)^2$ **16.** $\left(-\frac{4}{3}\right)^2$

17. $(0.3)^3$ **18.** $(0.2)^4$ **19.** $-(1.2)^2$ **20.** $-(1.1)^2$

For Problems 21–34, simplify each numerical expression.

21. $3^2 + 2^3 - 4^3$ **22.** $2^4 - 3^3 + 5^2$ **23.** $(-2)^3 - 2^4 - 3^2$

24. $(-3)^3 - 3^2 - 6^2$ **25.** $5(2)^2 - 4(2) - 1$ **26.** $7(-2)^2 - 6(-2) - 8$

27. $-2(3)^3 - 3(3)^2 + 4(3) - 6$ **28.** $5(-3)^3 - 4(-3)^2 + 6(-3) + 1$

29. $-7^2 - 6^2 + 5^2$ **30.** $-8^2 + 3^4 - 4^3$

31. $-3(-4)^2 - 2(-3)^3 + (-5)^2$ **32.** $-4(-3)^3 + 5(-2)^3 - (4)^2$

33. $\dfrac{-3(2)^4}{12} + \dfrac{5(-3)^3}{15}$ **34.** $\dfrac{4(2)^3}{16} - \dfrac{2(3)^2}{6}$

For Problems 35–46, use exponents to help express each algebraic expression in a more compact form. For example, $3 \cdot 5 \cdot x \cdot x \cdot y = 15x^2y$ and $(3x)(2x^2) = 6x^3$.

35. $9 \cdot x \cdot x$

36. $8 \cdot x \cdot x \cdot x \cdot y$

37. $3 \cdot 4 \cdot x \cdot y \cdot y$

38. $7 \cdot 2 \cdot a \cdot a \cdot b \cdot b \cdot b$

39. $-2 \cdot 9 \cdot x \cdot x \cdot x \cdot x \cdot y$

40. $-3 \cdot 4 \cdot x \cdot y \cdot z \cdot z$

41. $(5x)(3y)$

42. $(3x^2)(2y)$

43. $(6x^2)(2x^2)$

44. $(-3xy)(6xy)$

45. $(-4a^2)(-2a^3)$

46. $(-7a^3)(-3a)$

For Problems 47–54, simplify each expression by combining similar terms.

47. $3x^2 - 7x^2 - 4x^2$

48. $-2x^3 + 7x^3 - 4x^3$

49. $-12y^3 + 17y^3 - y^3$

50. $-y^3 + 8y^3 - 13y^3$

51. $7x^2 - 2y^2 - 9x^2 + 8y^2$

52. $5x^3 + 9y^3 - 8x^3 - 14y^3$

53. $\dfrac{2}{3}n^2 - \dfrac{1}{4}n^2 - \dfrac{3}{5}n^2$

54. $-\dfrac{1}{2}n^2 + \dfrac{5}{6}n^2 - \dfrac{4}{9}n^2$

For Problems 55–64, reduce each fraction to simplest form.

55. $\dfrac{9xy}{15x}$

56. $\dfrac{8x^2y}{14x}$

57. $\dfrac{22xy^2}{6xy^3}$

58. $\dfrac{18x^3y}{12xy^4}$

59. $\dfrac{7a^2b^3}{17a^3b}$

60. $\dfrac{9a^3b^3}{22a^4b^2}$

61. $\dfrac{-24abc^2}{32bc}$

62. $\dfrac{4a^2c^3}{-22b^2c^4}$

63. $\dfrac{-5x^4y^3}{-20x^2y}$

64. $\dfrac{-32xy^2z^4}{-48x^3y^3z}$

For Problems 65–82, perform the indicated operations and express answers in reduced form.

65. $\left(\dfrac{7x^2}{9y}\right)\left(\dfrac{12y}{21x}\right)$

66. $\left(\dfrac{3x}{8y^2}\right)\left(\dfrac{14xy}{9y}\right)$

67. $\left(\dfrac{5c}{a^2b^2}\right) \div \left(\dfrac{12c}{ab}\right)$

68. $\left(\dfrac{13ab^2}{12c}\right) \div \left(\dfrac{26b}{14c}\right)$

69. $\dfrac{6}{x} + \dfrac{5}{y^2}$

70. $\dfrac{8}{y} - \dfrac{6}{x^2}$

71. $\dfrac{5}{x^4} - \dfrac{7}{x^2}$

72. $\dfrac{9}{x} - \dfrac{11}{x^3}$

73. $\dfrac{3}{2x^3} + \dfrac{6}{x}$

74. $\dfrac{5}{3x^2} + \dfrac{6}{x}$

75. $\dfrac{-5}{4x^2} + \dfrac{7}{3x^2}$

76. $\dfrac{-8}{5x^3} + \dfrac{10}{3x^3}$

77. $\dfrac{11}{a^2} - \dfrac{14}{b^2}$

78. $\dfrac{9}{x^2} + \dfrac{8}{y^2}$

79. $\dfrac{1}{2x^3} - \dfrac{4}{3x^2}$

80. $\dfrac{2}{3x^3} - \dfrac{5}{4x}$

81. $\dfrac{3}{x} - \dfrac{4}{y} - \dfrac{5}{xy}$

82. $\dfrac{5}{x} + \dfrac{7}{y} - \dfrac{1}{xy}$

For Problems 83–92, evaluate each algebraic expression for the given values of the variables.

83. $4x^2 + 7y^2$ for $x = -2$ and $y = -3$

84. $5x^2 + 2y^3$ for $x = -4$ and $y = -1$

85. $3x^2 - y^2$ for $x = \dfrac{1}{2}$ and $y = -\dfrac{1}{3}$

86. $x^2 - 2y^2$ for $x = -\dfrac{2}{3}$ and $y = \dfrac{3}{2}$

87. $x^2 - 2xy + y^2$　for $x = -\dfrac{1}{2}$ and $y = 2$　　**88.** $x^2 + 2xy + y^2$　for $x = -\dfrac{3}{2}$ and $y = -2$

89. $-x^2$　for $x = -8$　　　　　　　　　**90.** $-x^3$　for $x = 5$

91. $-x^2 - y^2$　for $x = -3$ and $y = -4$　　**92.** $-x^2 + y^2$　for $x = -2$ and $y = 6$

93. $-a^2 - 3b^3$　for $a = -6$ and $b = -1$　　**94.** $-a^3 + 3b^2$　for $a = -3$ and $b = -5$

95. $y^2 - 3xy$　for $x = 0.4$ and $y = -0.3$　　**96.** $x^2 + 5xy$　for $x = -0.2$ and $y = -0.6$

2.5　Translating from English to Algebra

In order to use the tools of algebra for solving problems, we must be able to translate back and forth between the English language and the language of algebra. In this section we want to translate algebraic expressions to English phrases (word phrases) and English phrases to algebraic expressions.

Let's begin by considering the following translations from algebraic expressions to word phrases.

Algebraic expression	Word phrase
$x + y$	The sum of x and y
$x - y$	The difference of x and y
xy	The product of x and y
$\dfrac{x}{y}$	The quotient of x and y
$3x$	The product of 3 and x
$x^2 + y^2$	The sum of x squared and y squared
$2x - 3y$	The difference of two times x and three times y
$2xy$	The product of 2, x, and y
$2(x + y)$	Two times the quantity x plus y
$x - 3$	Three less than x

Now let's consider the reverse process, translating from word phrases to algebraic expressions. Part of the difficulty in translating from English to algebra is the fact that different word phrases translate to the same algebraic expression. So we need to become familiar with *different ways of saying the same thing*, especially pertaining to the four fundamental operations. The following examples should help to acquaint you with some of the phrases used in the basic operations.

$$\left.\begin{array}{l}\text{The sum of } x \text{ and } 4 \\ \text{The sum ``}x \text{ plus } 4\text{''}\end{array}\right] \quad \rightarrow x + 4$$

$$\begin{bmatrix} \text{The difference of } n \text{ and 5} \\ \text{The difference of 5 subtracted from } n \\ \text{The difference "} n \text{ minus 5"} \\ \text{The difference "} n \text{ less 5"} \end{bmatrix} \rightarrow n - 5$$

$$\begin{bmatrix} \text{The product of 4 and } y \\ \text{The product "4 times } y\text{"} \end{bmatrix} \rightarrow 4y$$

$$\begin{bmatrix} \text{The quotient of } n \text{ and 6} \\ \text{The quotient when } n \text{ is divided by 6} \\ \text{The quotient "} n \text{ divided by 6"} \\ \text{The quotient of 6 divided into } n \end{bmatrix} \rightarrow \frac{n}{6}$$

Often a word phrase indicates more than one operation. Furthermore, the standard vocabulary of sum, difference, product, and quotient may be replaced by other terminology. Study the following translations very carefully.

Word phrase	Algebraic expression
The sum of two times x and three times y	$2x + 3y$
The sum of the squares of a and b	$a^2 + b^2$
Five times x divided by y	$\dfrac{5x}{y}$
Two more than the square of x	$x^2 + 2$
Three less than the cube of b	$b^3 - 3$
Five less than the product of x and y	$xy - 5$
Three less x	$3 - x$
Nine minus the product of x and y	$9 - xy$
Four times the sum of x and 2	$4(x + 2)$
Six times the quantity w minus 4	$6(w - 4)$

Suppose you are told that the sum of two numbers is 12 and one of the numbers is 8. What is the other number? The other number is $12 - 8$ which equals 4. Now suppose that you are told that the product of two numbers is 56 and one of the numbers is 7. What is the other number? The other number is $56 \div 7$, which equals 8. The following examples illustrate the use of these addition-subtraction and multiplication-division relationships in a more general setting.

EXAMPLE 1   The sum of two numbers is 83 and one of the numbers is x. What is the other number?

Solution   Using the addition and subtraction relationship we can represent the other number by $83 - x$.   ■

EXAMPLE 2   The difference of two numbers is 14. The smaller number is n. What is the larger number?

Solution   Since the smaller number plus the difference must equal the larger number, we can represent the larger number by $n + 14$.   ■

EXAMPLE 3   The product of two numbers is 39 and one of the numbers is y. Represent the other number.

Solution   Using the multiplication and division relationship we can represent the other number by $\dfrac{39}{y}$.   ■

The "English statement" may not contain any key words such as sum, difference, product, or quotient. Instead, the statement may describe a physical situation; from this description you need to deduce the operations involved. Some suggestions for handling such situations are made in the following examples.

EXAMPLE 4   Arlene can type 70 words per minute. How many words can she type in m minutes?

Solution   In 10 minutes she would type $70(10) = 700$ words. In 50 minutes she would type $70(50) = 3500$ words. Thus, in m minutes she would type $70m$ words.   ■

Notice the use of some specific examples—$70(10) = 700$ and $70(50) = 3500$—to help formulate the general expression. This technique of first formulating some specific examples and then generalizing can be very effective.

EXAMPLE 5   Lynn has n nickels and d dimes. Express, in cents, this amount of money.

Solution   Three nickels and 8 dimes would be $5(3) + 10(8) = 95$ cents. Thus, n nickels and d dimes would be $5n + 10d$ cents.   ■

EXAMPLE 6   A train travels at the rate of r miles per hour. How far will it travel in 8 hours?

Solution   Suppose that a train travels at 50 miles per hour. Using the formula "distance equals rate times time," it would travel $50 \cdot 8 = 400$ miles. Therefore, at r miles per hour, it would travel $r \cdot 8$ miles. The expression $r \cdot 8$ is usually written as $8r$.   ■

EXAMPLE 7   The cost of a 5-pound box of candy is d dollars. How much is the cost per pound for the candy?

Solution   The price per pound is figured by dividing the total cost by the number of pounds.

Therefore, the price per pound is represented by $\dfrac{d}{5}$.   ■

The "English statement" to be translated to algebra may contain some geometric ideas. For example, suppose that we want to express, in inches, the length of a line segment that is f feet long. Since 1 foot = 12 inches, we can represent f feet by 12 times f, written as $12f$ inches.

Tables 2.1 and 2.2 contain some of the basic relationships pertaining to linear measurement in the English and metric systems, respectively. (Additional tables pertaining to both systems are located in the inside back cover of this book.)

Table 2.1	English System
	12 inches = 1 foot
	3 feet = 36 inches = 1 yard
	5280 feet = 1760 yards = 1 mile

Table 2.2	Metric System
	1 kilometer = 1000 meters
	1 hectometer = 100 meters
	1 dekameter = 10 meters
	1 decimeter = 0.1 meter
	1 centimeter = 0.01 meter
	1 millimeter = 0.001 meter

EXAMPLE 8 The distance between two cities is k kilometers. Express this distance in meters.

Solution Since 1 kilometer equals 1000 meters, we need to multiply k by 1000. Therefore, the distance, in meters, is represented by 1000k. ■

EXAMPLE 9 The length of a line segment is i inches. Express that length in yards.

Solution To change from inches to yards, we must divide by 36. Therefore, $\dfrac{i}{36}$ represents, in yards, the length of the line segment. ■

EXAMPLE 10 The width of a rectangle is w centimeters and the length is 5 centimeters less than twice the width. What is the length of the rectangle? What is the perimeter of the rectangle?

Solution The length of the rectangle can be represented by $2w - 5$. Now we can sketch a rectangle and record the given information. The perimeter of a rectangle is

the sum of the lengths of the four sides. Therefore, the perimeter is given by $2w + 2(2w - 5)$, which can be written as $2w + 4w - 10$ and then simplified to $6w - 10$. ■

Problem Set 2.5

For Problems 1–12, write a word phrase for each of the algebraic expressions. For example, *lw* can be expressed as "the product of *l* and *w*."

1. $a - b$ **2.** $x + y$ **3.** $\dfrac{1}{3}Bh$ **4.** $\dfrac{1}{2}bh$

5. $2(l + w)$ **6.** πr^2 **7.** $\dfrac{A}{w}$ **8.** $\dfrac{C}{\pi}$

9. $\dfrac{a + b}{2}$ **10.** $\dfrac{a - b}{4}$ **11.** $3y + 2$ **12.** $3(x - y)$

For Problems 13–36, translate each word phrase into an algebraic expression. For example, "the sum of *x* and 14" translates into *x* + 14.

13. The sum of l and w **14.** The difference of x and y

15. The product of a and b **16.** The product of $\dfrac{1}{3}$, B, and h

17. The quotient of d divided by t **18.** The quotient of r divided into d

19. The product of l, w, and h **20.** The product of π and the square of r

21. The difference of x subtracted from y **22.** The difference "x subtract y"

23. Two larger than the product of x and y **24.** Six plus the cube of x

25. Seven minus the square of y **26.** The quantity, x minus 2, cubed

27. The quantity, x minus y, divided by four   **28.** Eight less than x

29. Ten less x **30.** Nine times the quantity, n minus four

31. Ten times the quantity, n plus two

32. The sum of four times x and five times y

33. The difference of seven subtracted from the product of x and y

34. Three times the sum of n and 2

35. Twelve less than the product of x and y

36. Twelve less the product of x and y

For Problems 37–68, answer the question with an algebraic expression.

37. The sum of two numbers is 35 and one of the numbers is n. What is the other number?

38. The sum of two numbers is 100 and one of the numbers is x. What is the other number?

39. The difference of two numbers is 45 and the smaller number is n. What is the other number?

40. The product of two numbers is 25 and one of the numbers is x. What is the other number?

41. Janet is y years old. How old will she be in 10 years? $y + 10$

42. Hector is y years old. How old was he 5 years ago? $y - 5$

43. Debra is x years old and her mother is 3 years less than twice as old as Debra. How old is Debra's mother?

44. Jack is x years old and Dudley is 1 year more than three times as old as Jack. How old is Dudley?

45. Donna has d dimes and q quarters in her bank. How much money, in cents, does she have?

46. Andy has c cents that is all in dimes. How many dimes does he have?

47. A car travels d miles in t hours. What is the rate of the car?

48. If g gallons of gas cost d dollars, what is the price per gallon?

49. If p pounds of candy cost d dollars, what is the price per pound?

50. Sue can type x words per minute. How many words can she type in an hour?

51. Larry's annual salary is d dollars. What is his monthly salary?

52. Nancy's monthly salary is d dollars. What is her annual salary?

53. If n represents a whole number, what represents the next larger whole number?

54. If n represents an even number, what represents the next larger even number?

55. If n represents an odd number, what represents the next larger odd number?

56. Maria is y years old and her sister is twice as old. What represents the sum of their ages?

57. Willie is y years old and his father is 2 years less than twice Willie's age. What represents the sum of their ages?

58. Harriet has p pennies, n nickels, and d dimes. How much money, in cents, does she have?

59. The perimeter of a rectangle is y yards and f feet. What is the perimeter expressed in inches?

60. The perimeter of a triangle is m meters and c centimeters. What is the perimeter expressed in centimeters?

61. A rectangular plot of ground is f feet long. What is its length expressed in yards?

62. The height of a telephone pole is f feet. What is the height expressed in yards?

63. The width of a rectangle is w feet and its length is three times the width. What is the perimeter of the rectangle expressed in feet?

64. The width of a rectangle is w feet and its length is 1 foot more than twice its width. What is the perimeter of the rectangle expressed in feet?

65. The length of a rectangle is l inches and its width is 2 inches less than one-half of its length. What is the perimeter of the rectangle expressed in inches?

66. The length of a rectangle is l inches and its width is 3 inches more than one-third of its length. What is the perimeter of the rectangle expressed in inches?

67. The first side of a triangle is f feet long. The second side is 2 feet longer than the first side. The third side is twice as long as the second side. What is the perimeter of the triangle expressed in inches?

68. The first side of a triangle is y yards long. The second side is 3 yards shorter than the first side. The third side is 3 times as long as the second side. What is the perimeter of the triangle expressed in feet?

Chapter 2 Summary

(2.1) The property $\dfrac{a \cdot k}{b \cdot k} = \dfrac{a}{b}$ is used to express fractions in reduced form.

To **multiply** rational numbers in common fractional form we multiply numerators, multiply denominators, and express the result in reduced form.

To **divide** rational numbers in common fractional form we invert the divisor and multiply.

(2.2) **Addition** and **subtraction** of rational numbers in common fractional form are based on the following.

$$\frac{a}{b} + \frac{c}{b} = \frac{a+c}{b}; \qquad \text{(Addition)}$$

$$\frac{a}{b} - \frac{c}{b} = \frac{a-c}{b}. \qquad \text{(Subtraction)}$$

To add or subtract fractions that do not have a common denominator, we use the fundamental principle of fractions, $\frac{a}{b} = \frac{a \cdot k}{b \cdot k}$, and obtain equivalent fractions having a common denominator.

(2.3) To **add** or **subtract decimals**, we write the numbers in a column so that the decimal points are lined up, and then we add or subtract as we do with integers.

To **multiply decimals** we (1) multiply the numbers ignoring the decimal points, and then (2) insert the decimal point in the product so that the number of digits to the right of the decimal point in the product is equal to the sum of the numbers of digits to the right of the decimal point in each factor.

To **divide a decimal by a nonzero whole number** we (1) place the decimal point in the quotient directly above the decimal point in the dividend, and then (2) divide as with whole numbers, except that in the division process, zeros are placed in the quotient immediately to the right of the decimal point as needed to show the correct place value.

To **divide by a decimal** we change to an equivalent problem having a whole number divisor.

(2.4) Expressions of the form b^n, where

$$b^n = b \cdot b \cdot b \cdots b, \qquad (n \text{ factors of } b)$$

are read as "b to the nth power"; b is the *base* and n is the *exponent*.

(2.5) To translate English phrases to algebraic expressions we must be familiar with the standard vocabulary of **sum**, **difference**, **product**, and **quotient** as well as other terms used to express the same ideas.

Chapter 2 Review Problem Set

Find the value of each of the following. (Problems 1–10)

1. 2^6 **2.** $(-3)^3$ **3.** -4^2 **4.** $\left(\frac{3}{4}\right)^2$

5. $\left(\dfrac{1}{2} + \dfrac{2}{3}\right)^2$ **6.** $(0.6)^3$ **7.** $(0.12)^2$ **8.** $(0.06)^2$

9. $\left(-\dfrac{2}{3}\right)^3$ **10.** $\left(-\dfrac{1}{2}\right)^4$

Perform the indicated operations and express answers in reduced form. (Problems 11–20)

11. $\dfrac{3}{8} + \dfrac{5}{12}$

12. $\dfrac{9}{14} - \dfrac{3}{35}$

13. $\dfrac{2}{3} + \dfrac{-3}{5}$

14. $\dfrac{7}{x} + \dfrac{9}{2y}$

15. $\dfrac{5}{xy} - \dfrac{8}{x^2}$

16. $\left(\dfrac{7y}{8x}\right)\left(\dfrac{14x}{35}\right)$

17. $\left(\dfrac{6xy}{9y^2}\right) \div \left(\dfrac{15y}{18x^2}\right)$

18. $\left(\dfrac{-3x}{12y}\right)\left(\dfrac{8y}{-7x}\right)$

19. $\left(-\dfrac{4y}{3x}\right)\left(-\dfrac{3x}{4y}\right)$

20. $\left(\dfrac{6n}{7}\right)\left(\dfrac{9n}{8}\right)$

Simplify each of the following numerical expressions. (Problems 21–30)

21. $\dfrac{1}{6} + \dfrac{2}{3} \cdot \dfrac{3}{4} - \dfrac{5}{6} \div \dfrac{8}{6}$

22. $\dfrac{3}{4} \cdot \dfrac{1}{2} - \dfrac{4}{3} \cdot \dfrac{3}{2}$

23. $\dfrac{7}{9} \cdot \dfrac{3}{5} + \dfrac{7}{9} \cdot \dfrac{2}{5}$

24. $\dfrac{4}{5} \div \dfrac{1}{5} \cdot \dfrac{2}{3} - \dfrac{1}{4}$

25. $\dfrac{2}{3} \cdot \dfrac{1}{4} \div \dfrac{1}{2} + \dfrac{2}{3} \cdot \dfrac{1}{4}$

26. $0.48 + 0.72 - 0.35 - 0.18$

27. $0.81 + (0.6)(0.4) - (0.7)(0.8)$

28. $1.28 \div 0.8 - 0.81 \div 0.9 + 1.7$

29. $(0.3)^2 + (0.4)^2 - (0.6)^2$

30. $(1.76)(0.8) + (1.76)(0.2)$

Simplify each of the following algebraic expressions by combining similar terms. Express answers in reduced form when working with common fractions. (Problems 31–36)

31. $\dfrac{3}{8}x^2 - \dfrac{2}{5}y^2 - \dfrac{2}{7}x^2 + \dfrac{3}{4}y^2$

32. $0.24ab + 0.73bc - 0.82ab - 0.37bc$

33. $\dfrac{1}{2}x + \dfrac{3}{4}x - \dfrac{5}{6}x + \dfrac{1}{24}x$

34. $1.4a - 1.9b + 0.8a + 3.6b$

35. $\dfrac{2}{5}n + \dfrac{1}{3}n - \dfrac{5}{6}n$

36. $n - \dfrac{3}{4}n + 2n - \dfrac{1}{5}n$

Evaluate the following algebraic expressions for the given values of the variables. (Problems 37–42)

37. $\dfrac{1}{4}x - \dfrac{2}{5}y$ for $x = \dfrac{2}{3}$ and $y = -\dfrac{5}{7}$

38. $a^3 + b^2$ for $a = -\dfrac{1}{2}$ and $b = \dfrac{1}{3}$

39. $2x^2 - 3y^2$ for $x = 0.6$ and $y = 0.7$

40. $0.7w + 0.9z$ for $w = 0.4$ and $z = -0.7$

41. $\dfrac{3}{5}x - \dfrac{1}{3}x + \dfrac{7}{15}x - \dfrac{2}{3}x$ for $x = \dfrac{15}{17}$

42. $\dfrac{1}{3}n + \dfrac{2}{7}n - n$ for $n = 21$

Answer each of the following questions with an algebraic expression. (Problems 43–50)

43. The sum of two numbers is 72 and one of the numbers is n. What is the other number?

44. Joan has p pennies and d dimes. How much money, in cents, does she have?

45. Ellen types x words in an hour. What is her typing rate per minute?

46. Harry is y years old. His brother is 3 years less than twice as old as Harry. How old is Harry's brother?

47. Larry chose a number n. Cindy chose a number 3 more than 5 times the number chosen by Larry. What number did Cindy choose?

48. The height of a file cabinet is y yards and f feet. How tall is the file cabinet expressed in inches?

49. The length of a rectangular room is m meters. How long, in centimeters, is the room?

50. Corinne has n nickels, d dimes, and q quarters. How much money, in cents, does she have?

For Problems 51–60, translate each word phrase into an algebraic expression.

51. Five less than n

52. Five less n

53. Ten times the quantity x minus 2

54. Ten times x minus 2

55. The difference of x squared and y cubed

56. The quotient of d divided by r

57. x squared plus nine

58. x plus nine, the quantity squared

59. The sum of the cubes of x and y

60. Four less than the product of x and y

Cumulative Review Problem Set

For Problems 1–12, simplify each of the numerical expressions.

1. $16 - 18 - 14 + 21 - 14 + 19$

2. $7(-6) - 8(-6) + 4(-9)$

3. $6 - [3 - (10 - 12)]$

4. $-9 - 2[4 - (-10 + 6)] - 1$

5. $\dfrac{-7(-4) - 5(-6)}{-2}$

6. $\dfrac{5(-3) + (-4)(6) - 3(4)}{-3}$

7. $\dfrac{3}{4} + \dfrac{1}{3} \div \dfrac{4}{3} - \dfrac{1}{2}$

8. $\left(\dfrac{2}{3}\right)\left(-\dfrac{3}{4}\right) - \left(\dfrac{5}{6}\right)\left(\dfrac{4}{5}\right)$

9. $\left(\dfrac{1}{2} - \dfrac{2}{3}\right)^2$

10. -4^3

11. $\left(-\dfrac{1}{2}\right)^3 - \left(-\dfrac{3}{4}\right)^2$

12. $(.2)^2 - (.3)^3 + (.4)^2$

For Problems 13–20, evaluate each algebraic expression for the given values of the variables.

13. $3xy - 2x - 4y$ for $x = -6$ and $y = 7$

14. $-4x^2y - 2xy^2 + xy$ for $x = -2$ and $y = -4$

15. $\dfrac{5x - 2y}{3x}$ for $x = \dfrac{1}{2}$ and $y = -\dfrac{1}{3}$

16. $.2x - .3y + 2xy$ for $x = .1$ and $y = .3$

17. $-7x + 4y + 6x - 9y + x - y$ for $x = -.2$ and $y = .4$

18. $\dfrac{2}{3}x - \dfrac{3}{5}y + \dfrac{3}{4}x - \dfrac{1}{2}y$ for $x = \dfrac{6}{5}$ and $y = -\dfrac{1}{4}$

19. $\dfrac{1}{5}n - \dfrac{1}{3}n + n - \dfrac{1}{6}n$ for $n = \dfrac{1}{5}$

20. $-ab + \dfrac{1}{5}a - \dfrac{2}{3}b$ for $a = -2$ and $b = \dfrac{3}{4}$

For Problems 21–24, express each of the numbers as a product of prime factors.

21. 54   22. 78   23. 91   24. 153

For Problems 25–28, find the greatest common factor of the given numbers.

25. 42 and 70   26. 63 and 81   27. 28, 36, and 52   28. 48, 66, and 78

For Problems 29–32, find the least common multiple of the given numbers.

29. 20 and 28   30. 40 and 100   31. 12, 18, and 27   32. 16, 20, and 80

For Problems 33–38, simplify each algebraic expression by combining similar terms.

33. $\dfrac{2}{3}x - \dfrac{1}{4}y - \dfrac{3}{4}x - \dfrac{2}{3}y$

34. $-n - \dfrac{1}{2}n + \dfrac{3}{5}n + \dfrac{5}{6}n$

35. $3.2a - 1.4b - 6.2a + 3.3b$

36. $-(n - 1) + 2(n - 2) - 3(n - 3)$

37. $-x + 4(x - 1) - 3(x + 2) - (x + 5)$

38. $2a - 5(a + 3) - 2(a - 1) - 4a$

For Problems 39–46, perform the indicated operations and express answers in reduced form.

39. $\dfrac{5}{12} - \dfrac{3}{16}$

40. $\dfrac{3}{4} - \dfrac{5}{6} - \dfrac{7}{9}$

41. $\dfrac{5}{xy} - \dfrac{2}{x} + \dfrac{3}{y}$

42. $-\dfrac{7}{x^2} + \dfrac{9}{xy}$

43. $\left(\dfrac{7x}{9y}\right)\left(\dfrac{12y}{14}\right)$

44. $\left(-\dfrac{5a}{7b^2}\right)\left(-\dfrac{8ab}{15}\right)$

45. $\left(\dfrac{6x^2y}{11}\right) \div \left(\dfrac{9y^2}{22}\right)$

46. $\left(-\dfrac{9a}{8b}\right) \div \left(\dfrac{12a}{18b}\right)$

For Problems 47–50, answer each question with an algebraic expression.

47. Hector has p pennies, n nickels, and d dimes. How much money, in cents, does he have?

48. Ginny chose a number n. Penny chose a number 5 less than 4 times the number chosen by Ginny. What number did Penny choose?

49. The height of a flagpole is y yards, f feet, and i inches. How tall is the flagpole expressed in inches?

50. A rectangular room is x meters by y meters. What is its perimeter expressed in centimeters?

3 First Degree Equations of One Variable

Throughout this book we will follow a common theme of developing some skills; then using the skills to help solve equations and inequalities; and finally using equations and inequalities to solve applied problems. In this chapter we want to use the skills we have developed in the first two chapters to help us solve equations and begin our work with applied problems.

3.1 Solving First-Degree Equations

In the previous chapters we made statements like $3 + (-4) = -1$; $(-5)(-6) = 30$; and $-30 \div 10 = -3$. Each of those statements is an example of an *equation*—also called a statement of equality. The symbol, $=$, is called an *equal sign*. In general, $a = b$ states that a and b are two different names for the same thing. Thus, in our first example, $3 + (-4)$ and -1 are two different names for the same number. The statement $a \neq b$ (read "a is not equal to b") means that a and b do not name the same thing.

Numerical Equations may be true or false. For example, the equation $3 + 4 = 7$ is true, but the equation $7 + 4 = 13$ is false.

Algebraic Equations contain one or more variables. The following are examples of algebraic equations.

$$x + 3 = 4; \qquad 2x + 3y = 4;$$
$$2x - 7 = 12; \qquad 6ab = 7.$$
$$x^2 + 2x - 6 = 0;$$

An algebraic equation such as $x + 3 = 4$ is neither true nor false as it stands and is sometimes called an **open sentence**. Each time a number is substituted for x, the algebraic equation $x + 3 = 4$ becomes a numerical statement that is true or false. For example, if $x = 5$, then $x + 3 = 4$ becomes $5 + 3 = 4$, which is false. If $x = 1$, then $x + 3 = 4$ becomes $1 + 3 = 4$, which is a true statement. **Solving an equation** refers to the process of finding the number (or numbers) that makes an algebraic equation a true numerical statement. Such numbers are called the **solutions** or **roots** of the equation and are said to **satisfy** the equation. The set of all solutions of an equation is called its **solution set**. Thus, $\{1\}$ is the solution set of $x + 3 = 4$. A set consisting of no elements is called the **empty** or **null set** and denoted by $\varnothing$. For example, the solution set of $x + 3 = x + 2$ is $\varnothing$; that is, there are no real numbers that satisfy $x + 3 = x + 2$.

In this chapter we shall consider techniques for solving **first-degree equations of one variable**. This means that the equations contain only one variable and this variable has an exponent of one. The following are examples of first degree equations of one variable.

$$3x + 4 = 7; \qquad 8w + 7 = 5w - 4;$$
$$\frac{1}{2}y + 2 = 9; \qquad 7x + 2x - 1 = 4x - 1.$$

Equivalent equations are equations having the same solution set. For example,

$$5x - 4 = 3x + 8,$$
$$2x = 12,$$
$$x = 6,$$

are all equivalent equations. This can be verified by showing that 6 is the solution for all three equations.

The general procedure for solving an equation is to continue replacing the given equation with equivalent, but simpler, equations until an equation of the form **variable = constant** or **constant = variable** is obtained. Thus, in the preceding example, $5x - 4 = 3x + 8$ has been simplified to $2x = 12$, which has been further simplified to $x = 6$, from which the solution of 6 is obvious. The exact procedure for simplifying equations becomes our next concern.

Two properties of equality play an important role in the process of solving equations. The first of these is the **addition-subtraction property of equality**, which can be stated as follows.

Property 3.1

For all real numbers a, b, and c,

1. $a = b$ if and only if $a + c = b + c$;
2. $a = b$ if and only if $a - c = b - c$.

Property 3.1 states that *any number can be added to or subtracted from both sides of an equation and an equivalent equation is produced.* Consider the use of this property in the next four examples.

EXAMPLE 1   Solve $x - 8 = 3$.

Solution

$$x - 8 = 3$$
$$x - 8 + 8 = 3 + 8 \qquad \text{Add 8 to both sides.}$$
$$x = 11.$$

The solution set is $\{11\}$.   ■

Remark   It is true that a simple equation such as in Example 1 can be solved "by inspection." That is to say, we could think "some number minus 8 produces 3." Obviously, the number is 11. However, as the equations become more complex, the technique of solving by inspection becomes ineffective. So it is necessary to develop more formal techniques for solving equations. Therefore, we will begin developing such techniques even with very simple types of equations.

EXAMPLE 2   Solve $x + 14 = -8$.

Solution
$$x + 14 = -8$$
$$x + 14 - 14 = -8 - 14 \qquad \text{Subtract 14 from both sides.}$$
$$x = -22.$$

The solution set is $\{-22\}$. ■

EXAMPLE 3 Solve $n - \dfrac{1}{3} = \dfrac{1}{4}$.

Solution
$$n - \frac{1}{3} = \frac{1}{4}$$

$$n - \frac{1}{3} + \frac{1}{3} = \frac{1}{4} + \frac{1}{3} \qquad \text{Add } \frac{1}{3} \text{ to both sides.}$$

$$n = \frac{3}{12} + \frac{4}{12}$$

$$n = \frac{7}{12}.$$

The solution set is $\left\{\dfrac{7}{12}\right\}$. ■

EXAMPLE 4 Solve $0.72 = y + 0.35$.

Solution
$$0.72 = y + 0.35$$
$$0.72 - 0.35 = y + 0.35 - 0.35 \qquad \text{Subtract 0.35 from both sides.}$$
$$0.37 = y.$$

The solution set is $\{0.37\}$. ■

Note that in Example 4 the final equation is $0.37 = y$ instead of $y = 0.37$. Technically, the **symmetry property of equality** (if $a = b$, then $b = a$) would permit us to change from $0.37 = y$ to $y = 0.37$, but such a change is not necessary to determine that the solution is 0.37.

One other comment pertaining to Property 3.1 should be made at this time. Since subtracting a number is equivalent to adding its opposite, Property 3.1 could be stated only in terms of addition. Thus, to solve an equation such as in Example 4 we could add -0.35 to both sides rather than subtract 0.35 from both sides.

The other important property pertaining to solving equations is the **multiplication-division property of equality**.

> **Property 3.2**
>
> For all real numbers, a, b, and c, where $c \neq 0$,
>
>   **1.** $a = b$ if and only if $ac = bc$;
>
>   **2.** $a = b$ if and only if $\dfrac{a}{c} = \dfrac{b}{c}$.

Property 3.2 states that *an equivalent equation is obtained whenever both sides of a given equation are multiplied or divided by the same nonzero real number.* The following examples illustrate the use of this property.

EXAMPLE 5   Solve $\dfrac{3}{4} x = 6$.

Solution

$$\frac{3}{4} x = 6$$

$$\frac{4}{3}\left(\frac{3}{4} x\right) = \frac{4}{3}(6) \qquad \text{Multiply both sides by } \frac{4}{3} \text{ since } \left(\frac{4}{3}\right)\left(\frac{3}{4}\right) = 1.$$

$$x = 8.$$

The solution set is $\{8\}$.   ■

EXAMPLE 6   Solve $5x = 27$.

Solution

$$5x = 27$$

$$\frac{5x}{5} = \frac{27}{5} \qquad \text{Divide both sides by 5.}$$

$$x = \frac{27}{5}. \qquad \frac{27}{5} \text{ could be expressed as } 5\frac{2}{5} \text{ or } 5.4.$$

The solution set is $\left\{\dfrac{27}{5}\right\}$.   ■

EXAMPLE 7   Solve $-\dfrac{2}{3} p = \dfrac{1}{2}$.

Solution

$$-\frac{2}{3}p = \frac{1}{2}$$

$$-\frac{3}{2}\left(-\frac{2}{3}p\right) = \left(-\frac{3}{2}\right)\left(\frac{1}{2}\right) \qquad \text{Multiply both sides by } -\frac{3}{2}$$

$$\text{since } \left(-\frac{3}{2}\right)\left(-\frac{2}{3}\right) = 1.$$

$$p = -\frac{3}{4}.$$

The solution set is $\left\{-\frac{3}{4}\right\}$. ■

EXAMPLE 8 Solve $26 = -6x$.

Solution

$$26 = -6x$$

$$\frac{26}{-6} = \frac{-6x}{-6} \qquad \text{Divide both sides by } -6.$$

$$-\frac{26}{6} = x \qquad\qquad \frac{26}{-6} = -\frac{26}{6}.$$

$$-\frac{13}{3} = x. \qquad\qquad \text{Don't forget to reduce!}$$

The solution set is $\left\{-\frac{13}{3}\right\}$. ■

Looking back at Examples 5–8, you will notice that we divided both sides of the equation by the coefficient of the variable whenever it was an integer; otherwise we used the multiplication part of Property 3.2. Technically, we should realize that since dividing by a number is equivalent to multiplying by its reciprocal, Property 3.2 could be stated only in terms of multiplication. Thus, to solve an equation such as $5x = 27$, we could multiply both sides by $\frac{1}{5}$ instead of dividing both sides by 5.

Problem Set 3.1

Use the properties of equality to help solve each of the following equations.

1. $x + 9 = 17$
2. $x + 7 = 21$
3. $x + 11 = 5$
4. $x + 13 = 2$
5. $-7 = x + 2$
6. $-12 = x + 4$

7. $8 = n + 14$ **8.** $6 = n + 19$ **9.** $21 + y = 34$

10. $17 + y = 26$ **11.** $x - 17 = 31$ **12.** $x - 22 = 14$

13. $14 = x - 9$ **14.** $17 = x - 28$ **15.** $-26 = n - 19$

16. $-34 = n - 15$ **17.** $y - \dfrac{2}{3} = \dfrac{3}{4}$ **18.** $y - \dfrac{2}{5} = \dfrac{1}{6}$

19. $x + \dfrac{3}{5} = \dfrac{1}{3}$ **20.** $x + \dfrac{5}{8} = \dfrac{2}{5}$ **21.** $b + 0.19 = 0.46$

22. $b + 0.27 = 0.74$ **23.** $n - 1.7 = -5.2$ **24.** $n - 3.6 = -7.3$

25. $15 - x = 32$ **26.** $13 - x = 47$ **27.** $-14 - n = 21$

28. $-9 - n = 61$ **29.** $7x = -56$ **30.** $9x = -108$

31. $-6x = 102$ **32.** $-5x = 90$ **33.** $5x = 37$

34. $7x = 62$ **35.** $-18 = 6n$ **36.** $-52 = 13n$

37. $-26 = -4n$ **38.** $-56 = -6n$ **39.** $\dfrac{t}{9} = 16$

40. $\dfrac{t}{12} = 8$ **41.** $\dfrac{n}{-8} = -3$ **42.** $\dfrac{n}{-9} = -5$

43. $-x = 15$ **44.** $-x = -17$ **45.** $\dfrac{3}{4}x = 18$

46. $\dfrac{2}{3}x = 32$ **47.** $-\dfrac{2}{5}n = 14$ **48.** $-\dfrac{3}{8}n = 33$

49. $\dfrac{2}{3}n = \dfrac{1}{5}$ **50.** $\dfrac{3}{4}n = \dfrac{1}{8}$ **51.** $\dfrac{5}{6}n = -\dfrac{3}{4}$

52. $\dfrac{6}{7}n = -\dfrac{3}{8}$ **53.** $\dfrac{3x}{10} = \dfrac{3}{20}$ **54.** $\dfrac{5x}{12} = \dfrac{5}{36}$

55. $\dfrac{-y}{2} = \dfrac{1}{6}$ **56.** $\dfrac{-y}{4} = \dfrac{1}{9}$ **57.** $-\dfrac{4}{3}x = -\dfrac{9}{8}$

58. $-\dfrac{6}{5}x = -\dfrac{10}{14}$ **59.** $-\dfrac{5}{12} = \dfrac{7}{6}x$ **60.** $-\dfrac{7}{24} = \dfrac{3}{8}x$

61. $-\dfrac{5}{7}x = 1$ **62.** $-\dfrac{11}{12}x = -1$ **63.** $-4n = \dfrac{1}{3}$

64. $-6n = \dfrac{3}{4}$ **65.** $-8n = \dfrac{6}{5}$ **66.** $-12n = \dfrac{8}{3}$

3.2 Equations and Problem Solving

Many times, more than one property of equality needs to be used to help find the solution of an equation. Consider the following examples.

EXAMPLE 1 Solve $3x + 1 = 7$.

Solution

$$3x + 1 = 7$$

$$3x + 1 - 1 = 7 - 1 \qquad \text{Subtract 1 from both sides.}$$

$$3x = 6$$

$$\frac{3x}{3} = \frac{6}{3} \qquad \text{Divide both sides by 3.}$$

$$x = 2.$$

The potential solution can be *checked* by substituting it into the original equation to see if a true numerical statement is obtained.

Check: $\quad 3x + 1 = 7$

$$3(2) + 1 \overset{?}{=} 7$$

$$6 + 1 \overset{?}{=} 7$$

$$7 = 7.$$

Now we know that the solution set is $\{2\}$. ■

EXAMPLE 2 Solve $5x - 6 = 14$.

Solution

$$5x - 6 = 14$$

$$5x - 6 + 6 = 14 + 6 \qquad \text{Add 6 to both sides.}$$

$$5x = 20$$

$$\frac{5x}{5} = \frac{20}{5} \qquad \text{Divide both sides by 5.}$$

$$x = 4.$$

Check: $\quad 5x - 6 = 14$

$$5(4) - 6 \overset{?}{=} 14$$

$$20 - 6 \overset{?}{=} 14$$

$$14 = 14.$$

The solution set is $\{4\}$. ■

EXAMPLE 3 Solve $-3a + 4 = 22$.

Solution

$$-3a + 4 = 22$$

$$-3a + 4 - 4 = 22 - 4 \qquad \text{Subtract 4 from both sides.}$$

$$-3a = 18$$

$$\frac{-3a}{-3} = \frac{18}{-3} \qquad \text{Divide both sides by } -3.$$

$$a = -6.$$

Check:

$$-3a + 4 = 22$$
$$-3(-6) + 4 \overset{?}{=} 22$$
$$18 + 4 \overset{?}{=} 22$$
$$22 = 22.$$

The solution set is $\{-6\}$. ■

Notice that in Examples 1, 2, and 3 we used the addition-subtraction property first and then used the multiplication-division property. In general, this sequence of steps provides the easiest format for solving such equations. Perhaps you should convince yourself of that fact by doing Example 1 again, but this time using the multiplication-division property first and then the addition-subtraction property.

EXAMPLE 4　Solve $19 = 2n + 4$.

Solution

$$19 = 2n + 4$$
$$19 - 4 = 2n + 4 - 4 \qquad \text{Subtract 4 from both sides.}$$
$$15 = 2n$$

$$\frac{15}{2} = \frac{2n}{2} \qquad \text{Divide both sides by 2.}$$

$$\frac{15}{2} = n.$$

Check:　$19 = 2n + 4$

$$19 \overset{?}{=} 2\left(\frac{15}{2}\right) + 4$$
$$19 \overset{?}{=} 15 + 4$$
$$19 = 19.$$

The solution set is $\left\{\frac{15}{2}\right\}$. ■

■　Word Problems

In the last section of Chapter 2 we did some translating of English phrases into algebraic expressions. We are now ready to extend that idea to the translation of English *sentences* into algebraic *equations*. Such translations allow us to use the concepts of algebra to solve verbal problems. Let's consider some examples.

PROBLEM 1 A certain number added to 17 yields a sum of 29. What is the number?

Solution Let n represent the number to be found. The sentence "A certain number added to 17 yields a sum of 29" translates into the algebraic equation $17 + n = 29$. Solving this equation, we obtain

$$17 + n = 29$$
$$17 + n - 17 = 29 - 17$$
$$n = 12.$$

The solution is 12, which is the number asked for in the problem.   ■

The statement "let n represent the number to be found" is often referred to as **declaring the variable**. We need to choose a letter to use as a variable and indicate what it represents for a specific problem. This may seem like an insignificant idea, but as the problems become more complex, the process of declaring the variable becomes even more important. Furthermore, we are aware that we could solve a problem such as Problem 1 above without setting up an algebraic equation. However, as problems increase in difficulty, the translation from English to an algebraic equation becomes a key issue. Therefore, even with these relatively simple problems we must concentrate on the translation process.

PROBLEM 2 Six years ago Bill was 13 years old. How old is he now?

Solution Let y represent Bill's age now; therefore, $y - 6$ represents his age six years ago. Thus,

$$y - 6 = 13$$
$$y - 6 + 6 = 13 + 6$$
$$y = 19.$$

Bill is presently 19 years old.   ■

PROBLEM 3 Betty worked 8 hours Saturday and earned $32. How much did she earn per hour?

Solution A Let x represent the amount Betty earned per hour. The number of hours worked times the wage per hour yields the total earnings. Thus,

$$8x = 32$$
$$\frac{8x}{8} = \frac{32}{8}$$
$$x = 4.$$

Betty earned $4 per hour.

Solution B　　Let y represent the amount Betty earned per hour. The wage per hour equals the total wage divided by the number of hours. Thus,

$$y = \frac{32}{8}$$

$$y = 4.$$

Betty earned $4 per hour.　　　　　　　　　　　　　　　　■

　　　Sometimes more than one equation can be used to solve a problem. In Solution A the equation was set up in terms of multiplication, whereas in Solution B we were thinking in terms of division.

PROBLEM 4　　If 2 is subtracted from five-times-a-certain-number the result is 28. Find the number.

Solution　　Let n represent the number to be found. Translating the first sentence in the problem into an algebraic equation, we obtain

$$5n - 2 = 28.$$

Solving this equation yields

$$5n - 2 + 2 = 28 + 2$$
$$5n = 30$$
$$\frac{5n}{5} = \frac{30}{5}$$
$$n = 6.$$

The number to be found is 6.　　　　　　　　　　　　　■

PROBLEM 5　　A plumbing repair bill was $95. This included $17 for parts and an amount for 3 hours of labor. Find the hourly rate that was charged for labor.

Solution　　Let l represent the hourly charge for labor; then $3l$ represents the total cost for labor. Thus, the cost for labor plus the cost for parts is the total bill of $95; so we can proceed as follows.

cost for labor + cost for parts = $95
$$3l \quad + \quad 17 \quad = 95.$$

Solving this equation we obtain

$$3l = 78$$

$$\frac{3l}{3} = \frac{78}{3}$$

$$l = 26.$$

The charge for labor was $26 per hour.   ■

Problem Set 3.2

For Problems 1–40, solve each equation.

1. $2x + 5 = 13$ **2.** $3x + 4 = 19$ **3.** $5x + 2 = 32$

4. $7x + 3 = 24$ **5.** $3x - 1 = 23$ **6.** $2x - 5 = 21$

7. $4n - 3 = 41$ **8.** $5n - 6 = 19$ **9.** $6y - 1 = 16$

10. $4y - 3 = 14$ **11.** $2x + 3 = 22$ **12.** $3x + 1 = 21$

13. $10 = 3t - 8$ **14.** $17 = 2t + 5$ **15.** $5x + 14 = 9$

16. $4x + 17 = 9$ **17.** $18 - n = 23$ **18.** $17 - n = 29$

19. $-3x + 2 = 20$ **20.** $-6x + 1 = 43$ **21.** $7 + 4x = 29$

22. $9 + 6x = 23$ **23.** $16 = -2 - 9a$ **24.** $18 = -10 - 7a$

25. $-7x + 3 = -7$ **26.** $-9x + 5 = -18$ **27.** $17 - 2x = -19$

28. $18 - 3x = -24$ **29.** $-16 - 4x = 9$ **30.** $-14 - 6x = 7$

31. $-12t + 4 = 88$ **32.** $-16t + 3 = 67$ **33.** $14y + 15 = -33$

34. $12y + 13 = -15$ **35.** $32 - 16n = -8$ **36.** $-41 = 12n - 19$

37. $17x - 41 = -37$ **38.** $19y - 53 = -47$ **39.** $29 = -7 - 15x$

40. $49 = -5 - 14x$

For each of the following problems, (a) choose a variable and indicate what it represents in the problem, (b) set up an equation that represents the situation described, and (c) solve the equation.

41. Twelve added to a certain number is 21. What is the number?

42. A certain number added to 14 is 25. Find the number.

43. Nine subtracted from a certain number is 13. Find the number.

44. A certain number subtracted from 32 is 15. What is the number?

45. Two items cost $43. If one of the items costs $25, what is the cost of the other item?

46. Eight years ago Rosa was 22 years old. Find Rosa's present age.

47. Six years from now, Nora will be 41 years old. What is her present age?

48. The Chi's bought 8 pizzas for a total of $50. What was the price per pizza?

49. Chad worked 6 hours Saturday for a total of $39. How much per hour did he earn?

50. Jill worked 8 hours Saturday at $5.50 per hour. How much did she earn?

51. If 6 is added to three times a certain number, the result is 24. Find the number.

52. If 2 is subtracted from five times a certain number, the result is 38. Find the number.

53. Nineteen is four larger than three times a certain number. Find the number.

54. If 9 times a certain number is subtracted from 7, the result is 52. Find the number.

55. Forty-nine is equal to 6 less than 5 times a certain number. Find the number.

56. Seventy-one is equal to 2 more than 3 times a certain number. Find the number.

57. If one is subtracted from 6 times a certain number, the result is 47. Find the number.

58. Five less than four times a number equals 31. Find the number.

59. If eight times a certain number is subtracted from 27, the result is 3. Find the number.

60. Twenty is 22 less than six times a certain number. Find the number.

61. The price of a ring is $550. This price represents $50 less than twice the cost of the ring. Find the cost of the ring.

62. Todd is on a 1750 calorie-per-day diet plan. This plan permits 650 calories less than twice the number of calories permitted by Lerae's diet plan. How many calories are permitted by Lerae's plan?

63. The length of a rectangular shaped floor is 18 meters. This represents 2 meters less than five times the width of the floor. Find the width of the floor.

64. An executive is earning $45,000 per year. This represents $15,000 less than twice her salary four years ago. Find her salary four years ago.

65. The LM car dealership sold 32 cars during December of 1984. This represented 4 more than twice the number of cars they sold during December of 1983. How many cars did they sell during December of 1983?

66. A car repair bill was $119. This included $42 for parts and $22 for each hour of labor. Find the number of hours of labor.

67. An electric golf cart repair bill was $156. This included $36 for parts and $24 per hour for labor. How many hours of labor were involved?

68. Richard earned $305 last week for working 40 hours. Included in the $305 was a bonus of $25. How much was Richard's hourly pay?

3.3 More on Solving Equations and Problem Solving

As equations become more complex, additional steps are needed to solve them. This creates a need to carefully organize our work so as to minimize the chances for error. So let's begin this section with some suggestions for solving equations and then illustrate a *solution format* that is effective.

The process of solving first degree equations of one variable can be summarized as follows.

Step 1. Simplify both sides of the equation as much as possible.

Step 2. Use the addition or subtraction property of equality to isolate a term containing the variable on one side and a constant on the other side of the equation.

Step 3. Use the multiplication or division property of equality to make the coefficient of the variable one.

The following examples illustrate this step-by-step process for solving equations. Study them carefully and be sure that you understand each step taken in the solving process.

EXAMPLE 1 Solve $5y - 4 + 3y = 12$.

Solution

$$5y - 4 + 3y = 12$$
$$8y - 4 = 12 \qquad \text{Combine similar terms on the left side.}$$
$$8y - 4 + 4 = 12 + 4 \qquad \text{Add 4 to both sides.}$$
$$8y = 16$$

$$\frac{8y}{8} = \frac{16}{8} \qquad \text{Divide both sides by 8.}$$

$$y = 2.$$

The solution set is $\{2\}$. You can do the check alone now! ■

EXAMPLE 2 Solve $7x - 2 = 3x + 9$.

Solution Notice that both sides of the equation are in simplified form; thus, we can begin by applying the subtraction property of equality.

$$7x - 2 = 3x + 9$$
$$7x - 2 - 3x = 3x + 9 - 3x \qquad \text{Subtract } 3x \text{ from both sides.}$$
$$4x - 2 = 9$$
$$4x - 2 + 2 = 9 + 2 \qquad \text{Add 2 to both sides.}$$
$$4x = 11$$

$$\frac{4x}{4} = \frac{11}{4} \qquad \text{Divide both sides by 4.}$$

$$x = \frac{11}{4}.$$

The solution set is $\left\{\dfrac{11}{4}\right\}$. ■

EXAMPLE 3 Solve $5n + 12 = 9n - 16$.

Solution

$$5n + 12 = 9n - 16$$

$5n + 12 - 9n = 9n - 16 - 9n$ Subtract $9n$ from both sides.

$$-4n + 12 = -16$$

$-4n + 12 - 12 = -16 - 12$ Subtract 12 from both sides.

$$-4n = -28$$

$$\frac{-4n}{-4} = \frac{-28}{-4}$$ Divide both sides by -4.

$$n = 7.$$

The solution set is $\{7\}$.   ■

■ Word Problems

As we expand our skills for solving equations, we also expand our capabilities for solving verbal problems. There is no one definite procedure that will insure success at solving word problems, but the following suggestions can be helpful.

Suggestions for Solving Word Problems

1. Read the problem carefully, making certain that you understand the meanings of all the words. Be especially alert for any technical terms used in the statement of the problem.

2. Read the problem a second time (perhaps even a third time) to get an overview of the situation being described and to determine the known facts as well as what is to be found.

3. Sketch any figure, diagram, or chart that might be helpful in analyzing the problem.

4. Choose a meaningful variable to represent an unknown quantity in the problem (perhaps t, if time is an unknown quantity) and represent any other unknowns in terms of that variable.

5. Look for a *guideline* that can be used to set up an equation. A guideline might be a formula such as "distance equals rate times time" or a statement of a relationship such as "the sum of the two numbers is 28."

6. Form an equation containing the variable that translates the conditions of the guideline from English to algebra.

7. Solve the equation and use the solution to determine all facts requested in the problem.

8. **Check** all answers back into the **original statement of the problem**.

If you decide not to check an answer, at least use the *reasonableness-of-answer* idea as a partial check. That is to say, ask yourself the question: Is this answer reasonable? For example, if the problem involves two investments totaling $10,000, then an answer of $12,000 for one investment is certainly *not reasonable*.

Now let's consider some problems and illustrate the use of these suggestions.

PROBLEM 1 Find two consecutive even numbers whose sum is 74.

Solution To solve this problem we must know the meaning of the technical phrase "two consecutive even numbers." Two consecutive even numbers are two even numbers with one and only one whole number between them. For example, 2 and 4 are consecutive even numbers. Now we can proceed as follows. Let n represent the first even number; then $n + 2$ represents the next even number. Since their sum is 74, we can set up and solve the following equation.

$$n + n + 2 = 74$$
$$2n + 2 = 74$$
$$2n + 2 - 2 = 74 - 2$$
$$2n = 72$$
$$\frac{2n}{2} = \frac{72}{2}$$
$$n = 36.$$

If $n = 36$, then $n + 2 = 38$; thus, the numbers are 36 and 38.

Check: To check our answers for Problem 1, we must determine whether they satisfy the conditions stated in the original problem. Since 36 and 38 are two consecutive even numbers and $36 + 38 = 74$ (their sum is 74), we know that our answers are correct. ■

Suggestion 5 in our list of problem solving suggestions was to "look for a *guideline* that can be used to set up an equation." The guideline may not be explicitly stated in the problem but may be implied by the nature of the problem. Consider the following example.

PROBLEM 2 Barry sells bicycles on a salary-plus-commission basis. He receives a monthly salary of $300 and a commission of $15 for each bicycle that he sells. How many bicycles must he sell in a month to have a total monthly salary of $750?

Solution Let b represent the number of bicycles to be sold in a month. Then $15b$ represents his commission for those bicycles. The *guideline* "fixed salary plus commission equals total monthly salary" generates the following equation.

$$\text{fixed salary} + \text{commission} = \text{total monthly salary}$$
$$\downarrow \qquad \downarrow \qquad \qquad \downarrow$$
$$\$300 \quad + \quad 15b \quad = \quad \$750.$$

Solving this equation yields

$$300 + 15b - 300 = 750 - 300$$
$$15b = 450$$

$$\frac{15b}{15} = \frac{450}{15}$$

$$b = 30.$$

He must sell 30 bicycles per month. (Does this number check?)   ■

■ Geometric Problems

Sometimes the guideline for setting up an equation to solve a problem is based on a geometric relationship. There are several basic geometric relationships pertaining to angle measure. Let's state some of these relationships and then consider some problems involving them.

1. **Two** angles for which the sum of their measure is 90° (the symbol ° indicates degrees) are called **complementary angles**.

2. **Two** angles for which the sum of their measure is 180° are called **supplementary angles**.

3. The sum of the measures of the three angles of a triangle is 180°.

PROBLEM 3 One of two complementary angles is 14° larger than the other. Find the measure of each of the angles.

Solution If we let a represent the measure of the smaller angle, then $a + 14$ represents the measure of the larger angle. Since they are complementary angles their sum is 90° and we can proceed as follows.

$$a + a + 14 = 90$$
$$2a + 14 = 90$$
$$2a + 14 - 14 = 90 - 14$$
$$2a = 76$$
$$\frac{2a}{2} = \frac{76}{2}$$
$$a = 38.$$

If $a = 38$, then $a + 14 = 52$, and the angles are of measure 38° and 52°.   ■

PROBLEM 4 Find the measures of the three angles of a triangle if the second is three times the first and the third is twice the second.

Solution If we let a represent the measure of the smallest angle, then $3a$ and $2(3a)$ represent the measures of the other two angles. Therefore, we can set up and solve the following equation.

$$a + 3a + 2(3a) = 180$$
$$a + 3a + 6a = 180$$
$$10a = 180$$
$$\frac{10a}{10} = \frac{180}{10}$$
$$a = 18.$$

If $a = 18$, then $3a = 54$ and $2(3a) = 108$. So the angles have measures of $18°$, $54°$, and $108°$. ■

Problem Set 3.3

Solve each of the following equations.

1. $2x + 7 + 3x = 32$
2. $3x + 9 + 4x = 30$
3. $7x - 4 - 3x = -36$
4. $8x - 3 - 2x = -45$
5. $3y - 1 + 2y - 3 = 4$
6. $y + 3 + 2y - 4 = 6$
7. $5n - 2 - 8n = 31$
8. $6n - 1 - 10n = 51$
9. $-2n + 1 - 3n + n - 4 = 7$
10. $-n + 7 - 2n + 5n - 3 = -6$
11. $3x + 4 = 2x - 5$
12. $5x - 2 = 4x + 6$
13. $5x - 7 = 6x - 9$
14. $7x - 3 = 8x - 13$
15. $6x + 1 = 3x - 8$
16. $4x - 10 = x + 17$
17. $7y - 3 = 5y + 10$
18. $8y + 4 = 5y - 4$
19. $8n - 2 = 11n - 7$
20. $7n - 10 = 9n - 13$
21. $-2x - 7 = -3x + 10$
22. $-4x + 6 = -5x - 9$
23. $-3x + 5 = -5x - 8$
24. $-4x + 7 = -6x + 4$
25. $-7 - 6x = 9 - 9x$
26. $-10 - 7x = 14 - 12x$
27. $2x - 1 - x = 3x - 5$
28. $3x - 4 - 4x = 5 - 5x + 3x$
29. $5n - 4 - n = -3n - 6 + n$
30. $4x - 3 + 2x = 8x - 3 - x$
31. $-7 - 2n - 6n = 7n - 5n + 12$
32. $-3n + 6 + 5n = 7n - 8n - 9$

Solve each of the following verbal problems by setting up and solving an algebraic equation.

33. The sum of a number plus four times the number is 85. What is the number?
34. A number subtracted from three times the number yields 68. Find the number.
35. Find two consecutive odd numbers whose sum is 72.
36. Find two consecutive even numbers whose sum is 94.
37. Find three consecutive even numbers whose sum is 114.
38. Find three consecutive odd numbers whose sum is 159.

39. Two more than three times a certain number is the same as four less than seven times the number. Find the number.

40. One more than five times a certain number is equal to eleven less than nine times the number. What is the number?

41. The sum of a number and five times the number equals eighteen less than three times the number. Find the number.

42. One of two supplementary angles is five times as large as the other. Find the measure of each angle.

43. One of two complementary angles is 6 less than twice the other angle. Find the measure of each angle.

44. If two angles are complementary and the difference of their measures is 62°, find the measure of each angle.

45. If two angles are supplementary and the larger angle is 20° less than three times the smaller angle, find the measure of each angle.

46. Find the measures of the three angles of a triangle if the largest is 14° less than three times the smallest and the other angle is 4° more than the smallest.

47. One of the angles of a triangle has a measure of 40°. Find the measures of the other two angles if the difference of their measures is 10°.

48. Larry sold encyclopedias one summer on a salary-plus-commission basis. He was given a flat salary of $1800 for the summer plus a commission of $75 for each set of encyclopedias that he sold. His total earnings for the summer were $2550. How many sets of encyclopedias did he sell?

49. Daniel sold some stock at $35 per share. This was $17 a share less than twice what he paid for it. What price did he pay for each share of the stock?

50. Becky sold some stock at $37 per share. This was $14 a share less than three times what she paid for it. What price did she pay for the stock?

51. Bob is paid two times his normal hourly rate for each hour worked over 40 hours in a week. Last week he earned $504 for 48 hours of work. What is his hourly wage?

52. On an algebra test, the highest grade was 9 points less than three times the lowest grade. The sum of the two grades was 135. Find the lowest and highest grades on the test.

53. At a university-sponsored concert, there were three times as many females as males. A total of 600 people attended the concert. How many males and how many females attended?

54. A triangular lot is enclosed with 135 yards of fencing. The longest side of the lot is 5 yards more than twice the length of the shortest side. The other side is 10 yards longer than the shortest side. Find the lengths of the three sides of the lot.

55. Melton and Sanchez were opposing candidates in a school board election. Sanchez received 10 more than twice the number of votes that Melton received. If there was a total of 1030 votes cast, how many votes did Sanchez receive?

56. The plans for an apartment complex provided for 260 apartments, each apartment having one, two, or three bedrooms. There were to be 25 more three-bedroom than one-bedroom apartments, and the number of two-bedroom apartments was to be 15 less than three times the number of one-bedroom apartments. How many apartments of each type were in the plans?

57. A board 20 feet long is to be cut into two pieces so that one piece is 8 feet longer than the other. How long should the shorter piece be?

58. A copper wire 110 centimeters long was bent in the shape of a rectangle. The length of the rectangle was 10 centimeters more than twice the width. Find the dimensions of the rectangle.

59. Seventy-eight yards of fencing was purchased to enclose a rectangular yard. The length of the yard is 1 yard less than 3 times its width. Find the length and width of the yard.

■ Miscellaneous Problems

60. Solve each of the following equations.

(a) $7x - 3 = 4x - 3$

(b) $-x - 4 + 3x = 2x - 7$

(c) $-3x + 9 - 2x = -5x + 9$

(d) $5x - 3 = 6x - 7 - x$

(e) $7x + 4 = -x + 4 + 8x$

(f) $3x - 2 - 5x = 7x - 2 - 5x$

(g) $-6x - 8 = 6x + 4$

(h) $-8x + 9 = -8x + 5$

3.4 Equations Involving Parentheses and Fractional Forms

The distributive property is frequently used in this section as we expand our techniques for solving equations. Recall that in symbolic form the distributive property states that $a(b + c) = ab + ac$. Consider the following examples that illustrate using this property to *remove parentheses*. Pay special attention to the last two examples involving a negative number in front of the parentheses.

$$3(x + 2) = \boxed{3 \cdot x + 3 \cdot 2} = 3x + 6;$$

$$5(y - 3) = \boxed{5 \cdot y - 5 \cdot 3} = 5y - 15; \quad (a(b - c) = ab - ac)$$

$$2(4x + 7) = \boxed{2(4x) + 2(7)} = 8x + 14;$$

$$-1(n + 4) = \boxed{(-1)(n) + (-1)(4)} = -n - 4;$$

$$-6(x - 2) = \boxed{(-6)(x) - (-6)(2)} = -6x + 12.$$

$$\downarrow$$

Do this step
mentally!

It is often necessary to solve equations in which the variable is part of an expression enclosed in parentheses. The distributive property is used to remove the parentheses; we then proceed in the usual way. Consider the following examples. (Notice that we are beginning to show *only the major steps* when solving an equation.)

EXAMPLE 1 Solve $3(x + 2) = 23$.

Solution

$$3(x + 2) = 23$$

$$3x + 6 = 23 \qquad \text{Applied distributive property to left side.}$$

$$3x = 17 \qquad \text{Subtracted 6 from both sides.}$$

$$x = \frac{17}{3}. \qquad \text{Divided both sides by 3.}$$

The solution set is $\left\{\dfrac{17}{3}\right\}$. ■

EXAMPLE 2 Solve $4(x + 3) = 2(x - 6)$.

Solution

$$4(x + 3) = 2(x - 6)$$

$$4x + 12 = 2x - 12 \qquad \text{Applied distributive property on each side.}$$

$$2x + 12 = -12 \qquad \text{Subtracted } 2x \text{ from both sides.}$$

$$2x = -24 \qquad \text{Subtracted 12 from both sides.}$$

$$x = -12. \qquad \text{Divided both sides by 2.}$$

The solution set is $\{-12\}$. ■

It may be necessary to remove more than one set of parentheses and then to use the distributive property again to combine similar terms. Consider the following two examples.

EXAMPLE 3 Solve $5(w + 3) + 3(w + 1) = 14$.

Solution

$$5(w + 3) + 3(w + 1) = 14$$

$$5w + 15 + 3w + 3 = 14 \qquad \text{Applied distributive property.}$$

$$8w + 18 = 14 \qquad \text{Combined similar terms.}$$

$$8w = -4 \qquad \text{Subtracted 18 from both sides.}$$

$$w = -\frac{4}{8} \qquad \text{Divided both sides by 8.}$$

$$w = -\frac{1}{2} \qquad \text{Reduced!}$$

The solution set is $\left\{-\dfrac{1}{2}\right\}$. ■

EXAMPLE 4 Solve $6(x - 7) - 2(x - 4) = 13$.

Solution

$$6(x - 7) - 2(x - 4) = 13 \quad \text{(Be careful with this sign!)}$$
$$6x - 42 - 2x + 8 = 13 \qquad \text{Distributive property.}$$
$$4x - 34 = 13 \qquad \text{Combined similar terms.}$$
$$4x = 47 \qquad \text{Added 34 to both sides.}$$
$$x = \frac{47}{4}. \qquad \text{Divided both sides by 4.}$$

The solution set is $\left\{\dfrac{47}{4}\right\}$. ■

In a previous section we solved equations like $x - \dfrac{2}{3} = \dfrac{3}{4}$ by adding $\dfrac{2}{3}$ to both sides. If an equation contains several fractions, then it is usually easier to *clear the equation of all fractions* by multiplying both sides by the least common denominator of all of the denominators. Perhaps several examples will clarify this idea.

EXAMPLE 5 Solve $\dfrac{1}{2}x + \dfrac{2}{3} = \dfrac{5}{6}$.

Solution

$$\frac{1}{2}x + \frac{2}{3} = \frac{5}{6}$$

$$6\left(\frac{1}{2}x + \frac{2}{3}\right) = 6\left(\frac{5}{6}\right) \qquad \text{6 is the LCD of 2, 3, and 6.}$$

$$6\left(\frac{1}{2}x\right) + 6\left(\frac{2}{3}\right) = 6\left(\frac{5}{6}\right) \qquad \text{Distributive property.}$$

$$3x + 4 = 5 \qquad \text{Note how the equation has been }$$
$$\textit{cleared of all fractions.}$$

$$3x = 1$$

$$x = \frac{1}{3}.$$

The solution set is $\left\{\dfrac{1}{3}\right\}$. ■

EXAMPLE 6 Solve $\dfrac{5n}{6} - \dfrac{1}{4} = \dfrac{3}{8}$.

Solution

$$\frac{5n}{6} - \frac{1}{4} = \frac{3}{8} \qquad \text{Remember } \frac{5n}{6} = \frac{5}{6}n.$$

$$24\left(\frac{5n}{6} - \frac{1}{4}\right) = 24\left(\frac{3}{8}\right) \qquad \text{24 is the LCD of 6, 4, and 8.}$$

$$24\left(\frac{5n}{6}\right) - 24\left(\frac{1}{4}\right) = 24\left(\frac{3}{8}\right) \qquad \text{Distributive property.}$$

$$20n - 6 = 9$$

$$20n = 15$$

$$n = \frac{15}{20} = \frac{3}{4}.$$

The solution set is $\left\{\frac{3}{4}\right\}$.   ■

Many of the ideas presented in this section are used to help solve the following equations. Study the solutions carefully and be sure that you can supply reasons for each step. It might be helpful to cover up the solutions and to try to solve the equations on your own.

EXAMPLE 7   Solve $\dfrac{x + 3}{2} + \dfrac{x + 4}{5} = \dfrac{3}{10}$.

Solution

$$\frac{x + 3}{2} + \frac{x + 4}{5} = \frac{3}{10}$$

$$10\left(\frac{x + 3}{2} + \frac{x + 4}{5}\right) = 10\left(\frac{3}{10}\right) \qquad \text{10 is the LCD of 2, 5, and 10.}$$

$$10\left(\frac{x + 3}{2}\right) + 10\left(\frac{x + 4}{5}\right) = 10\left(\frac{3}{10}\right) \qquad \text{Distributive property.}$$

$$5(x + 3) + 2(x + 4) = 3$$

$$5x + 15 + 2x + 8 = 3$$

$$7x + 23 = 3$$

$$7x = -20$$

$$x = -\frac{20}{7}.$$

The solution set is $\left\{-\dfrac{20}{7}\right\}$.   ■

EXAMPLE 8   Solve $\dfrac{x - 1}{4} - \dfrac{x - 2}{6} = \dfrac{2}{3}$.

Solution

$$\frac{x-1}{4} - \frac{x-2}{6} = \frac{2}{3}$$

$$12\left(\frac{x-1}{4} - \frac{x-2}{6}\right) = 12\left(\frac{2}{3}\right) \qquad \text{12 is the LCD of 4, 6, and 3.}$$

$$12\left(\frac{x-1}{4}\right) - 12\left(\frac{x-2}{6}\right) = 12\left(\frac{2}{3}\right) \qquad \text{Distributive property.}$$

$$3(x-1) - 2(x-2) = 8$$

$$3x - 3 - 2x + 4 = 8 \qquad \text{Be careful with this sign.}$$

$$x + 1 = 8$$

$$x = 7.$$

The solution set is $\{7\}$.   ■

■ Word Problems

We are now ready to solve some verbal problems using equations of the different types presented in this section. Again, it might be helpful for you to attempt to solve the problems on your own before looking at the book's approach.

PROBLEM 1   Loretta has 19 coins consisting of quarters and nickels amounting to $2.35. How many coins of each kind does she have?

Solution   Let q represent the number of quarters. Then $19 - q$ represents the number of nickels. We can use the following guideline to help set up an equation.

$$\begin{array}{c} \text{value of quarters} \\ \text{in cents} \end{array} + \begin{array}{c} \text{value of nickels} \\ \text{in cents} \end{array} = \begin{array}{c} \text{total value} \\ \text{in cents} \end{array}$$

$$\downarrow \qquad\qquad \downarrow \qquad\qquad \downarrow$$

$$25q \qquad + \quad 5(19 - q) \quad = \quad 235.$$

Solving the equation we obtain

$$25q + 95 - 5q = 235$$

$$20q + 95 = 235$$

$$20q = 140$$

$$q = 7.$$

If $q = 7$, then $19 - q = 12$; so she has 7 quarters and 12 nickels.   ■

PROBLEM 2   Find a number such that 4-less-than-two-thirds-the-number is equal to one-sixth the number.

Solution   Let n represent the number. Then $\frac{2}{3}n - 4$ represents 4 less than two-thirds the number, and $\frac{1}{6}n$ represents one-sixth the number.

$$\frac{2}{3}n - 4 = \frac{1}{6}n$$

$$6\left(\frac{2}{3}n - 4\right) = 6\left(\frac{1}{6}n\right)$$

$$4n - 24 = n$$

$$3n - 24 = 0$$

$$3n = 24$$

$$n = 8.$$

The number is 8.   ■

PROBLEM 3   Lance is paid $1\frac{1}{2}$ times his normal hourly rate for each hour worked over 40 hours in a week. Last week he worked 50 hours and earned \$462. What is his normal hourly rate?

Solution   Let x represent his normal hourly rate. Then $\frac{3}{2}x$ represents $1\frac{1}{2}$ times his normal hourly rate. The following guideline can be used to help set up the equation.

$$\underset{\downarrow}{\substack{\text{regular wages for}\\ \text{first 40 hours}}} + \underset{\downarrow}{\substack{\text{wages for 10 hours}\\ \text{of overtime}}} = \underset{\downarrow}{\substack{\text{total}\\ \text{wages}}}$$

$$40x \quad + \quad 10\left(\frac{3}{2}x\right) \quad = \quad 462.$$

Solving this equation we obtain

$$40x + 15x = 462$$

$$55x = 462$$

$$x = 8.40.$$

His normal hourly rate is \$8.40.   ■

PROBLEM 4   Find three consecutive whole numbers such that the sum of the first plus twice the second plus three times the third is 134.

Solution   Let n represent the first whole number. Then $n + 1$ represents the second whole number and $n + 2$ represents the third whole number.

$$n + 2(n + 1) + 3(n + 2) = 134$$
$$n + 2n + 2 + 3n + 6 = 134$$
$$6n + 8 = 134$$
$$6n = 126$$
$$n = 21.$$

The numbers are 21, 22, and 23.   ■

Problem Set 3.4

Solve each of the following equations.

1. $7(x + 2) = 21$
2. $4(x + 4) = 24$
3. $5(x - 3) = 35$
4. $6(x - 2) = 18$
5. $-3(x + 5) = 12$
6. $-5(x - 6) = -15$
7. $4(n - 6) = 5$
8. $3(n + 4) = 7$
9. $6(n + 7) = 8$
10. $8(n - 3) = 12$
11. $-10 = -5(t - 8)$
12. $-16 = -4(t + 7)$
13. $5(x - 4) = 4(x + 6)$
14. $7(x + 3) = 6(x - 8)$
15. $8(x + 1) = 9(x - 2)$
16. $4(x - 7) = 5(x + 2)$
17. $8(t + 5) = 6(t - 6)$
18. $7(t - 5) = 5(t + 3)$
19. $3(2t + 1) = 4(3t - 2)$
20. $2(3t - 1) = 3(3t - 2)$
21. $-2(x - 6) = -(x - 9)$
22. $-(x + 7) = -2(x + 10)$
23. $-3(t - 4) - 2(t + 4) = 9$
24. $5(t - 4) - 3(t - 2) = 12$
25. $3(n - 10) - 5(n + 12) = -86$
26. $4(n + 9) - 7(n - 8) = 83$
27. $3(x + 1) + 4(2x - 1) = 5(2x + 3)$
28. $4(x - 1) + 5(x + 2) = 3(x - 8)$
29. $-(x + 2) + 2(x - 3) = -2(x - 7)$
30. $-2(x + 6) + 3(3x - 2) = -3(x - 4)$
31. $5(2x - 1) - (3x + 4) = 4(x + 3) - 27$
32. $3(4x + 1) - 2(2x + 1) = -2(x - 1) - 1$
33. $-(a - 1) - (3a - 2) = 6 + 2(a - 1)$
34. $3(2a - 1) - 2(5a + 1) = 4(3a + 4)$
35. $3(x - 1) + 2(x - 3) = -4(x - 2) + 10(x + 4)$
36. $-2(x - 4) - (3x - 2) = -2 + (-6x + 2)$
37. $3 - 7(x - 1) = 9 - 6(2x + 1)$
38. $8 - 5(2x + 1) = 2 - 6(x - 3)$
39. $\dfrac{3}{4}x - \dfrac{2}{3} = \dfrac{5}{6}$
40. $\dfrac{1}{2}x - \dfrac{4}{3} = -\dfrac{5}{6}$
41. $\dfrac{5}{6}x + \dfrac{1}{4} = -\dfrac{9}{4}$
42. $\dfrac{3}{8}x + \dfrac{1}{6} = -\dfrac{7}{12}$

43. $\dfrac{1}{2}x - \dfrac{3}{5} = \dfrac{3}{4}$

44. $\dfrac{1}{4}x - \dfrac{2}{5} = \dfrac{5}{6}$

45. $\dfrac{n}{3} + \dfrac{5n}{6} = \dfrac{1}{8}$

46. $\dfrac{n}{6} + \dfrac{3n}{8} = \dfrac{5}{12}$

47. $\dfrac{5y}{6} - \dfrac{3}{5} = \dfrac{2y}{3}$

48. $\dfrac{3y}{7} + \dfrac{1}{2} = \dfrac{y}{4}$

49. $\dfrac{h}{6} + \dfrac{h}{8} = 1$

50. $\dfrac{h}{4} + \dfrac{h}{3} = 1$

51. $\dfrac{x + 2}{3} + \dfrac{x + 3}{4} = \dfrac{13}{3}$

52. $\dfrac{x - 1}{4} + \dfrac{x + 2}{5} = \dfrac{39}{20}$

53. $\dfrac{x - 1}{5} - \dfrac{x + 4}{6} = -\dfrac{13}{15}$

54. $\dfrac{x + 1}{7} - \dfrac{x - 3}{5} = \dfrac{4}{5}$

55. $\dfrac{x + 8}{2} - \dfrac{x + 10}{7} = \dfrac{3}{4}$

56. $\dfrac{x + 7}{3} - \dfrac{x + 9}{6} = \dfrac{5}{9}$

57. $\dfrac{x - 2}{8} - 1 = \dfrac{x + 1}{4}$

58. $\dfrac{x - 4}{2} + 3 = \dfrac{x - 2}{4}$

59. $\dfrac{x + 1}{4} = \dfrac{x - 3}{6} + 2$

60. $\dfrac{x + 3}{5} = \dfrac{x - 6}{2} + 1$

Solve each of the following problems by setting up and solving an appropriate algebraic equation.

61. Find two consecutive whole numbers such that the smaller plus four times the larger equals 39.

62. Find two consecutive whole numbers such that the smaller subtracted from five times the larger equals 57.

63. Find three consecutive whole numbers such that twice the sum of the two smallest numbers is 10 more than three times the largest number.

64. Find four consecutive whole numbers such that the sum of the first three equals the fourth number.

65. The sum of two numbers is 17. If twice the smaller is 1 more than the larger, find the numbers.

66. The sum of two numbers is 53. If three times the smaller is 1 less than the larger, find the numbers.

67. Find a number such that 20 more than one-third of the number equals three-fourths of the number.

68. The sum of three-eighths of a number and five-sixths of the same number is 29. Find the number.

69. The difference of two numbers is 6. One-half of the larger number is 5 larger than one-third of the smaller. Find the numbers.

70. The difference of two numbers is 16. Three-fourths of the larger number is 14 larger than one-half of the smaller. Find the numbers.

71. A board 20 feet long is cut into two pieces. Four times the length of the shorter piece is 4 feet less than three times the length of the longer piece. Find the length of each piece.

72. Ellen is paid "time and a half" for each hour over 40 hours worked in a week. Last week she worked 44 hours and earned $299. What is her normal hourly rate?

73. Lucy has 69 coins consisting of nickels, dimes, and quarters. She has 5 more dimes than nickels and the number of quarters is 4 more than three times the number of nickels. How many coins of each kind does she have?

74. Julian has 44 coins consisting of pennies and nickels. If the number of nickels is 2 more than twice the number of pennies, find the number of coins of each kind.

75. Max has a collection of 210 coins consisting of nickels, dimes, and quarters. He has twice as many dimes as nickels and 10 more quarters than dimes. How many coins of each kind does he have?

76. Ginny has a collection of 425 coins consisting of pennies, nickels, and dimes. She has 50 more nickels than pennies and 25 more dimes than nickels. How many coins of each kind does she have?

77. Mary has 18 coins consisting of dimes and quarters amounting to $3.30. How many coins of each kind does she have?

78. Ike has some nickels and dimes amounting to $2.90. The number of dimes is 1 less than twice the number of nickels. How many coins of each kind does he have?

79. A collection of pennies, nickels, and dimes totals $30.75. There are twice as many nickels as pennies and three times as many dimes as pennies. How many coins of each kind are in the collection?

80. Tickets for a concert were priced at $3 for students and $5 for nonstudents. There were 1500 tickets sold for a total of $5000. How many student tickets were sold?

81. The supplement of an angle is 30° more than twice its complement. Find the measure of the angle.

82. The sum of the measure of an angle and three times its complement is 202°. Find the measure of the angle.

83. In triangle ABC, the measure of angle A is 2° less than one-fifth of the measure of angle C. The measure of angle B is 5° less than one-half of the measure of angle C. Find the measures of the three angles of the triangle.

84. If one-fourth of the complement of an angle plus one-fifth of the supplement of the angle equals 36°, find the measure of the angle.

85. The supplement of an angle is 10° less than three times its complement. Find the size of the angle.

86. In triangle ABC, the measure of angle C is eight times the measure of angle A and the measure of angle B is 10° more than the measure of angle C. Find the measure of each angle of the triangle.

■ Miscellaneous Problems

87. Solve each of the following equations.

(a) $-2(x - 1) = -2x + 2$

(b) $3(x + 4) = 3x - 4$

(c) $5(x - 1) = -5x - 5$

(d) $\dfrac{x - 3}{3} + 4 = 3$

(e) $\dfrac{x+2}{3} + 1 = \dfrac{x-2}{3}$ (f) $\dfrac{x-1}{5} - 2 = \dfrac{x-11}{5}$

(g) $4(x-2) - 2(x+3) = 2(x+6)$ (h) $5(x+3) - 3(x-5) = 2(x+15)$

(i) $7(x-1) + 4(x-2) = 15(x-1)$

88. Find three consecutive integers such that the sum of the smallest and largest is equal to twice the middle integer.

3.5 Ratios, Proportions, and Percents

In Figure 3.1, as gear A revolves 4 times, gear B will revolve 3 times. We say that the *gear ratio* of A to B is 4 to 3 or the gear ratio of B to A is 3 to 4.

Mathematically, a **ratio** *is the comparison of two numbers by division.* The gear ratio of A to B above can be written as

$$4 \text{ to } 3 \quad \text{or} \quad 4:3 \quad \text{or} \quad \dfrac{4}{3}.$$

Ratios are commonly expressed as fractions in reduced form. For example, if there are 7500 females and 5000 males at a certain university, then the ratio of females to males is $\dfrac{7500}{5000} = \dfrac{3}{2}.$

Figure 3.1

A statement of equality between two ratios is called a **proportion**. For example,

$$\dfrac{2}{3} = \dfrac{8}{12}$$

is a proportion that states that the ratios $\frac{2}{3}$ and $\frac{8}{12}$ are equal. In the general proportion

$$\frac{a}{b} = \frac{c}{d} \qquad (b \neq 0 \quad \text{and} \quad d \neq 0),$$

if we multiply both sides of the equation by the common denominator, bd, we obtain

$$(bd)\left(\frac{a}{b}\right) = (bd)\left(\frac{c}{d}\right)$$

$$ad = bc.$$

The products ad and bc are sometimes referred to as *cross products* and therefore we can state that *in a proportion, the cross products are equal*. This fact can be used to help solve algebraic equations that are in the form of a proportion, as the next examples illustrate.

EXAMPLE 1 Solve $\frac{x}{20} = \frac{3}{4}$.

Solution

$$\frac{x}{20} = \frac{3}{4}$$

$$4x = 60 \qquad \text{Cross products are equal.}$$

$$x = 15.$$

The solution set is $\{15\}$. ■

EXAMPLE 2 Solve $\frac{x-3}{5} = \frac{x+2}{4}$.

Solution

$$\frac{x-3}{5} = \frac{x+2}{4} \qquad\qquad 1-24$$

$$4(x-3) = 5(x+2) \qquad \text{Cross products are equal.}$$

$$4x - 12 = 5x + 10 \qquad \text{Distributive property.}$$

$$-12 = x + 10 \qquad \text{Subtracted } 4x \text{ from both sides.}$$

$$-22 = x. \qquad \text{Subtracted 10 from both sides.}$$

The solution set is $\{-22\}$. ■

If a variable appears in one or both of the denominators, then a proper restriction should be made to avoid division by zero, as the next example illustrates.

EXAMPLE 3   Solve $\dfrac{7}{a-2} = \dfrac{4}{a+3}$.

Solution

$$\frac{7}{a-2} = \frac{4}{a+3}, \qquad a \neq 2 \text{ and } a \neq -3$$

$7(a+3) = 4(a-2)$ Cross products are equal.

$7a + 21 = 4a - 8$ Distributive property.

$3a + 21 = -8$ Subtracted $4a$ from both sides.

$3a = -29$ Subtracted 21 from both sides.

$$a = -\frac{29}{3}. \qquad \text{Divided both sides by 3.}$$

The solution set is $\left\{ -\dfrac{29}{3} \right\}$.   ■

■ Problem Solving Using Proportions

Some word problems can be conveniently set up and solved using the concepts of ratio and proportion. Consider the following examples.

PROBLEM 1   On a certain map, 1 inch represents 20 miles. If two cities are $6\frac{1}{2}$ inches apart on the map, find the number of miles between the cities.

Solution

Let m represent the number of miles between the two cities. Now let's set up a proportion for which one ratio compares distances, in inches, on the map and the other ratio compares *corresponding* distance, in miles, on land.

$$\frac{1}{6\frac{1}{2}} = \frac{20}{m}.$$

Solving this equation by equating the cross products produces

$$m(1) = \left(6\frac{1}{2}\right)(20)$$

$$m = \left(\frac{13}{2}\right)(20) = 130.$$

The distance between the two cities is 130 miles.   ■

PROBLEM 2 A sum of $1750 is to be divided between two people in the ratio of 3 to 4. How much does each person receive?

Solution Let d represent the amount of money to be received by one person. Then $1750 - d$ represents the amount for the other person. The following proportion can be set up and solved.

$$\frac{d}{1750 - d} = \frac{3}{4}$$

$$4d = 3(1750 - d)$$
$$4d = 5250 - 3d$$
$$7d = 5250$$
$$d = 750.$$

If $d = 750$, then $1750 - d = 1000$; therefore, one person receives $750 and the other person receives $1000.   ■

■ Percent

The word *percent* means "per one hundred" and is expressed using the symbol %. For example, 7 percent is written as 7% and means $\frac{7}{100}$ or 0.07. In other words, percent is a special kind of ratio, namely, one in which the denominator is always 100.

Proportions provide a convenient basis for changing common fractions to percents. Consider the following examples.

EXAMPLE 4 Express $\frac{7}{20}$ as a percent.

Solution We are asking "what number compares to 100 as 7 compares to 20?" Therefore, if we let n represent that number, the following proportion can be set up and solved.

$$\frac{n}{100} = \frac{7}{20}$$

$$20n = 700$$
$$n = 35.$$

Thus, $\frac{7}{20} = \frac{35}{100} = 35\%$.   ■

EXAMPLE 5 Express $\frac{5}{6}$ as a percent.

Solution

$$\frac{n}{100} = \frac{5}{6}$$

$$6n = 500$$

$$n = \frac{500}{6} = \frac{250}{3} = 83\frac{1}{3}.$$

Therefore, $\frac{5}{6} = 83\frac{1}{3}\%$.   ■

■ Some Basic Percent Problems

What is 8% of 35? Fifteen percent of what number is 24? Twenty-one is what percent of 70? These are often referred to as the three basic types of percent problems. Each of these problems can be easily solved by translating into and solving a simple algebraic equation.

PROBLEM 3   What is 8% of 35?

Solution   Let n represent the number to be found. The "is" refers to equality and "of" means multiplication. Thus, the question translates into

$$n = (8\%)(35),$$

which can be solved as follows.

$$n = (0.08)(35)$$
$$= 2.8.$$

Therefore, 2.8 is 8% of 35.   ■

PROBLEM 4   Fifteen percent of what number is 24?

Solution   Let n represent the number to be found.

$$(15\%)(n) = 24$$
$$0.15n = 24$$
$$15n = 2400 \qquad \text{Multiplied both sides by 100.}$$
$$n = 160.$$

Therefore, 15% of 160 is 24.   ■

PROBLEM 5 Twenty-one is what percent of 70?

Solution Let r represent the percent to be found.

$$21 = r(70)$$

$$\frac{21}{70} = r$$

$$\frac{3}{10} = r \qquad \text{Reduce!}$$

$$\frac{30}{100} = r \qquad \text{Changed } \frac{3}{10} \text{ to } \frac{30}{100}.$$

$$30\% = r.$$

Therefore, 21 is 30% of 70. ■

PROBLEM 6 Seventy-two is what percent of 60?

Solution Let r represent the percent to be found.

$$72 = r(60)$$

$$\frac{72}{60} = r$$

$$\frac{6}{5} = r$$

$$\frac{120}{100} = r$$

$$120\% = r.$$

Therefore, 72 is 120% of 60. ■

As mentioned before, it is helpful to get into the habit of checking answers for "reasonableness." We would also suggest that many times you can alert yourself to a computational error by *estimating* the answer before you actually do the problem. For example, prior to doing Problem 6 we may have estimated as follows. Since 72 is larger than 60, we know that the answer has to be greater than 100%. Furthermore, 1.5 (or 150%) times 60 equals 90. Therefore, we can estimate the answer to be somewhere between 100% and 150%. That may seem like a rather rough estimate, but many times such an estimate will detect a computational error.

Problem Set 3.5

For Problems 1–24, solve each of the equations treating them as proportions.

1. $\dfrac{x}{6} = \dfrac{3}{2}$

2. $\dfrac{x}{9} = \dfrac{5}{3}$

3. $\dfrac{5}{12} = \dfrac{n}{24}$

4. $\dfrac{7}{8} = \dfrac{n}{16}$

5. $\dfrac{x}{3} = \dfrac{5}{2}$

6. $\dfrac{x}{7} = \dfrac{4}{3}$

7. $\dfrac{x-2}{4} = \dfrac{x+4}{3}$

8. $\dfrac{x-6}{7} = \dfrac{x+9}{8}$

9. $\dfrac{x+1}{6} = \dfrac{x+2}{4}$

10. $\dfrac{x-2}{6} = \dfrac{x-6}{8}$

11. $\dfrac{-4}{x+2} = \dfrac{-3}{x-7}$

12. $\dfrac{-9}{x+1} = \dfrac{-8}{x+5}$

13. $\dfrac{-1}{x-7} = \dfrac{5}{x-1}$

14. $\dfrac{3}{x-10} = \dfrac{-2}{x+6}$

15. $\dfrac{3}{2x-1} = \dfrac{2}{3x+2}$

16. $\dfrac{1}{4x+3} = \dfrac{2}{5x-3}$

17. $\dfrac{n+1}{n} = \dfrac{8}{7}$

18. $\dfrac{5}{6} = \dfrac{n}{n+1}$

19. $\dfrac{n}{150-n} = \dfrac{1}{2}$

20. $\dfrac{n}{200-n} = \dfrac{3}{5}$

21. $\dfrac{300-n}{n} = \dfrac{3}{2}$

22. $\dfrac{80-n}{n} = \dfrac{7}{9}$

23. $\dfrac{-1}{5x-1} = \dfrac{-2}{3x+7}$

24. $\dfrac{-3}{2x-5} = \dfrac{-4}{x-3}$

For Problems 25–36, use proportions to change each common fraction to a percent.

25. $\dfrac{11}{20}$

26. $\dfrac{17}{20}$

27. $\dfrac{3}{5}$

28. $\dfrac{7}{25}$

29. $\dfrac{1}{6}$

30. $\dfrac{5}{7}$

31. $\dfrac{3}{8}$

32. $\dfrac{1}{16}$

33. $\dfrac{3}{2}$

34. $\dfrac{5}{4}$

35. $\dfrac{12}{5}$

36. $\dfrac{13}{6}$

For Problems 37–48, answer the question by setting up and solving an appropriate equation.

37. What is 7% of 38?

38. What is 35% of 52?

39. 15% of what number is 6.3?

40. 55% of what number is 38.5?

41. 76 is what percent of 95?

42. 72 is what percent of 120?

43. What is 120% of 50?

44. What is 160% of 70?

45. 46 is what percent of 40?

46. 26 is what percent of 20?

47. 160% of what number is 144?

48. 220% of what number is 66?

For Problems 49–65, solve each problem using a proportion.

49. A blueprint has a scale of 1 inch represents 6 feet. Find the dimensions of a rectangular room that measures $2\dfrac{1}{2}$ inches by $3\dfrac{1}{4}$ inches on the blueprint.

50. On a certain map, 1 inch represents 15 miles. If two cities are 7 inches apart on the map, find the number of miles between the cities.

51. If a car can travel 264 miles using 12 gallons of gasoline, how far will it go on 15 gallons?

52. Jesse used 10 gallons of gasoline to drive 170 miles. How much gasoline will he need to travel 238 miles?

53. If the ratio of the length of a rectangle to its width is $\frac{5}{2}$ and the width is 24 centimeters, find its length.

54. If the ratio of the width of a rectangle to its length is $\frac{4}{5}$ and the length is 45 centimeters, find its width.

55. A salt water solution is made by dissolving 3 pounds of salt in 10 gallons of water. At this rate, how many pounds of salt are needed for 25 gallons of water?

56. If a home valued at $50,000 is assessed $900 in real estate taxes, at the same rate how much are the taxes on a home assessed at $60,000?

57. If 20 pounds of fertilizer will cover 1500 square feet of lawn, how many pounds are needed for 2500 square feet?

58. A flu epidemic is affecting 6 out of every 10 college students in a certain part of the country. At this rate, how many students will be affected at a university of 15,000 students?

59. A preelection poll indicated that 3 out of every 7 eligible voters were going to vote in an upcoming election. At this rate, how many people would be expected to vote in a city of 210,000?

60. A board 28 feet long is cut into two pieces whose lengths are in the ratio of 2 to 5. Find the lengths of the two pieces.

61. The perimeter of a rectangle is 50 inches. If the ratio of its length to its width is 3 to 2, find the dimensions of the rectangle.

62. The ratio of male students to female students at a certain university is 5 to 4. If there is a total of 6975 students, find the number of male students and the number of female students.

63. An investment of $500 earns $45 in a year. At the same rate, how much additional money must be invested to raise the earnings to $72 per year?

64. A sum of $1250 is to be divided between two people in the ratio of 2 to 3. How much does each person receive?

65. An inheritance of $180,000 is to be divided between a child and the local cancer fund in the ratio of 5 to 1. How much money will the child receive?

■ Miscellaneous Problems

Solve each of the following equations. Don't forget that division by zero is undefined.

66. $\dfrac{3}{x-2} = \dfrac{6}{2x-4}$

67. $\dfrac{8}{2x+1} = \dfrac{4}{x-3}$

68. $\dfrac{5}{x-3} = \dfrac{10}{x-6}$

69. $\dfrac{6}{x-1} = \dfrac{5}{x-1}$

3.6 More on Percents and Problem Solving

The equation $x + 0.35 = 0.72$ can be solved by subtracting 0.35 from both sides of the equation. Another technique for solving equations containing decimals is to "clear the equation of all decimals" by multiplying both sides by an appropriate power of 10. The following examples illustrate the use of the "clear equation of all decimals" strategy in a variety of situations.

EXAMPLE 1 Solve $0.5x = 14$.

Solution

$$0.5x = 14$$
$$5x = 140 \qquad \text{Multiplied both sides by 10.}$$
$$x = 28. \qquad \text{Divided both sides by 5.}$$

The solution set is $\{28\}$. ■

EXAMPLE 2 Solve $x + 0.07x = 0.13$.

Solution

$$x + 0.07x = 0.13$$
$$100(x + 0.07x) = 100(0.13) \qquad \text{Multiply both sides by 100.}$$
$$100(x) + 100(0.07x) = 100(0.13) \qquad \text{Distributive property.}$$
$$100x + 7x = 13$$
$$107x = 13$$
$$x = \frac{13}{107}.$$

The solution set is $\left\{\dfrac{13}{107}\right\}$. ■

EXAMPLE 3 Solve $0.08y + 0.09y = 3.4$.

Solution

$$0.08y + 0.09y = 3.4$$
$$8y + 9y = 340 \qquad \text{Multiplied both sides by 100.}$$
$$17y = 340$$
$$y = 20.$$

The solution set is $\{20\}$. ■

EXAMPLE 4 Solve $0.10t = 560 - 0.12(t + 1000)$.

Solution

$$0.10t = 560 - 0.12(t + 1000)$$
$$10t = 56{,}000 - 12(t + 1000) \qquad \text{Multiplied both sides by 100.}$$

$$10t = 56{,}000 - 12t - 12{,}000 \qquad \text{Distributive property.}$$
$$22t = 44{,}000$$
$$t = 2000.$$

The solution set is $\{2000\}$.   ■

■ Problems Involving Percents

Many consumer problems can be solved using an "equation approach." For example, the following general guideline indicates a basic relationship involving discount sales.

original selling price − discount = discount sale price

Let's consider some examples using our algebraic techniques along with this basic guideline.

PROBLEM 1 Amy bought a dress at a 30% discount sale for $35. What was the original price of the dress?

Solution Let p represent the original price of the dress. We can use the basic discount relationship as a guideline for setting up an algebraic equation.

original selling price − discount = discount sale price
$$(100\%)(p) \quad - (30\%)(p) = \quad \$35.$$

Solving this equation we obtain

$$(100\%)(p) - (30\%)(p) = 35$$
$$(70\%)(p) = 35$$
$$0.7p = 35$$
$$7p = 350$$
$$p = 50.$$

The original price of the dress was $50.   ■

PROBLEM 2 Find the cost of a $30 pair of jogging shoes that is on sale for 20% off.

Solution Let x represent the discount sale price.

original price − discount = sale price
$$\$30 \quad - (20\%)(\$30) = \quad x.$$

Solving this equation we obtain

$$30 - (20\%)(30) = x$$
$$30 - (0.2)(30) = x$$
$$30 - 6 = x$$
$$24 = x.$$

The sale price is $24.   ■

Another basic relationship pertaining to consumer problems is expressed in the following equation.

selling price = cost + profit

Profit (also called *markup*, *markon*, *margin*, and *margin of profit*) may be stated in different ways. It may be stated as a percent of the selling price, a percent of the cost, or simply in terms of dollars and cents. Let's consider some problems for which the profit is either a percent of the selling price or a percent of the cost.

PROBLEM 3   A retailer has some shirts that cost him $20 each. He wants to sell them at a profit of 60% of the cost. What selling price should be marked on the shirts?

Solution   Let s represent the selling price. The basic relationship "selling price equals cost plus profit" can be used as a guideline.

selling price = cost + profit (% of cost)
$$\downarrow \qquad\quad \downarrow \qquad\qquad \downarrow$$
$$s \qquad = \$20 + \quad (60\%)(20).$$

Solving this equation we obtain

$$s = 20 + (60\%)(20)$$
$$s = 20 + (.6)(20)$$
$$s = 20 + 12$$
$$s = 32.$$

The selling price should be $32.   ■

PROBLEM 4   Kathrin bought a painting for $120 and later decided to resell it. She made a profit of 40% of the selling price. What did she receive for the painting?

Solution   We can use the same basic relationship as a guideline except this time the profit is a percent of the selling price. Let s represent the selling price.

$$\text{selling price} = \text{cost} + \text{profit (\% of selling price)}$$
$$\downarrow \qquad \downarrow \qquad \downarrow$$
$$s \qquad = 120 + \qquad (40\%)(s).$$

Solving this equation we obtain

$$s = 120 + (40\%)(s)$$
$$s = 120 + .4s$$
$$.6s = 120$$
$$s = \frac{120}{.6} = 200.$$

She received $200 for the painting.   ■

Certain types of investment problems can be solved by translating into algebraic equations as these next examples illustrate.

PROBLEM 5 A woman invests a total of $5000. Part of it is invested at 10% and the remainder at 12%. Her total yearly interest from the two investments is $560. How much does she have invested at each rate?

Solution Let x represent amount invested at 12%. Then $5000 - x$ represents amount invested at 10%. The following guideline can be used.

$$\begin{array}{ccc}
\text{interest earned from} & + & \text{interest earned from} & = & \text{total interest} \\
\text{12\% investment} & & \text{10\% investment} & & \text{earned} \\
\downarrow & & \downarrow & & \downarrow \\
(12\%)(x) & & + (10\%)(\$5000 - x) = & & \$560.
\end{array}$$

Solving this equation yields

$$(12\%)(x) + (10\%)(5000 - x) = 560$$
$$0.12x + 0.10(5000 - x) = 560$$
$$12x + 10(5000 - x) = 56{,}000$$
$$12x + 50{,}000 - 10x = 56{,}000$$
$$2x + 50{,}000 = 56{,}000$$
$$2x = 6000$$
$$x = 3000.$$
$$\text{Therefore } 5000 - x = 2000.$$

$3000 was invested at 12% and $2000 at 10%.   ■

PROBLEM 6   An investor invests a certain amount of money at 9%. Then he finds a better deal and invests $5000 more than that amount at 12%. His yearly income from the two investments was $1650. How much did he invest at each rate?

Solution   Let x represent the amount invested at 9%. Then $x + 5000$ represents the amount invested at 12%.

$$(9\%)(x) + (12\%)(x + 5000) = 1650$$
$$0.09x + 0.12(x + 5000) = 1650$$
$$9x + 12(x + 5000) = 165{,}000$$
$$9x + 12x + 60{,}000 = 165{,}000$$
$$21x + 60{,}000 = 165{,}000$$
$$21x = 105{,}000$$
$$x = 5000.$$
$$\text{Therefore } x + 5000 = 10{,}000.$$

He invested $5000 at 9% and $10,000 at 12%.   ■

Problem Set 3.6

For Problems 1–22, solve each of the equations.

1. $x - 0.36 = 0.75$ **2.** $x - 0.15 = 0.42$ **3.** $x + 7.6 = 14.2$

4. $x + 11.8 = 17.1$ **5.** $0.62 - y = 0.14$ **6.** $7.4 - y = 2.2$

7. $0.7t = 56$ **8.** $1.3t = 39$ **9.** $x = 3.36 - 0.12x$

10. $x = 5.3 - 0.06x$ **11.** $s = 35 + 0.3s$ **12.** $s = 40 + 0.5s$

13. $s = 42 + 0.4s$ **14.** $s = 24 + 0.6s$

15. $0.07x + 0.08(x + 600) = 78$ **16.** $0.06x + 0.09(x + 200) = 63$

17. $0.09x + 0.1(2x) = 130.5$ **18.** $0.11x + 0.12(3x) = 188$

19. $0.08x + 0.11(500 - x) = 50.5$ **20.** $0.07x + 0.09(2000 - x) = 164$

21. $0.09x = 550 - 0.11(5400 - x)$ **22.** $0.08x = 580 - 0.1(6000 - x)$

For Problems 23–46, set up an equation and solve each problem.

23. Tom bought a pair of trousers at a 30% discount sale for $35. What was the original price of the trousers?

24. Louise bought a dress for $140, which represented a 20% discount of the original price. What was the original price of the dress?

25. Find the cost of a $48 sweater that is on sale for 25% off.

26. Byron purchased a bicycle at a 10% discount sale for $121.50. What was the original price of the bicycle?

27. Jack bought a $32 putter on sale for 35% off. How much did he pay for the putter?

28. Mindy bought a 13-inch portable color TV for 20% off of the list price. The list price was $349.95. What did she pay for the TV?

29. Pierre bought a coat for $126 that was listed for $180. What rate of discount did he receive?

30. Phoebe paid $32 for a pair of gold shoes that were listed for $40. What rate of discount did she receive?

31. A retailer has some ties that cost him $5 each. He wants to sell them at a profit of 70% of the cost. What should be the selling price of the ties?

32. A retailer has some blouses that cost her $25 each. She wants to sell them at a profit of 80% of the cost. What price should she charge for the blouses?

★ 33. The owner of a pizza parlor wants to make a profit of 55% of the cost for each pizza sold. If it costs $3 to make a pizza, at what price should it be sold?

34. Produce in a food market usually has a high markup because of loss due to spoilage. If a head of lettuce costs a retailer $.40, at what price should it be sold to realize a profit of 130% of the cost?

35. Jewelry has a very high markup rate. If a ring costs a jeweler $400, at what price should it be sold to gain a profit of 60% of the selling price?

36. If a box of candy costs a retailer $2.50 and he wants to make a profit of 50% based on the selling price, what price should he charge for the candy?

37. If the cost for a pair of shoes for a retailer is $32 and he sells them for $44.80, what is his rate of profit based on the cost?

38. A retailer has some skirts that cost her $24. If she sells them for $31.20, find her rate of profit based on the cost.

39. Lou invested a certain amount of money at 9% interest and $250 more than that amount at 10%. His total yearly interest was $101. How much did he invest at each rate?

40. A total of $12,000 is to be invested, part of it at 12% interest and the remainder at 14%. If the total yearly interest amounts to $1580, how much is invested at each rate?

41. A total of $1200 is invested, part of it at 9% interest and the remainder at 12%. If the total yearly interest amounts to $129, how much is invested at each rate?

42. Sally invested a certain sum of money at 9%, twice that sum at 10%, and three times that sum at 11%. Her total yearly interest from all three investments was $310. How much did she invest at each rate?

43. If $2000 is invested at 8% interest, how much additional money must be invested at 11% interest so that the total return for both investments averages 10%?

44. Fawn invested a certain amount of money at 10% interest and $250 more than that amount at 11%. Her total yearly interest was $153.50. How much did she invest at each rate?

★ 45. A sum of $2300 is invested, part of it at 10% interest and the remainder at 12%. If the interest earned by the 12% investment is $100 more than the interest earned by the 10% investment, find the amount invested at each rate.

46. If $3000 is invested at 9% interest, how much additional money must be invested at 12% so that the total return for both investments averages 11%?

47. How can $5400 be invested, part of it at 8% and the remainder at 10%, so that the two investments will produce the same amount of interest?

48. A sum of $6000 is invested, part of it at 9% interest and the remainder at 11%. If the interest earned by the 9% investment is $160 less than the interest earned by the 11% investment, find the amount invested at each rate.

■ Miscellaneous Problems

49. A retailer buys an item for $40, resells it for $50, and claims that he is only making a 20% profit. Is his claim correct?

50. A store has a special discount sale of 40% off on all items. It also advertises an additional 10% off on items bought in quantities of a dozen or more. How much will it cost to buy a dozen items of some particular kind that regularly sell for $5 per item? (Be careful, a 40% discount followed by a 10% discount is not equal to a 50% discount.)

51. Some people use the following formula for determining the selling price of an item when the profit is based on a percent of the selling price.

$$\text{selling price} = \frac{\text{cost}}{100\% - \text{percent of profit}}.$$

Show how this formula is developed.

Chapter 3 Summary

(3.1) **Numerical equations** may be true or false. **Algebraic equations** (open sentences) contain one or more variables. **Solving an equation** refers to the process of finding the number (or numbers) that makes an algebraic equation a true statement. A **first-degree equation of one variable** is an equation containing only one variable and this variable has an exponent of one.

Properties 3.1 and 3.2 provide the basis for solving equations. Be sure that you can use these properties to solve a variety of equations presented in this chapter.

(3.2) It is often necessary to use both the addition and multiplication properties of equality to solve an equation.

Be sure to *declare your variable* as you translate English sentences into algebraic equations.

(3.3) Keep the following suggestions in mind as you solve word problems.

  1. Read the problem carefully.

  2. Sketch any figure, diagram, or chart that might be helpful.

  3. Choose a meaningful variable.

  4. Look for a *guideline*.

  5. Form an equation.

  6. Solve the equation.

  7. Check your answers.

(3.4) The **distributive property** is used to *remove parentheses.*

If an equation contains several fractions, then it is usually advisable to *clear the equation of all fractions* by multiplying both sides by the least common denominator of all of the denominators in the equation.

(3.5) A **ratio** is the comparison of two numbers by division.

A statement of equality between two ratios is a **proportion**. In a proportion, the *cross products* are equal. That is to say,

$$\text{if } \frac{a}{b} = \frac{c}{d}, \quad \text{then } ad = bc, \quad (b \neq 0 \quad \text{and} \quad d \neq 0).$$

The cross product property can be used to solve equations that are in the form of a proportion.

A variety of word problems can be conveniently set up and solved using proportions.

The concept of **percent** means "per one hundred" and is therefore a special ratio, namely, one having a denominator of 100. Proportions provide a convenient basis for changing common fractions to percents.

(3.6) To solve equations containing decimals, you can clear the equation of all decimals by multiplying both sides by an appropriate power of ten.

Many consumer problems involve the concept of *percent* and can be solved using an equation approach. The basic relationships *selling price equals cost plus profit* and *original selling price minus discount equals discount sale price* are used frequently.

Chapter 3 Review Problem Set

For Problems 1–25, solve each of the equations.

1. $9x - 2 = -29$

2. $-3 = -4y + 1$

3. $7 - 4x = 10$

4. $6y - 5 = 4y + 13$

5. $4n - 3 = 7n + 9$

6. $7(y - 4) = 4(y + 3)$

7. $2(x + 1) + 5(x - 3) = 11(x - 2)$

8. $-3(x + 6) = 5x - 3$

9. $\dfrac{2}{5}n - \dfrac{1}{2}n = \dfrac{7}{10}$

10. $\dfrac{3n}{4} + \dfrac{5n}{7} = \dfrac{1}{14}$

11. $\dfrac{x - 3}{6} + \dfrac{x + 5}{8} = \dfrac{11}{12}$

12. $\dfrac{n}{2} - \dfrac{n - 1}{4} = \dfrac{3}{8}$

13. $0.5x + 0.7x = 1.7$

14. $0.07t + 0.12(t - 3) = 0.59$

15. $-2(x - 4) = -3(x + 8)$

16. $3x - 4x - 2 = 7x - 14 - 9x$

17. $5(n - 1) - 4(n + 2) = -3(n - 1) + 3n + 5$

18. $\dfrac{9}{x - 3} = \dfrac{8}{x + 4}$

19. $\dfrac{-3}{x - 1} = \dfrac{-4}{x + 2}$

20. $-(t - 3) - (2t + 1) = 3(t + 5) - 2(t + 1)$

21. $\dfrac{2x - 1}{3} = \dfrac{3x + 2}{2}$

22. $0.1x + 0.12(1700 - x) = 188$

23. $x - 0.25x = 12$

24. $0.2(x - 3) = 14$

25. $3(2t - 4) + 2(3t + 1) = -2(4t + 3) - (t - 1)$

Set up an equation and solve each of the following problems (26–43).

26. Three-fourths of a number equals 18. Find the number.

27. Nineteen is two less than three times a certain number. Find the number.

28. The difference of two numbers is 21. If 12 is the smaller number, find the other number.

29. One subtracted from nine times a certain number is the same as fifteen added to seven times the number. Find the number.

30. Fifteen percent of a certain number is 6. Find the number.

31. The sum of two numbers is 40. Six times the smaller number equals four times the larger. Find the numbers.

32. Eighteen is what percent of 30?

33. Find a number such that two less than two-thirds of the number is one more than one-half of the number.

34. The sum of two numbers is 96 and their ratio is 5 to 7. Find the numbers.

35. Miriam has 30 coins consisting of nickels and dimes amounting to $2.60. How many coins of each kind does she have?

36. Russ has a bunch of nickels, dimes, and quarters amounting to $15.40. The number of dimes is one more than three times the number of nickels and the number of. quarters is twice the number of dimes. How many coins of each kind does he have?

37. The supplement of an angle is $14°$ more than three times the complement of the angle. Find the measure of the angle.

38. The ratio of the complement of an angle to the supplement of the angle is 7 to 16. Find the measure of the angle.

39. If a car uses 18 gallons of gasoline for a 369-mile trip, at the same rate of consumption how many gallons will it use on a 615-mile trip?

40. A sum of $2100 is invested, part of it at 9% interest and the remainder at 11%. If the interest earned by the 11% investment is $51 more than the interest from the 9% investment, find the amount invested at each rate.

41. A retailer has some sweaters that cost $28 each. At what price should the sweaters be sold to obtain a profit of 30% of the selling price?

42. Robin bought a dress on sale for $39 and the original price of the dress was $60. What percent of discount did she receive?

43. Pam rented a car from a rental agency that charges $25 a day and $0.20 per mile. She kept the car for 3 days and her bill was $215. How many miles did she drive during that 3-day period?

4 Formulas, Problem Solving, and Inequalities

In this chapter we shall expand some of the ideas from the previous chapter: solving equations and using equations to solve word problems. We shall also consider some techniques for solving inequalities.

4.1   Formulas

To find the distance traveled in 3 hours at a rate of 50 miles per hour, we multiply the rate by the time. Thus, the distance is $50(3) = 150$ miles. The rule *distance equals rate times time* is commonly stated as a **formula**: $d = rt$. Formulas are simply rules stated in symbolic language and usually expressed as equations. Thus, the formula, $d = rt$, is an equation involving three variables, d, r, and t.

As we work with formulas it is often necessary to solve for some specific variable when numerical values for the remaining variables are known. Consider the following examples.

EXAMPLE 1   Solve $d = rt$ for r if $d = 330$ and $t = 6$.

Solution   Substituting 330 for d and 6 for t in the given formula, we obtain

$$330 = r(6).$$

Solving this equation yields

$$330 = 6r$$
$$55 = r. \qquad \blacksquare$$

EXAMPLE 2   Solve $C = \dfrac{5}{9}(F - 32)$ for F if $C = 10$. (This formula expresses the relationship between the Fahrenheit and Celsius temperature scales.)

Solution   Substituting 10 for C, we obtain

$$10 = \frac{5}{9}(F - 32).$$

Solving this equation produces

$$\frac{9}{5}(10) = \frac{9}{5}\left(\frac{5}{9}\right)(F - 32) \qquad \text{Multiply both sides by } \frac{9}{5}.$$
$$18 = F - 32$$
$$50 = F. \qquad \blacksquare$$

Sometimes when using a formula it may be convenient to change its form by using the properties of equality. For example, the formula, $d = rt$, can be changed as follows.

$$d = rt$$

$$\frac{d}{r} = \frac{rt}{r} \qquad \text{Divide both sides by } r.$$

$$\frac{d}{r} = t.$$

We say that the formula, $d = rt$, **has been solved for the variable t**. The formula can also be **solved for r** as follows.

$$d = rt$$

$$\frac{d}{t} = \frac{rt}{t} \qquad \text{Divide both sides by } t.$$

$$\frac{d}{t} = r.$$

■ Geometric Formulas

There are several formulas in geometry that are used quite often. Let's briefly review them at this time and then they will be used periodically throughout the remainder of the text. (These formulas along with some others are also listed in the inside front cover of this text.)

$$A = lw \qquad P = 2l + 2w$$

    A area
    P perimeter
    l length
    w width

Rectangle

Triangle

$$A = \frac{1}{2}bh$$

A area
b base
h altitude (height)

Trapezoid

$$A = \frac{1}{2}h(b_1 + b_2)$$

A area
b_1, b_2 bases
h altitude

Parallelogram

$A = bh$

A area
b base
h altitude (height)

Circle

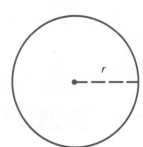

$A = \pi r^2 \quad C = 2\pi r$

A area
C circumference
r radius

Sphere

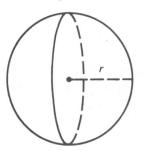

$S = 4\pi r^2 \quad V = \dfrac{4}{3}\pi r^3$

S surface area
V volume
r radius

Prism

$V = Bh$

V volume
B area of base
h altitude (height)

Pyramid

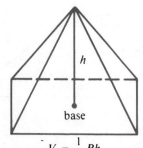

$V = \dfrac{1}{3} Bh$

V volume
B area of base
h altitude (height)

Right circular cylinder

$V = \pi r^2 h \quad S = 2\pi r^2 + 2\pi rh$

V volume
S total surface area
r radius
h altitude (height)

Right circular cone

$V = \dfrac{1}{3}\pi r^2 h \quad S = \pi r^2 + \pi rs$

V volume
S total surface area
r radius
h altitude (height)
s slant height

EXAMPLE 3 Find the total surface area of a right circular cylinder having a radius of 10 inches and a height of 14 inches.

Solution Substituting 10 for r and 14 for h in the formula for finding total surface area of a right circular cylinder we obtain

$$
\begin{aligned}
S &= 2\pi r^2 + 2\pi rh \\
&= 2\pi(10)^2 + 2\pi(10)(4) \\
&= 200\pi + 280\pi \\
&= 480\pi.
\end{aligned}
$$

The total surface area is 480π square inches. ■

EXAMPLE 4 Solve $C = 2\pi r$ for r.

Solution

$$C = 2\pi r$$

$$\frac{C}{2\pi} = \frac{2\pi r}{2\pi} \qquad \text{Divide both sides by } 2\pi.$$

$$\frac{C}{2\pi} = r.$$ ■

EXAMPLE 5 Solve $V = \dfrac{1}{3} Bh$ for h.

Solution

$$V = \frac{1}{3} Bh$$

$$3(V) = 3\left(\frac{1}{3} Bh\right) \qquad \text{Multiply both sides by 3.}$$

$$3V = Bh$$

$$\frac{3V}{B} = \frac{Bh}{B} \qquad \text{Divide both sides by } B.$$

$$\frac{3V}{B} = h.$$ ■

EXAMPLE 6 Solve $P = 2l + 2w$ for w.

Solution

$$P = 2l + 2w$$

$$P - 2l = 2l + 2w - 2l \qquad \text{Subtract } 2l \text{ from both sides.}$$

$$P - 2l = 2w$$

$$\frac{P - 2l}{2} = \frac{2w}{2} \qquad \text{Divide both sides by 2.}$$

$$\frac{P - 2l}{2} = w. \qquad\qquad\qquad\qquad\qquad\qquad ■$$

In Chapter 8 you will be working with equations containing two variables. At times you will need to solve for one variable in terms of the other variable; that is, change the form of the equation as we have been doing with formulas. The following examples illustrate once again how the properties of equality are used for such situations.

EXAMPLE 7 Solve $3x + y = 4$ for x.

Solution

$$3x + y = 4$$
$$3x + y - y = 4 - y \qquad \text{Subtract } y \text{ from both sides.}$$
$$3x = 4 - y$$

$$\frac{3x}{3} = \frac{4 - y}{3} \qquad \text{Divide both sides by 3.}$$

$$x = \frac{4 - y}{3}. \qquad\qquad\qquad\qquad\qquad ■$$

EXAMPLE 8 Solve $4x - 5y = 7$ for y.

Solution

$$4x - 5y = 7$$
$$4x - 5y - 4x = 7 - 4x \qquad \text{Subtract } 4x \text{ from both sides.}$$
$$-5y = 7 - 4x$$

$$\frac{-5y}{-5} = \frac{7 - 4x}{-5} \qquad \text{Divide both sides by } -5.$$

$$y = \frac{7 - 4x}{-5}\left(\frac{-1}{-1}\right) \qquad \begin{array}{l} \text{Multiply numerator and denominator} \\ \text{of fraction on the right by } -1. \end{array}$$

$$y = \frac{4x - 7}{5}. \qquad \begin{array}{l} \text{(We commonly do this so that the} \\ \text{denominator is positive.)} \end{array} \qquad ■$$

EXAMPLE 9 Solve $y = mx + b$ for m.

Solution

$$y = mx + b$$
$$y - b = mx + b - b \qquad \text{Subtract } b \text{ from both sides.}$$
$$y - b = mx$$

$$\frac{y-b}{x} = \frac{mx}{x}$$ Divide both sides by x.

$$\frac{y-b}{x} = m.$$ ■

Problem Set 4.1

For Problems 1–10, solve for the specified variable using the given facts.

1. Solve $d = rt$ for t if $d = 336$ and $r = 48$.

2. Solve $d = rt$ for r if $d = 486$ and $t = 9$.

3. Solve $i = Prt$ for P if $i = 200$, $r = 0.08$, and $t = 5$.

4. Solve $i = Prt$ for t if $i = 540$, $P = 750$, and $r = 0.09$.

5. Solve $F = \frac{9}{5}C + 32$ for C if $F = 68$.

6. Solve $C = \frac{5}{9}(F - 32)$ for F if $C = 15$.

7. Solve $V = \frac{1}{3}Bh$ for B if $V = 112$ and $h = 7$.

8. Solve $V = \frac{1}{3}Bh$ for h if $V = 216$ and $B = 54$.

9. Solve $A = P + Prt$ for t if $A = 652$, $P = 400$, and $r = 0.07$.

10. Solve $A = P + Prt$ for P if $A = 1032$, $r = 0.06$, and $t = 12$.

For Problems 11–32, use the geometric formulas given in this section to help solve the problems.

11. Find the perimeter of a rectangle that is 14 centimeters long and 9 centimeters wide.

12. If the perimeter of a rectangle is 80 centimeters and its length is 24 centimeters, find its width.

13. If the perimeter of a rectangle is 108 inches and its length is $3\frac{1}{4}$ feet, find its width in inches.

14. How many yards of fencing would it take to enclose a rectangular plot of ground that is 69 feet long and 42 feet wide?

15. A dirt path 4 feet wide surrounds a rectangular garden that is 38 feet long and 17 feet wide. Find the area of the dirt path.

16. Find the area of a cement walk 3 feet wide that surrounds a rectangular plot of ground 86 feet long and 42 feet wide.

17. Suppose that paint costs $2.00 per liter and that one liter will cover 9 square meters of surface. We are going to paint (on one side only) 50 pieces of wood of the same size, that are rectangular in shape, having a length of 60 centimeters and a width of 30 centimeters. What will the cost for the paint be?

18. A lawn is in the shape of a triangle with one side 130 feet long and the altitude to that side 60 feet long. Will one sack of fertilizer, which covers 4000 square feet, be enough to fertilize the lawn?

19. Find the length of an altitude of a trapezoid having bases of 8 inches and 20 inches and an area of 98 square inches.

20. A flower garden is in the shape of a trapezoid having bases of measure 6 yards and 10 yards. The distance between the bases is 4 yards. Find the area of the garden.

21. The diameter of a metal washer is 4 centimeters. The diameter of the hole is 2 centimeters. How many square centimeters of metal are there in 50 washers? Express the answer in terms of π.

22. Find the area of a circular plot of ground having a radius of length 14 meters. Use $3\frac{1}{7}$ as an approximation for π.

23. Find the area of a circular region having a diameter of 1 yard. Express the answer in terms of π.

24. Find the area of a circular region if the circumference is 12π units. Express the answer in terms of π.

25. Find the total surface area and volume of a sphere having a radius 9 inches long. Express the answers in terms of π.

26. A circular pool is 34 feet in diameter and has a flagstone walk around it that is 3 feet wide. Find the area of the walk. Express the answer in terms of π.

27. Find the volume and total surface area of a right circular cylinder having a radius of 8 feet and a height of 18 feet. Express the answers in terms of π.

28. Find the total surface area and volume of a sphere having a diameter 12 centimeters long. Express the answers in terms of π.

29. If the volume of a right circular cone is 324π cubic inches and a radius of the base is 9 inches long, find the height of the cone.

30. Find the volume and total surface area of a tin can if the radius of a base is 3 centimeters and the height of the can is 10 centimeters. Express the answers in terms of π.

31. If the total surface area of a right circular cone is 65π square feet and a radius of the base is 5 feet long, find the slant height of the cone.

32. If the total surface area of a right circular cylinder is 104π square meters and a radius of the base is 4 meters long, find the height of the cylinder.

For Problems 33–44, solve each of the formulas for the indicated variable.

33. $V = Bh$ for $h = \frac{V}{B}$ (Volume of prism)

34. $A = lw$ for l (Area of a rectangle)

35. $V = \frac{1}{3}Bh$ for B (Volume of a pyramid)

36. $A = \frac{1}{2}bh$ for h (Area of a triangle)

37. $P = 2l + 2w$ for w (Perimeter of a rectangle)

38. $V = \pi r^2 h$ for h (Volume of a cylinder)

39. $V = \frac{1}{3}\pi r^2 h$ for h (Volume of a cone)

40. $i = Prt$ for t (Simple interest formula)

41. $F = \dfrac{9}{5}C + 32$ for C (Celsius to Fahrenheit)

42. $A = P + Prt$ for P (Simple interest formula)

43. $A = 2\pi r^2 + 2\pi rh$ for h (Surface area of a cylinder)

44. $C = \dfrac{5}{9}(F - 32)$ for F (Fahrenheit to Celsius)

For Problems 45–60, solve each equation for the indicated variable.

45. $3x + 7y = 9$ for x **46.** $5x + 2y = 12$ for x **47.** $9x - 6y = 13$ for y

48. $3x - 5y = 19$ for y **49.** $-2x + 11y = 14$ for x **50.** $-x + 14y = 17$ for x

51. $y = -3x - 4$ for x **52.** $y = -7x + 10$ for x **53.** $\dfrac{x-2}{4} = \dfrac{y-3}{6}$ for y

54. $\dfrac{x+1}{3} = \dfrac{y-5}{2}$ for y **55.** $ax - by - c = 0$ for y **56.** $ax + by = c$ for y

57. $\dfrac{x+6}{2} = \dfrac{y+4}{5}$ for x **58.** $\dfrac{x-3}{6} = \dfrac{y-4}{8}$ for x **59.** $m = \dfrac{y-b}{x}$ for y

60. $y = mx + b$ for x

4.2 Problem Solving

Let's begin this section by restating the suggestions for solving word problems offered in Section 3.3.

Suggestions for Solving Word Problems

1. Read the problem carefully, making certain that you understand the meanings of all the words. Be especially alert for any technical terms used in the statement of the problem.

2. Read the problem a second time (perhaps even a third time) to get an overview of the situation being described and to determine the known facts as well as what is to be found.

3. Sketch any figure, diagram, or chart that might be helpful in analyzing the problem.

4. Choose a meaningful variable to represent an unknown quantity in the problem (perhaps t, if time is the unknown quantity); represent any other unknowns in terms of that variable.

5. Look for a guideline that can be used to set up an equation. A guideline might be a formula such as "selling price equals cost plus profit," or a relationship such as "interest earned from a 9% investment plus interest earned from a 10% investment equals total amount of interest earned."

6. Form an equation containing the variable that translates the conditions of the guideline from English to algebra.

7. Solve the equation and use the solution to determine all facts requested in the problem.

8. **Check all answers back into the original statement of the problem.**

Again we emphasize the importance of suggestion number 5. Determining the guideline to follow when setting up the equation is a vital part of the analysis of a problem. Sometimes the guideline is a formula, such as one of the formulas presented in the previous section and accompanying problem set. Let's consider a problem of that type.

PROBLEM 1 How long will it take $500 to double itself if it is invested at 8% simple interest?

Solution Let's use the basic simple interest formula, $i = Prt$, where i represents interest, P is the principal (money invested), r is the rate (percent), and t is the time in years. For $500 to "double itself" means that we want $500 to earn another $500 in interest. Thus, using $i = Prt$ as a guideline we can proceed as follows.

$$i = Prt.$$
$$500 = 500(8\%)(t)$$

Now let's solve this equation.

$$500 = 500(.08)(t)$$
$$1 = .08t$$
$$100 = 8t$$
$$\frac{100}{8} = t$$
$$12\frac{1}{2} = t.$$

It will take $12\frac{1}{2}$ years. ■

If the problem involves a geometric formula, then a sketch of the figure is helpful for recording the given information and analyzing the problem. The next example illustrates this idea.

PROBLEM 2 The length of a football field is 40 feet more than twice its width and the perimeter of the field is 1040 feet. Find the length and width of the field.

Solution   Since the length is stated in terms of the width, we can let w represent the width and then $2w + 40$ represents the length.

$$2w + 40$$

A guideline for this problem is the perimeter formula, $P = 2l + 2w$. Thus, the following equation can be set up and solved.

$$P = 2l + 2w$$
$$1040 = 2(2w + 40) + 2w$$
$$1040 = 4w + 80 + 2w$$
$$1040 = 6w + 80$$
$$960 = 6w$$
$$160 = w.$$

If $w = 160$, then $2w + 40 = 2(160) + 40 = 360$. Thus, the football field is 360 feet long and 160 feet wide.   ■

Sometimes formulas are used within the analysis of a problem but do not serve as the main guideline for setting up the equation. For example, uniform motion problems involve the formula $d = rt$, but the main guideline for setting up an equation for such problems is usually a statement about either *times*, *rates*, or *distances*. Let's consider an example to illustrate this.

PROBLEM 3 Jerry leaves city A on a moped traveling toward city B at 18 miles per hour. At the same time, Cindy leaves city B on a moped traveling toward city A at 23 miles per hour. The distance between the two cities is 123 miles. How long will it take before Jerry and Cindy meet on their mopeds?

Remark   Before presenting a solution to this problem we want to emphasize a point made earlier. *Estimating* an answer before attempting to solve a problem may, in fact, help in determining the guideline to be used. To estimate an answer for Problem 3, suppose both Jerry and Cindy travel at 20 miles per hour. Then together they travel 40 miles per hour; so it should take approximately 3 hours to travel 123 miles.

Solution

First, let's sketch a diagram.

Let t represent the time that Jerry travels. Then t also represents the time that Cindy travels.

distance Jerry travels + distance Cindy travels = total distance
$$\downarrow \qquad\qquad \downarrow \qquad\qquad \downarrow$$
$$18t \qquad + \qquad 23t \qquad = \qquad 123.$$

Solving this equation yields

$$18t + 23t = 123$$
$$41t = 123$$
$$t = 3.$$

They both travel for 3 hours.   ■

Some people find it helpful to use a chart to organize the known and unknown facts in a uniform motion problem. Let's illustrate this with an example.

PROBLEM 4   A car leaves a town at 60 kilometers per hour. How long will it take a second car traveling at 75 kilometers per hour to catch the first car if the second car leaves 1 hour later?

Solution

Let t represent the time of the second car. Then $t + 1$ represents the time of the first car since it travels one hour longer. The information of the problem can now be recorded in a chart as follows.

	rate	time	distance
first car	60	$t + 1$	$60(t + 1)$
second car	75	t	$75t$

$d = rt$

Since the second car is to overtake the first car, the distances must be equal.

distance of second car equals distance of first car
$$\downarrow \qquad\qquad\qquad \downarrow$$
$$75t \qquad = \qquad 60(t + 1).$$

Solving this equation yields

$$75t = 60(t + 1)$$
$$75t = 60t + 60$$
$$15t = 60$$
$$t = 4.$$

The second car should overtake the first car in 4 hours. (Check the answer!)   ■

We would like to offer one bit of advice at this time. Don't become discouraged if solving word problems is giving you trouble. Problem solving is not a skill that can be developed overnight. It takes time, patience, hard work, and an open mind. Keep giving it your best shot and gradually you should become more confident in your approach to such problems. Furthermore, we realize that some (perhaps many) of these problems may not seem "practical" to you. However, keep in mind that the real issue here is that of developing problem solving techniques. From the previous examples the ideas of finding and using a guideline, sketching a figure to record information and help in the analysis, estimating an answer before attempting to solve the problem, and using a chart to record information are the key issues.

Problem Set 4.2

Set up an equation and solve each of the following problems. Keep in mind the suggestions offered in this section.

1. How long will it take $750 to double itself if it is invested at 8% simple interest?
2. How long will it take $1000 to double itself if it is invested at 10% simple interest?
3. How long will it take $800 to triple itself if it is invested at 10% simple interest?
4. How long will it take $500 to earn $750 in interest if it is invested at 6% simple interest?
5. The length of a rectangle is three times its width. If the perimeter of the rectangle is 112 inches, find its length and width.
6. The width of a rectangle is one-half of its length. If the perimeter of the rectangle is 54 feet, find its length and width.
7. The length of a rectangle is 2 centimeters less than three times its width. The perimeter of the rectangle is 92 centimeters. Find the length and width of the rectangle.
8. The length of a rectangle is 1 meter more than five times its width. The perimeter of the rectangle is 98 meters. Find the length and width of the rectangle.
9. The width of a rectangle is 3 inches less than one-half of its length. If the perimeter of the rectangle is 42 inches, find the area of the rectangle.
10. The width of a rectangle is 1 foot more than one-third of its length. If the perimeter of the rectangle is 74 feet, find the area of the rectangle.
11. The perimeter of a triangle is 100 feet. The longest side is 3 feet less than twice the shortest

side and the third side is 7 feet longer than the shortest side. Find the lengths of the sides of the triangle.

12. A triangular plot of ground has a perimeter of 54 yards. The longest side is twice the shortest side and the third side is two yards longer than the shortest side. Find the lengths of the sides of the triangle.

13. The second side of a triangle is 1 centimeter more than three times the first side. The third side is 2 centimeters longer than the second side. If the perimeter is 46 centimeters, find the length of each side of the triangle.

14. The second side of a triangle is 3 meters less than twice the first side. The third side is 4 meters longer than the second side. If the perimeter is 58 meters, find the length of each side of the triangle.

15. The perimeter of an equilateral triangle is 4 centimeters more than the perimeter of a square and the length of a side of the triangle is 4 centimeters more than the length of a side of the square. Find the length of a side of the equilateral triangle. (An equilateral triangle has three sides of the same length.)

16. A square and an equilateral triangle have the same perimeter. Each side of the equilateral triangle is 6 centimeters longer than each side of the square. Find the length of each side of the square. (An equilateral triangle has three sides of the same length.)

17. Suppose that the length of a radius of a circle is the same as the length of a side of a square. If the circumference of the circle is 15.96 centimeters longer than the perimeter of the square, find the length of a radius of the circle. (Use 3.14 as an approximation for π).

18. The circumference of a circle is 2.24 centimeters more than six times the length of a radius. Find the radius of the circle. (Use 3.14 as an approximation for π).

19. Sandy leaves a town traveling in her car at a rate of 45 miles per hour. One hour later, Monica leaves the same town traveling the same route at a rate of 50 miles per hour. How long will it take Monica to overtake Sandy?

20. Two cars start from the same place traveling in opposite directions. One car travels 4 miles per hour faster than the other car. Find their speeds if after 5 hours they are 520 miles apart.

21. The distance between city A and city B is 325 miles. A freight train leaves city A and travels toward city B at 40 miles per hour. At the same time, a passenger train leaves city B and travels toward city A at 90 miles per hour. How long will it take the two trains to meet?

22. Kirk starts jogging at 5 miles per hour. One-half hour later Nancy starts jogging on the same route at 7 miles per hour. How long will it take Nancy to catch Kirk?

23. A car leaves a town at 40 miles per hour. Two hours later a second car leaves the town traveling the same route and overtakes the first car in 5 hours and 20 minutes. How fast was the second car traveling?

24. Two airplanes leave St. Louis at the same time and fly in opposite directions. If one travels at 500 kilometers per hour and the other at 600 kilometers per hour, how long will it take for them to be 1925 kilometers apart?

25. Two trains leave at the same time, one traveling east and the other traveling west. At the end of $9\frac{1}{2}$ hours they are 1292 miles apart. If the rate of the train traveling east is 8 miles per hour faster than the other train, find their rates.

26. Dawn starts on a 58 mile trip on her moped at 20 miles per hour. After a time the motor "kicks out" and she pedals the remainder of the trip at 12 miles per hour. The entire trip took $3\frac{1}{2}$ hours. How far had Dawn traveled when the motor on the moped quit running?

27. Jeff leaves home and rides his bicycle out into the country for 3 hours. On his return trip along the same route, it takes him three-quarters of an hour longer. If his rate on the return trip was 2 miles per hour slower than on the trip out into the country, find the total roundtrip distance.

28. In $1\frac{1}{4}$ hours more time, Rita, riding her bicycle at 12 miles per hour, rode 2 miles further than Sonya, who was riding her bicycle at 16 miles per hour. How long did each girl ride?

4.3 More About Problem Solving

Let's begin this section by considering another important but often overlooked facet of problem solving. That is the idea of *looking back* over your solution and considering some of the following questions.

1. Is your answer to the problem a reasonable answer? Does it agree with the estimated answer you arrived at before doing the problem?

2. Have you checked your answer by substituting it back into the conditions stated in the problem?

3. Do you now see another plan that could be used to solve the problem? Perhaps there is even another guideline that could be used.

4. Do you now see that this problem is closely related to another problem that you have previously solved?

5. Have you "tucked away for future reference" the technique used to solve this problem?

Looking back over the solution of a newly solved problem can often lay important groundwork for solving problems in the future.

Now let's consider three problems that are often referred to as mixture problems. There is no basic formula that applies for all of these problems, but the suggestion that "you think in terms of a pure substance" is often helpful in setting up a guideline. Let's illustrate what we mean by that suggestion. (A phrase such as "30% solution of acid" means that 30% of the amount of solution is acid and the remaining 70% is water.)

PROBLEM 1 How many milliliters of pure acid must be added to 150 milliliters of a 30% solution of acid to obtain a 40% solution?

Solution The key idea for solving such a problem is to recognize the following guideline.

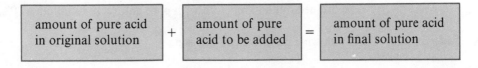

$$\boxed{\begin{array}{c}\text{amount of pure acid} \\ \text{in original solution}\end{array}} + \boxed{\begin{array}{c}\text{amount of pure} \\ \text{acid to be added}\end{array}} = \boxed{\begin{array}{c}\text{amount of pure acid} \\ \text{in final solution}\end{array}}$$

Letting x represent the amount of pure acid to be added, the guideline translates into $(30\%)(150) + x = (40\%)(150 + x)$. Solving this equation yields

$$45 + x = 60 + 0.40x$$
$$0.60x = 15$$
$$60x = 1500$$
$$x = 25.$$

We must add 25 mL of pure acid. (Perhaps you should check this answer!).   ■

PROBLEM 2 Suppose that you have a supply of a 30% solution of alcohol and a 70% solution. How many quarts of each should be mixed to produce a 20-quart solution that is 40% alcohol?

Solution A guideline similar to the one in Problem 1 can be used.

pure alcohol in 30% solution	+	pure alcohol in 70% solution	=	pure alcohol in 40% solution

Let x represent the amount of 30% solution. Then $20 - x$ represents the amount of 70% solution. Now using the guideline, we translate to

$$(30\%)(x) + (70\%)(20 - x) = (40\%)(20).$$

Solving this equation we obtain

$$0.30x + 0.70(20 - x) = 8$$
$$30x + 70(20 - x) = 800$$
$$30x + 1400 - 70x = 800$$
$$-40x = -600$$
$$x = 15.$$

Therefore, $20 - x = 5$. We should mix 15 quarts of the 30% solution with 5 quarts of the 70% solution.   ■

PROBLEM 3 A 4-gallon radiator is full and contains a 40% solution of antifreeze. How much needs to be drained out and replaced with pure antifreeze to obtain a 70% solution?

Solution The following guideline can be used.

Let x represent the amount of pure antifreeze to be added. Then x also represents the amount of the 40% solution to be drained out. Thus, the guideline translates into the following equation.

$$(40\%)(4) - (40\%)(x) + x = (70\%)(4).$$

Solving this equation we obtain

$$.4(4) - .4x + x = .7(4)$$
$$1.6 + .6x = 2.8$$
$$.6x = 1.2$$
$$x = 2.$$

Therefore, we must drain out 2 gallons of the 40% solution and then add 2 gallons of pure antifreeze. (Checking this answer would be a worthwhile exercise for you!). ■

Now let's consider a problem where the process of representing the various unknown quantities in terms of one variable is really a key issue in solving the problem.

PROBLEM 4 Jody is 6 years younger than her sister Cathy and in 7 years she will be three-fourths as old as Cathy. Find their present ages.

Solution By letting x represent Cathy's present age we can represent all of the unknown quantities as follows:

$$x: \qquad \text{Cathy's present age}$$
$$x - 6: \qquad \text{Jody's present age}$$
$$x + 7: \qquad \text{Cathy's age in 7 years}$$
$$x - 6 + 7 \text{ or } x + 1: \qquad \text{Jody's age in 7 years}$$

The statement that Jody's age in 7 years will be three-fourths of Cathy's age then serves as the guideline. So we can set up and solve the following equation.

$$x + 1 = \frac{3}{4}(x + 7)$$

$$4x + 4 = 3(x + 7)$$
$$4x + 4 = 3x + 21$$
$$x = 17.$$

Therefore, Cathy's present age is 17 and Jody's present age is $17 - 6 = 11$.   ■

Problem Set 4.3

Set up an equation and solve each of the following problems.

1. How many milliliters of pure acid must be added to 100 milliliters of a 10% acid solution to obtain a 20% solution?

2. How many liters of pure alcohol must be added to 20 liters of a 40% solution to obtain a 60% solution?

3. How many centiliters of distilled water must be added to 10 centiliters of a 50% acid solution to obtain a 20% acid solution?

4. How many milliliters of distilled water must be added to 50 milliliters of a 40% acid solution to reduce it to a 10% acid solution?

5. We want to mix some 30% alcohol solution with some 50% alcohol solution to obtain 10 quarts of a 35% solution. How many quarts of each kind should we use?

6. We have a 20% alcohol solution and a 50% solution. How many pints must be used from each to obtain 8 pints of a 30% solution?

7. How much water needs to be removed from 20 gallons of a 30% salt solution to change it to a 40% salt solution?

8. How much water needs to be removed from 30 liters of a 20% salt solution to change it to a 50% salt solution?

9. A 12-quart radiator contains a 20% solution of antifreeze. How much solution needs to be drained out and replaced with pure antifreeze to obtain a 40% solution of antifreeze?

10. A tank contains 50 gallons of a 40% solution of antifreeze. How much solution needs to be drained out and replaced with pure antifreeze to obtain a 50% solution?

11. How many gallons of a 15% salt solution must be mixed with 8 gallons of a 20% salt solution to obtain a 17% salt solution?

12. How many liters of a 10% salt solution must be mixed with 15 liters of a 40% salt solution to obtain a 20% salt solution?

13. Thirty ounces of a punch containing 10% grapefruit juice is added to 50 ounces of punch containing 20% grapefruit juice. Find the percent of grapefruit juice in the resulting mixture.

14. Suppose that 20 gallons of a 20% salt solution is mixed with 30 gallons of a 25% salt solution. What is the percent of salt in the resulting solution?

15. Suppose that the perimeter of a square equals the perimeter of a rectangle. The width of the rectangle is 9 inches less than twice the side of the square and the length of the rectangle is 3 inches less than twice the side of the square. Find the dimensions of the square and the rectangle.

16. The perimeter of a triangle is 40 centimeters. The longest side is 1 centimeter more than twice the shortest side. The other side is 2 centimeters shorter than the longest side. Find the lengths of the three sides.

17. Butch starts walking from point A at 2 miles per hour. One-half hour later, Dick starts walking from point A at $3\frac{1}{2}$ miles per hour and follows the same route. How long will it take Dick to catch up with Butch?

18. Karen, riding her bicycle at 15 miles per hour, rode 10 miles further than Michelle, who was riding her bicycle at 14 miles per hour. Karen rode for 30 minutes longer than Michelle. How long did each girl ride her bicycle?

19. Pam is half as old as her brother Bill. Six years ago Bill was four times older than Pam. How old is each now?

20. The sum of the present ages of Tom and his father is 100 years. Ten years ago Tom's father was three times as old as Tom was at that time. Find their present ages.

21. The difference of the present ages of Abby and her mother is 21 years. In ten years Abby's mother's age will be three years less than twice Abby's age at that time. Find their present ages.

22. Kelly has a coin collection in which a penny is twice as old as a dime. In 10 years, the age of the penny will be $1\frac{1}{2}$ times the age of the dime at that time. Find the present ages of the penny and the dime.

23. The sum of the ages of our two cars is 11 years. In five years, the age of the older car will be three years more than twice the age of the newer car at that time. Find the present ages of the two cars.

24. Mitch is three-fourths as old as his sister Candy. In 8 years he will be five-sixths as old as Candy. Find their present ages.

25. Steve rode his bicycle from one town to another at 14 miles per hour. If he had increased his speed to 16 miles per hour he could have made the trip in one-half hour less time. How far apart are the towns?

4.4 Inequalities

Just as we use the symbol $=$ to represent *is equal to*, we also use the symbols $<$ and $>$ to represent *is less than* and *is greater than*, respectively. The following are examples of **statements of inequality**. Notice that the first four are true statements and the last two are false.

$$6 + 4 > 7;$$
$$8 - 2 < 14;$$
$$4 \cdot 8 > 4 \cdot 6;$$
$$5 \cdot 2 < 5 \cdot 7;$$
$$5 + 8 > 19;$$
$$9 - 2 < 3.$$

Algebraic inequalities contain one or more variables. The following are examples of algebraic inequalities.

$$x + 3 > 4;$$
$$2x - 1 < 6;$$
$$x^2 + 2x - 1 > 0;$$
$$2x + 3y < 7;$$
$$7ab < 9.$$

An algebraic inequality such as $x + 1 > 2$ is neither true or false as it stands and is called an **open sentence**. Each time a number is substituted for x, the algebraic inequality $x + 1 > 2$ becomes a numerical statement that is either true or false. For example, if $x = 0$, then $x + 1 > 2$ becomes $0 + 1 > 2$, which is false. If $x = 2$, then $x + 1 > 2$ becomes $2 + 1 > 2$, which is true. **Solving an inequality** refers to the process of finding the numbers that make an algebraic inequality a true numerical statement. Such numbers, called the **solutions of the inequality**, are said to satisfy the inequality. The set of all solutions of an inequality is called its **solution set**. Solution sets for inequalities are often stated using **set builder notation**. For example, the solution set for $x + 1 > 2$ is the set of real numbers greater than 1 and can be expressed as $\{x \mid x > 1\}$. The set builder notation $\{x \mid x > 1\}$ is read as "the set of all x such that x is greater than 1." Solution sets for inequalities are sometimes graphed on a number line. For example, the solution set for $\{x \mid x > 1\}$ can be pictured on a number line as follows.

The unshaded circle around 1 indicates that 1 is not a solution; the shaded portion indicates that all numbers to the right of (greater than) 1 are solutions. The shaded portion of the number line is referred to as the *graph* of the solution set $\{x \mid x > 1\}$.

The following examples contain some simple algebraic inequalities, their solution sets, and the graphs of the solution sets. Look them over very carefully because they also introduce some additional symbolism.

Algebraic inequality	Solution set	Graph of solution set
$x < 2$	$\{x \mid x < 2\}$	
$x > -1$	$\{x \mid x > -1\}$	
$1 > x$	$\{x \mid x < 1\}$	
$3 < x$	$\{x \mid x > 3\}$	
$x \geq 1$ ($\geq$ is read "greater than or equal to")	$\{x \mid x \geq 1\}$	
$x \leq 2$ ($\leq$ is read "less than or equal to")	$\{x \mid x \leq 2\}$	

The general process for solving inequalities closely parallels that for solving equations. We continue to replace the given inequality with equivalent, but simpler, inequalities. For example,

$$2x + 1 > 9; \tag{1}$$
$$2x > 8; \tag{2}$$
$$x > 4 \tag{3}$$

are all equivalent inequalities; that is, they have the same solutions. Thus, to solve (1) we can solve (3), which is obviously all numbers greater than 4. The exact procedure for simplifying inequalities is primarily based on two properties and they become our topics of discussion at this time. The first of these is the **addition-subtraction property of inequality**.

Property 4.1

For all real numbers a, b, and c,

   1. $a > b$ if and only if $a + c > b + c$;

   2. $a > b$ if and only if $a - c > b - c$.

Property 4.1 states that "any number can be added to or subtracted from both sides of an inequality and an equivalent inequality is produced." The property is stated in terms of $>$, but analogous properties exist for $<, \geq$, and $\leq$. Consider the use of this property in the next three examples.

EXAMPLE 1 Solve $x - 3 > -1$ and graph the solutions.

Solution

$$x - 3 > -1$$
$$x - 3 + 3 > -1 + 3 \qquad \text{Add 3 to both sides.}$$
$$x > 2$$

The solution set is $\{x \mid x > 2\}$ and it can be graphed as follows.

EXAMPLE 2 Solve $x + 4 \leq 5$ and graph the solutions.

Solution

$$x + 4 \leq 5$$
$$x + 4 - 4 \leq 5 - 4 \qquad \text{Subtract 4 from both sides.}$$
$$x \leq 1.$$

The solution set is $\{x \mid x \leq 1\}$ and it can be graphed as follows.

EXAMPLE 3 Solve $5 > 6 + x$ and graph the solutions.

Solution

$$5 > 6 + x$$
$$5 - 6 > 6 + x - 6 \qquad \text{Subtract 6 from both sides.}$$
$$-1 > x.$$

Since $-1 > x$ is equivalent to $x < -1$, the solution set is $\{x \mid x < -1\}$ and it can be graphed as follows.

Now let's look at some numerical examples to see what happens when both sides of an inequality are multiplied or divided by some number.

$$4 > 3 \longrightarrow \underline{5}(4) > \underline{5}(3);$$
$$-2 > -3 \longrightarrow \underline{4}(-2) > \underline{4}(-3);$$
$$6 > 4 \longrightarrow \frac{6}{\underline{2}} > \frac{4}{\underline{2}};$$
$$8 > -2 \longrightarrow \frac{8}{\underline{4}} > \frac{-2}{\underline{4}}.$$

Notice that multiplying or dividing both sides of an inequality by a positive number produces an inequality of the same sense. This means that if the original inequality is "greater than," then the new inequality is "greater than," and if the original is "less than", then the resulting inequality is "less than."

Now note what happens when we multiply or divide both sides by a negative number.

$$3 < 5 \quad \text{but } \underline{-2}(3) > \underline{-2}(5);$$
$$-4 < 1 \quad \text{but } \underline{-5}(-4) > \underline{-5}(1);$$
$$4 > 2 \quad \text{but } \frac{4}{\underline{-2}} < \frac{2}{\underline{-2}};$$
$$-3 > -6 \quad \text{but } \frac{-3}{\underline{-3}} < \frac{-6}{\underline{-3}}.$$

Multiplying or dividing both sides of an inequality *by a negative number reverses the sense of the inequality.* The following property summarizes these ideas.

Property 4.2

(a) For all real numbers a, b, and c, *with $c > 0$,*

  1. $a > b$ if and only if $ac > bc$;

  2. $a > b$ if and only if $\dfrac{a}{c} > \dfrac{b}{c}$.

(b) For all real numbers a, b, and c, *with $c < 0$,*

  1. $a > b$ if and only if $ac < bc$;

  2. $a > b$ if and only if $\dfrac{a}{c} < \dfrac{b}{c}$.

Similar properties hold if each inequality is reversed or if $>$ is replaced with $\geq$ and $<$ with $\leq$. For example, if $a \leq b$ and $c < 0$, then $ac \geq bc$ and $\dfrac{a}{c} \geq \dfrac{b}{c}$.

Observe the use of Property 4.2 in the next three examples.

EXAMPLE 4 Solve $2x > 4$.

Solution

$2x > 4$

$\dfrac{2x}{2} > \dfrac{4}{2}$ Divide both sides by 2.

$x > 2$.

The solution set is $\{x \mid x > 2\}$.   ■

EXAMPLE 5 Solve $\dfrac{3}{4}x \le \dfrac{1}{5}$.

Solution

$\dfrac{3}{4}x \le \dfrac{1}{5}$

$\dfrac{4}{3}\left(\dfrac{3}{4}x\right) \le \dfrac{4}{3}\left(\dfrac{1}{5}\right)$ Multiply both sides by $\dfrac{4}{3}$.

$x \le \dfrac{4}{15}$.

The solution set is $\left\{x \mid x \le \dfrac{4}{15}\right\}$.   ■

EXAMPLE 6 Solve $-3x > 9$.

Solution

$-3x > 9$

$\dfrac{-3x}{-3} < \dfrac{9}{-3}$ Divide both sides by -3, which reverses the sense.

$x < -3$

The solution set is $\{x \mid x < -3\}$.   ■

As mentioned earlier, many of the same techniques used to solve equations may be used to solve inequalities. However, we must be extremely careful when applying Property 4.2. Study the following examples and notice the similarities between solving equations and solving inequalities.

EXAMPLE 7 Solve $4x - 3 > 9$.

Solution

$$4x - 3 > 9$$

$$4x - 3 + 3 > 9 + 3 \qquad \text{Add 3 to both sides.}$$

$$4x > 12$$

$$\frac{4x}{4} > \frac{12}{4} \qquad \text{Divide both sides by 4.}$$

$$x > 3.$$

The solution set is $\{x \mid x > 3\}$.

EXAMPLE 8 Solve $-3n + 5 < 11$.

Solution

$$-3n + 5 < 11$$

$$-3n + 5 - 5 < 11 - 5 \qquad \text{Subtract 5 from both sides.}$$

$$-3n < 6$$

$$\frac{-3n}{-3} > \frac{6}{-3} \qquad \text{Divide both sides by } -3, \text{ which reverses the sense.}$$

$$n > -2.$$

The solution set is $\{n \mid n > -2\}$. ■

Problem Set 4.4

For Problems 1–10, determine whether each numerical inequality is *true* or *false*.

1. $2(3) - 4(5) < 5(3) - 2(-1) + 4$

2. $5 + 6(-3) - 8(-4) > 17$

3. $\frac{2}{3} - \frac{3}{4} + \frac{1}{6} > \frac{1}{5} + \frac{3}{4} - \frac{7}{10}$

4. $\frac{1}{2} + \frac{1}{3} < \frac{1}{3} + \frac{1}{4}$

5. $\left(-\frac{1}{2}\right)\left(\frac{4}{9}\right) > \left(\frac{3}{5}\right)\left(-\frac{1}{3}\right)$

6. $\left(\frac{5}{6}\right)\left(\frac{8}{12}\right) < \left(\frac{3}{7}\right)\left(\frac{14}{15}\right)$

7. $\frac{3}{4} + \frac{2}{3} \div \frac{1}{5} > \frac{2}{3} + \frac{1}{2} \div \frac{3}{4}$

8. $1.9 - 2.6 - 3.4 < 2.5 - 1.6 - 4.2$

9. $0.16 + 0.34 > 0.23 + 0.17$

10. $(0.6)(1.4) > (0.9)(1.2)$

For Problems 11–22, state the solution set and graph it on a number line.

11. $x > -2$

12. $x > -4$

13. $x \leq 3$

14. $x \leq 0$

15. $2 < x$

16. $-3 \leq x$

17. $-2 \geq x$

18. $1 > x$

19. $-x > 1$

20. $-x < 2$

21. $-2 < -x$

22. $-1 > -x$

For Problems 23–60, solve each of the inequalities.

23. $x + 6 < -14$ **24.** $x + 7 > -15$ **25.** $x - 4 \geq -13$ **26.** $x - 3 \leq -12$

27. $4x > 36$ **28.** $3x < 51$ **29.** $6x < 20$ **30.** $8x > 28$

31. $-5x > 40$ **32.** $-4x < 24$ **33.** $-7n \leq -56$ **34.** $-9n \geq -63$

35. $48 > -14n$ **36.** $36 < -8n$ **37.** $16 < 9 + n$ **38.** $19 > 27 + n$

39. $3x + 2 > 17$ **40.** $2x + 5 < 19$ **41.** $4x - 3 \leq 21$ **42.** $5x - 2 \geq 28$

43. $-2x - 1 \geq 41$ **44.** $-3x - 1 \leq 35$ **45.** $6x + 2 < 18$ **46.** $8x + 3 > 25$

47. $3 > 4x - 2$ **48.** $7 < 6x - 3$ **49.** $-2 < -3x + 1$

50. $-6 > -2x + 4$ **51.** $-38 \geq -9t - 2$ **52.** $36 \geq -7t + 1$

53. $5x - 4 - 3x > 24$ **54.** $7x - 8 - 5x < 38$ **55.** $4x + 2 - 6x < -1$

56. $6x + 3 - 8x > -3$ **57.** $-5 \geq 3t - 4 - 7t$ **58.** $6 \leq 4t - 7t - 10$

59. $-x - 4 - 3x > 5$ **60.** $-3 - x - 3x < 10$

■ Miscellaneous Problems

Solve each of the following inequalities.

61. $x + 3 < x - 4$ **62.** $x - 4 < x + 6$ **63.** $2x + 4 > 2x - 7$

64. $5x + 2 > 5x + 7$ **65.** $3x - 4 - 3x > 6$ **66.** $-2x + 7 + 2x > 1$

67. $-5 \leq -4x - 1 + 4x$ **68.** $-7 \geq 5x - 2 - 5x$

4.5 Inequalities, Compound Inequalities, and Problem Solving

Let's begin this section by solving three inequalities using the same basic steps as we did with equations. Again, be careful when applying the multiplication and division properties of inequality.

EXAMPLE 1 Solve $5x + 8 \leq 3x - 10$.

Solution

$$5x + 8 \leq 3x - 10$$

$$5x + 8 - 3x \leq 3x - 10 - 3x \qquad \text{Subtract } 3x \text{ from both sides.}$$

$$2x + 8 \leq -10$$

$$2x + 8 - 8 \leq -10 - 8 \qquad \text{Subtract 8 from both sides.}$$

$$2x \leq -18$$

$$\frac{2x}{2} \leq \frac{-18}{2} \qquad \text{Divide both sides by 2.}$$

$$x \leq -9.$$

The solution set is $\{x \mid x \leq -9\}$. ■

EXAMPLE 2 Solve $4(x + 3) + 3(x - 4) \geq 2(x - 1)$.

Solution

$$4(x + 3) + 3(x - 4) \geq 2(x - 1)$$
$$4x + 12 + 3x - 12 \geq 2x - 2 \qquad \text{Distributive property.}$$
$$7x \geq 2x - 2 \qquad \text{Combined similar terms.}$$
$$7x - 2x \geq 2x - 2 - 2x \qquad \text{Subtract } 2x \text{ from both sides.}$$
$$5x \geq -2$$
$$\frac{5x}{5} \geq \frac{-2}{5} \qquad \text{Divide both sides by 5.}$$
$$x \geq -\frac{2}{5}.$$

The solution set is $\left\{ x \mid x \geq -\frac{2}{5} \right\}$. ■

EXAMPLE 3 Solve $-\frac{3}{2}n + \frac{1}{6}n < \frac{3}{4}$.

Solution

$$-\frac{3}{2}n + \frac{1}{6}n < \frac{3}{4}$$

$$12\left(-\frac{3}{2}n + \frac{1}{6}n\right) < 12\left(\frac{3}{4}\right) \qquad \begin{array}{l}\text{Multiply both sides by 12, which is} \\ \text{the LCD of all denominators.}\end{array}$$

$$12\left(-\frac{3}{2}n\right) + 12\left(\frac{1}{6}n\right) < 12\left(\frac{3}{4}\right) \qquad \text{Distributive property.}$$

$$-18n + 2n < 9$$
$$-16n < 9$$
$$\frac{-16n}{-16} > \frac{9}{-16} \qquad \begin{array}{l}\text{Divide both sides by } -16, \text{ which} \\ \text{reverses the inequality.}\end{array}$$
$$n > -\frac{9}{16}.$$

The solution set is $\left\{ n \mid n > -\frac{9}{16} \right\}$. ■

Checking solutions for an inequality presents a problem. Obviously, we cannot check all of the infinitely many solutions for a particular inequality. However, by checking at least one solution, especially when the multiplication property has been used, we might catch a mistake of forgetting to reverse an inequality sign. In

Example 3 we are claiming that "all numbers greater than $-\dfrac{9}{16}$ will satisfy the original inequality." Let's check one number, say 0.

$$-\frac{3}{2}n + \frac{1}{6}n = \frac{3}{4}$$

$$-\frac{3}{2}(0) + \frac{1}{6}(0) \overset{?}{<} \frac{3}{4}$$

$$0 < \frac{3}{4}.$$

Therefore, 0 satisfies the original inequality. Had we forgotten to reverse the inequality sign when both sides were divided by -16, then our answer would have been $n < -\dfrac{9}{16}$ and such an error would be detected by the check.

■ Compound Statements

The words *and* and *or* are used in mathematics to form **compound statements**. The following are examples of some compound numerical statements using *and*. Such statements are called **conjunctions**. We agree to call a conjunction true only if all of its component parts are true.

1. $3 + 6 = 9$ and $4 < 7$ (True)
2. $4 > 2$ and $0 > -2$ (True)
3. $8 + 7 = 15$ and $2 < 1$ (False)
4. $9 > 14$ and $13 < 18$ (False)
5. $5 > 8$ and $-6 > 0$ (False)

Compound statements using *or* are called **disjunctions**. A disjunction is true if at least one of its component parts is true. Said another way, a disjunction is false only if all of its component parts are false. Consider the following disjunctions.

1. $17 > 14$ or $18 < 25$ (True)
2. $19 > 12$ or $13 > 17$ (True)
3. $4 + 6 = 8$ or $0 < 1$ (True)
4. $7 < -4$ or $9 < 2$ (False)

Now let's consider finding solutions for some compound statements involving algebraic inequalities. Keep in mind that the previous discussion regarding labeling conjunctions and disjunctions true or false forms the basis for our reasoning.

EXAMPLE 4 Graph the solutions for $x > 2$ and $x < 5$.

Solution   The key word is *and*, so both inequalities need to be satisfied. Thus, all numbers between 2 and 5 are solutions and can be graphed as follows.

■

EXAMPLE 5   Graph the solutions for $x < -1$ or $x > 3$.

Solution   The key word is *or*, so any numbers satisfying either (or both) inequalities are solutions. Thus, all numbers greater than 3 along with all numbers less than -1 are solutions and can be graphed as follows.

■

EXAMPLE 6   Graph the solutions for $x \geq -2$ and $x < 1$.

Solution   Since it is an *and* statement, both inequalities need to be satisfied. Thus, all numbers between -2 and 1, including -2 (but not including 1), are solutions. Notice on the graph that we shade the circle at -2, but do not shade the circle at 1.

■

EXAMPLE 7   Graph the solutions for $x > 1$ or $x > 4$.

Solution   Since it is an *or* statement, numbers that satisfy either inequality (or both) are solutions. Thus, all numbers greater than 1 are included and the graph is as follows.

■

Let's consider this section by considering some word problems that translate into inequality statements. The *suggestions for solving word problems* given in Section 4.2 continue to apply here except that the situation being described in a problem will translate into an inequality instead of an equation.

PROBLEM 1   Debbie had scores of 95, 82, 93, and 84 on her first four exams of the semester. What score must she obtain on the fifth exam to have an average of 90 or better for the five exams?

Solution   Let s represent the score needed on the fifth exam. Since the average is computed by adding all fives scores and dividing by 5 (the number of scores), we have the following inequality to solve.

$$\frac{95 + 82 + 93 + 84 + s}{5} \geq 90.$$

Solving this inequality, we obtain

$$\frac{354 + s}{5} \geq 90 \qquad\qquad \text{Simplified numerator of left side.}$$

$$5\left(\frac{354 + s}{5}\right) \geq 5(90) \qquad\qquad \text{Multiply both sides by 5.}$$

$$354 + s \geq 450$$
$$354 + s - 354 \geq 450 - 354 \qquad \text{Subtract 354 from both sides.}$$
$$s \geq 96.$$

She must receive a score of 96 or better on the fifth exam.   ■

PROBLEM 2   The Cubs have won 40 baseball games and have lost 62 games. They have 60 more games to play. To win more than 50% of all of their games, how many of the 60 games remaining must they win?

Solution   Let w represent the number of games they must win out of the 60 games remaining. Since they are playing a total of $40 + 62 + 60 = 162$ games, to win more than 50% of their games they would need to win more than 81 games. Thus, we have the following inequality.

$$w + 40 > 81.$$

Solving this yields

$$w > 41.$$

They need to win at least 42 of the 60 games remaining.   ■

Problem Set 4.5

Solve each of the following inequalities.

1. $3x + 4 > x + 8$ **2.** $5x + 3 < 3x + 11$ **3.** $7x - 2 < 3x - 6$

4. $8x - 1 > 4x - 21$ **5.** $6x + 7 > 3x - 3$ **6.** $7x + 5 < 4x - 12$

7. $5n - 2 \leq 6n + 9$ **8.** $4n - 3 \geq 5n + 6$ **9.** $2t + 9 \geq 4t - 13$

10. $6t + 14 \leq 8t - 16$ **11.** $-3x - 4 < 2x + 7$ **12.** $-x - 2 > 3x - 7$

13. $-4x + 6 > -2x + 1$ **14.** $-6x + 8 < -4x + 5$ **15.** $5(x - 2) \leq 30$

16. $4(x + 1) \geq 16$ **17.** $2(n + 3) > 9$ **18.** $3(n - 2) < 7$

19. $-3(y - 1) < 12$ **20.** $-2(y + 4) > 18$ **21.** $-2(x + 6) > -17$

22. $-3(x - 5) < -14$ **23.** $3(x - 2) < 2(x + 1)$ **24.** $5(x + 3) > 4(x - 2)$

25. $4(x + 3) > 6(x - 5)$ **26.** $6(x - 1) < 8(x + 5)$

27. $3(x - 4) + 2(x + 3) < 24$ **28.** $2(x + 1) + 3(x + 2) > -12$

29. $5(n + 1) - 3(n - 1) > -9$ **30.** $4(n - 5) - 2(n - 1) < 13$

31. $\dfrac{1}{2}n - \dfrac{2}{3}n \geq -7$ **32.** $\dfrac{3}{4}n + \dfrac{1}{6}n \leq 1$

33. $\dfrac{3}{4}n - \dfrac{5}{6}n < \dfrac{3}{8}$ **34.** $\dfrac{2}{3}n - \dfrac{1}{2}n > \dfrac{1}{4}$

35. $\dfrac{3x}{5} - \dfrac{2}{3} > \dfrac{x}{10}$ **36.** $\dfrac{5x}{4} + \dfrac{3}{8} < \dfrac{7x}{12}$

37. $n \geq 3.4 + 0.15n$ **38.** $x \geq 2.1 + 0.3x$

39. $0.09t + 0.1(t + 200) > 77$ **40.** $0.07t + 0.08(t + 100) > 38$

41. $0.06x + 0.08(250 - x) \geq 19$ **42.** $0.08x + 0.09(2x) \leq 130$

43. $\dfrac{x - 1}{2} + \dfrac{x + 3}{5} > \dfrac{1}{10}$ **44.** $\dfrac{x + 3}{4} + \dfrac{x - 5}{7} < \dfrac{1}{28}$

45. $\dfrac{x + 2}{6} - \dfrac{x + 1}{5} < -2$ **46.** $\dfrac{x - 6}{8} - \dfrac{x + 2}{7} > -1$

47. $\dfrac{n + 3}{3} + \dfrac{n - 7}{2} > 3$ **48.** $\dfrac{n - 4}{4} + \dfrac{n - 2}{3} < 4$

49. $\dfrac{x - 3}{7} - \dfrac{x - 2}{4} \leq \dfrac{9}{14}$ **50.** $\dfrac{x - 1}{5} - \dfrac{x + 2}{6} \geq \dfrac{7}{15}$

Graph the solutions for each of the following compound inequalities.

51. $x > -1$ and $x < 2$ **52.** $x > 1$ and $x < 4$ **53.** $x < -2$ or $x > 1$

54. $x < 0$ or $x > 3$ **55.** $x > -2$ and $x \leq 2$ **56.** $x \geq -1$ and $x < 3$

57. $x > -1$ and $x > 2$ **58.** $x < -2$ and $x < 3$ **59.** $x > -4$ or $x > 0$

60. $x < 2$ or $x < 4$ **61.** $x > 3$ and $x < -1$ **62.** $x < -3$ and $x > 6$

63. $x \leq 0$ or $x \geq 2$ **64.** $x \leq -2$ or $x \geq 1$ **65.** $x > -4$ or $x < 3$

66. $x > -1$ or $x < 2$

Solve each of the following problems by setting up and solving an appropriate inequality.

67. Five more than 3 times a number is greater than 26. Find the numbers satisfying this relationship.

68. Fourteen increased by twice a number is less than or equal to 3 times the number. Find the numbers satisfying this relationship.

69. The perimeter of a rectangle is to be no greater than 70 inches and the length of the rectangle must be 20 inches. Find the largest possible value for the width of the rectangle.

70. One side of a triangle is three times as long as another side. The third side is 15 centimeters long. If the perimeter of the triangle is to be no greater than 75 centimeters, find the largest lengths that the other two sides can be.

71. Sue bowled 132 and 160 in her first two games. What must she bowl in the third game to have an average of at least 150 for the three games?

72. Mike has scores of 87, 81, and 74 on his first three algebra tests. What score must he make on the 4th test to have an average of 85 or better for the four tests?

73. Lance has scores of 96, 90, and 94 on his first three algebra exams. What must he average on the last two exams to have an average of better than 92 for all five exams?

74. The Mets have won 45 baseball games and lost 55 games. They have 62 more games to play. To win more than 50% of all their games, how many of the 62 games remaining must they win?

75. Mona has $1000 to invest. If she invests $500 at 8% interest, at what rate must she invest the other $500 so that the two investments together yield more than $100 of yearly interest?

76. The average height of the two forwards and the center of a basketball team is 6 feet and 8 inches. What must the average height of the two guards be so that the team average is at least 6 feet and 4 inches?

77. Scott shot rounds of 82, 84, 78, and 79 on the first four days of the golf tournament. What must he shoot on the fifth day of the tournament to average 80 or less for the five days?

78. Lance has $500 to invest. If he invests $300 at 9% interest, at what rate must he invest the other $200 so that the two investments together yield more than $47 in yearly interest?

4.6   Equations and Inequalities Involving Absolute Value

In Chapter 1 we used the concept of absolute value to describe addition and multiplication of integers. We defined the absolute value of a number to be the distance between the number and zero on a number line. For example, the absolute value of 3 is 3. The absolute value of -3 is 3. The absolute value of 0 is 0. Symbolically, absolute value is denoted with vertical bars.

$$|3| = 3; \qquad |-3| = 3; \qquad |0| = 0.$$

In general, we can say that the absolute value of any number, except 0, is positive.

Interpreting absolute value as distance on a number line allows us to solve a variety of equations and inequalities involving absolute value. First, let's consider some equations.

EXAMPLE 1   Solve and graph the solutions for $|x| = 2$.

Solution   Thinking in terms of "distance between the number and zero" we see that x must

be 2 or -2. Thus,

$$|x| = 2$$
$$x = 2 \quad \text{or} \quad x = -2.$$

The solution set is $\{-2, 2\}$ and its graph is as follows.

EXAMPLE 2 Solve and graph the solutions for $|x + 4| = 1$.

Solution The number, $x + 4$, must be 1 or -1. Thus,

$$|x + 4| = 1$$
$$x + 4 = 1 \quad \text{or} \quad x + 4 = -1$$
$$x = -3 \quad \text{or} \quad x = -5.$$

The solution set is $\{-5, -3\}$ and its graph is as follows.

EXAMPLE 3 Solve and graph the solutions for $|3x - 2| = 4$.

Solution The number, $3x - 2$, must be 4 or -4. Thus,

$$|3x - 2| = 4$$
$$3x - 2 = 4 \quad \text{or} \quad 3x - 2 = -4$$
$$3x = 6 \quad \text{or} \quad 3x = -2$$
$$x = 2 \quad \text{or} \quad x = -\frac{2}{3}.$$

The solution set is $\left\{-\frac{2}{3}, 2\right\}$ and its graph is as follows.

The *distance interpretation* for absolute value also provides a good basis for solving inequalities involving absolute value. Consider the following examples.

EXAMPLE 4 Solve and graph the solutions for $|x| < 2$.

Solution   The number, x, must be *less than two units away from zero*. Thus,

$$|x| < 2$$
$$x > -2 \quad \text{and} \quad x < 2.$$

The solution set is $\{x \mid x > -2 \text{ and } x < 2\}$ and its graph is as follows.

EXAMPLE 5 Solve and graph the solutions for $|x - 1| < 2$.

Solution   The number, $x - 1$, must be *less than two units away from zero*. Thus,

$$|x - 1| < 2$$
$$x - 1 > -2 \quad \text{and} \quad x - 1 < 2$$
$$x > -1 \quad \text{and} \quad x < 3.$$

The solution set is $\{x \mid x > -1 \text{ and } x < 3\}$ and its graph is as follows.

EXAMPLE 6 Solve and graph the solutions for $|2x + 5| \leq 1$.

Solution   The number, $2x + 5$, must be *equal to or less than one unit away from zero*. Therefore,

$$|2x + 5| \leq 1$$
$$2x + 5 \geq -1 \quad \text{and} \quad 2x + 5 \leq 1$$
$$2x \geq -6 \quad \text{and} \quad 2x \leq -4$$
$$x \geq -3 \quad \text{and} \quad x \leq -2.$$

The solution set is $\{x \mid x \geq -3 \text{ and } x \leq -2\}$ and its graph is as follows.

EXAMPLE 7 Solve and graph the solutions for $|x| > 2$.

Solution The number, x, must be *greater than two units away from zero*. Thus,

$$|x| > 2$$
$$x < -2 \quad \text{or} \quad x > 2.$$

The solution set is $\{x \mid x < -2 \text{ or } x > 2\}$ and its graph is as follows.

$$-5 \;\; -4 \;\; -3 \;\; -2 \;\; -1 \quad 0 \quad 1 \quad 2 \quad 3 \quad 4 \quad 5$$

■

EXAMPLE 8 Solve and graph the solutions for $|3x - 1| > 4$.

Solution The number, $3x - 1$, must be *greater than four units away from zero*. Thus.

$$|3x - 1| > 4$$
$$3x - 1 < -4 \quad \text{or} \quad 3x - 1 > 4$$
$$3x < -3 \quad \text{or} \quad 3x > 5$$
$$x < -1 \quad \text{or} \quad x > \frac{5}{3}.$$

The solution set is $\left\{ x \mid x < -1 \text{ or } x > \frac{5}{3} \right\}$ and its graph is as follows.

$$-5 \;\; -4 \;\; -3 \;\; -2 \;\; -1 \quad 0 \quad 1 \quad 2 \quad 3 \quad 4 \quad 5$$

■

The solutions for equations and inequalities such as $|3x - 7| = -4$, $|x + 5| < -3$, and $|2x - 3| > -7$ can be found *by inspection*. Notice that in each of these examples the right side is a negative number. Therefore, using the fact that the **absolute value of any number is nonnegative**, we can reason as follows.

$|3x - 7| = -4$ has *no solutions* since the absolute value of a number cannot be negative.

$|x + 5| < -3$ has *no solutions* since we cannot obtain an absolute value less than -3.

$|2x - 3| > -7$ is *satisfied by all real numbers* since the absolute value of $2x - 3$, regardless of what number is substituted for x, will always be greater than -7.

Problem Set 4.6

For Problems 1–26, solve and graph the solutions for each of the following.

1. $|x| = 4$ **2.** $|x| = 3$ **3.** $|x| < 1$

4. $|x| < 4$ **5.** $|x| \geq 2$ **6.** $|x| \geq 1$

7. $|x + 2| = 1$ **8.** $|x + 3| = 2$ **9.** $|x - 1| = 2$

10. $|x - 2| = 1$ **11.** $|x - 2| \leq 2$ **12.** $|x + 1| \leq 3$

13. $|x + 1| > 3$ **14.** $|x - 3| > 1$ **15.** $|2x + 1| = 3$

16. $|3x - 1| = 5$ **17.** $|5x - 2| = 4$ **18.** $|4x + 3| = 8$

19. $|2x - 3| \geq 1$ **20.** $|2x + 1| \geq 3$ **21.** $|4x + 3| < 2$

22. $|5x - 2| < 8$ **23.** $|3x + 6| = 0$ **24.** $|4x - 3| = 0$

25. $|3x - 2| > 0$ **26.** $|2x + 7| < 0$

For Problems 27–42, solve each of the following.

27. $|3x - 1| = 17$ **28.** $|4x + 3| = 27$ **29.** $|2x + 1| > 9$

30. $|3x - 4| > 20$ **31.** $|3x - 5| < 19$ **32.** $|5x + 3| < 14$

33. $|-3x - 1| = 17$ **34.** $|-4x + 7| = 26$ **35.** $|4x - 7| \leq 31$

36. $|5x - 2| \leq 21$ **37.** $|5x + 3| \geq 18$ **38.** $|2x - 11| \geq 4$

39. $|-x - 2| < 4$ **40.** $|-x - 5| < 7$ **41.** $|-2x + 1| > 6$

42. $|-3x + 2| > 8$

For Problems 43–50, solve each of the following *by inspection*.

43. $|7x| = 0$ **44.** $|3x - 1| = -4$ **45.** $|x - 6| > -4$

46. $|3x + 1| > -3$ **47.** $|x + 4| < -7$ **48.** $|5x - 2| < -2$

49. $|x + 6| \leq 0$ **50.** $|x + 7| > 0$

■ Miscellaneous Problems

A conjunction such as "$x > -2$ and $x < 4$" can be written in a more compact form $-2 < x < 4$, which is read as "-2 is less than x and x is less than 4." In other words, x is clamped between -2 and 4. The compact form is very convenient for solving conjunctions as follows.

$$-3 < 2x - 1 < 5$$
$$-2 < \quad 2x \quad < 6 \qquad \text{Add 1 to the left side, middle, and right side.}$$
$$-1 < \quad x \quad < 3 \qquad \text{Divide through by 2.}$$

Thus, the solution set can be expressed as $\{x \mid -1 < x < 3\}$.

For Problems 51–60, solve the compound inequalities using the compact form.

51. $-2 < x - 6 < 8$ **52.** $-1 < x + 3 < 9$ **53.** $1 \leq 2x + 3 \leq 11$

54. $-2 \le 3x - 1 \le 14$ **55.** $-4 < \dfrac{x-1}{3} < 2$ **56.** $2 < \dfrac{x+1}{4} < 5$

57. $|x + 4| < 3$ **58.** $|x - 6| < 5$ **59.** $|2x - 5| < 7$

60. $|3x + 2| < 14$

Chapter 4 Summary

(4.1) **Formulas** are rules stated in symbolic form. A formula such as $P = 2l + 2w$ can be solved for $l\left(l = \dfrac{P - 2w}{2} \right)$ or for $w\left(w = \dfrac{P - 2l}{2} \right)$ by applying the properties of equality.

(4.2) and (4.3) Don't forget the following suggestions for solving word problems.

1. Read the problem carefully.
2. Sketch a figure, diagram, or chart that might be helpful to organize the facts.
3. Choose a meaningful variable.
4. Look for a guideline.
5. Form an equation or inequality.
6. Solve the equation or inequality.
7. Check your answers.

(4.3) and (4.4) Properties 4.1 and 4.2 provide the basis for solving inequalities. Be sure that you can use these properties to solve a variety of inequalities as presented in this chapter.

Many of the same techniques used to solve equations can also be used to solve inequalities *but* we must be very careful when multiplying or dividing both sides of an inequality by the same number. *Don't forget* that when multiplying or dividing both sides of an inequality by a *negative number*, the resulting inequality *reverses sense*.

(4.5) The words **and** and **or** are used to form compound inequalities.

To solve inequalities involving **and** we must satisfy all of the conditions. Thus, the compound inequality $x > 1$ **and** $x < 3$ is satisfied by all numbers between 1 and 3.

To solve inequalities involving **or** we must satisfy one or more of the conditions. Thus, the compound inequality $x < -1$ **or** $x > 2$ is satisfied by (*a*) all numbers less than -1, or (*b*) all numbers greater than 2, or (*c*) both (*a*) and (*b*).

(4.6) Interpreting **absolute value** to mean "the distance between a number and zero on a number line" allows us to solve a variety of equations and inequalities involving absolute value.

Chapter 4 Review Problem Set

1. Solve $P = 2l + 2w$ for w if $P = 50$ and $l = 19$.

2. Solve $F = \dfrac{9}{5}C + 32$ for C if $F = 77$.

3. Solve $A = P + Prt$ for t.

4. Solve $2x - 3y = 13$ for x.

5. Find the area of a trapezoid having one base 8 inches long and the other base 14 inches long, if the altitude between the two bases is 7 inches.

6. If the area of a triangle is 27 square centimeters and the length of one side is 9 centimeters, find the length of the altitude to that side.

7. If the total surface area of a right circular cylinder is 152π square feet and a radius of a base is 4 feet long, find the height of the cylinder.

Solve each of the following inequalities.

8. $3x - 2 > 10$

9. $-2x - 5 < 3$

10. $2x - 9 \geq x + 4$

11. $3x + 1 \leq 5x - 10$

12. $6(x - 3) > 4(x + 13)$

13. $2(x + 3) + 3(x - 6) < 14$

14. $\dfrac{2n}{5} - \dfrac{n}{4} < \dfrac{3}{10}$

15. $\dfrac{n + 4}{5} + \dfrac{n - 3}{6} > \dfrac{7}{15}$

16. $s \geq 4.5 + 0.25s$

17. $0.07t + 0.09(500 - t) \geq 43$

18. $-16 < 8 + 2y - 3y$

19. $-24 > 5x - 4 - 7x$

20. $-3(n - 4) > 5(n + 2) + 3n$

21. $-4(n - 2) - (n - 1) < -4(n + 6)$

22. $\dfrac{3}{4}n - 6 \leq \dfrac{2}{3}n + 4$

23. $\dfrac{1}{2}n - \dfrac{1}{3}n - 4 \geq \dfrac{3}{5}n + 2$

24. $-12 > -4(x - 1) + 2$

25. $36 < -3(x + 2) - 1$

Graph the solutions for each of the following compound inequalities.

26. $x > -3$ and $x < 2$

27. $x < -1$ or $x > 4$

28. $x < 2$ or $x > 0$

29. $x > 1$ and $x > 0$

Solve and graph the solutions for each of the following.

30. $|3x - 5| = 7$

31. $|x - 4| < 1$

32. $|2x - 1| \geq 3$

33. $|3x - 2| \leq 4$

34. $|2x - 1| = 9$

35. $|5x - 2| \geq 6$

Set up an equation or an inequality and solve each of the following problems.

36. The length of a rectangle is 5 meters more than twice the width. The perimeter of the rectangle is 46 meters. Find the length and width.

37. Two airplanes leave Chicago at the same time and fly in opposite directions. If one travels at 350 miles per hour and the other at 400 miles per hour, how long will it take them to be 1125 miles apart?

38. How many liters of pure alcohol must be added to 10 liters of a 70% solution to obtain a 90% solution?

39. Monica has scores of 83, 89, 78, and 86 on her first four exams. What score must she receive on the fifth exam so that her average for all five exams is 85 or better?

40. One angle of a triangle has a measure of 47°. Of the other two angles, one of them is 3° less than three times the other angle. Find the measures of the two remaining angles.

41. Connie rides out into the country on her bicycle at a rate of 10 miles per hour. An hour later Jay leaves from the same place that Connie did and rides his bicycle along the same route at 12 miles per hour. How long will it take Jay to catch Connie?

42. How many gallons of a 10% salt solution must be mixed with 12 gallons of a 15% salt solution to obtain a 12% salt solution?

43. Susan's average score for her first three psychology exams is 84. What must she get on the 4th exam so that her average for the four exams is 85 or better?

44. Suppose that 20 ounces of a punch containing 20% orange juice is added to 30 ounces of punch containing 30% orange juice. Find the percent of orange juice in the resulting mixture.

45. The present ages of Bill and Dan are 20 years and 30 years, respectively. Will there be or was there ever a time when Bill's age is (was) one-half of Dan's age?

Cumulative Review Problem Set

For Problems 1–10, simplify each algebraic expression by combining similar terms.

1. $7x - 9x - 14x$

2. $-10a - 4 + 13a + a - 2$

3. $5(x - 3) + 7(x + 6)$

4. $3(x - 1) - 4(2x - 1)$

5. $-3n - 2(n - 1) + 5(3n - 2) - n$

6. $6n + 3(4n - 2) - 2(2n - 3) - 5$

7. $\dfrac{1}{2}x - \dfrac{3}{4}x + \dfrac{2}{3}x - \dfrac{1}{6}x$

8. $\dfrac{1}{3}n - \dfrac{4}{15}n + \dfrac{5}{6}n - n$

9. $0.4x + 0.7x - 0.8x + x$

10. $0.5(x - 2) + 0.4(x + 3) - 0.2x$

For Problems 11–20, evaluate each of the algebraic expressions for the given values of the variables.

11. $5x - 7y + 2xy$ for $x = -2$ and $y = 5$

12. $2ab - a + 6b$ for $a = 3$ and $b = -4$

13. $-3(x - 1) + 2(x + 6)$ for $x = -5$

14. $5(n + 3) - (n + 4) - n$ for $n = 7$

15. $\dfrac{3x - 2y}{2x - 3y}$ for $x = 3$ and $y = -6$

16. $\dfrac{3}{4}n - \dfrac{1}{3}n + \dfrac{5}{6}n$ for $n = -\dfrac{2}{3}$

17. $2a^2 - 4b^2$ for $a = 0.2$ and $b = -0.3$

18. $x^2 - 3xy - 2y^2$ for $x = \dfrac{1}{2}$ and $y = \dfrac{1}{4}$

19. $5x - 7y - 8x + 3y$ for $x = 9$ and $y = -8$

20. $\dfrac{3a - b - 4a + 3b}{a - 6b - 4b - 3a}$ for $a = -1$ and $b = 3$

For Problems 21–26, evaluate each of the expressions.

21. 3^4

22. -2^6

23. $\left(\dfrac{2}{3}\right)^3$

24. $\left(-\dfrac{1}{2}\right)^5$

25. $\left(\dfrac{1}{2}+\dfrac{1}{3}\right)^2$

26. $\left(\dfrac{3}{4}-\dfrac{7}{8}\right)^3$

For Problems 27–40, solve each of the equations.

27. $-5x + 2 = 22$

28. $3x - 4 = 7x + 4$

29. $7(n - 3) = 5(n + 7)$

30. $2(x - 1) - 3(x - 2) = 12$

31. $\dfrac{2}{5}x - \dfrac{1}{3} = \dfrac{1}{3}x + \dfrac{1}{2}$

32. $\dfrac{t - 2}{4} + \dfrac{t + 3}{3} = \dfrac{1}{6}$

33. $\dfrac{2n - 1}{5} - \dfrac{n + 2}{4} = 1$

34. $0.09x + 0.12(500 - x) = 54$

35. $|2x - 7| = 3$

36. $|3x + 5| = 7$

37. $-5(n - 1) - (n - 2) = 3(n - 1) - 2n$

38. $\dfrac{-2}{x - 1} = \dfrac{-3}{x + 4}$

39. $0.2x + 0.1(x - 4) = 0.7x - 1$

40. $-(t - 2) + (t - 4) = 2\left(t - \dfrac{1}{2}\right) - 3\left(t + \dfrac{1}{3}\right)$

For Problems 41–50, solve each of the inequalities.

41. $4x - 6 > 3x + 1$

42. $-3x - 6 < 12$

43. $-2(n - 1) \leq 3(n - 2) + 1$

44. $\dfrac{2}{7}x - \dfrac{1}{4} \geq \dfrac{1}{4}x + \dfrac{1}{2}$

45. $0.08t + 0.1(300 - t) > 28$

46. $-4 > 5x - 2 - 3x$

47. $|3x - 1| < 5$

48. $|2x + 3| > 4$

49. $\dfrac{2}{3}n - 2 \geq \dfrac{1}{2}n + 1$

50. $-3 < -2(x - 1) - x$

5 Exponents and Polynomials

Certain arithmetic skills are necessary before you can use arithmetic to solve problems. A similar situation exists in algebra. There are basic algebraic skills needed in order to use algebra to solve certain kinds of problems. This chapter focuses primarily on the development of some algebraic skills that will be useful when we study a class of algebraic expressions called polynomials.

5.1 Addition and Subtraction of Polynomials

In the previous chapters, algebraic expressions such as $4x$, $5y$, $-6ab$, $7x^2$, and $-9xy^2z^3$ were called *terms*. Recall that a term is an indicated product that may contain any number of factors. The variables involved in a term are called *literal factors* and the numerical factor is called the *numerical coefficient* of the term. Thus, in $-6ab$, the a and b are literal factors and the numerical coefficient is -6. Terms having the same literal factors are similar, or like, terms.

Terms containing variables with only whole numbers as exponents are called **monomials**. The previously listed terms, $4x$, $5y$, $-6ab$, $7x^2$, and $-9xy^2z^3$ are all monomials. (We shall work with some algebraic expressions later, such as $7x^{-1}y^{-1}$ and $4a^{-2}b^{-3}$, which are not monomials.)

The degree of a monomial is the sum of the exponents of the literal factors.

$4xy$ is of degree 2;

$5x$ is of degree 1;

$14a^2b$ is of degree 3;

$-17xy^2z^3$ is of degree 6;

$-9y^4$ is of degree 4.

If the monomial contains only one variable, then the exponent of the variable is the degree of the monomial. Any nonzero constant term is said to be of degree zero.

A **polynomial** is a monomial or a finite sum (or difference) of monomials. The **degree of a polynomial** is the degree of the term with the highest degree in the polynomial. Some special classifications of polynomials are made according to the number of terms. We call a one-term polynomial a **monomial**, a two-term polynomial a **binomial**, and a three-term polynomial a **trinomial**. The following examples illustrate some of this terminology.

The polynomial $5x^3y^4$ is a monomial of degree 7;

The polynomial $4x^2y - 3xy$ is a binomial of degree 3;

The polynomial $5x^2 - 6x + 4$ is a trinomial of degree 2;

The polynomial $9x^4 - 7x^3 + 6x^2 + x - 2$ is given no special name, but is of degree 4.

▪ Adding Polynomials

In the preceding chapters you have already worked many problems involving addition and subtraction of polynomials. For example, simplifying $4x^2 + 6x + 7x^2 - 2x$ to $11x^2 + 4x$ by combining similar terms can actually be considered the addition problem $(4x^2 + 6x) + (7x^2 - 2x)$. At this time we simply want to review and extend some of those ideas.

EXAMPLE 1   Add $5x^2 + 7x - 2$ and $9x^2 - 12x + 13$.

Solution   The horizontal format is most commonly used for such work. Thus,

$$(5x^2 + 7x - 2) + (9x^2 - 12x + 13) = (5x^2 + 9x^2) + (7x - 12x) + (-2 + 13)$$
$$= 14x^2 - 5x + 11.$$ ■

The commutative, associative, and distributive properties provide the basis for rearranging, regrouping, and combining similar terms.

EXAMPLE 2   Add $5x - 1$, $3x + 4$, and $9x - 7$.

Solution
$$(5x - 1) + (3x + 4) + (9x - 7) = (5x + 3x + 9x) + (-1 + 4 + (-7))$$
$$= 17x - 4.$$ ■

■ Subtracting Polynomials

Recall from Chapter 1 that $a - b = a + (-b)$. Subtraction is defined as *adding the opposite*. This same idea extends to polynomials in general. The opposite of a polynomial can be formed by taking the opposite of each term. For example, the opposite of $2x^2 - 7x + 3$ is $-2x^2 + 7x - 3$. Symbolically, this is expressed as

$$-(2x^2 - 7x + 3) = -2x^2 + 7x - 3.$$

Now consider the following subtraction problems.

EXAMPLE 3   Subtract $2x^2 + 9x - 3$ from $5x^2 - 7x - 1$.

Solution   Using the horizontal format, we have

$$(5x^2 - 7x - 1) - (2x^2 + 9x - 3) = (5x^2 - 7x - 1) + (-2x^2 - 9x + 3)$$
$$= (5x^2 - 2x^2) + (-7x - 9x) + (-1 + 3)$$
$$= 3x^2 - 16x + 2.$$ ■

EXAMPLE 4   Subtract $-8y^2 - y + 5$ from $2y^2 + 9$.

Solution
$$(2y^2 + 9) - (-8y^2 - y + 5) = (2y^2 + 9) + (8y^2 + y - 5)$$
$$= (2y^2 + 8y^2) + (y) + (9 - 5)$$
$$= 10y^2 + y + 4.$$ ■

Later, when dividing polynomials, we will need to subtract polynomials using a vertical format; therefore, let's consider two such examples.

EXAMPLE 5   Subtract $3x^2 + 5x - 2$ from $9x^2 - 7x - 1$.

Solution

$9x^2 - 7x - 1$ Notice which polynomial goes on the bottom and
$3x^2 + 5x - 2$ the alignment of similar terms in columns.

Now we can "mentally form the opposite of the bottom polynomial" and add.

$9x^2 - 7x - 1$
$3x^2 + 5x - 2$
$6x^2 - 12x + 1$ The opposite of $3x^2 + 5x - 2$ is $-3x^2 - 5x + 2$.   ■

EXAMPLE 6   Subtract $15y^3 + 5y^2 + 3$ from $13y^3 + 7y - 1$.

Solution

$13y^3 + 7y - 1$ The similar terms are arranged in columns.
$15y^3 + 5y^2 + 3$
$-2y^3 - 5y^2 + 7y - 4$ We mentally formed the opposite of the
bottom polynomial and added.   ■

The distributive property along with the properties $a = 1(a)$ and $-a = -1(a)$ can be used when adding and subtracting polynomials. The next examples illustrate this approach.

EXAMPLE 7   Perform the indicated operations.

$$(3x - 4) + (2x - 5) - (7x - 1).$$

Solution

$(3x - 4) + (2x - 5) - (7x - 1)$
$= 1(3x - 4) + 1(2x - 5) - 1(7x - 1)$
$= 1(3x) - 1(4) + 1(2x) - 1(5) - 1(7x) - 1(-1)$
$= 3x - 4 + 2x - 5 - 7x + 1$
$= 3x + 2x - 7x - 4 - 5 + 1.$
$= -2x - 8.$ ■

Certainly some of the steps can be done mentally and our format may be as illustrated in the next example.

EXAMPLE 8   Perform the indicated operations.

$$(-y^2 + 5y - 2) - (-2y^2 + 8y + 6) + (4y^2 - 2y - 5).$$

Solution

$$(-y^2 + 5y - 2) - (-2y^2 + 8y + 6) + (4y^2 - 2y - 5)$$
$$= -y^2 + 5y - 2 + 2y^2 - 8y - 6 + 4y^2 - 2y - 5$$
$$= -y^2 + 2y^2 + 4y^2 + 5y - 8y - 2y - 2 - 6 - 5$$
$$= 5y^2 - 5y - 13.$$ ■

When using the horizontal format, as in Examples 7 and 8, parentheses are used to indicate a quantity. In Example 7 the quantities $(3x - 4)$ and $(2x - 5)$ are to be added and from this result we are to subtract the quantity $(7x - 1)$. Brackets, [ ], and braces, { }, are also sometimes used as grouping symbols, especially if there is a need to indicate quantities within quantities. Removal of grouping symbols can be accomplished by performing the operations indicated, starting with the innermost set of symbols. Let's consider two examples of this type.

EXAMPLE 9 Perform the indicated operations.

$$3x - [2x + (3x - 1)].$$

Solution First, we need to add the quantities $2x$ and $(3x - 1)$.

$$3x - [2x + (3x - 1)] = 3x - [2x + 3x - 1]$$
$$= 3x - [5x - 1].$$

Now we need to subtract the quantity $[5x - 1]$ from $3x$.

$$3x - [5x - 1] = 3x - 5x + 1$$
$$= -2x + 1.$$ ■

EXAMPLE 10 Perform the indicated operations.

$$8 - [7x - [2 + (x - 1)] + 4x]$$

Solution Starting with the innermost set of grouping symbols (the parentheses) we can proceed as follows.

$$8 - [7x - [2 + (x - 1)] + 4x] = 8 - [7x - [2 + x - 1] + 4x]$$
$$= 8 - [7x - [x + 1] + 4x]$$
$$= 8 - [7x - x - 1 + 4x]$$
$$= 8 - [10x - 1]$$
$$= 8 - 10x + 1$$
$$= -10x + 9.$$ ■

Problem Set 5.1

For Problems 1–8, determine the degree of each polynomial.

1. $7x^2y + 6xy$

2. $4xy - 7x$

3. $5x^2 - 9$

4. $8x^2y^2 - 2xy^2 - x$

5. $5x^3 - x^2 - x + 3$

6. $8x^4 - 2x^2 + 6$

7. $5xy$

8. $-7x + 4$

For Problems 9–18, add the polynomials.

9. $3x + 4$ and $5x + 7$

10. $3x - 5$ and $2x - 9$

11. $-5y - 3$ and $9y + 13$

12. $x^2 - 2x - 1$ and $-2x^2 + x + 4$

13. $-2x^2 + 7x - 9$ and $4x^2 - 9x - 14$

14. $3a^2 + 4a - 7$ and $-3a^2 - 7a + 10$

15. $5x - 2$, $3x - 7$, and $9x - 10$

16. $-x - 4$, $8x + 9$, and $-7x - 6$

17. $2x^2 - x + 4$, $-5x^2 - 7x - 2$, and $9x^2 + 3x - 6$

18. $-3x^2 + 2x - 6$, $6x^2 + 7x + 3$, and $-4x^2 - 9$

For Problems 19–30, subtract the polynomials using a horizontal format.

19. $7x + 1$ from $12x + 6$

20. $10x + 3$ from $14x + 13$

21. $5x - 2$ from $3x - 7$

22. $7x - 2$ from $2x + 3$

23. $-x - 1$ from $-4x + 6$

24. $-3x + 2$ from $-x - 9$

25. $x^2 - 7x + 2$ from $3x^2 + 8x - 4$

26. $2x^2 + 6x - 1$ from $8x^2 - 2x + 6$

27. $-2n^2 - 3n + 4$ from $3n^2 - n + 7$

28. $3n^2 - 7n - 9$ from $-4n^2 + 6n + 10$

29. $-4x^3 - x^2 + 6x - 1$ from $-7x^3 + x^2 + 6x - 12$

30. $-4x^2 + 6x - 2$ from $-3x^3 + 2x^2 + 7x - 1$

For Problems 31–40, subtract the polynomials using a vertical format.

31. $3x - 2$ from $12x - 4$

32. $-4x + 6$ from $7x - 3$

33. $-5a - 6$ from $-3a + 9$

34. $7a - 11$ from $-2a - 1$

35. $8x^2 - x + 6$ from $6x^2 - x + 11$

36. $3x^2 - 2$ from $-2x^2 + 6x - 4$

37. $-2x^3 - 6x^2 + 7x - 9$ from $4x^3 + 6x^2 + 7x - 14$

38. $4x^3 + x - 10$ from $3x^2 - 6$

39. $2x^2 - 6x - 14$ from $4x^3 - 6x^2 + 7x - 2$

40. $3x - 7$ from $7x^3 + 6x^2 - 5x - 4$

For Problems 41–60, perform the indicated operations.

41. $(5x + 3) - (7x - 2) + (3x + 6)$

42. $(3x - 4) + (9x - 1) - (14x - 7)$

43. $(-x - 1) - (-2x + 6) + (-4x - 7)$

44. $(-3x + 6) + (-x - 8) - (-7x + 10)$

45. $(x^2 - 7x - 4) + (2x^2 - 8x - 9) - (4x^2 - 2x - 1)$

46. $(3x^2 + x - 6) - (8x^2 - 9x + 1) - (7x^2 + 2x - 6)$

47. $(-x^2 - 3x + 4) + (-2x^2 - x - 2) - (-4x^2 + 7x + 10)$

48. $(-3x^2 - 2) + (7x^2 - 8) - (9x^2 - 2x - 4)$

49. $(3a - 2b) - (7a + 4b) - (6a - 3b)$ **50.** $(5a + 7b) + (-8a - 2b) - (5a + 6b)$

51. $(n - 6) - (2n^2 - n + 4) + (n^2 - 7)$ **52.** $(3n + 4) - (n^2 - 9n + 10) - (-2n + 4)$

53. $7x + [3x - (2x - 1)]$ **54.** $-6x + [-2x - (5x + 2)]$

55. $-7n - [4n - (6n - 1)]$ **56.** $9n - [3n - (5n + 4)]$

57. $(5a - 1) - [3a + (4a - 7)]$ **58.** $(-3a + 4) - [-7a + (9a - 1)]$

59. $13x - [5x - [4x - (x - 6)]]$ **60.** $-10x - [7x - [3x - (2x - 3)]]$

61. Subtract $5x - 3$ from the sum of $4x - 2$ and $7x + 6$.

62. Subtract $7x + 5$ from the sum of $9x - 4$ and $-3x - 2$.

63. Subtract the sum of $-2n - 5$ and $-n + 7$ from $-8n + 9$.

64. Subtract the sum of $7n - 11$ and $-4n - 3$ from $13n - 4$.

5.2 Multiplying Monomials

In Section 2.4 we used exponents and some of the basic properties of real numbers to simplify algebraic expressions into a more compact form. For example,

$$(3x)(4xy) = 3 \cdot 4 \cdot x \cdot x \cdot y$$
$$= 12x^2y.$$

Actually, we were **multiplying monomials**, and it is this topic that we will pursue further at this time.

Multiplying monomials is made easier by using some basic properties of exponents. These properties are the direct result of the definition of an exponent. The following examples lead into the first property.

$$x^2 \cdot x^3 = (x \cdot x)(x \cdot x \cdot x) = x^5;$$
$$a^3 \cdot a^4 = (a \cdot a \cdot a)(a \cdot a \cdot a \cdot a) = a^7;$$
$$b \cdot b^2 = (b)(b \cdot b) = b^3.$$

In general,

$$b^n \cdot b^m = \underbrace{(b \cdot b \cdot b \cdots b)}_{\substack{n \text{ factors} \\ \text{of } b}}\underbrace{(b \cdot b \cdot b \cdots b)}_{\substack{m \text{ factors} \\ \text{of } b}}$$

$$= \underbrace{b \cdot b \cdot b \cdots b}_{\substack{(n + m) \text{ factors} \\ \text{of } b}}$$

$$= b^{n+m}.$$

Property 5.1

If b is any real number and n and m are positive integers, then

$$b^n \cdot b^m = b^{n+m}.$$

Property 5.1 states that when multiplying powers with the same base, add exponents.

EXAMPLE 1 Multiply each of the following.

(a) $x^4 \cdot x^3$ (b) $a^8 \cdot a^7$

Solution (a) $x^4 \cdot x^3 = x^{4+3} = x^7$.

(b) $a^8 \cdot a^7 = a^{8+7} = a^{15}$.   ■

Another property of exponents is illustrated by the following examples.

$$(x^2)^3 = x^2 \cdot x^2 \cdot x^2 = x^{2+2+2} = x^6;$$
$$(a^3)^2 = a^3 \cdot a^3 = a^{3+3} = a^6;$$
$$(b^3)^4 = b^3 \cdot b^3 \cdot b^3 \cdot b^3 = b^{3+3+3+3} = b^{12}.$$

In general,

$$(b^n)^m = \underbrace{b^n \cdot b^n \cdot b^n \cdots b^n}_{m \text{ factors of } b^n}$$

$$= \overbrace{b^{n+n+n+\cdots+n}}^{m \text{ of these } n\text{'s}}$$

$$= b^{mn}.$$

Property 5.2

If b is any real number and m and n are positive integers, then

$$(b^n)^m = b^{mn}.$$

Property 5.2 states that when raising a power to a power, multiply exponents.

EXAMPLE 2 Raise each of the following to the indicated power.

(a) $(x^4)^3$ (b) $(a^5)^6$

Solution   (a) $(x^4)^3 = x^{3 \cdot 4} = x^{12}$.

(b) $(a^5)^6 = a^{6 \cdot 5} = a^{30}$.   ■

The third property of exponents to be used in this section pertains to raising a monomial to a power.

$$(2x)^3 = (2x)(2x)(2x) = 2 \cdot 2 \cdot 2 \cdot x \cdot x \cdot x = 2^3 \cdot x^3;$$
$$(3a^4)^2 = (3a^4)(3a^4) = 3 \cdot 3 \cdot a^4 \cdot a^4 = (3)^2(a^4)^2;$$
$$(-2xy^5)^2 = (-2xy^5)(-2xy^5) = (-2)(-2)(x)(x)(y^5)(y^5) = (-2)^2(x)^2(y^5)^2.$$

In general,

$$(ab)^n = \underbrace{ab \cdot ab \cdot ab \cdots \cdot ab}_{n \text{ factors of } ab}$$

$$= \underbrace{(a \cdot a \cdot a \cdots a)}_{n \text{ factors of } a}\underbrace{(b \cdot b \cdot b \cdots b)}_{n \text{ factors of } b}$$

$$= a^n b^n.$$

Property 5.3

If a and b are real numbers and n is a positive integer, then

$$(ab)^n = a^n b^n.$$

Property 5.3 states that when raising a monomial to a power, raise each factor to that power.

EXAMPLE 3 Raise each of the following to the indicated power.

(a) $(2x^2y^3)^4$ (b) $(-3ab^5)^3$

Solution   (a) $(2x^2y^3)^4 = (2)^4(x^2)^4(y^3)^4 = 16x^8y^{12}$.

(b) $(-3ab^5)^3 = (-3)^3(a^1)^3(b^5)^3 = -27a^3b^{15}$.   ■

Consider the following examples in which the properties of exponents are used to help simplify the process of multiplying monomials.

1. $(3x^3)(5x^4) = 3 \cdot 5 \cdot x^3 \cdot x^4$
    $\qquad = 15x^7. \qquad (x^3 \cdot x^4 = x^{3+4} = x^7)$

2. $(-4a^2b^3)(6ab^2) = -4 \cdot 6 \cdot a^2 \cdot a \cdot b^3 \cdot b^2$
    $\qquad = -24a^3b^5.$

3. $(xy)(7xy^5) = 1 \cdot 7 \cdot x \cdot x \cdot y \cdot y^5$ The numerical coefficient of xy is 1.
$$= 7x^2y^6.$$

4. $\left(\dfrac{3}{4}x^2y^3\right)\left(\dfrac{1}{2}x^3y^5\right) = \dfrac{3}{4} \cdot \dfrac{1}{2} \cdot x^2 \cdot x^3 \cdot y^3 \cdot y^5$
$$= \dfrac{3}{8}x^5y^8.$$

Raising a monomial to a power is also a simple process when using the properties of exponents. Study the following examples.

5. $(2x^3)^4 = (2)^4(x^3)^4$ by using $(ab)^n = a^nb^n$
$ = (2)^4(x^{12})$ by using $(b^n)^m = b^{mn}$
$ = 16x^{12}.$

6. $(-2a^4)^5 = (-2)^5(a^4)^5$
$ = -32a^{20}.$

7. $\left(\dfrac{2}{5}x^2y^3\right)^3 = \left(\dfrac{2}{5}\right)^3(x^2)^3(y^3)^3$
$$\phantom{\left(\dfrac{2}{5}x^2y^3\right)^3} = \dfrac{8}{125}x^6y^9.$$

8. $(0.2a^6b^7)^2 = (0.2)^2(a^6)^2(b^7)^2$
$ = 0.04a^{12}b^{14}.$

Sometimes problems involve first raising monomials to a power and then multiplying the resulting monomials, as in the following examples.

9. $(3x^2)^3(2x^3)^2 = (3)^3(x^2)^3(2)^2(x^3)^2$
$ = (27)(x^6)(4)(x^6)$
$ = 108x^{12}.$

10. $(-x^2y^3)^5(-2x^2y)^2 = (-1)^5(x^2)^5(y^3)^5(-2)^2(x^2)^2(y)^2$
$ = (-1)(x^{10})(y^{15})(4)(x^4)(y^2)$
$ = -4x^{14}y^{17}.$

The distributive property along with the properties of exponents form a basis for finding the product of a monomial and a polynomial. The following examples illustrate these ideas.

11. $(3x)(2x^2 + 6x + 1) = (3x)(2x^2) + (3x)(6x) + (3x)(1)$
$ = 6x^3 + 18x^2 + 3x.$

12. $(5a^2)(a^3 - 2a^2 - 1) = (5a^2)(a^3) - (5a^2)(2a^2) - (5a^2)(1)$
$ = 5a^5 - 10a^4 - 5a^2.$

13. $(-2xy)(6x^2y - 3xy^2 - 4y^3)$
$ = (-2xy)(6x^2y) - (-2xy)(3xy^2) - (-2xy)(4y^3)$
$ = -12x^3y^2 + 6x^2y^3 + 8xy^4.$

Once you feel comfortable with this process, you may want to perform most of the work mentally and simply write down the final result. See if you understand the following examples.

14. $3x(2x + 3) = 6x^2 + 9x.$

15. $-4x(2x^2 - 3x - 1) = -8x^3 + 12x^2 + 4x.$

16. $ab(3a^2b - 2ab^2 - b^3) = 3a^3b^2 - 2a^2b^3 - ab^4.$

Problem Set 5.2

For Problems 1–30, multiply using the properties of exponents to help with the manipulation.

1. $(5x)(9x)$

2. $(7x)(8x)$

3. $(3x^2)(7x)$

4. $(9x)(4x^3)$

5. $(-3xy)(2xy)$

6. $(6xy)(-3xy)$

7. $(-2x^2y)(-7x)$

8. $(-5xy^2)(-4y)$

9. $(4a^2b^2)(-12ab)$

10. $(-3a^3b)(13ab^2)$

11. $(-xy)(-5x^3)$

12. $(-7y^2)(-x^2y)$

13. $(8ab^2c)(13a^2c)$

14. $(9abc^3)(14bc^2)$

15. $(5x^2)(2x)(3x^3)$

16. $(4x)(2x^2)(6x^4)$

17. $(4xy)(-2x)(7y^2)$

18. $(5y^2)(-3xy)(5x^2)$

19. $(-2ab)(-ab)(-3b)$

20. $(-7ab)(-4a)(-ab)$

21. $(6cd)(-3c^2d)(-4d)$

22. $(2c^3d)(-6d^3)(-5cd)$

23. $\left(\frac{2}{3}xy\right)\left(\frac{3}{5}x^2y^4\right)$

24. $\left(-\frac{5}{6}x\right)\left(\frac{8}{3}x^2y\right)$

25. $\left(-\frac{7}{12}a^2b\right)\left(\frac{8}{21}b^4\right)$

26. $\left(-\frac{9}{5}a^3b^4\right)\left(-\frac{15}{6}ab^2\right)$

27. $(0.4x^5)(0.7x^3)$

28. $(-1.2x^4)(0.3x^2)$

29. $(-4ab)(1.6a^3b)$

30. $(-6a^2b)(-1.4a^2b^4)$

For Problems 31–46, raise each monomial to the indicated power. Use the properties of exponents to help with the manipulation.

31. $(2x^4)^2$

32. $(3x^3)^2$

33. $(-3a^2b^3)^2$

34. $(-8a^4b^5)^2$

35. $(3x^2)^3$

36. $(2x^4)^3$

37. $(-4x^4)^3$

38. $(-3x^3)^3$

39. $(9x^4y^5)^2$

40. $(8x^6y^4)^2$

41. $(2x^2y)^4$

42. $(2x^2y^3)^5$

43. $(-3a^3b^2)^4$

44. $(-2a^4b^2)^4$

45. $(-x^2y)^6$

46. $(-x^2y^3)^7$

For Problems 47–60, multiply by using the distributive property.

47. $5x(3x + 2)$

48. $7x(2x + 5)$

49. $3x^2(6x - 2)$

50. $4x^2(7x - 2)$

51. $-4x(7x^2 - 4)$

52. $-6x(9x^2 - 5)$

53. $2x(x^2 - 4x + 6)$

54. $3x(2x^2 - x + 5)$

55. $-6a(3a^2 - 5a - 7)$

56. $-8a(4a^2 - 9a - 6)$

57. $7xy(4x^2 - x + 5)$

58. $5x^2y(3x^2 + 7x - 9)$

59. $-xy(9x^2 - 2x - 6)$

60. $xy^2(6x^2 - x - 1)$

For Problems 61–70, remove parentheses by multiplying and then simplify by combining similar terms. For example,

$$3(x - y) + 2(x - 3y) = 3x - 3y + 2x - 6y$$
$$= 5x - 9y.$$

61. $5(x + 2y) + 4(2x + 3y)$ **62.** $3(2x + 5y) + 2(4x + y)$

63. $4(x - 3y) - 3(2x - y)$ **64.** $2(5x - 3y) - 5(x + 4y)$

65. $2x(x^2 - 3x - 4) + x(2x^2 + 3x - 6)$ **66.** $3x(2x^2 - x + 5) - 2x(x^2 + 4x + 7)$

67. $3[2x - (x - 2)] - 4(x - 2)$ **68.** $2[3x - (2x + 1)] - 2(3x - 4)$

69. $-4(3x + 2) - 5[2x - (3x + 4)]$ **70.** $-5(2x - 1) - 3[x - (4x - 3)]$

For Problems 71–80, perform the indicated operations and simplify.

71. $(3x)^2(2x^3)$ **72.** $(-2x)^3(4x^5)$ **73.** $(-3x)^3(-4x)^2$

74. $(3xy)^2(2x^2y)^4$ **75.** $(5x^2y)^2(xy^2)^3$ **76.** $(-x^2y)^3(6xy)^2$

77. $(-a^2bc^3)^3(a^3b)^2$ **78.** $(ab^2c^3)^4(-a^2b)^3$ **79.** $(-2x^2y^2)^4(-xy^3)^3$

80. $(-3xy)^3(-x^2y^3)^4$

■ Miscellaneous Problems

For Problems 81–90, find each of the indicated products. Assume that the variables in the exponents represent positive integers. For example,

$$(x^{2n})(x^{4n}) = x^{2n + 4n} = x^{6n}.$$

81. $(x^n)(x^{3n})$ **82.** $(x^{2n})(x^{5n})$ **83.** $(x^{2n-1})(x^{3n+2})$

84. $(x^{5n+2})(x^{n-1})$ **85.** $(x^3)(x^{4n-5})$ **86.** $(x^{6n-1})(x^4)$

87. $(2x^n)(3x^{2n})$ **88.** $(4x^{3n})(-5x^{7n})$ **89.** $(-6x^{2n+4})(5x^{3n-4})$

90. $(-3x^{5n-2})(-4x^{2n+2})$

5.3 Multiplying Polynomials

In general, to go from multiplying a monomial and a polynomial to multiplying two polynomials requires the use of the distributive property. Consider the following examples.

EXAMPLE 1   Find the product of $(x + 3)$ and $(y + 4)$.

Solution

$$(x + 3)(y + 4) = x(y + 4) + 3(y + 4)$$
$$= x(y) + x(4) + 3(y) + 3(4).$$
$$= xy + 4x + 3y + 12.$$

■

Notice that each term of the first polynomial is multiplied times each term of the second polynomial.

EXAMPLE 2 Find the product of $(x - 2)$ and $(y + z + 5)$.

Solution
$$(x - 2)(y + z + 5) = x(y) + x(z) + x(5) - 2(y) - 2(z) - 2(5)$$
$$= xy + xz + 5x - 2y - 2z - 10.$$ ■

Frequently, multiplying polynomials will produce similar terms that can be combined to simplify the resulting polynomial.

EXAMPLE 3 Multiply $(x + 3)(x + 2)$.

Solution
$$(x + 3)(x + 2) = x(x + 2) + 3(x + 2)$$
$$= x^2 + 2x + 3x + 6$$
$$= x^2 + 5x + 6.$$ ■

EXAMPLE 4 Multiply $(x - 4)(x + 9)$.

Solution
$$(x - 4)(x + 9) = x(x + 9) - 4(x + 9)$$
$$= x^2 + 9x - 4x - 36$$
$$= x^2 + 5x - 36.$$ ■

EXAMPLE 5 Multiply $(x + 4)(x^2 + 3x + 2)$.

Solution
$$(x + 4)(x^2 + 3x + 2) = x(x^2 + 3x + 2) + 4(x^2 + 3x + 2)$$
$$= x^3 + 3x^2 + 2x + 4x^2 + 12x + 8$$
$$= x^3 + 7x^2 + 14x + 8.$$ ■

EXAMPLE 6 Multiply $(2x - y)(3x^2 - 2xy + 4y^2)$.

Solution
$$(2x - y)(3x^2 - 2xy + 4y^2) = 2x(3x^2 - 2xy + 4y^2) - y(3x^2 - 2xy + 4y^2)$$
$$= 6x^3 - 4x^2y + 8xy^2 - 3x^2y + 2xy^2 - 4y^3$$
$$= 6x^3 - 7x^2y + 10xy^2 - 4y^3.$$ ■

Perhaps the most frequently used type of multiplication problem is the product of two binomials. It will be a big help later if you can become proficient at multiplying binomials without showing all of the intermediate steps. This is quite easy to do by developing a three-step shortcut pattern as demonstrated by the following examples.

EXAMPLE 7 Multiply $(x + 5)(x + 7)$.

Solution

$$(x + 5)(x + 7) = x^2 + 12x + 35.$$

Step 1. Multiply $x \cdot x$.

Step 2. Multiply $5 \cdot x$ and $7 \cdot x$ and combine.

Step 3. Multiply $5 \cdot 7$. ■

EXAMPLE 8 Multiply $(x - 8)(x + 3)$.

Solution

$$(x - 8)(x + 3) = x^2 - 5x - 24.$$ ■

EXAMPLE 9 Multiply $(3x + 2)(2x - 5)$.

Solution

$$(3x + 2)(2x - 5) = 6x^2 - 11x - 10.$$ ■

Now see if *you* can use the pattern to find the following products.

$(x + 3)(x + 7) = $ _____

$(3x + 1)(2x + 5) = $ _____

$(x - 2)(x - 3) = $ _____

$(4x + 5)(x - 2) = $ _____

Your answers should be $x^2 + 10x + 21$, $6x^2 + 17x + 5$, and $x^2 - 5x + 6$, and $4x^2 - 3x - 10$.

Keep in mind that the shortcut pattern discussed above applies only to finding the product of two binomials. For other situations, such as finding the product of a binomial and a trinomial, we would suggest showing the intermediate steps as follows.

$$(x + 3)(x^2 + 6x - 7) = x(x^2) + x(6x) - x(7) + 3(x^2) + 3(6x) - 3(7)$$
$$= x^3 + 6x^2 - 7x + 3x^2 + 18x - 21$$
$$= x^3 + 9x^2 + 11x - 21.$$

Perhaps the first step could be omitted and the form shortened as follows.

$$(x - 4)(x^2 - 5x - 6) = x^3 - 5x^2 - 6x - 4x^2 + 20x + 24$$
$$= x^3 - 9x^2 + 14x + 24.$$

Remember that you are multiplying each term of the first polynomial times each term of the second polynomial and combining similar terms.

Exponents are also used to indicate repeated multiplication of polynomials. For example, $(x + 4)(x + 4)$ can be written as $(x + 4)^2$. Thus, to square a binomial we simply write it as the product of two equal binomials and apply the shortcut pattern.

$$(x + 4)^2 = (x + 4)(x + 4) = x^2 + 8x + 16;$$
$$(x - 5)^2 = (x - 5)(x - 5) = x^2 - 10x + 25;$$
$$(2x + 3)^2 = (2x + 3)(2x + 3) = 4x^2 + 12x + 9.$$

When squaring binomials be careful not to forget the middle term. That is to say, $(x + 3)^2 \neq x^2 + 3^2$; instead, $(x + 3)^2 = (x + 3)(x + 3) = x^2 + 6x + 9$.

The following example suggests a format to use when cubing a binomial.

$$(x + 4)^3 = (x + 4)(x + 4)(x + 4)$$
$$= (x + 4)(x^2 + 8x + 16)$$
$$= x(x^2 + 8x + 16) + 4(x^2 + 8x + 16)$$
$$= x^3 + 8x^2 + 16x + 4x^2 + 32x + 64$$
$$= x^3 + 12x^2 + 48x + 64.$$

■ Special Product Patterns

When multiplying binomials, there are some special patterns that occur that you should recognize. These patterns can be used to find products and will also be used later when factoring polynomials. Each of the patterns will be stated in general terms followed by examples to illustrate the use of each pattern.

PATTERN

$$(a + b)^2 = (a + b)(a + b) = a^2 + 2ab + b^2$$

square of first term of binomial
+ twice the product of the two terms of binomial +
square of second term of binomial

Examples

$$(x + 4)^2 = x^2 + 8x + 16.$$
$$(2x + 3y)^2 = 4x^2 + 12xy + 9y^2.$$
$$(5a + 7b)^2 = 25a^2 + 70ab + 49b^2.$$

PATTERN

$$(a - b)^2 = (a - b)(a - b) = a^2 - \quad 2ab \quad + \quad b^2$$

square of
first term — product of + second term
of binomial the two terms of binomial
of binomial

twice the

square of

Examples

$(x - 8)^2 = x^2 - 16x + 64.$

$(3x - 4y)^2 = 9x^2 - 24xy + 16y^2.$

$(4a - 9b)^2 = 16a^2 - 72ab + 81b^2.$

PATTERN

$$(a + b)(a - b) = a^2 - \quad b^2$$

square of   square of
first term   — second term
of binomials   of binomials

Examples

$(x + 7)(x - 7) = x^2 - 49.$

$(2x + y)(2x - y) = 4x^2 - y^2.$

$(3a - 2b)(3a + 2b) = 9a^2 - 4b^2.$

Problem Set 5.3

For Problems 1–10, find the indicated products by applying the distributive property. For example,

$$(x + 1)(y + 5) = x(y) + x(5) + 1(y) + 1(5)$$
$$= xy + 5x + y + 5.$$

1. $(x + 2)(y + 3)$ **2.** $(x + 3)(y + 6)$ **3.** $(x - 4)(y + 1)$

4. $(x - 5)(y + 7)$ **5.** $(x - 5)(y - 6)$ **6.** $(x - 7)(y - 9)$

7. $(x + 2)(y + z + 1)$ **8.** $(x + 4)(y - z + 4)$ **9.** $(2x + 3)(3y + 1)$

10. $(3x - 2)(2y - 5)$

For Problems 11–36, find the indicated products by applying the distributive property and combining similar terms. Use the following format to show your work.

$$(x + 3)(x + 8) = x(x) + x(8) + 3(x) + 3(8)$$
$$= x^2 + 8x + 3x + 24$$
$$= x^2 + 11x + 24.$$

11. $(x + 3)(x + 7)$ 12. $(x + 4)(x + 2)$ 13. $(x + 8)(x - 3)$

14. $(x + 9)(x - 6)$ 15. $(x - 7)(x + 1)$ 16. $(x - 10)(x + 8)$

17. $(n - 4)(n - 6)$ 18. $(n - 3)(n - 7)$ 19. $(3n + 1)(n + 6)$

20. $(4n + 3)(n + 6)$ 21. $(5x - 2)(3x + 7)$ 22. $(3x - 4)(7x + 1)$

23. $(x + 3)(x^2 + 4x + 9)$ 24. $(x + 2)(x^2 + 6x + 2)$ 25. $(x + 4)(x^2 - x - 6)$

26. $(x + 5)(x^2 - 2x - 7)$ 27. $(x - 5)(2x^2 + 3x - 7)$

28. $(x - 4)(3x^2 + 4x - 6)$ 29. $(2a - 1)(4a^2 - 5a + 9)$

30. $(3a - 2)(2a^2 - 3a - 5)$ 31. $(3a + 5)(a^2 - a - 1)$

32. $(5a + 2)(a^2 + a - 3)$ 33. $(x^2 + 2x + 3)(x^2 + 5x + 4)$

34. $(x^2 - 3x + 4)(x^2 + 5x - 2)$ 35. $(x^2 - 6x - 7)(x^2 + 3x - 9)$

36. $(x^2 - 5x - 4)(x^2 + 7x - 8)$

For Problems 37–80, find the indicated products by using the shortcut pattern for multiplying binomials.

 Foil

37. $(x + 2)(x + 9)$ 38. $(x + 3)(x + 8)$ 39. $(x + 6)(x - 2)$

40. $(x + 8)(x - 6)$ 41. $(x + 3)(x - 11)$ 42. $(x + 4)(x - 10)$

43. $(n - 4)(n - 3)$ 44. $(n - 5)(n - 9)$ 45. $(n + 6)(n + 12)$

46. $(n + 8)(n + 13)$ 47. $(y + 3)(y - 7)$ 48. $(y + 2)(y - 12)$

49. $(y - 7)(y - 12)$ 50. $(y - 4)(y - 13)$ 51. $(x - 5)(x + 7)$

52. $(x - 1)(x + 9)$ 53. $(x - 14)(x + 8)$ 54. $(x - 15)(x + 6)$

55. $(a + 10)(a - 9)$ 56. $(a + 7)(a - 6)$ 57. $(2a + 1)(a + 6)$

58. $(3a + 2)(a + 4)$ 59. $(5x - 2)(x + 7)$ 60. $(2x - 3)(x + 8)$

61. $(3x - 7)(2x + 1)$ 62. $(5x - 6)(4x + 3)$ 63. $(4a + 3)(3a - 4)$

64. $(5a + 4)(4a - 5)$ 65. $(6n - 5)(2n - 3)$ 66. $(4n - 3)(6n - 7)$

67. $(7x - 4)(2x + 3)$ 68. $(8x - 5)(3x + 7)$ 69. $(5 - x)(9 - 2x)$

70. $(4 - 3x)(2 + x)$ 71. $(-2x + 3)(4x - 5)$ 72. $(-3x + 1)(9x - 2)$

73. $(-3x - 1)(3x - 4)$ 74. $(-2x - 5)(4x + 1)$ 75. $(8n + 3)(9n - 4)$

76. $(6n + 5)(9n - 7)$ 77. $(3 - 2x)(9 - x)$ 78. $(5 - 4x)(4 - 5x)$

79. $(-4x + 3)(-5x - 2)$ 80. $(-2x + 7)(-7x - 3)$

For Problems 81–110, use one of the appropriate patterns $(a + b)^2 = a^2 + 2ab + b^2$, $(a - b)^2 = a^2 - 2ab + b^2$, or $(a + b)(a - b) = a^2 - b^2$ to help find the indicated products.

 Patterns

81. $(x + 7)^2$ 82. $(x + 9)^2$ 83. $(5x - 2)(5x + 2)$

84. $(6x + 1)(6x - 1)$ 85. $(x - 1)^2$ 86. $(x - 4)^2$

87. $(3x + 7)^2$ 88. $(2x + 9)^2$ 89. $(2x - 3)^2$

90. $(4x - 5)^2$ 91. $(2x + 3y)(2x - 3y)$ 92. $(3a - b)(3a + b)$

93. $(1 - 5n)^2$ 94. $(2 - 3n)^2$ 95. $(3x + 4y)^2$

96. $(2x + 5y)^2$ 97. $(3 + 4y)^2$ 98. $(7 + 6y)^2$

99. $(1 + 7n)(1 - 7n)$ 100. $(2 + 9n)(2 - 9n)$ 101. $(4a - 7b)^2$

102. $(6a - b)^2$ **103.** $(x + 8y)^2$ **104.** $(x + 6y)^2$

105. $(5x - 11y)(5x + 11y)$ **106.** $(7x - 9y)(7x + 9y)$ **107.** $x(8x + 1)(8x - 1)$

108. $3x(5x + 7)(5x - 7)$ **109.** $-2x(4x + y)(4x - y)$ **110.** $-4x(2 - 3x)(2 + 3x)$

For Problems 111–118, find the indicated products. Don't forget that $(x + 2)^3$ means $(x + 2)(x + 2)(x + 2)$.

111. $(x + 2)^3$ **112.** $(x + 4)^3$ **113.** $(x - 3)^3$ **114.** $(x - 1)^3$

115. $(2n + 1)^3$ **116.** $(3n + 2)^3$ **117.** $(3n - 2)^3$ **118.** $(4n - 3)^3$

■ Miscellaneous Problems

119. The following two patterns result from cubing binomials.

$$(a + b)^3 = a^3 + 3a^2b + 3ab^2 + b^3.$$
$$(a - b)^3 = a^3 - 3a^2b + 3ab^2 - b^3.$$

Use these patterns and redo Problems 111–118.

120. Find a pattern for the expansion of $(a + b)^4$. Then use the pattern to expand $(x + 2)^4$, $(x + 3)^4$, and $(2x + 1)^4$.

5.4 Dividing by Monomials

Developing an effective process for dividing by a monomial relies on yet another property of exponents. This property is also a direct consequence of the definition of exponent and is illustrated by the following examples.

$$\frac{x^5}{x^2} = \frac{x \cdot x \cdot x \cdot x \cdot x}{x \cdot x} = x^3;$$

$$\frac{a^4}{a^3} = \frac{a \cdot a \cdot a \cdot a}{a \cdot a \cdot a} = a;$$

$$\frac{y^7}{y^3} = \frac{y \cdot y \cdot y \cdot y \cdot y \cdot y \cdot y}{y \cdot y \cdot y} = y^4;$$

$$\frac{x^4}{x^4} = \frac{x \cdot x \cdot x \cdot x}{x \cdot x \cdot x \cdot x} = 1;$$

$$\frac{y^3}{y^3} = \frac{y \cdot y \cdot y}{y \cdot y \cdot y} = 1.$$

Property 5.4

If b is any nonzero real number and n and m are positive integers, then

1. $\dfrac{b^n}{b^m} = b^{n-m}$ when $n > m$;

2. $\dfrac{b^n}{b^m} = 1$ when $n = m$.

Applying Property 5.4 to the above examples yields

$$\frac{x^5}{x^2} = x^{5-2} = x^3;$$

$$\frac{a^4}{a^3} = a^{4-3} = a^1; \quad \text{(usually written as } a\text{)}$$

$$\frac{y^7}{y^3} = y^{7-3} = y^4;$$

$$\frac{x^4}{x^4} = 1;$$

$$\frac{y^3}{y^3} = 1.$$

(The situation when $n < m$ will be discussed in a later section.)

Property 5.4 along with our knowledge of dividing integers provides the basis for dividing a monomial by another monomial. Consider the following examples.

$$\frac{16x^5}{2x^3} = 8x^{5-3} = 8x^2; \qquad\qquad \frac{-81a^{12}}{-9a^4} = 9a^{12-4} = 9a^8;$$

$$\frac{-35x^9}{5x^4} = -7x^{9-4} = -7x^5; \qquad \frac{45x^4}{9x^4} = 5 \quad \left(\frac{x^4}{x^4} = 1\right);$$

$$\frac{56y^6}{-7y^2} = -8y^{6-2} = -8y^4; \qquad \frac{54x^3y^7}{-6xy^5} = -9x^{3-1}y^{7-5} = -9x^2y^2.$$

Recall that $\dfrac{a}{c} + \dfrac{b}{c} = \dfrac{a+b}{c}$. This same property —except viewed as $\dfrac{a+b}{c} = \dfrac{a}{c} + \dfrac{b}{c}$— serves as the basis for dividing a polynomial by a monomial. Consider the following examples.

$$\frac{25x^3 + 10x^2}{5x} = \frac{25x^3}{5x} + \frac{10x^2}{5x} = 5x^2 + 2x;$$

$$\frac{-35x^8 - 28x^6}{7x^3} = \frac{-35x^8}{7x^3} - \frac{28x^6}{7x^3} = -5x^5 - 4x^3. \qquad \left(\frac{a-b}{c} = \frac{a}{c} - \frac{b}{c}\right).$$

To divide a polynomial by a monomial we simply divide each term of the polynomial by the monomial. Here are some further examples using this idea.

$$\frac{12x^3y^2 - 14x^2y^5}{-2xy} = \frac{12x^3y^2}{-2xy} - \frac{14x^2y^5}{-2xy} = -6x^2y + 7xy^4;$$

$$\frac{48ab^5 + 64a^2b}{-16ab} = \frac{48ab^5}{-16ab} + \frac{64a^2b}{-16ab} = -3b^4 - 4a;$$

$$\frac{33x^6 - 24x^5 - 18x^4}{3x} = \frac{33x^6}{3x} - \frac{24x^5}{3x} - \frac{18x^4}{3x}$$

$$= 11x^5 - 8x^4 - 6x^3.$$

As with many skills, once you feel comfortable with the process, you may want to mentally perform some of the steps. Your work could take on the following format.

$$\frac{24x^4y^5 - 56x^3y^9}{8x^2y^3} = 3x^2y^2 - 7xy^6;$$

$$\frac{13a^2b - 12ab^2}{-ab} = -13a + 12b.$$

Problem Set 5.4

For Problems 1–24, divide the monomials.

1. $\dfrac{x^{10}}{x^2}$

2. $\dfrac{x^{12}}{x^5}$

3. $\dfrac{4x^3}{2x}$

4. $\dfrac{8x^5}{4x^3}$

5. $\dfrac{-16n^6}{2n^2}$

6. $\dfrac{-54n^8}{6n^4}$

7. $\dfrac{72x^3}{-9x^3}$

8. $\dfrac{84x^5}{-7x^5}$

9. $\dfrac{65x^2y^3}{5xy}$

10. $\dfrac{70x^3y^4}{5x^2y}$

11. $\dfrac{-91a^4b^6}{-13a^3b^4}$

12. $\dfrac{-72a^5b^4}{-12ab^2}$

13. $\dfrac{18x^2y^6}{xy^2}$

14. $\dfrac{24x^3y^4}{x^2y^2}$

15. $\dfrac{32x^6y^2}{-x}$

16. $\dfrac{54x^5y^3}{-y^2}$

17. $\dfrac{-96x^5y^7}{12y^3}$ 18. $\dfrac{-84x^4y^9}{14x^4}$ 19. $\dfrac{-ab}{ab}$ 20. $\dfrac{6ab}{-ab}$

21. $\dfrac{56a^2b^3c^5}{4abc}$ 22. $\dfrac{60a^3b^2c}{15a^2c}$ 23. $\dfrac{-80xy^2z^6}{-5xyz^2}$ 24. $\dfrac{-90x^3y^2z^8}{-6xy^2z^4}$

For Problems 25–50, perform the following divisions of polynomials by monomials.

25. $\dfrac{8x^4 + 12x^5}{2x^2}$ 26. $\dfrac{12x^3 + 16x^6}{4x}$

27. $\dfrac{9x^6 - 24x^4}{3x^3}$ 28. $\dfrac{35x^8 - 45x^6}{5x^4}$

29. $\dfrac{-28n^5 + 36n^2}{4n^2}$ 30. $\dfrac{-42n^6 + 54n^4}{6n^4}$

31. $\dfrac{35x^6 - 56x^5 - 84x^3}{7x^2}$ 32. $\dfrac{27x^7 - 36x^5 - 45x^3}{3x}$

33. $\dfrac{-24n^8 + 48n^5 - 78n^3}{-6n^3}$ 34. $\dfrac{-56n^9 + 84n^6 - 91n^2}{-7n^2}$

35. $\dfrac{-60a^7 - 96a^3}{-12a}$ 36. $\dfrac{-65a^8 - 78a^4}{-13a^2}$

37. $\dfrac{27x^2y^4 - 45xy^4}{-9xy^3}$ 38. $\dfrac{-40x^4y^7 + 64x^5y^8}{-8x^3y^4}$

39. $\dfrac{48a^2b^2 + 60a^3b^4}{-6ab}$ 40. $\dfrac{45a^3b^4 - 63a^2b^6}{-9ab^2}$

41. $\dfrac{12a^2b^2c^2 - 52a^2b^3c^5}{-4a^2bc}$ 42. $\dfrac{48a^3b^2c + 72a^2b^4c^5}{-12ab^2c}$

43. $\dfrac{9x^2y^3 - 12x^3y^4}{-xy}$ 44. $\dfrac{-15x^3y + 27x^2y^4}{xy}$

45. $\dfrac{-42x^6 - 70x^4 + 98x^2}{14x^2}$ 46. $\dfrac{-48x^8 - 80x^6 + 96x^4}{16x^4}$

47. $\dfrac{15a^3b - 35a^2b - 65ab^2}{-5ab}$ 48. $\dfrac{-24a^4b^2 + 36a^3b - 48a^2b}{-6ab}$

49. $\dfrac{-xy + 5x^2y^3 - 7x^2y^6}{xy}$ 50. $\dfrac{-9x^2y^3 - xy + 14xy^4}{-xy}$

5.5 Dividing by Binomials

Perhaps the easiest way to explain the process of dividing a polynomial by a binomial is to work a few examples and describe the step-by-step procedure as we go along.

EXAMPLE 1   Divide $x^2 + 5x + 6$ by $x + 2$.

Solution   Step 1. Use the conventional long division format from arithmetic and arrange both the dividend and the divisor in descending powers of the variable.

$$x + 2\overline{)x^2 + 5x + 6}$$

Step 2. Find the first term of the quotient by dividing the first term of the dividend by the first term of the divisor.

$$\begin{array}{r} x \phantom{{}+5x+6} \\ x + 2\overline{)x^2 + 5x + 6} \end{array} \qquad \left(\dfrac{x^2}{x} = x\right)$$

Step 3. Multiply the entire divisor by the term of the quotient found in Step 2 and place this product so as to be subtracted from the dividend.

$$\begin{array}{r} x \phantom{{}+5x+6} \\ x + 2\overline{)x^2 + 5x + 6} \\ x^2 + 2x \end{array} \qquad (x(x + 2) = x^2 + 2x)$$

Step 4. Subtract.

Remember to add the opposite!

$$\begin{array}{r} x \phantom{{}+5x+6} \\ x + 2\overline{)x^2 + 5x + 6} \\ \longrightarrow x^2 + 2x \\ \hline 3x + 6 \end{array}$$

Step 5. Repeat the process beginning with Step 2; use the polynomial that resulted from the subtraction in Step 4 as a new dividend.

$$\begin{array}{r} x + 3 \\ x + 2\overline{)x^2 + 5x + 6} \\ x^2 + 2x \\ \hline 3x + 6 \\ 3x + 6 \end{array} \qquad \left(\dfrac{3x}{x} = 3\right)$$
$$(3(x + 2) = 3x + 6)$$

Thus, $(x^2 + 5x + 6) \div (x + 2) = x + 3$; this can be checked by multiplying $(x + 2)$ and $(x + 3)$.

$$(x + 2)(x + 3) = x^2 + 5x + 6.$$ ■

A division problem such as $(x^2 + 5x + 6) \div (x + 2)$ can also be written as $\dfrac{x^2 + 5x + 6}{x + 2}$. Using this format the final result for Example 1 could be expressed as $\dfrac{x^2 + 5x + 6}{x + 2} = x + 3$. (Technically, the restriction $x \neq -2$ should be made to avoid division by zero.)

In general, to check a division problem we can multiply the divisor times the quotient and add the remainder. This can be expressed as

Dividend = (Divisor)(Quotient) + Remainder.

Sometimes the remainder is expressed as a fractional part of the divisor. The relationship then becomes

$$\frac{\text{Dividend}}{\text{Divisor}} = \text{Quotient} + \frac{\text{Remainder}}{\text{Divisor}}.$$

EXAMPLE 2 Divide $2x^2 - 3x - 20$ by $x - 4$.

Solution

Step 1. $x - 4\overline{)2x^2 - 3x - 20}$

Step 2. $\begin{array}{r} 2x \\ x - 4\overline{)2x^2 - 3x - 20} \end{array}$ $\left(\dfrac{2x^2}{x} = 2x\right)$

Step 3. $\begin{array}{r} 2x \\ x - 4\overline{)2x^2 - 3x - 20} \\ 2x^2 - 8x \end{array}$ $(2x(x - 4) = 2x^2 - 8x)$

Step 4. $\begin{array}{r} 2x \\ x - 4\overline{)2x^2 - 3x - 20} \\ \underline{2x^2 - 8x} \\ 5x - 20 \end{array}$

Step 5. $\begin{array}{r} 2x \ + 5 \\ x - 4\overline{)2x^2 - 3x - 20} \\ \underline{2x^2 - 8x} \\ 5x - 20 \\ 5x - 20 \end{array}$ $\left(\dfrac{5x}{x} = 5\right)$

$(5(x - 4) = 5x - 20)$

Check: $(x - 4)(2x + 5) = 2x^2 - 3x - 20.$

Therefore, $\dfrac{2x^2 - 3x - 20}{x - 4} = 2x + 5.$ ■

Now let's continue to think in terms of the step-by-step division process but organize our work in the typical long division format.

EXAMPLE 3 Divide $12x^2 + x - 6$ by $3x - 2$.

Solution

$$\begin{array}{r} 4x + 3 \\ 3x - 2\overline{)12x^2 + \cdot x - 6} \\ \underline{12x^2 - 8x} \\ 9x - 6 \\ \underline{9x - 6} \end{array}$$

Check: $(3x - 2)(4x + 3) = 12x^2 + x - 6.$

Therefore, $\dfrac{12x^2 + x - 6}{3x - 2} = 4x + 3.$ ■

Each of the next three examples illustrates another point regarding the division process. Study them carefully; then you should be ready to work the exercises in the next problem set.

EXAMPLE 4 Perform the division $(7x^2 - 3x - 4) \div (x - 2)$.

Solution

$$
\begin{array}{r}
7x \; + 11 \\
x - 2 \overline{)\, 7x^2 - 3x - 4} \\
\underline{7x^2 - 14x} \\
11x - 4 \\
\underline{11x - 22} \\
18 \longleftarrow \text{a remainder of } 18
\end{array}
$$

Check: Just as in arithmetic, we check by *adding* the remainder to the product of the divisor and quotient.

$$(x - 2)(7x + 11) + 18 \overset{?}{=} 7x^2 - 3x - 4$$
$$7x^2 - 3x - 22 + 18 \overset{?}{=} 7x^2 - 3x - 4$$
$$7x^2 - 3x - 4 = 7x^2 - 3x - 4.$$

Therefore, $\dfrac{7x^2 - 3x - 4}{x - 2} = 7x + 11 + \dfrac{18}{x - 2}.$ ■

EXAMPLE 5 Perform the division $\dfrac{x^3 - 8}{x - 2}.$

Solution

$$
\begin{array}{r}
x^2 + 2x \; + 4 \\
x - 2 \overline{)\, x^3 + 0x^2 + 0x - 8} \\
\underline{x^3 - 2x^2} \\
2x^2 + 0x - 8 \\
\underline{2x^2 - 4x} \\
4x - 8 \\
\underline{4x - 8}
\end{array}
$$

$\longleftarrow$ Notice the inserting of x squared and x terms with zero coefficients.

Check:
$$(x - 2)(x^2 + 2x + 4) \overset{?}{=} x^3 - 8$$
$$x^3 + 2x^2 + 4x - 2x^2 - 4x - 8 \overset{?}{=} x^3 - 8$$
$$x^3 - 8 = x^3 - 8.$$

Therefore, $\dfrac{x^3 - 8}{x - 2} = x^2 + 2x + 4.$ ■

EXAMPLE 6 Perform the division $\dfrac{x^3 + 5x^2 - 3x - 4}{x^2 + 2x}.$

Solution

$$\begin{array}{r} x\ +3 \\ x^2 + 2x\overline{)x^3 + 5x^2 - 3x - 4} \\ \underline{x^3 + 2x^2} \\ 3x^2 - 3x - 4 \\ \underline{3x^2 + 6x} \\ -9x - 4 \end{array}$$ ←—— a remainder of $-9x - 4$

The division process is stopped when the degree of the remainder is less than the degree of the divisor.

Check: $(x^2 + 2x)(x + 3) + (-9x - 4) \stackrel{?}{=} x^3 + 5x^2 - 3x - 4$

$$x^3 + 3x^2 + 2x^2 + 6x - 9x - 4 \stackrel{?}{=} x^3 + 5x^2 - 3x - 4$$

$$x^3 + 5x^2 - 3x - 4 = x^3 + 5x^2 - 3x - 4.$$

Therefore, $\dfrac{x^3 + 5x^2 - 3x - 4}{x^2 + 2x} = x + 3 + \dfrac{-9x - 4}{x^2 + 2x}$. ■

Problem Set 5.5

Perform the following divisions.

1. $(x^2 + 16x + 48) \div (x + 4)$

2. $(x^2 + 15x + 54) \div (x + 6)$

3. $(x^2 - 5x - 14) \div (x - 7)$

4. $(x^2 + 8x - 65) \div (x - 5)$

5. $(x^2 + 11x + 28) \div (x + 3)$

6. $(x^2 + 11x + 15) \div (x + 2)$

7. $(x^2 - 4x - 39) \div (x - 8)$

8. $(x^2 - 9x - 30) \div (x - 12)$

9. $(5n^2 - n - 4) \div (n - 1)$

10. $(7n^2 - 61n - 90) \div (n - 10)$

11. $(8y^2 + 53y - 19) \div (y + 7)$

12. $(6y^2 + 47y - 72) \div (y + 9)$

13. $(20x^2 - 31x - 7) \div (5x + 1)$

14. $(27x^2 + 21x - 20) \div (3x + 4)$

15. $(6x^2 + 25x + 8) \div (2x + 7)$

16. $(12x^2 + 28x + 27) \div (6x + 5)$

17. $(2x^3 - x^2 - 2x - 8) \div (x - 2)$

18. $(3x^3 - 7x^2 - 26x + 24) \div (x - 4)$

19. $(5n^3 + 11n^2 - 15n - 9) \div (n + 3)$

20. $(6n^3 + 29n^2 - 6n - 5) \div (n + 5)$

21. $(n^3 - 40n + 24) \div (n - 6)$

22. $(n^3 - 67n - 24) \div (n + 8)$

23. $(x^3 - 27) \div (x - 3)$

24. $(x^3 + 8) \div (x + 2)$

25. $\dfrac{27x^3 - 64}{3x - 4}$

26. $\dfrac{8x^3 + 27}{2x + 3}$

27. $\dfrac{1 + 3n^2 - 2n}{n + 2}$

28. $\dfrac{x + 5 + 12x^2}{3x - 2}$

29. $\dfrac{9t^2 + 3t + 4}{-1 + 3t}$

30. $\dfrac{4n^2 + 6n - 1}{4 + 2n}$

31. $\dfrac{6n^3 - 5n^2 - 7n + 4}{2n - 1}$

32. $\dfrac{21n^3 + 23n^2 - 9n - 10}{3n + 2}$

33. $\dfrac{4x^3 + 23x^2 - 30x + 32}{x + 7}$

34. $\dfrac{5x^3 - 12x^2 + 13x - 14}{x - 1}$

35. $(x^3 + 2x^2 - 3x - 1) \div (x^2 - 2x)$

36. $(x^3 - 6x^2 - 2x + 1) \div (x^2 + 3x)$

37. $(2x^3 - 4x^2 + x - 5) \div (x^2 + 4x)$

38. $(2x^3 - x^2 - 3x + 5) \div (x^2 + x)$

39. $(x^4 - 16) \div (x + 2)$

40. $(x^4 - 81) \div (x - 3)$

5.6 Zero and Negative Integers as Exponents

Thus far in this text we have used only positive integers as exponents. The following definition and properties have served as a basis for our work with exponents.

Definition 5.1

If n is a positive integer and b is any real number, then,

$$b^n = \underbrace{bbb \cdots b}_{n \text{ factors of } b}.$$

Property 5.5

If m and n are positive integers and a and b are real numbers, except $b \neq 0$ whenever it appears in a denominator, then

1. $b^n \cdot b^m = b^{n+m}$;

2. $(b^n)^m = b^{mn}$;

3. $(ab)^n = a^n b^n$;

4. $\left(\dfrac{a}{b}\right)^n = \dfrac{a^n}{b^n}$;   (Part 4 has not been previously stated.)

5. $\dfrac{b^n}{b^m} = b^{n-m}$ when $n > m$;

   $\dfrac{b^n}{b^m} = 1$ when $n = m$.

■ Zero as an Exponent

Zero and the negative integers can also be used as exponents. First, let's consider the use of 0 as an exponent. We want to use 0 as an exponent in such a way that the basic properties of exponents will continue to hold. So consider the example $x^4 \cdot x^0$.

If part 1 of Property 5.5 is to hold, then

$$x^4 \cdot x^0 = x^{4+0} = x^4.$$

Note that x^0 *acts like* 1 since $x^4 \cdot x^0 = x^4$. This suggests the following definition.

Definition 5.2

If b is a nonzero real number, then

$$b^0 = 1.$$

According to Definition 5.2 the following statements are all true.

$$4^0 = 1;$$
$$(-628)^0 = 1;$$
$$\left(\frac{4}{7}\right)^0 = 1;$$
$$n^0 = 1; \qquad (n \neq 0)$$
$$(x^2 y^5)^0 = 1. \qquad (x \neq 0) \text{ and } y \neq 0)$$

■ Negative Integers as Exponents

A similar line of reasoning indicates how negative integers should be used as exponents. Consider the example $x^3 \cdot x^{-3}$.

If part 1 of Property 5.5 is to hold, then

$$x^3 \cdot x^{-3} = x^{3+(-3)} = x^0 = 1.$$

Thus, x^{-3} must be the reciprocal of x^3 since their product is 1; that is,

$$x^{-3} = \frac{1}{x^3}.$$

This kind of reasoning suggests the following definition.

Definition 5.3

If n is a positive integer and b is a nonzero real number, then

$$b^{-n} = \frac{1}{b^n}.$$

According to Definition 5.3 the following statements are all true.

$$x^{-6} = \frac{1}{x^6};$$

$$2^{-3} = \frac{1}{2^3} = \frac{1}{8};$$

$$10^{-2} = \frac{1}{10^2} = \frac{1}{100} \quad \text{or} \quad 0.01;$$

$$\frac{1}{x^{-4}} = \frac{1}{\frac{1}{x^4}} = x^4;$$

$$\left(\frac{2}{3}\right)^{-2} = \frac{1}{\left(\frac{2}{3}\right)^2} = \frac{1}{\frac{4}{9}} = \frac{9}{4}.$$

It can be verified (we will not do so in this text) that all of the parts of Property 5.5 hold for *all integers*. In fact, part 5 can be replaced with the following statement.

Replacement for Part 5 of Property 5.5

$$\frac{b^n}{b^m} = b^{n-m} \quad \text{for all integers } n \text{ and } m.$$

The following examples illustrate the use of this new property. In each example the original expression has been simplified and the final result expressed using only positive exponents.

$$\frac{x^2}{x^5} = x^{2-5} = x^{-3} = \frac{1}{x^3};$$

$$\frac{a^{-3}}{a^{-7}} = a^{-3-(-7)} = a^{-3+7} = a^4;$$

$$\frac{y^{-5}}{y^{-2}} = y^{-5-(-2)} = y^{-5+2} = y^{-3} = \frac{1}{y^3};$$

$$\frac{x^{-6}}{x^{-6}} = x^{-6-(-6)} = x^{-6+6} = x^0 = 1.$$

The properties of exponents provide a basis for simplifying certain types of numerical expressions, as the following examples illustrate.

$$2^{-4} \cdot 2^6 = 2^{-4+6} = 2^2 = 4;$$

$$10^5 \cdot 10^{-6} = 10^{5+(-6)} = 10^{-1} = \frac{1}{10} \quad \text{or} \quad .1;$$

$$\frac{10^2}{10^{-2}} = 10^{2-(-2)} = 10^{2+2} = 10^4 = 10000;$$

$$(2^{-3})^{-2} = 2^{-3(-2)} = 2^6 = 64.$$

Having the use of all integers as exponents also expands the type of work that we can do with algebraic expressions. In each of the following examples a given expression has been simplified and we expressed the final result using only positive exponents.

$$x^8 x^{-2} = x^{8+(-2)} = x^6;$$

$$a^{-4}a^{-3} = a^{-4+(-3)} = a^{-7} = \frac{1}{a^7};$$

$$(y^{-3})^4 = y^{-3(4)} = y^{-12} = \frac{1}{y^{12}};$$

$$(x^{-2}y^4)^{-3} = (x^{-2})^{-3}(y^4)^{-3} = x^6 y^{-12} = \frac{x^6}{y^{12}};$$

$$\left(\frac{x^{-1}}{y^2}\right)^{-2} = \frac{(x^{-1})^{-2}}{(y^2)^{-2}} = \frac{x^2}{y^{-4}} = x^2 y^4;$$

$$(4x^{-2})(3x^{-1}) = 12x^{-2+(-1)} = 12x^{-3} = \frac{12}{x^3};$$

$$\left(\frac{12x^{-6}}{6x^{-2}}\right)^{-2} = (2x^{-6-(-2)})^{-2} = (2x^{-4})^{-2}$$

$$= (2)^{-2}(x^{-4})^{-2}$$

$$= \left(\frac{1}{2^2}\right)(x^8) = \frac{x^8}{4}.$$

■ Scientific Notation

Many scientific applications of mathematics involve the use of very large and very small numbers. For example,

1. the speed of light is approximately 29,979,200,000 centimeters per second;

2. a light year —the distance light travels in one year— is approximately 5,865,696,000,000 miles;

3. a millimicron equals 0.000000001 of a meter.

Working with numbers of this type in standard form is quite cumbersome. It is much more convenient to represent very small and very large numbers in **scientific notation**, sometimes called scientific form. A number is expressed in scientific notation when it is written as a product of a number between 1 and 10 and an integral power of 10. Symbolically, a number written in scientific notation has the form

$$(N)(10)^k$$

where N is a number between 1 and 10, written in decimal form, and k is an integer. For example, 621 can be written as $(6.21)(10^2)$ and 0.0023 can be written as $(2.3)(10^{-3})$.

To switch from ordinary notation to scientific notation, you can use the following procedure.

Write the given number as the product of a number between 1 and 10 and an integral power of 10. The exponent of 10 is determined by counting the number of places that the decimal point was moved in going from the original number to the number between 1 and 10. This exponent is (a) negative if the original number is less than 1, (b) positive if the original number is greater than 10, and (c) zero if the original number itself is between 1 and 10.

Thus, we can write

$0.000179 = (1.79)(10^{-4});$ (According to part (a) of the rule.)

$8175 = (8.175)(10^3);$ (According to part (b).)

$3.14 = (3.14)(10^0).$ (According to part (c).)

The applications given earlier can be expressed in scientific notation as follows.

Speed of Light $29,979,200,000 = (2.99792)(10^{10})$ centimeters per second.

Light Year $5,865,696,000,000 = (5.865696)(10^{12})$ miles.

Metric Units   A millimicron is $0.000000001 = (1)(10^{-9})$ of a meter.

To switch from scientific notation to ordinary decimal notation you can use the following procedure.

Move the decimal point the number of places indicated by the exponent of 10. The decimal point is moved to the right if the exponent is positive and to the left if it is negative.

Thus, we can write

$(4.71)(10^4) = 47,100;$ Two zeros are needed for place value purposes.

$(1.78)(10^{-2}) = 0.0178.$ One zero is needed for place value purposes.

The use of scientific notation along with the properties of exponents can make some arithmetic problems much easier to evaluate. The following examples illustrate this point.

EXAMPLE 1   Evaluate $(4,000)(0.000012)$.

Solution

$$(4,000)(0.000012) = (4)(10^3)(1.2)(10^{-5})$$
$$= (4)(1.2)(10^3)(10^{-5})$$
$$= (4.8)(10^{-2})$$
$$= 0.048.$$

■

EXAMPLE 2 Evaluate $\dfrac{960,000}{0.032}$.

Solution

$$\frac{960,000}{0.032} = \frac{(9.6)(10^5)}{(3.2)(10^{-2})}$$
$$= (3)(10^7) \qquad \left(\frac{10^5}{10^{-2}} = 10^{5-(-2)} = 10^7\right)$$
$$= 30,000,000.$$

■

EXAMPLE 3 Evaluate $\dfrac{(6,000)(0.00008)}{(40,000)(0.006)}$.

Solution

$$\frac{(6,000)(0.00008)}{(40,000)(0.006)} = \frac{(6)(10^3)(8)(10^{-5})}{(4)(10^4)(6)(10^{-3})}$$
$$= \frac{(48)(10^{-2})}{(24)(10^1)}$$
$$= (2)(10^{-3}) \qquad \left(\frac{10^{-2}}{10^1} = 10^{-2-1} = 10^{-3}\right)$$
$$= 0.002.$$

■

Problem Set 5.6

For Problems 1–30, evaluate each numerical expression.

1. 3^{-2}

2. 2^{-5}

3. 4^{-3}

4. 5^{-2}

5. $\left(\dfrac{3}{2}\right)^{-1}$

6. $\left(\dfrac{3}{4}\right)^{-2}$

7. $\dfrac{1}{2^{-4}}$

8. $\dfrac{1}{3^{-1}}$

9. $\left(-\dfrac{4}{3}\right)^{0}$

10. $\left(-\dfrac{1}{2}\right)^{-3}$

11. $\left(-\dfrac{2}{3}\right)^{-3}$

12. $(-16)^0$

13. $(-2)^{-2}$

14. $(-3)^{-2}$

15. $-(3^{-2})$

16. $-(2^{-2})$

17. $\dfrac{1}{\left(\dfrac{3}{4}\right)^{-3}}$

18. $\dfrac{1}{\left(\dfrac{3}{2}\right)^{-4}}$

19. $2^6 \cdot 2^{-9}$

20. $3^5 \cdot 3^{-2}$

21. $3^6 \cdot 3^{-3}$ **22.** $2^{-7} \cdot 2^2$ **23.** $\dfrac{10^2}{10^{-1}}$ **24.** $\dfrac{10^1}{10^{-3}}$

25. $\dfrac{10^{-1}}{10^2}$ **26.** $\dfrac{10^{-2}}{10^{-2}}$ **27.** $(2^{-1} \cdot 3^{-2})^{-1}$ **28.** $(3^{-1} \cdot 4^{-2})^{-1}$

29. $\left(\dfrac{4^{-1}}{3}\right)^{-2}$ **30.** $\left(\dfrac{3}{2^{-1}}\right)^{-3}$

For Problems 31–84, simplify each algebraic expression and express answers using positive exponents only.

31. $x^6 x^{-1}$ **32.** $x^{-2} x^7$ **33.** $n^{-4} n^2$

34. $n^{-8} n^3$ **35.** $a^{-2} a^{-3}$ **36.** $a^{-4} a^{-6}$

37. $(2x^3)(4x^{-2})$ **38.** $(5x^{-4})(6x^7)$ **39.** $(3x^{-6})(9x^2)$

40. $(8x^{-8})(4x^2)$ **41.** $(5y^{-1})(-3y^{-2})$ **42.** $(-7y^{-3})(9y^{-4})$

43. $(8x^{-4})(12x^4)$ **44.** $(-3x^{-2})(-6x^2)$ **45.** $\dfrac{x^7}{x^{-3}}$

46. $\dfrac{x^2}{x^{-4}}$ **47.** $\dfrac{n^{-1}}{n^3}$ **48.** $\dfrac{n^{-2}}{n^5}$ **49.** $\dfrac{4n^{-1}}{2n^{-3}}$

50. $\dfrac{12n^{-2}}{3n^{-5}}$ **51.** $\dfrac{-24x^{-6}}{8x^{-2}}$ **52.** $\dfrac{56x^{-5}}{-7x^{-1}}$ **53.** $\dfrac{-52y^{-2}}{-13y^{-2}}$

54. $\dfrac{-91y^{-3}}{-7y^{-3}}$ **55.** $(x^{-3})^{-2}$ **56.** $(x^{-1})^{-5}$ **57.** $(x^2)^{-2}$

58. $(x^3)^{-1}$ **59.** $(x^3 y^4)^{-1}$ **60.** $(x^4 y^{-2})^{-2}$ **61.** $(x^{-2} y^{-1})^3$

62. $(x^{-3} y^{-4})^2$ **63.** $(2n^{-2})^3$ **64.** $(3n^{-1})^4$ **65.** $(4n^3)^{-2}$

66. $(2n^2)^{-3}$ **67.** $(3a^{-2})^4$ **68.** $(5a^{-1})^2$ **69.** $(5x^{-1})^{-2}$

70. $(4x^{-2})^{-2}$ **71.** $(2x^{-2} y^{-1})^{-1}$ **72.** $(3x^2 y^{-3})^{-2}$ **73.** $\left(\dfrac{x^2}{y}\right)^{-1}$

74. $\left(\dfrac{y^2}{x^3}\right)^{-2}$ **75.** $\left(\dfrac{a^{-1}}{b^2}\right)^{-4}$ **76.** $\left(\dfrac{a^3}{b^{-2}}\right)^{-3}$ **77.** $\left(\dfrac{x^{-1}}{y^{-3}}\right)^{-2}$

78. $\left(\dfrac{x^{-3}}{y^{-4}}\right)^{-1}$ **79.** $\left(\dfrac{x^2}{x^3}\right)^{-1}$ **80.** $\left(\dfrac{x^4}{x}\right)^{-2}$ **81.** $\left(\dfrac{2x^{-1}}{x^{-2}}\right)^{-3}$

82. $\left(\dfrac{3x^{-2}}{x^{-5}}\right)^{-1}$ **83.** $\left(\dfrac{18x^{-1}}{9x}\right)^{-2}$ **84.** $\left(\dfrac{35x^2}{7x^{-1}}\right)^{-1}$

For Problems 85–94, write each number in scientific notation. For example $786 = (7.86)(10^2)$.

85. 321   **86.** 74   **87.** 8000

88. 500   **89.** 0.00246   **90.** 0.017

91. 0.0000179   **92.** 0.00000049   **93.** 87,000,000

94. 623,000,000,000

For Problems 95–106, write each number in standard decimal form. For example, $(1.4)(10^3) = 1400$.

95. $(8)(10^3)$ **96.** $(6)(10^2)$ **97.** $(5.21)(10^4)$ **98.** $(7.2)(10^3)$

99. $(1.14)(10^7)$ **100.** $(5.64)(10^8)$ **101.** $(7)(10^{-2})$ **102.** $(8.14)(10^{-1})$

103. $(9.87)(10^{-4})$ **104.** $(4.37)(10^{-5})$ **105.** $(8.64)(10^{-6})$ **106.** $(3.14)(10^{-7})$

For Problems 107–118, use scientific notation and the properties of exponents to help evaluate each numerical expression.

107. $(0.007)(120)$

108. $(0.0004)(13)$

109. $(5,000,000)(0.00009)$

110. $(800,000)(0.0000006)$

111. $\dfrac{6000}{0.0015}$

112. $\dfrac{480}{0.012}$

113. $\dfrac{0.00086}{4300}$

114. $\dfrac{0.0057}{30000}$

115. $\dfrac{0.00039}{0.0013}$

116. $\dfrac{0.0000082}{0.00041}$

117. $\dfrac{(0.0008)(0.07)}{(20,000)(0.0004)}$

118. $\dfrac{(0.006)(600)}{(0.00004)(30)}$

Chapter 5 Summary

(5.1)   Terms containing variables with only whole numbers as exponents are called **monomials**. A **polynomial** is a monomial or a finite sum (or difference) of monomials. Polynomials of one term, two terms, and three terms are called **monomials, binomials**, and **trinomials**, respectively.

Addition and subtraction of polynomials are based on the distributive property and the idea of combining similar terms.

(5.2) and (5.3)   The following properties of exponents serve as a basis for multiplying polynomials.

    1. $b^n \cdot b^m = b^{n+m}$.

    2. $(b^n)^m = b^{mn}$.

    3. $(ab)^n = a^n b^n$.

(5.4)   The following properties of exponents serve as a basis for dividing monomials.

    1. $\dfrac{b^n}{b^m} = b^{n-m}$ when $n > m$.

    2. $\dfrac{b^n}{b^m} = 1$ when $n = m$.

Dividing a polynomial by a monomial is based on the property $\dfrac{a+b}{c} = \dfrac{a}{c} + \dfrac{b}{c}$.

(5.5)   To review dividing a polynomial by a binomial, return to Section 5.5 and study the examples carefully.

(5.6)   The following two definitions are used to extend our work with exponents to include zero and the negative integers.

Definition 5.2   If b is a nonzero real number, then $b^0 = 1$.

Definition 5.3   If n is a positive integer and b is a nonzero real number, then

$$b^{-n} = \frac{1}{b^n}.$$

The following properties of exponents are true for all integers.

1. $b^n \cdot b^m = b^{n+m}$.
2. $(b^n)^m = b^{mn}$.
3. $(ab)^n = a^n b^n$.
4. $\left(\dfrac{a}{b}\right)^n = \dfrac{a^n}{b^n}$.   ($b \neq 0$ whenever it appears in a denominator.)
5. $\dfrac{b^n}{b^m} = b^{n-m}$.

To represent a number in scientific notation, express it as the product of a number between 1 and 10 and a power of 10.

Chapter 5 Review Problem Set

Test 3 — Thursday after Spring Break

Perform the following additions and subtractions.

1. $(5x^2 - 6x + 4) + (3x^2 - 7x - 2)$
2. $(7y^2 + 9y - 3) - (4y^2 - 2y + 6)$
3. $(2x^2 + 3x - 4) + (4x^2 - 3x - 6) - (3x^2 - 2x - 1)$
4. $(-3x^2 - 2x + 4) - (x^2 - 5x - 6) - (4x^2 + 3x - 8)$

Remove parentheses and combine similar terms.

5. $5(2x - 1) + 7(x + 3) - 2(3x + 4)$
6. $3(2x^2 - 4x - 5) - 5(3x^2 - 4x + 1)$
7. $6(y^2 - 7y - 3) - 4(y^2 + 3y - 9)$
8. $3(a - 1) - 2(3a - 4) - 5(2a + 7)$
9. $-(a + 4) + 5(-a - 2) - 7(3a - 1)$
10. $-2(3n - 1) - 4(2n + 6) + 5(3n + 4)$
11. $3(n^2 - 2n - 4) - 4(2n^2 - n - 3)$
12. $-5(-n^2 + n - 1) + 3(4n^2 - 3n - 7)$

Find the indicated products.

13. $(5x^2)(7x^4)$
14. $(-6x^3)(9x^5)$
15. $(-4xy^2)(-6x^2 y^3)$
16. $(2a^3 b^4)(-3ab^5)$
17. $(2a^2 b^3)^3$
18. $(-3xy^2)^2$
19. $5x(7x + 3)$
20. $(-3x^2)(8x - 1)$

Find the indicated products. Be sure to simplify answers.

21. $(x + 9)(x + 8)$

22. $(3x + 7)(x + 1)$

23. $(x - 5)(x + 2)$

24. $(y - 4)(y - 9)$

25. $(2x - 1)(7x + 3)$

26. $(4a - 7)(5a + 8)$

27. $(3a - 5)^2$

28. $(x + 6)(2x^2 + 5x - 4)$

29. $(5n - 1)(6n + 5)$

30. $(3n + 4)(4n - 1)$

31. $(2n + 1)(2n - 1)$

32. $(4n - 5)(4n + 5)$

33. $(2a + 7)^2$

34. $(3a + 5)^2$

35. $(x - 2)(x^2 - x + 6)$

36. $(2x - 1)(x^2 + 4x + 7)$

37. $(a + 5)^3$

38. $(a - 6)^3$

39. $(x^2 - x - 1)(x^2 + 2x + 5)$

40. $(n^2 + 2n + 4)(n^2 - 7n - 1)$

Perform the following divisions.

41. $\dfrac{36x^4 y^5}{-3xy^2}$

42. $\dfrac{-56a^5 b^7}{-8a^2 b^3}$

43. $\dfrac{-18x^4 y^3 - 54x^6 y^2}{6x^2 y^2}$

44. $\dfrac{-30a^5 b^{10} + 39a^4 b^8}{-3ab}$

45. $\dfrac{56x^4 - 40x^3 - 32x^2}{4x^2}$

46. $(x^2 + 9x - 1) \div (x + 5)$

47. $(21x^2 - 4x - 12) \div (3x + 2)$

48. $(2x^3 - 3x^2 + 2x - 4) \div (x - 2)$

Evaluate each of the following.

49. $3^2 + 2^2$

50. $(3 + 2)^2$

51. 2^{-4}

52. $(-5)^0 = 1$

53. $-5^0 = -1$

54. $\dfrac{1}{3^{-2}}$

55. $\left(\dfrac{3}{4}\right)^{-2}$

56. $\dfrac{1}{\left(\dfrac{1}{4}\right)^{-1}}$

57. $\dfrac{1}{(-2)^{-3}}$

58. $2^{-1} + 3^{-2}$

59. $3^0 + 2^{-2}$

60. $(2 + 3)^{-2}$

Simplify each of the following and express answers using positive exponents only.

61. $x^5 x^{-8}$

62. $(3x^5)(4x^{-2})$

63. $\dfrac{x^{-4}}{x^{-6}}$

64. $\dfrac{x^{-6}}{x^{-4}}$

65. $\dfrac{24a^5}{3a^{-1}}$

66. $\dfrac{48n^{-2}}{12n^{-1}}$

all of these ?

67. $(x^{-2} y)^{-1}$

68. $(a^2 b^{-3})^{-2}$

69. $(2x)^{-1}$

70. $(3n^2)^{-2}$

71. $(2n^{-1})^{-3}$

72. $(4ab^{-1})(-3a^{-1} b^2)$

Write each of the following as a standard decimal fraction.

73. $(6.1)(10^2)$

6.10

74. $(5.6)(10^4)$

5.6000

75. $(8)(10^{-2})$

.08

76. $(9.2)(10^{-4})$

0.00092

Write each of the following in scientific notation.

77. 9000 $(9)(10^3)$ **78.** 47 $(4.7)(10)$ **79.** 0.047 $(4.7)(10^{-2})$ **80.** 0.00021 $(2.1)(10^{-4})$

Use scientific notation and the properties of exponents to help evaluate each of the following.

81. (0.00004)(12,000) 0.48 **82.** (0.0021)(2000) 4.2

83. $\dfrac{0.0056}{0.0000028}$ 2000 **84.** $\dfrac{0.00078}{39,000}$ 0.00000002

81. $(4)(10^{-5})\ (1.2)(10^4)$

82. $(2.1)(10^{-3})\ (2)(10^3)$

83. $(5.6)(10^{-3}) \div (2.8)(10^{-6})$

84. $(7.8)(10^{-4}) \div 3.9(10^4)$

6 Factoring and Solving Equations

The distributive property has played an important role in combining similar terms and multiplying polynomials. In this chapter we shall see yet another use of the distributive property as we learn how to **factor polynomials**. Being able to factor polynomials will allow us to solve other kinds of equations, which will in turn permit us to solve more kinds of word problems.

6.1 Factoring by Using the Distributive Property

In Chapter 1 we found the *greatest common factor* of two or more whole numbers by inspection or by using the prime factored form of the numbers. For example, by inspection we see that the greatest common factor of 8 and 12 is 4. This means that 4 is the largest whole number that is a factor of both 8 and 12. If it was difficult to determine the highest common factor by inspection then we used the prime factorization technique as follows.

$$42 = 2 \cdot 3 \cdot 7; \qquad 70 = 2 \cdot 5 \cdot 7.$$

From this we see that $2 \cdot 7 = 14$ is the greatest common factor of 42 and 70.

It is meaningful to extend the concept of greatest common factor to monomials. Consider the following example.

EXAMPLE 1   Find the greatest common factor of $8x^2$ and $12x^3$.

Solution
$$8x^2 = 2 \cdot 2 \cdot 2 \cdot x \cdot x;$$
$$12x^3 = 2 \cdot 2 \cdot 3 \cdot x \cdot x \cdot x.$$

Therefore, the greatest common factor is $2 \cdot 2 \cdot x \cdot x = 4x^2$.   ■

By the greatest common factor of two or more monomials we mean the monomial with the largest numerical coefficient and highest power of the variables that is a factor of the given monomials.

EXAMPLE 2   Find the greatest common factor of $16x^2y$, $24x^3y^2$, and $32xy$.

Solution
$$16x^2y = 2 \cdot 2 \cdot 2 \cdot 2 \cdot x \cdot x \cdot y;$$
$$24x^3y^2 = 2 \cdot 2 \cdot 2 \cdot 3 \cdot x \cdot x \cdot x \cdot y \cdot y;$$
$$32xy = 2 \cdot 2 \cdot 2 \cdot 2 \cdot 2 \cdot x \cdot y.$$

Therefore, the greatest common factor is $2 \cdot 2 \cdot 2 \cdot x \cdot y = 8xy$.   ■

You have used the distributive property to multiply a polynomial by a monomial. For example,

$$3x(x + 2) = 3x^2 + 6x.$$

Suppose that you start with $3x^2 + 6x$ and want to express it in factored form. This can be done by using the distributive property in the form $ab + ac = a(b + c)$.

$$3x^2 + 6x = 3x(x) + 3x(2) \qquad 3x \text{ is the greatest common factor of } 3x^2 \text{ and } 6x.$$
$$= 3x(x + 2). \qquad \text{Use the distributive property.}$$

The next four examples further illustrate this process of **factoring out the greatest common monomial factor**.

EXAMPLE 3 Factor $12x^3 - 8x^2$.

Solution
$$12x^3 - 8x^2 = 4x^2(3x) - 4x^2(2)$$
$$= 4x^2(3x - 2). \qquad ab - ac = a(b - c).$$ ■

EXAMPLE 4 Factor $12x^2y + 18xy^2$.

Solution
$$12x^2y + 18xy^2 = 6xy(2x) + 6xy(3y)$$
$$= 6xy(2x + 3y).$$ ■

EXAMPLE 5 Factor $24x^3 + 30x^4 - 42x^5$.

Solution
$$24x^3 + 30x^4 - 42x^5 = 6x^3(4) + 6x^3(5x) - 6x^3(7x^2)$$
$$= 6x^3(4 + 5x - 7x^2).$$ ■

EXAMPLE 6 Factor $9x^2 + 9x$.

Solution
$$9x^2 + 9x = 9x(x) + 9x(1)$$
$$= 9x(x + 1).$$ ■

We want to emphasize the point made prior to Example 3. It is important to realize that we are factoring out the *greatest* common monomial factor. An expression such as $9x^2 + 9x$ in Example 6 could be factored as $9(x^2 + x)$, $3(3x^2 + 3x)$, $3x(3x + 3)$, or even as $\frac{1}{2}(18x^2 + 18x)$, but it is the form $9x(x + 1)$ that we want to obtain. This can be accomplished by factoring out the greatest common monomial factor; we sometimes refer to this process as **factoring completely**. A polynomial with integral coefficients is in completely factored form if:

1. it is expressed as a product of polynomials with integral coefficients;
2. no polynomial, other than a monomial, within the factored form can be further factored into polynomials with integral coefficients.

Thus, $9(x^2 + x)$, $3(3x^2 + 3x)$, and $3x(3x + 3)$ are not completely factored because they violate part 2. The form $\frac{1}{2}(18x^2 + 18x)$ violates both parts 1 and 2.

Sometimes there may be a **common binomial factor** rather than a common monomial factor. For example, each of the two terms of $x(y + 2) + z(y + 2)$ has a binomial factor of $(y + 2)$. Thus, we can factor $(y + 2)$ from each term and our result is as follows.

$$x(y + 2) + z(y + 2) = (y + 2)(x + z).$$

Consider a few more examples involving a common binomial factor.

$$a(b + c) - d(b + c) = (b + c)(a - d);$$
$$x(x + 2) + 3(x + 2) = (x + 2)(x + 3);$$
$$x(x + 5) - 4(x + 5) = (x + 5)(x - 4).$$

It may be that the original polynomial exhibits no apparent common monomial or binomial factor, as is the case with

$$ab + 3a + bc + 3c.$$

However, by factoring a from the first two terms and c from the last two terms, it can be expressed as

$$ab + 3a + bc + 3c = a(b + 3) + c(b + 3).$$

Now a common binomial factor of $(b + 3)$ is obvious and we can proceed as before.

$$a(b + 3) + c(b + 3) = (b + 3)(a + c)$$

This factoring process is called **factoring by grouping**. Let's consider two more examples of factoring by grouping.

$$x^2 - x + 5x - 5 = x(x - 1) + 5(x - 1)$$
Factor x from first two terms and 5 from last two terms.

$$= (x - 1)(x + 5).$$
Factor common binomial factor of $(x - 1)$ from both terms.

$$6x^2 - 4x - 3x + 2 = 2x(3x - 2) - 1(3x - 2)$$
Factor $2x$ from first two terms and -1 from last two terms.

$$= (3x - 2)(2x - 1).$$
Factor common binomial factor of $(3x - 2)$ from both terms.

■ Back to Solving Equations

Suppose we are told that the product of two numbers is zero. What do we know about the numbers? Do you agree we can conclude that at least one of the numbers must be zero? The following property formalizes this idea.

Property 6.1

For all real numbers a and b,

$$ab = 0 \quad \text{if and only if } a = 0 \text{ or } b = 0.$$

Property 6.1 provides us with another technique for solving equations.

EXAMPLE 7 Solve $x^2 + 6x = 0$.

Solution

$x^2 + 6x = 0$

$x(x + 6) = 0$

$x = 0 \quad \text{or} \quad x + 6 = 0$ Property 6.1.

$x = 0 \quad \text{or} \qquad x = -6.$

The solution set is $\{-6, 0\}$. (Be sure to check both of these in the original equation.) ■

EXAMPLE 8 Solve $x^2 = 12x$.

Solution

$x^2 = 12x$

$x^2 - 12x = 0$ Added $-12x$ to both sides.

$x(x - 12) = 0$

$x = 0 \quad \text{or} \quad x - 12 = 0$ Property 6.1.

$x = 0 \quad \text{or} \qquad x = 12.$

The solution set is $\{0, 12\}$. ■

EXAMPLE 9 Solve $4x^2 - 3x = 0$.

Solution

$4x^2 - 3x = 0$

$x(4x - 3) = 0$

$x = 0 \quad \text{or} \quad 4x - 3 = 0$

$x = 0 \quad \text{or} \qquad 4x = 3$

$x = 0 \quad \text{or} \qquad x = \dfrac{3}{4}.$

The solution set is $\left\{ 0, \dfrac{3}{4} \right\}$. ■

Remark   Notice in Example 8 we *did not* divide both sides of the original equation by x. This would cause us to lose the solution of 0.

As always, each time that we extend our equation solving capabilities, we also gain more techniques for solving problems. Let's consider a geometric problem that can be solved using the ideas from this section.

PROBLEM 1   The area of a square is numerically equal to twice its perimeter. Find the length of a side of the square.

Solution   Sketch a square and let s represent the length of each side. Then the area is represented by s^2 and the perimeter by $4s$. Thus,

$$s^2 = 2(4s)$$
$$s^2 = 8s$$
$$s^2 - 8s = 0$$
$$s(s - 8) = 0$$
$$s = 0 \quad \text{or} \quad s - 8 = 0$$
$$s = 0 \quad \text{or} \qquad s = 8.$$

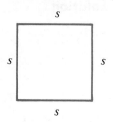

Since 0 is not a reasonable answer to the problem, the solution is 8. (Be sure to check this solution in the original statement of the problem!)   ■

Problem Set 6.1

For Problems 1–10, find the greatest common factor of the given expressions.

1. $24y$ and $30xy$

2. $32x$ and $40xy$

3. $60x^2y$ and $84xy^2$

4. $72x^3$ and $63x^2$

5. $42ab^3$ and $70a^2b^2$

6. $48a^2b^2$ and $96ab^4$

7. $6x^3$, $8x$, and $24x^2$

8. $72xy$, $36x^2y$, and $84xy^2$

9. $16a^2b^2$, $40a^2b^3$, and $56a^3b^4$

10. $70a^3b^3$, $42a^2b^4$, and $49ab^5$

For Problems 11–46, factor completely each polynomial.

11. $8x + 12y$

12. $18x + 24y$

13. $14xy - 21y$

14. $24x - 40xy$

15. $18x^2 + 45x$

16. $12x + 28x^3$

17. $12xy^2 - 30x^2y$

18. $28x^2y^2 - 49x^2y$

19. $36a^2b - 60a^3b^4$

20. $65ab^3 - 45a^2b^2$

21. $16xy^3 + 25x^2y^2$

22. $12x^2y^2 + 29x^2y$

23. $64ab - 72cd$ 24. $45xy - 72zw$ 25. $9a^2b^4 - 27a^2b$

26. $7a^3b^5 - 42a^2b^6$ 27. $52x^4y^2 + 60x^6y$ 28. $70x^5y^3 - 42x^8y^2$

29. $40x^2y^2 + 8x^2y$ 30. $84x^2y^3 + 12xy^3$ 31. $12x + 15xy + 21x^2$

32. $30x^2y + 40xy + 55y$ 33. $2x^3 - 3x^2 + 4x$ 34. $x^4 + x^3 + x^2$

35. $44y^5 - 24y^3 - 20y^2$ 36. $14a - 18a^3 - 26a^5$

37. $14a^2b^3 + 35ab^2 - 49a^3b$ 38. $24a^3b^2 + 36a^2b^4 - 60a^4b^3$

39. $x(y + 1) + z(y + 1)$ 40. $a(c + d) + 2(c + d)$

41. $a(b - 4) - c(b - 4)$ 42. $x(y - 6) - 3(y - 6)$

43. $x(x + 3) + 6(x + 3)$ 44. $x(x - 7) + 9(x - 7)$

45. $2x(x + 1) - 3(x + 1)$ 46. $4x(x + 8) - 5(x + 8)$

For Problems 47–60, use the process of *factoring by grouping* to factor each of the following.

47. $5x + 5y + bx + by$ 48. $7x + 7y + zx + zy$ 49. $bx - by - cx + cy$

50. $2x - 2y - ax + ay$ 51. $ac + bc + a + b$ 52. $x + y + ax + ay$

53. $x^2 + 5x + 12x + 60$ 54. $x^2 + 3x + 7x + 21$ 55. $x^2 - 2x - 8x + 16$

56. $x^2 - 4x - 9x + 36$ 57. $2x^2 + x - 10x - 5$ 58. $3x^2 + 2x - 18x - 12$

59. $6n^2 - 3n - 8n + 4$ 60. $20n^2 + 8n - 15n - 6$

For Problems 61–80, solve each equation.

61. $x^2 - 8x = 0$ 62. $x^2 - 12x = 0$ 63. $x^2 + x = 0$

64. $x^2 + 7x = 0$ 65. $n^2 = 5n$ 66. $n^2 = -2n$

67. $2y^2 - 3y = 0$ 68. $4y^2 - 7y = 0$ 69. $7x^2 = -3x$

70. $5x^2 = -2x$ 71. $3n^2 + 15n = 0$ 72. $6n^2 - 24n = 0$

73. $4x^2 = 6x$ 74. $12x^2 = 8x$ 75. $7x - x^2 = 0$

76. $9x - x^2 = 0$ 77. $13x = x^2$ 78. $15x = -x^2$

79. $5x = -2x^2$ 80. $7x = -5x^2$

For Problems 81–87, set up an equation and solve each problem.

81. The square of a number equals nine times that number. Find the number.

82. Four times the square of a number equals twenty times that number. What is the number?

83. The area of a square is numerically equal to five times its perimeter. Find the length of a side of the square.

84. The area of a square is 14 times as large as the area of a triangle. One side of the triangle is 7 inches long and the altitude to that side is the same length as a side of the square. Find the length of a side of the square. Also find the areas of both figures and be sure that your answer checks.

85. Suppose that the area of a circle is numerically equal to the perimeter of a square and that the length of a radius of the circle is equal to the length of a side of the square. Find the length of a side of the square. Express your answer in terms of π.

86. One side of a parallelogram, an altitude to that side, and one side of a rectangle all have the same measure. If an adjacent side of the rectangle is 20 centimeters long and the area of the rectangle is twice the area of the parallelogram, find the area of both figures.

87. The area of a rectangle is twice the area of a square. If the rectangle is 6 inches long and the width of the rectangle is the same as the length of a side of the square, find the dimensions of both the rectangle and the square.

■ Miscellaneous Problems

88. The total surface area of a right circular cylinder is given by the formula $A = 2\pi r^2 + 2\pi rh$, where r represents the radius of a base and h represents the height of the cylinder. For computational purposes it may be more convenient to change the form of the right side of the formula by factoring it.

$$A = 2\pi r^2 + 2\pi rh$$
$$= 2\pi r(r + h).$$

Use $A = 2\pi r(r + h)$ to find the total surface area of each of the following cylinders. Also, use $\dfrac{22}{7}$ as an approximation for π.

(a) $r = 7$ centimeters and $h = 12$ centimeters;

(b) $r = 14$ meters and $h = 20$ meters;

(c) $r = 3$ feet and $h = 4$ feet;

(d) $r = 5$ yards and $h = 9$ yards.

89. The formula $A = P + Prt$ yields the total amount of money accumulated (A) when P dollars is invested at r percent simple interest for t years. For computational purposes it may be convenient to change the right side of the formula by factoring.

$$A = P + Prt$$
$$= P(1 + rt).$$

Use $A = P(1 + rt)$ to find the total amount of money accumulated for each of the following investments.

(a) $100 at 8% for 2 years;

(b) $200 at 9% for 3 years;

(c) $500 at 10% for 5 years;

(d) $1000 at 10% for 10 years.

For Problems 90–93, solve each equation for the indicated variable.

90. $ax + bx = c$ for x

91. $b^2x^2 - cx = 0$ for x

92. $5ay^2 = by$ for y

93. $y + ay - by - c = 0$ for y

6.2 Factoring the Difference of Two Squares

In Section 5.3 we noted some special multiplication patterns. One of these patterns was

$$(a - b)(a + b) = a^2 - b^2.$$

This same pattern can be viewed as the factoring pattern that follows.

Difference of Two Squares

$$a^2 - b^2 = (a - b)(a + b).$$

Applying the pattern is a fairly simple process, as these next examples illustrate. The steps inside the box are often performed mentally.

$$
\begin{aligned}
x^2 - 36 &= \boxed{(x)^2 - (6)^2} = (x - 6)(x + 6); \\
4x^2 - 25 &= \boxed{(2x)^2 - (5)^2} = (2x - 5)(2x + 5); \\
9x^2 - 16y^2 &= \boxed{(3x)^2 - (4y)^2} = (3x - 4y)(3x + 4y); \\
64 - y^2 &= \boxed{(8)^2 - (y)^2} = (8 - y)(8 + y).
\end{aligned}
$$

Since multiplication is commutative, the order of writing the factors is not important. For example, $(x - 6)(x + 6)$ can also be written as $(x + 6)(x - 6)$.

You must be careful not to assume an analogous factoring pattern for the *sum* of two squares; it does not exist. For example, $x^2 + 4 \neq (x + 2)(x + 2)$ since $(x + 2)(x + 2) = x^2 + 4x + 4$. We say that the **sum of two squares is not factorable using integers**. The phrase "using integers" is necessary because $x^2 + 4$ could be written as $\dfrac{1}{2}(2x^2 + 8)$, but such *factoring* is of no help. Furthermore, we do not consider $(1)(x^2 + 4)$ as factoring $x^2 + 4$.

It is possible that both the technique of *factoring out a common monomial factor* and the pattern *difference of two squares* can be applied to the same polynomial. In general, it is best to first look for a common monomial factor.

EXAMPLE 1   Factor $2x^2 - 50$.

Solution

$$
\begin{aligned}
2x^2 - 50 &= 2(x^2 - 25) && \text{Common factor of 2.} \\
&= 2(x - 5)(x + 5). && \text{Difference of squares.}
\end{aligned}
$$

■

By expressing $2x^2 - 50$ as $2(x - 5)(x + 5)$ we say that it has been **factored completely**. That means, the factors 2, $x - 5$, and $x + 5$ cannot be factored any further using integers.

EXAMPLE 2 Factor completely $18y^3 - 8y$.

Solution

$$18y^3 - 8y = 2y(9y^2 - 4) \qquad \text{Common factor of } 2y.$$
$$= 2y(3y - 2)(3y + 2). \qquad \text{Difference of squares.} \qquad ■$$

Sometimes it is possible to apply the difference-of-squares pattern more than once.

EXAMPLE 3 Factor completely $x^4 - 16$.

Solution

$$x^4 - 16 = (x^2 + 4)(x^2 - 4)$$
$$= (x^2 + 4)(x + 2)(x - 2). \qquad ■$$

The following examples should help you to summarize our factoring ideas thus far.

$5x^2 + 20 = 5(x^2 + 4)$;

$25 - y^2 = (5 - y)(5 + y)$;

$3 - 3x^2 = 3(1 - x^2) = 3(1 + x)(1 - x)$;

$36x^2 - 49y^2 = (6x - 7y)(6x + 7y)$;

$a^2 + 9$ is not factorable using integers;

$9x + 17y$ is not factorable using integers.

■ Solving Equations

Each time that we pick up a new factoring technique we also develop more power for solving equations. Let's consider how the difference-of-squares factoring pattern can be used to help solve certain kinds of equations.

EXAMPLE 4 Solve $x^2 = 25$.

Solution

$$x^2 = 25$$
$$x^2 - 25 = 0$$
$$(x + 5)(x - 5) = 0$$
$$x + 5 = 0 \quad \text{or} \quad x - 5 = 0 \qquad \text{Remember: } ab = 0 \text{ if and only if}$$
$$x = -5 \quad \text{or} \qquad x = 5. \qquad a = 0 \text{ or } b = 0.$$

The solution set is $\{-5, 5\}$. Check these answers! $\qquad ■$

EXAMPLE 5 Solve $9x^2 = 25$.

Solution

$$9x^2 = 25$$
$$9x^2 - 25 = 0$$
$$(3x + 5)(3x - 5) = 0$$
$$3x + 5 = 0 \quad \text{or} \quad 3x - 5 = 0$$
$$3x = -5 \quad \text{or} \quad 3x = 5$$
$$x = -\frac{5}{3} \quad \text{or} \quad x = \frac{5}{3}.$$

The solution set is $\left\{-\dfrac{5}{3}, \dfrac{5}{3}\right\}$. ■

EXAMPLE 6 Solve $5y^2 = 20$.

Solution

$$5y^2 = 20$$
$$\frac{5y^2}{5} = \frac{20}{5} \qquad \text{Divide both sides by 5.}$$
$$y^2 = 4$$
$$y^2 - 4 = 0$$
$$(y + 2)(y - 2) = 0$$
$$y + 2 = 0 \quad \text{or} \quad y - 2 = 0$$
$$y = -2 \quad \text{or} \quad y = 2.$$

The solution set is $\{-2, 2\}$. Check them! ■

EXAMPLE 7 Solve $x^3 - 9x = 0$.

Solution

$$x^3 - 9x = 0$$
$$x(x^2 - 9) = 0$$
$$x(x - 3)(x + 3) = 0$$
$$x = 0 \quad \text{or} \quad x - 3 = 0 \quad \text{or} \quad x + 3 = 0$$
$$x = 0 \quad \text{or} \quad x = 3 \quad \text{or} \quad x = -3.$$

The solution set is $\{-3, 0, 3\}$. ■

The more we know about solving equations, the better off we are when we solve word problems.

PROBLEM 1 The combined area of two squares is 20 square centimeters. Each side of one square is twice as long as a side of the other square. Find the lengths of the sides of each square.

Solution Let's sketch two squares and label the sides of the smaller square s. Then the sides of the larger square are $2s$. Since the sum of the areas of the two squares is 20 square centimeters, we can set up and solve the following equation.

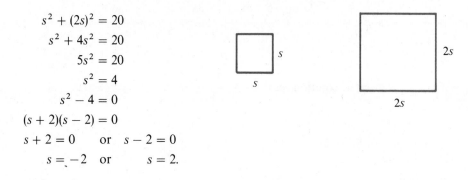

$$s^2 + (2s)^2 = 20$$
$$s^2 + 4s^2 = 20$$
$$5s^2 = 20$$
$$s^2 = 4$$
$$s^2 - 4 = 0$$
$$(s + 2)(s - 2) = 0$$
$$s + 2 = 0 \quad \text{or} \quad s - 2 = 0$$
$$s = -2 \quad \text{or} \quad s = 2.$$

Since s represents the length of a side of a square, the solution -2 must be disregarded. Thus, one square has sides of length 2 centimeters and the other square has sides of length $2(2) = 4$ centimeters.   ■

Problem Set 6.2

For Problems 1–12, use the difference-of-squares pattern to factor each polynomial.

1. $x^2 - 1$ **2.** $x^2 - 25$ **3.** $x^2 - 100$ **4.** $x^2 - 121$

5. $x^2 - 4y^2$ **6.** $x^2 - 36y^2$ **7.** $9x^2 - y^2$ **8.** $49y^2 - 64x^2$

9. $36a^2 - 25b^2$ **10.** $4a^2 - 81b^2$ **11.** $1 - 4n^2$ **12.** $4 - 9n^2$

For Problems 13–40, factor completely each polynomial. Indicate any that are not factorable using integers. Don't forget to first look for a common monomial factor.

13. $5x^2 - 20$ **14.** $7x^2 - 7$ **15.** $8x^2 + 32$ **16.** $12x^2 + 60$

17. $2x^2 - 18y^2$ **18.** $8x^2 - 32y^2$ **19.** $x^3 - 25x$ **20.** $2x^3 - 2x$

21. $x^2 + 9y^2$ **22.** $18x - 42y$ **23.** $45x^2 - 36xy$ **24.** $16x^2 + 25y^2$

25. $36 - 4x^2$ **26.** $75 - 3x^2$ **27.** $4a^4 + 16a^2$ **28.** $9a^4 + 81a^2$

29. $x^4 - 81$ **30.** $16 - x^4$ **31.** $x^4 + x^2$ **32.** $x^5 + 2x^3$

33. $3x^3 + 48x$ **34.** $6x^3 + 24x$ **35.** $5x - 20x^3$ **36.** $4x - 36x^3$

37. $4x^2 - 64$ **38.** $9x^2 - 9$ **39.** $75x^3y - 12xy^3$ **40.** $32x^3y - 18xy^3$

For Problems 41–64, solve each equation.

41. $x^2 = 9$ 42. $x^2 = 1$ 43. $4 = n^2$ 44. $144 = n^2$

45. $9x^2 = 16$ 46. $4x^2 = 9$ 47. $n^2 - 121 = 0$ 48. $n^2 - 81 = 0$

49. $25x^2 = 4$ 50. $49x^2 = 36$ 51. $3x^2 = 75$ 52. $7x^2 = 28$

53. $3x^3 - 48x = 0$ 54. $x^3 - x = 0$ 55. $n^3 = 16n$ 56. $2n^3 = 8n$

57. $5 - 45x^2 = 0$ 58. $3 - 12x^2 = 0$ 59. $4x^3 - 400x = 0$ 60. $2x^3 - 98x = 0$

61. $64x^2 = 81$ 62. $81x^2 = 25$ 63. $36x^3 = 9x$ 64. $64x^3 = 4x$

For Problems 65–76, set up an equation and solve the problem.

65. Forty-nine less than the square of a number equals zero. Find the number.

66. The cube of a number equals nine times the number. Find the number.

67. Five times the cube of a number equals eighty times the number. Find the number.

68. Ten times the square of a number equals 40. Find the number.

69. The sum of the areas of two squares is 234 square inches. Each side of the larger square is 5 times the length of a side of the smaller square. Find the length of a side of each square.

70. The difference of the areas of two squares is 75 square feet. Each side of the larger square is twice the length of a side of the smaller square. Find the length of a side of each square.

71. The length of a rectangle is $2\dfrac{1}{2}$ times its width and the area of the rectangle is 160 square centimeters. Find the length and width of the rectangle.

72. The width of a rectangle is three-fourths of its length and the area of the rectangle is 108 square meters. Find the length and width of the rectangle.

73. The sum of the areas of two circles is 80π square meters. Find the length of a radius of each circle if one of them is twice as long as the other.

74. The area of a triangle is 98 square feet. If one side of the triangle and the altitude to that side are of equal length, find the length.

75. The total surface area of a right circular cylinder is 100π square centimeters. If a radius of the base and the altitude of the cylinder are of the same length, find the length of a radius.

76. The total surface area of a right circular cone is 128π square feet. If the slant height of the cone is equal in length to a radius of the base, find the length of a radius.

■ Miscellaneous Problems

The following patterns can be used to factor the sum and difference of two cubes.

$$a^3 + b^3 = (a + b)(a^2 - ab + b^2).$$
$$a^3 - b^3 = (a - b)(a^2 + ab + b^2).$$

Consider the following examples.

$$x^3 + 8 = (x)^3 + (2)^3 = (x + 2)(x^2 - 2x + 4).$$
$$x^3 - 1 = (x)^3 - (1)^3 = (x - 1)(x^2 + x + 1).$$

Use the sum and difference-of-cubes patterns to factor each of the following.

77. $x^3 + 1$ **78.** $x^3 - 8$ **79.** $n^3 - 27$ **80.** $n^3 + 64$

81. $8x^3 + 27y^3$ **82.** $27a^3 - 64b^3$ **83.** $1 - 8x^3$ **84.** $1 + 27a^3$

85. $x^3 + 8y^3$ **86.** $8x^3 - y^3$ **87.** $a^3b^3 - 1$ **88.** $27x^3 - 8y^3$

89. $8 + n^3$ **90.** $125x^3 + 8y^3$ **91.** $27n^3 - 125$ **92.** $64 + x^3$

6.3   Factoring Trinomials of the Form $x^2 + bx + c$

Expressing a trinomial as the product of two binomials is one of the most common types of factoring used in algebra. In this section we want to consider trinomials where the coefficient of the squared term is 1; that is, trinomials of the form $x^2 + bx + c$.

Again, to develop a factoring technique we first look at some multiplication ideas. Consider the product $(x + r)(x + s)$ using the distributive property to show how each term of the resulting trinomial is formed.

$$(x + r)(x + s) = x(x) + \underbrace{x(s) + r(x)} + r(s).$$
$$\qquad\qquad\quad x^2 \;+\; (s + r)x \;+\; rs$$

Notice that the coefficient of the middle term is the *sum* of r and s and the last term is the *product* of r and s. These two relationships are used in the following examples.

EXAMPLE 1   Factor $x^2 + 7x + 12$.

Solution   We need to complete the following with two numbers whose sum is 7 and whose product is 12.

$$x^2 + 7x + 12 = (x + \underline{\quad})(x + \underline{\quad}).$$

This can be done by setting up a small table as follows.

Product	Sum
$1(12) = 12$	$1 + 12 = 13$
$2(6) = 12$	$2 + 6 \ = 8$
$3(4) = 12$	$3 + 4 \ = 7$

The bottom line contains the numbers that we need. Thus,

$$x^2 + 7x + 12 = (x + 3)(x + 4).$$

■

EXAMPLE 2 Factor $x^2 - 11x + 24$.

Solution We need two numbers whose product is 24 and whose sum is -11.

Product	Sum
$(-1)(-24) = 24$	$-1 + (-24) = -25$
$(-2)(-12) = 24$	$-2 + (-12) = -14$
$(-3)(-8) = 24$	$-3 + (-8) = -11$
$(-4)(-6) = 24$	$-4 + (-6) = -10$

The third line contains the numbers that we want.

$$x^2 - 11x + 24 = (x - 3)(x - 8).$$

■

EXAMPLE 3 Factor $x^2 + 3x - 10$.

Solution We need two numbers whose product is -10 and whose sum is 3.

Product	Sum
$1(-10) = -10$	$1 + (-10) = -9$
$-1(10) = -10$	$-1 + 10 = 9$
$2(-5) = -10$	$2 + (-5) = -3$
$-2(5) = -10$	$-2 + 5 = 3$

The bottom line is the key line. Thus,

$$x^2 + 3x - 10 = (x + 5)(x - 2).$$

■

EXAMPLE 4 Factor $x^2 - 2x - 8$.

Solution We need two numbers whose product is -8 and whose sum is -2.

Product	Sum
$1(-8) = -8$	$1 + (-8) = -7$
$-1(8) = -8$	$-1 + 8 = 7$
$2(-4) = -8$	$2 + (-4) = -2$
$-2(4) = -8$	$-2 + 4 = 2$

The third line has the desired information.

$$x^2 - 2x - 8 = (x - 4)(x + 2).$$

■

The tables in the four previous examples have been used to illustrate one way of organizing your thoughts for such problems. We have shown complete tables. That is, for Example 4 the bottom line was included even though the desired numbers were obtained in the third line. If you use such tables, then keep in mind that as soon as the desired numbers are obtained, the table need not be completed any further. Furthermore, many times you may be able to find the numbers without using a table. The key ideas are the product and sum relationships.

EXAMPLE 5 Factor $x^2 - 13x + 12$.

Solution

Product	Sum
$(-1)(-12) = 12$	$(-1) + (-12) = -13$

We need not complete the table.

$$x^2 - 13x + 12 = (x - 1)(x - 12).$$ ■

In the next example, we refer to the concept of absolute value. Recall that the absolute value refers to the number, disregarding the sign. For example,

$$|4| = 4 \quad \text{and} \quad |-4| = 4.$$

EXAMPLE 6 Factor $x^2 - x - 56$.

Solution Notice that the coefficient of the middle term is -1. Therefore, we are looking for two numbers whose product is -56; since their sum is -1, the absolute value of the negative number must be one larger than the absolute value of the positive number. The numbers are -8 and 7 and we have

$$x^2 - x - 56 = (x - 8)(x + 7).$$ ■

EXAMPLE 7 Factor $x^2 + 10x + 12$.

Solution

Product	Sum
$1(12) = 12$	$1 + 12 = 13$
$2(6) = 12$	$2 + 6 = 8$
$3(4) = 12$	$3 + 4 = 7$

Since the table is complete and no two factors of 12 produce a sum of 10, we conclude that

$$x^2 + 10x + 12$$

is not factorable using integers.   ■

In a problem such as Example 7 we need to be sure that all possibilities have been tried before concluding that the trinomial is not factorable.

■ Back to Solving Equations

The property "$ab = 0$ if and only if $a = 0$ or $b = 0$" continues to play an important role as we solve equations involving the factoring ideas of this section. Consider the following examples.

EXAMPLE 8 Solve $x^2 + 8x + 15 = 0$.

Solution

$$x^2 + 8x + 15 = 0$$
$$(x + 3)(x + 5) = 0 \qquad \text{Factor the left side.}$$
$$x + 3 = 0 \quad \text{or} \quad x + 5 = 0 \qquad \text{Use: } ab = 0 \text{ if and only if}$$
$$\hspace{10cm} a = 0 \text{ or } b = 0.$$
$$x = -3 \quad \text{or} \quad x = -5.$$

The solution set is $\{-5, -3\}$. ■

EXAMPLE 9 Solve $x^2 + 5x - 6 = 0$.

Solution

$$x^2 + 5x - 6 = 0$$
$$(x + 6)(x - 1) = 0$$
$$x + 6 = 0 \quad \text{or} \quad x - 1 = 0$$
$$x = -6 \quad \text{or} \quad x = 1.$$

The solution set is $\{-6, 1\}$. ■

EXAMPLE 10 Solve $y^2 - 4y = 45$.

Solution

$$y^2 - 4y = 45$$
$$y^2 - 4y - 45 = 0$$
$$(y - 9)(y + 5) = 0$$
$$y - 9 = 0 \quad \text{or} \quad y + 5 = 0$$
$$y = 9 \quad \text{or} \quad y = -5.$$

The solution set is $\{-5, 9\}$. ■

Don't forget that we can always check to be absolutely sure of our solutions. Let's check the solutions for Example 10.

If $y = 9$, then $y^2 - 4y = 45$ becomes

$$9^2 - 4(9) \overset{?}{=} 45$$
$$81 - 36 \overset{?}{=} 45$$
$$45 = 45.$$

If $y = -5$, then $y^2 - 4y = 45$ becomes

$$(-5)^2 - 4(-5) \overset{?}{=} 45$$
$$25 + 20 \overset{?}{=} 45$$
$$45 = 45.$$

The more we know about factoring and solving equations the more word problems we can solve.

PROBLEM 1 Find two consecutive integers whose product is 72.

Solution Let n represent one integer. Then $n + 1$ represents the next integer.

$$n(n + 1) = 72 \qquad \text{The product of the two integers is 72.}$$
$$n^2 + n = 72$$
$$n^2 + n - 72 = 0$$
$$(n + 9)(n - 8) = 0$$
$$n + 9 = 0 \quad \text{or} \quad n - 8 = 0$$
$$n = -9 \quad \text{or} \qquad n = 8.$$
If $n = -9$ then $n + 1 = -9 + 1 = -8$.
If $n = 8$ then $n + 1 = 8 + 1 = 9$.

Thus, the consecutive integers are -9 and -8 or 8 and 9.   ■

PROBLEM 2 A rectangular plot is 6 meters longer than it is wide. The area of the plot is 16 square meters. Find the length and width of the plot.

Solution Let w represent the width of the plot. Then $w + 6$ represents the length.

$w + 6$

Using the area formula ($A = lw$), we obtain

$$w(w + 6) = 16$$
$$w^2 + 6w = 16$$
$$w^2 + 6w - 16 = 0$$
$$(w + 8)(w - 2) = 0$$
$$w + 8 = 0 \quad \text{or} \quad w - 2 = 0$$
$$w = -8 \quad \text{or} \quad w = 2.$$

The solution of -8 is not possible for the width of a rectangle, so the plot is 2 meters wide and its length ($w + 6$) is 8 meters.   ■

The Pythagorean Theorem, an important theorem pertaining to right triangles, can also serve as a guideline for solving certain types of problems. The Pythagorean Theorem states that **in any right triangle, the square of the longest side** (*called the* hypotenuse) **is equal to the sum of the squares of the other two sides** (*called* legs). Let's use this relationship to help solve a problem.

PROBLEM 3 The lengths of the three sides of a right triangle are represented by consecutive whole numbers. Find the lengths of the three sides.

Solution Let s represent the length of the shortest leg. Then $s + 1$ represents the length of the other leg and $s + 2$ represents the length of the hypotenuse.
Using the Pythagorean Theorem as a guideline, we obtain the following equation.

$$\overbrace{s^2 + (s + 1)^2}^{\text{sum of squares of two legs}} = \overbrace{(s + 2)^2}^{\text{square of hypotenuse}}.$$

Solving this equation yields

$$s^2 + s^2 + 2s + 1 = s^2 + 4s + 4$$
$$2s^2 + 2s + 1 = s^2 + 4s + 4$$
$$s^2 + 2s + 1 = 4s + 4$$

$$s^2 - 2s + 1 = 4$$
$$s^2 - 2s - 3 = 0$$
$$(s - 3)(s + 1) = 0$$
$$s - 3 = 0 \quad \text{or} \quad s + 1 = 0$$
$$s = 3 \quad \text{or} \quad s = -1.$$

The solution of -1 is not possible for the length of a side, so the shortest side (s) is of length 3. The other two sides $(s + 1$ and $s + 2)$ have lengths of 4 and 5. ■

Problem Set 6.3

For Problems 1–30, factor completely each trinomial. Indicate any that are not factorable using integers.

1. $x^2 + 10x + 24$ **2.** $x^2 + 9x + 14$ **3.** $x^2 + 13x + 40$

4. $x^2 + 11x + 24$ **5.** $x^2 - 11x + 18$ **6.** $x^2 - 5x + 4$

7. $n^2 - 11n + 28$ **8.** $n^2 - 7n + 10$ **9.** $n^2 + 6n - 27$

10. $n^2 + 3n - 18$ **11.** $n^2 - 6n - 40$ **12.** $n^2 - 4n - 45$

13. $t^2 + 12t + 24$ **14.** $t^2 + 20t + 96$ **15.** $x^2 - 18x + 72$

16. $x^2 - 14x + 32$ **17.** $x^2 + 5x - 66$ **18.** $x^2 + 11x - 42$

19. $y^2 - y - 72$ **20.** $y^2 - y - 30$ **21.** $x^2 + 21x + 80$

22. $x^2 + 21x + 90$ **23.** $x^2 + 6x - 72$ **24.** $x^2 - 8x - 36$

25. $x^2 - 10x - 48$ **26.** $x^2 - 12x - 64$ **27.** $x^2 + 3xy - 10y^2$

28. $x^2 - 4xy - 12y^2$ **29.** $a^2 - 4ab - 32b^2$ **30.** $a^2 + 3ab - 54b^2$

For Problems 31–50, solve each equation.

31. $x^2 + 10x + 21 = 0$ **32.** $x^2 + 9x + 20 = 0$ **33.** $x^2 - 9x + 18 = 0$

34. $x^2 - 9x + 8 = 0$ **35.** $x^2 - 3x - 10 = 0$ **36.** $x^2 - x - 12 = 0$

37. $n^2 + 5n - 36 = 0$ **38.** $n^2 + 3n - 18 = 0$ **39.** $n^2 - 6n - 40 = 0$

40. $n^2 - 8n - 48 = 0$ **41.** $t^2 + t - 56 = 0$ **42.** $t^2 + t - 72 = 0$

43. $x^2 - 16x + 28 = 0$ **44.** $x^2 - 18x + 45 = 0$ **45.** $x^2 + 11x = 12$

46. $x^2 + 8x = 20$ **47.** $x(x - 10) = -16$ **48.** $x(x - 12) = -35$

49. $-x^2 - 2x + 24 = 0$ **50.** $-x^2 + 6x + 16 = 0$

For Problems 51–68, set up an equation and solve each problem.

51. Find two consecutive integers whose product is 56.

52. Find two consecutive odd whole numbers whose product is 63.

53. Find two consecutive even whole numbers whose product is 168.

54. One number is 2 larger than another number. The sum of their squares is 100. Find the numbers.

55. Find four consecutive integers such that the product of the two larger integers is 22 less than twice the product of the two smaller integers.

56. Find three consecutive integers such that the product of the two smaller integers is two more than ten times the largest integer.

57. One number is 3 smaller than another number. The square of the larger number is 9 larger than ten times the smaller number. Find the numbers.

58. The area of the floor of a rectangular room is 84 square feet. The length of the room is 5 feet more than its width. Find the length and width of the room.

59. The width of a rectangle is 3 inches less than its length. The area is numerically 6 less than twice the perimeter. Find the length and width of the rectangle.

60. The sum of the areas of a square and a rectangle is 64 square centimeters. The length of the rectangle is 4 centimeters more than a side of the square and the width of the rectangle is 2 centimeters more than a side of the square. Find the dimensions of the square and the rectangle.

61. The perimeter of a rectangle is 30 centimeters and the area is 54 square centimeters. Find the length and width of the rectangle. (*Hint*: let w represent the width; then $15 - w$ represents the length.)

62. The perimeter of a rectangle is 44 inches and its area is 120 square inches. Find the length and width of the rectangle.

63. An apple orchard contains 84 trees. The number of trees per row is 5 more than the number of rows. Find the number of rows.

64. A room contains 54 chairs. The number of rows is three less than the number of chairs per row. Find the number of rows.

65. One leg of a right triangle is 7 feet shorter than the other leg. The hypotenuse is 2 feet longer than the longer leg. Find the lengths of all three sides of the right triangle.

66. One leg of a right triangle is 7 meters longer than the other leg. The hypotenuse is 1 meter longer than the longer leg. Find the lengths of all three sides of the right triangle.

67. The length of one leg of a right triangle is 2 inches less than the length of the other leg. If the length of the hypotenuse is 10 inches, find the length of each leg.

68. The length of one leg of a right triangle is 3 centimeters more than the length of the other leg. The length of the hypotenuse is 15 centimeters. Find the lengths of the two legs.

■ Miscellaneous Problems

For Problems 69–72, factor each trinomial assuming that all variables appearing as exponents represent positive integers.

69. $x^{2a} + 10x^a + 24$

70. $x^{2a} + 13x^a + 40$

71. $x^{2a} - 2x^a - 8$

72. $x^{2a} + 6x^a - 27$

73. Suppose that we want to factor $n^2 + 26n + 168$ so that we can solve the equation $n^2 + 26n + 168 = 0$. We need to find two positive integers whose product is 168 and whose sum is 26. Since the constant term, 168, is rather large, let's look at it in prime factored form.

$$168 = 2 \cdot 2 \cdot 2 \cdot 3 \cdot 7$$

Now we can mentally form two numbers by using all of these factors in different combinations. Using two 2's and the 3 in one number and the other 2 and the 7 in another number produces $2 \cdot 2 \cdot 3 = 12$ and $2 \cdot 7 = 14$. Therefore, we can solve the given equation as follows.

$$n^2 + 26n + 168 = 0$$
$$(n + 12)(n + 14) = 0$$
$$n + 12 = 0 \quad \text{or} \quad n + 14 = 0$$
$$n = -12 \quad \text{or} \quad n = -14$$

The solution set is $\{-14, -12\}$.
Solve each of the following equations.

(a) $n^2 + 30n + 216 = 0$ **(b)** $n^2 + 35n + 294 = 0$ **(c)** $n^2 - 40n + 384 = 0$

(d) $n^2 - 40n + 375 = 0$ **(e)** $n^2 + 6n - 432 = 0$ **(f)** $n^2 - 16n - 512 = 0$

6.4 Factoring Trinomials of the Form $ax^2 + bx + c$

Now let's consider factoring trinomials where the coefficient of the squared term is not 1. First, let's illustrate an informal trial and error technique that works quite well for certain types of trinomials. This technique simply relies on our knowledge of multiplication of binomials.

EXAMPLE 1   Factor $2x^2 + 7x + 3$.

Solution   By looking at the first term, $2x^2$, and the positive signs of the other two terms, we know that the binomials are of the form

$$(2x + \underline{\quad})(x + \underline{\quad}).$$

Since the factors of the constant term, 3, are 1 and 3, we have only two possibilities to try.

$$(2x + 3)(x + 1) \quad \text{or} \quad (2x + 1)(x + 3)$$

By checking the middle term of both of these products we find that the second one yields the correct middle term of $7x$. Therefore,

$$2x^2 + 7x + 3 = (2x + 1)(x + 3).$$ ■

EXAMPLE 2   Factor $6x^2 - 17x + 5$.

Solution   First, note that $6x^2$ can be written as $2x \cdot 3x$ or $6x \cdot x$. Secondly, since the middle term of the trinomial is negative and the last term is positive, we know that the binomials are of the form

$$(2x - \underline{\hspace{1em}})(3x - \underline{\hspace{1em}}) \quad \text{or} \quad (6x - \underline{\hspace{1em}})(x - \underline{\hspace{1em}}).$$

Since the factors of the constant term, 5, are 1 and 5, the following possibilities exist.

$$(2x - 5)(3x - 1); \quad (2x - 1)(3x - 5);$$
$$(6x - 5)(x - 1); \quad (6x - 1)(x - 5).$$

By checking the middle term for each of these products we find that the product $(2x - 5)(3x - 1)$ produces the desired term of $-17x$. Therefore,

$$6x^2 - 17x + 5 = (2x - 5)(3x - 1). \qquad ■$$

EXAMPLE 3   Factor $4x^2 - 4x - 15$.

Solution   First, note that $4x^2$ can be written as $4x \cdot x$ or $2x \cdot 2x$. Secondly, the last term, -15, can be written as $(1)(-15)$, $(-1)(15)$, $(3)(-5)$, or $(-3)(5)$. Thus, we can generate the possibilities for the binomial factors as follows.

Using 1 and -15	Using -1 and 15
$(4x - 15)(x + 1)$	$(4x - 1)(x + 15)$
$(4x + 1)(x - 15)$	$(4x + 15)(x - 1)$
$(2x + 1)(2x - 15)$	$(2x - 1)(2x + 15)$

Using 3 and -5	Using -3 and 5
$(4x + 3)(x - 5)$	$(4x - 3)(x + 5)$
$(4x - 5)(x + 3)$	$(4x + 5)(x - 3)$
$\checkmark\ (2x - 5)(2x + 3)$	$(2x + 5)(2x - 3)$

By checking the middle term of each of these products we find that the product indicated with a check mark produces the desired middle term of $-4x$. Therefore,

$$4x^2 - 4x - 15 = (2x - 5)(2x + 3). \qquad ■$$

Let's pause for a moment and look back over Examples 1, 2, and 3. Obviously, Example 3 created the most difficulty because of the many possibilities that had to be considered. We have suggested one possible format for considering the possibilities but as you practice such problems you may develop a format of your own that works better for you. Regardless of the format that you use, the key idea is to organize your work so that all possibilities are being considered. Let's look at another example.

EXAMPLE 4   Factor $4x^2 + 6x + 9$.

Solution   First, note that $4x^2$ can be written as $4x \cdot x$ or $2x \cdot 2x$. Secondly, since the middle term is positive and the last term is positive, we know that the binomials are of the form

$$(4x + \underline{\hspace{1cm}})(x + \underline{\hspace{1cm}}) \quad \text{or} \quad (2x + \underline{\hspace{1cm}})(2x + \underline{\hspace{1cm}}).$$

Since 9 can be written as $9 \cdot 1$ or $3 \cdot 3$, we have only the five following possibilities to try.

$$(4x + 9)(x + 1); \quad (4x + 1)(x + 9);$$
$$(4x + 3)(x + 3); \quad (2x + 1)(2x + 9);$$
$$(2x + 3)(2x + 3).$$

Trying all of these possibilities, we find that none of them yields a middle term of $6x$. Therefore, $4x^2 + 6x + 9$ is *not factorable* using integers.   ■

Remark   Example 4 illustrates the importance of organizing your work so that *all* possibilities are tried before you conclude that a particular trinomial is not factorable.

■ Another Approach

There is another more systematic technique that you may wish to use with some trinomials. It is an extension of the method used in the previous section. Recall that at the beginning of Section 6.3 we looked at the following product.

$$(x + r)(x + s) = x(x) + x(s) + r(x) + r(s)$$
$$= x^2 + (s + r)x + rs$$

$$\overset{\nearrow}{\text{sum of } r \text{ and } s} \quad \overset{\nwarrow}{\text{product of } r \text{ and } s}$$

Now let's look at the following product.

$$(px + r)(qx + s) = px(qx) + px(s) + r(qx) + r(s)$$
$$= (pq)x^2 + (ps + rq)x + rs.$$

Notice that the product of the coefficient of the x^2 term, (pq), and the constant term, (rs), is $pqrs$. Likewise, the product of the two coefficients of x, (ps and rq), is also $pqrs$. Therefore, the two coefficients of x must have a sum of $ps + rq$ and a product of $pqrs$. This may seen a little confusing, but the next few examples illustrate how easy it is to apply.

EXAMPLE 5 Factor $3x^2 + 14x + 8$.

Solution

$$3x^2 + 14x + 8. \qquad \text{sum of } 14$$

product of $3 \cdot 8 = 24$

We need to find two integers whose sum is 14 and whose product is 24. Obviously, 2 and 12 satisfy these conditions. Therefore, we will express the middle term of the trinomial, $14x$, as $2x + 12x$ and proceed as follows.

$$\begin{aligned}
3x^2 + 14x + 8 &= 3x^2 + 2x + 12x + 8 \\
&= x(3x + 2) + 4(3x + 2) \\
&= (3x + 2)(x + 4).
\end{aligned}$$

■

EXAMPLE 6 Factor $16x^2 - 26x + 3$.

Solution

$$16x^2 - 26x + 3. \qquad \text{sum of } -26$$

product of $16(3) = 48$

We need two integers whose sum is -26 and whose product is 48. The integers -2 and -24 satisfy these conditions and allow us to express the middle term, $-26x$, as $-2x - 24x$. Then, we can factor as follows.

$$\begin{aligned}
16x^2 - 26x + 3 &= 16x^2 - 2x - 24x + 3 \\
&= 2x(8x - 1) - 3(8x - 1) \\
&= (8x - 1)(2x - 3).
\end{aligned}$$

■

EXAMPLE 7 Factor $6x^2 - 5x - 6$.

Solution

$$6x^2 - 5x - 6. \qquad \text{sum of } -5$$

product of $6(-6) = -36$

We need two integers whose product is -36 and whose sum is -5. Furthermore, since the sum is negative, the absolute value of the negative number must be greater than the absolute value of the positive number. A little searching will determine that the numbers are -9 and 4. Thus, we can express the middle term of $-5x$ as $-9x + 4x$ and proceed as follows.

$$\begin{aligned}
6x^2 - 5x - 6 &= 6x^2 - 9x + 4x - 6 \\
&= 3x(2x - 3) + 2(2x - 3) \\
&= (2x - 3)(3x + 2).
\end{aligned}$$

■

Now that we have shown you two possible techniques for factoring trinomials of the form $ax^2 + bx + c$, the ball is in your court. Practice may not make you perfect at factoring, but it will surely help. We are not promoting one technique over the other; that is an individual choice. Many people find the trial and error technique presented first very useful if the number of possibilities for the factors is fairly small. However, as the list of possibilities grows, the second technique does have the advantage of being systematic. So perhaps having both techniques at your fingertips is your best bet.

■ Now We Can Solve More Equations

Being able to factor certain trinomials of the form $ax^2 + bx + c$ provides us with greater equation solving capabilities. Consider the following examples.

EXAMPLE 8   Solve $3x^2 + 17x + 10 = 0$.

Solution

$$3x^2 + 17x + 10 = 0$$
$$(x + 5)(3x + 2) = 0$$

Factoring $3x^2 + 17x + 10$ as $(x + 5)(3x + 2)$ may require some extra work on scratch paper.

$$x + 5 = 0 \quad \text{or} \quad 3x + 2 = 0 \,, \quad ab = 0 \text{ if and only if } a = 0 \text{ or } b = 0.$$
$$x = -5 \quad \text{or} \quad 3x = -2$$
$$x = -5 \quad \text{or} \quad x = -\frac{2}{3}.$$

The solution set is $\left\{ -5, -\frac{2}{3} \right\}$. Check them!   ■

EXAMPLE 9   Solve $24x^2 + 2x - 15 = 0$.

Solution

$$24x^2 + 2x - 15 = 0$$
$$(4x - 3)(6x + 5) = 0$$
$$4x - 3 = 0 \quad \text{or} \quad 6x + 5 = 0$$
$$4x = 3 \quad \text{or} \quad 6x = -5$$
$$x = \frac{3}{4} \quad \text{or} \quad x = -\frac{5}{6}.$$

The solution set is $\left\{ -\frac{5}{6}, \frac{3}{4} \right\}$.   ■

Problem Set 6.4

For Problems 1–50, factor completely each of the trinomials. Indicate any that are not factorable using integers.

1. $3x^2 + 7x + 2$
2. $2x^2 + 9x + 4$
3. $6x^2 + 19x + 10$
4. $12x^2 + 19x + 4$
5. $4x^2 - 25x + 6$
6. $5x^2 - 22x + 8$
7. $12x^2 - 31x + 20$
8. $8x^2 - 30x + 7$
9. $5y^2 - 33y - 14$
10. $3y^2 - 2y - 8$
11. $2n^2 + 13n - 24$
12. $4n^2 + 17n - 15$
13. $2x^2 + x + 7$
14. $7x^2 + 19x + 10$
15. $18x^2 + 45x + 7$
16. $10x^2 + x - 5$
17. $7x^2 - 30x + 8$
18. $6x^2 - 17x + 12$
19. $8x^2 + 2x - 21$
20. $9x^2 + 15x - 14$
21. $9t^2 - 15t - 14$
22. $12t^2 - 20t - 25$
23. $12y^2 + 79y - 35$
24. $9y^2 + 52y - 12$
25. $6n^2 + 2n - 5$
26. $20n^2 - 27n + 9$
27. $14x^2 + 55x + 21$
28. $15x^2 + 34x + 15$
29. $20x^2 - 31x + 12$
30. $8t^2 - 3t - 4$
31. $16n^2 - 8n - 15$
32. $25n^2 - 20n - 12$
33. $24x^2 - 50x + 25$
34. $24x^2 - 41x + 12$
35. $2x^2 + 25x + 72$
36. $2x^2 + 23x + 56$
37. $21a^2 + a - 2$
38. $14a^2 + 5a - 24$
39. $12a^2 - 31a - 15$
40. $10a^2 - 39a - 4$
41. $4x^2 + 12x + 9$
42. $9x^2 - 12x + 4$
43. $6x^2 - 5xy + y^2$
44. $12x^2 + 13xy + 3y^2$
45. $20x^2 + 7xy - 6y^2$
46. $8x^2 - 6xy - 35y^2$
47. $5x^2 - 32x + 12$
48. $3x^2 - 35x + 50$
49. $8x^2 - 55x - 7$
50. $12x^2 - 67x - 30$

For Problems 51–80, solve each equation.

51. $2x^2 + 13x + 6 = 0$
52. $3x^2 + 16x + 5 = 0$
53. $12x^2 + 11x + 2 = 0$
54. $15x^2 + 56x + 20 = 0$
55. $3x^2 - 25x + 8 = 0$
56. $4x^2 - 31x + 21 = 0$
57. $15n^2 - 41n + 14 = 0$
58. $6n^2 - 31n + 40 = 0$
59. $6t^2 + 37t - 35 = 0$
60. $2t^2 + 15t - 27 = 0$
61. $16y^2 - 18y - 9 = 0$
62. $9y^2 - 15y - 14 = 0$
63. $9x^2 - 6x - 8 = 0$
64. $12n^2 + 28n - 5 = 0$
65. $10x^2 - 29x + 10 = 0$
66. $4x^2 - 16x + 15 = 0$
67. $6x^2 + 19x = -10$
68. $12x^2 + 17x = -6$
69. $16x(x + 1) = 5$
70. $5x(5x + 2) = 8$
71. $35n^2 - 34n - 21 = 0$
72. $18n^2 - 3n - 28 = 0$
73. $4x^2 - 45x + 50 = 0$
74. $7x^2 - 65x + 18 = 0$
75. $7x^2 + 46x - 21 = 0$
76. $2x^2 + 7x - 30 = 0$
77. $12x^2 - 43x - 20 = 0$
78. $14x^2 - 13x - 12 = 0$
79. $18x^2 + 55x - 28 = 0$
80. $24x^2 + 17x - 20 = 0$

■ Miscellaneous Problems

81. Consider the following approach to factoring $20x^2 + 39x + 18$.

$$20x^2 + 39x + 18. \quad \text{sum of 39}$$

product of $20(18) = 360$

We need two integers whose sum is 39 and whose product is 360. To help find these integers, let's prime factor 360.

$$360 = 2 \cdot 2 \cdot 2 \cdot 3 \cdot 3 \cdot 5.$$

Now by grouping these factors in various ways we find that $2 \cdot 2 \cdot 2 \cdot 3 = 24$, $3 \cdot 5 = 15$, and $24 + 15 = 39$. So the numbers are 15 and 24 and the middle term of the given trinomial, $39x$, can be expressed as $15x + 24x$. Therefore, we can complete the factoring as follows.

$$\begin{aligned} 20x^2 + 39x + 18 &= 20x^2 + 15x + 24x + 18 \\ &= 5x(4x + 3) + 6(4x + 3) \\ &= (4x + 3)(5x + 6). \end{aligned}$$

Factor each of the following trinomials.

(a) $20x^2 + 41x + 20$ (b) $24x^2 - 79x + 40$

(c) $30x^2 + 23x - 40$ (d) $36x^2 + 65x - 36$

6.5 Factoring, Solving Equations, and Problem Solving

Before summarizing our work with factoring techniques let's look at two more special factoring patterns. These patterns emerge when multiplying binomials. Consider the following examples.

$$(x + 5)^2 = (x + 5)(x + 5) = x^2 + 10x + 25;$$
$$(2x + 3)^2 = (2x + 3)(2x + 3) = 4x^2 + 12x + 9;$$
$$(4x + 7)^2 = (4x + 7)(4x + 7) = 16x^2 + 56x + 49.$$

In general, $(a + b)^2 = (a + b)(a + b) = a^2 + 2ab + b^2$.

$$(x - 6)^2 = (x - 6)(x - 6) = x^2 - 12x + 36;$$
$$(3x - 4)^2 = (3x - 4)(3x - 4) = 9x^2 - 24x + 16;$$
$$(5x - 2)^2 = (5x - 2)(5x - 2) = 25x^2 - 20x + 4.$$

In general, $(a - b)^2 = (a - b)(a - b) = a^2 - 2ab + b^2$. Thus, we have the following patterns.

Perfect Square Trinomials

$$a^2 + 2ab + b^2 = (a + b)^2;$$
$$a^2 - 2ab + b^2 = (a - b)^2.$$

Trinomials of the form $a^2 + 2ab + b^2$ or $a^2 - 2ab + b^2$ are called **perfect square trinomials**. They are easily recognized by the nature of their terms. For example, $9x^2 + 30x + 25$ is a perfect square trinomial because

1. the first term is a square; $(3x)^2$
2. the last term is a square; $(5)^2$
3. the middle term is twice the product of the quantities being squared in the first and last terms. $2(3x)(5)$

Likewise, $25x^2 - 40xy + 16y^2$ is a perfect square trinomial because

1. the first term is a square; $(5x)^2$
2. the last term is a square; $(4y)^2$
3. the middle term is twice the product of the quantities being squared in the first and last terms. $2(5x)(4y)$

Once we know that we have a perfect square trinomial, then the factoring process follows immediately from the two basic patterns.

$$9x^2 + 30x + 25 = (3x + 5)^2;$$
$$25x^2 - 40xy + 16y^2 = (5x - 4y)^2.$$

Here are some additional examples of perfect square trinomials and their factored form.

$$
\begin{array}{rcl}
x^2 - 16x + 64 = & (x)^2 - 2(x)(8) + (8)^2 & = (x - 8)^2; \\
16x^2 - 56x + 49 = & (4x)^2 - 2(4x)(7) + (7)^2 & = (4x - 7)^2; \\
25x^2 + 20xy + 4y^2 = & (5x)^2 + 2(5x)(2y) + (2y)^2 & = (5x + 2y)^2; \\
1 + 6y + 9y^2 = & (1)^2 + 2(1)(3y) + (3y)^2 & = (1 + 3y)^2; \\
4m^2 - 4mn + n^2 = & (2m)^2 - 2(2m)(n) + (n)^2 & = (2m - n)^2.
\end{array}
$$

Perhaps you will want to do this step mentally after you feel comfortable with the process.

We have considered some basic factoring techniques in this chapter one at a time, but we must be able to apply them as needed in a variety of situations. So, let's first summarize the techniques and then consider some examples illustrating them.

In this chapter we have discussed factoring

1. by using the distributive property to factor out the greatest common monomial or binomial factor;
2. by grouping;
3. by applying the difference-of-squares pattern;

4. by applying the perfect-square-trinomial pattern;

5. of trinomials of the form $x^2 + bx + c$ into the product of two binomials;

6. of trinomials of the form $ax^2 + bx + c$ into the product of two binomials.

As a general guideline, **always look for a greatest common monomial factor first**, and then proceed with the other factoring techniques.

In each of the following examples we have factored completely whenever possible. Study them carefully and notice the factoring techniques used.

1. $2x^2 + 12x + 10 = 2(x^2 + 6x + 5)$
$$= 2(x + 1)(x + 5).$$

2. $4x^2 + 36 = 4(x^2 + 9).$

    Remember that the sum of two squares is not factorable using integers unless there is a common factor.

3. $4t^2 + 20t + 25 = (2t + 5)^2.$

    If you fail to recognize a perfect trinomial square, no harm is done. Simply proceed to factor into the product of two binomials and then you will recognize that the two binomials are the same.

4. $x^2 - 3x - 8$ is not factorable using integers. This can be determined from the following table.

Product	Sum
$1(-8) = -8$	$1 + (-8) = -7$
$-1(8) = -8$	$-1 + 8 = 7$
$2(-4) = -8$	$2 + (-4) = -2$
$-2(4) = -8$	$-2 + 4 = 2$

No two factors of -8 produce a sum of -3.

5. $6y^2 - 13y - 28 = (2y - 7)(3y + 4)$. The binomial factors can be found as follows.

$(y +$ ___ $)(6y -$ ___ $)$

    or $1 \cdot 28$ or $28 \cdot 1$

$(y -$ ___ $)(6y +$ ___ $)$ $2 \cdot 14$ or $14 \cdot 2$

    or $4 \cdot 7$ or $(7 \cdot 4)$

$(2y -$ ___ $)(3y +$ ___ $)$

    or

$(2y +$ ___ $)(3y -$ ___ $).$

6. $32x^2 - 50y^2 = 2(16x^2 - 25y^2)$
$$= 2(4x + 5y)(4x - 5y).$$

■ Solving Equations by Factoring

Throughout this chapter each time that we considered a new factoring technique we used that technique to help solve some equations. Success at solving equations relies on being able to recognize which technique works for a particular type of equation.

EXAMPLE 1 Solve $x^2 = 25x$.

Solution

$$x^2 = 25x$$
$$x^2 - 25x = 0$$
$$x(x - 25) = 0$$
$$x = 0 \quad \text{or} \quad x - 25 = 0$$
$$x = 0 \quad \text{or} \quad x = 25.$$

The solution set is $\{0, 25\}$. Check them! ■

EXAMPLE 2 Solve $x^3 - 36x = 0$.

Solution

$$x^3 - 36x = 0$$
$$x(x^2 - 36) = 0$$
$$x(x + 6)(x - 6) = 0$$
$$x = 0 \quad \text{or} \quad x + 6 = 0 \quad \text{or} \quad x - 6 = 0$$

If $abc = 0$ then $a = 0$ or $b = 0$ or $c = 0$.

$$x = 0 \quad \text{or} \quad x = -6 \quad \text{or} \quad x = 6.$$

The solution set is $\{-6, 0, 6\}$. Do they check? ■

EXAMPLE 3 Solve $10x^2 - 13x - 3 = 0$.

Solution

$$10x^2 - 13x - 3 = 0$$
$$(5x + 1)(2x - 3) = 0$$
$$5x + 1 = 0 \quad \text{or} \quad 2x - 3 = 0$$
$$5x = -1 \quad \text{or} \quad 2x = 3$$
$$x = -\frac{1}{5} \quad \text{or} \quad x = \frac{3}{2}.$$

The solution set is $\left\{-\frac{1}{5}, \frac{3}{2}\right\}$. Do they check? ■

EXAMPLE 4 Solve $4x^2 - 28x + 49 = 0$.

Solution

$$4x^2 - 28x + 49 = 0$$
$$(2x - 7)^2 = 0$$
$$(2x - 7)(2x - 7) = 0$$
$$2x - 7 = 0 \quad \text{or} \quad 2x - 7 = 0$$
$$2x = 7 \quad \text{or} \quad 2x = 7$$
$$x = \frac{7}{2} \quad \text{or} \quad x = \frac{7}{2}.$$

The solution set is $\left\{ \frac{7}{2} \right\}$. ■

Pay special attention to the next example. We need to change the form of the original equation before the property "$ab = 0$ if and only if $a = 0$ or $b = 0$" can be applied. The uniqueness of this property is that we have an indicated product set equal to zero.

EXAMPLE 5 Solve $(x + 1)(x + 4) = 40$.

Solution

$$(x + 1)(x + 4) = 40$$
$$x^2 + 5x + 4 = 40$$
$$x^2 + 5x - 36 = 0$$
$$(x + 9)(x - 4) = 0$$
$$x + 9 = 0 \quad \text{or} \quad x - 4 = 0$$
$$x = -9 \quad \text{or} \quad x = 4.$$

The solution set is $\{-9, 4\}$. Check them! ■

EXAMPLE 6 Solve $2n^2 + 16n - 40 = 0$.

Solution

$$2n^2 + 16n - 40 = 0$$
$$2(n^2 + 8n - 20) = 0$$
$$n^2 + 8n - 20 = 0 \qquad \text{Multiplied both sides of equation by } \frac{1}{2}.$$
$$(n + 10)(n - 2) = 0$$
$$n + 10 = 0 \quad \text{or} \quad n - 2 = 0$$
$$n = -10 \quad \text{or} \quad n = 2.$$

The solution set is $\{-10, 2\}$. Do they check? ■

■ Problem Solving

The preface of this book states that a common thread throughout the book is "to learn a skill," then "to use that skill to help solve equations," and then "to use equations to help solve problems." This thread should be very apparent in this chapter. Developing our factoring skills has provided us with more ways of solving equations, which in turn gives us more power for solving word problems. Let's conclude the chapter by solving a few more problems.

PROBLEM 1 Find two numbers whose product is 65 if one of the numbers is 3 more than twice the other number.

Solution Let n represent one of the numbers; then $2n + 3$ represents the other number. Since their product is 65, we can set up and solve the following equation.

$$n(2n + 3) = 65$$
$$2n^2 + 3n - 65 = 0$$
$$(2n + 13)(n - 5) = 0$$
$$2n + 13 = 0 \quad \text{or} \quad n - 5 = 0$$
$$2n = -13 \quad \text{or} \quad n = 5$$
$$n = -\frac{13}{2} \quad \text{or} \quad n = 5$$

If $n = -\dfrac{13}{2}$, then $2n + 3 = 2\left(-\dfrac{13}{2}\right) + 3 = -10$.

If $n = 5$, then $2n + 3 = 2(5) + 3 = 13$.

Thus, the numbers are $-\dfrac{13}{2}$ and -10, or 5 and 13. ■

PROBLEM 2 The area of a triangular sheet of paper is 14 square inches. One side of the triangle is 3 inches longer than the altitude to that side. Find the length of the one side and the length of the altitude to that side.

Solution Let h represent the altitude to the side. Then $h + 3$ represents the side of the triangle.

$h + 3$

Since the formula for finding the area of a triangle is $A = \dfrac{1}{2}bh$, we have

$$\frac{1}{2}h(h + 3) = 14$$

$$h(h + 3) = 28 \qquad \text{Multiplied both sides by 2.}$$
$$h^2 + 3h = 28$$
$$h^2 + 3h - 28 = 0$$
$$(h + 7)(h - 4) = 0$$
$$h + 7 = 0 \quad \text{or} \quad h - 4 = 0$$
$$h = -7 \quad \text{or} \qquad h = 4.$$

The solution of -7 is not reasonable. Thus, the altitude is 4 inches and the length of the side to which that altitude is drawn is 7 inches.

PROBLEM 3   A strip of uniform width is shaded along both sides and both ends of a rectangular poster 12 inches by 16 inches. How wide is the strip if one-half of the poster is shaded?

Solution   If we let x represent the width of the strip, then the following diagram depicts the situation. The area of the strip is one-half of the area of the poster; therefore, it is $\dfrac{1}{2}(12)(16) = 96$ square inches. Furthermore, the area of the strip around the poster can be represented by "the area of the poster minus the area of the unshaded portion of the poster." Thus, we can set up and solve the following equation.

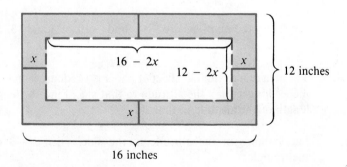

$$
\underbrace{16(12)}_{\text{area of poster}} \; \underbrace{- (16 - 2x)(12 - 2x)}_{\substack{\text{area of unshaded}\\\text{portion}}} = \underbrace{96}_{\text{area of strip}}
$$

$$192 - (192 - 56x + 4x^2) = 96$$
$$192 - 192 + 56x - 4x^2 = 96$$

$$-4x^2 + 56x - 96 = 0$$
$$x^2 - 14x + 24 = 0$$
$$(x - 12)(x - 2) = 0$$
$$x - 12 = 0 \quad \text{or} \quad x - 2 = 0$$
$$x = 12 \quad \text{or} \quad x = 2.$$

Obviously, the strip cannot be 12 inches wide since the total width of the poster is 12 inches. Thus, the solution of 12 must be disregarded, and the strip is 2 inches wide. ■

Problem Set 6.5

For Problems 1–12, factor each of the perfect square trinomials.

1. $x^2 + 4x + 4$ **2.** $x^2 + 18x + 81$ **3.** $x^2 - 10x + 25$

4. $x^2 - 24x + 144$ **5.** $9n^2 + 12n + 4$ **6.** $25n^2 + 30n + 9$

7. $16a^2 - 8a + 1$ **8.** $36a^2 - 84a + 49$ **9.** $4 + 36x + 81x^2$

10. $1 - 4x + 4x^2$ **11.** $16x^2 - 24xy + 9y^2$ **12.** $64x^2 + 16xy + y^2$

For Problems 13–40, factor each polynomial completely. Indicate any that are not factorable using integers.

13. $2x^2 + 17x + 8$ **14.** $x^2 + 19x$ **15.** $2x^3 - 72x$

16. $30x^2 - x - 1$ **17.** $n^2 - 7n - 60$ **18.** $4n^3 - 100n$

19. $3a^2 - 7a - 4$ ∩F   **20.** $a^2 + 7a - 30$ **21.** $8x^2 + 72$

22. $3y^3 - 36y^2 + 96y$ **23.** $9x^2 + 30x + 25$ **24.** $5x^2 - 5x - 6$

25. $15x^2 + 65x + 70$ **26.** $4x^2 - 20xy + 25y^2$ **27.** $24x^2 + 2x - 15$

28. $9x^2y - 27xy$ **29.** $xy + 5y - 8x - 40$ **30.** $xy - 3y + 9x - 27$

31. $20x^2 + 31xy - 7y^2$ **32.** $2x^2 - xy - 36y^2$ **33.** $24x^2 + 18x - 81$

34. $30x^2 + 55x - 50$ **35.** $12x^2 + 6x + 30$ **36.** $24x^2 - 8x + 32$

37. $5x^4 - 80$ **38.** $3x^5 - 3x$ **39.** $x^2 + 12xy + 36y^2$

40. $4x^2 - 28xy + 49y^2$

For Problems 41–70, solve each equation.

41. $4x^2 - 20x = 0$ **42.** $-3x^2 - 24x = 0$ **43.** $x^2 - 9x - 36 = 0$

44. $x^2 + 8x - 20 = 0$ **45.** $-2x^3 + 8x = 0$ **46.** $4x^3 - 36x = 0$

47. $6n^2 - 29n - 22 = 0$ **48.** $30n^2 - n - 1 = 0$ **49.** $(3n - 1)(4n - 3) = 0$

50. $(2n - 3)(7n + 1) = 0$ **51.** $(n - 2)(n + 6) = -15$ **52.** $(n + 3)(n - 7) = -25$

53. $2x^2 = 12x$ **54.** $-3x^2 = 15x$ **55.** $t^3 - 2t^2 - 24t = 0$

56. $2t^3 - 16t^2 - 18t = 0$ **57.** $12 - 40x + 25x^2 = 0$ **58.** $12 - 7x - 12x^2 = 0$

59. $n^2 - 28n + 192 = 0$ **60.** $n^2 + 33n + 270 = 0$ **61.** $(3n + 1)(n + 2) = 12$

62. $(2n + 5)(n + 4) = -1$ **63.** $x^3 = 6x^2$ **64.** $x^3 = -4x^2$

65. $9x^2 - 24x + 16 = 0$ **66.** $25x^2 + 60x + 36 = 0$ **67.** $x^3 + 10x^2 + 25x = 0$

68. $x^3 - 18x^2 + 81x = 0$ **69.** $24x^2 + 17x - 20 = 0$ **70.** $24x^2 + 74x - 35 = 0$

For Problems 71–88, set up an equation and solve each problem.

71. Find two numbers whose product is 15 such that one of the numbers is 7 more than four times the other number.

72. Find two numbers whose product is 12 such that one of the numbers is 4 less than eight times the other number.

73. Find two numbers whose product is -1. One of the numbers is 3 more than twice the other number.

74. The sum of the squares of three consecutive integers is 110. Find the integers.

75. One number is 1 more than twice another number. The sum of the squares of the two numbers is 97. Find the numbers.

76. One number is 1 less than three times another number. If the product of the two numbers is 102, find the numbers.

77. A room contains 54 chairs. The number of chairs per row is 3 less than twice the number of rows. Find the number of rows and the number of chairs per row.

78. An apple orchard contains 85 trees. The number of trees in each row is 3 less than 4 times the number of rows. Find the number of rows and the number of trees per row.

79. The combined area of two squares is 360 square feet. Each side of the larger square is 3 times as long as a side of the smaller square. How big is each square?

80. The area of a rectangular slab of sidewalk is 45 square feet. Its length is 3 more than 4 times its width. Find the length and width of the slab.

81. The length of a rectangle is 1 centimeter more than twice its width and the area of the rectangle is 55 square centimeters. Find the length and width of the rectangle.

82. The length of a rectangle is three times its width. If the length is increased by 2 inches and the width increased by 1 inch, the newly formed rectangle has an area of 70 square inches. Find the length and width of the original rectangle.

83. The area of a triangle is 51 square inches. One side of the triangle is 1 inch less than three times the length of the altitude to that side. Find the length of that side and the length of the altitude to that side.

84. Suppose that a square and a rectangle have equal areas. Furthermore, suppose that the length of the rectangle is twice the length of a side of the square and the width of the rectangle is 4 centimeters less than the length of a side of the square. Find the dimensions of both figures.

85. A strip of uniform width is to be cut off of both sides and both ends of a sheet of paper that is 8 inches by 11 inches in order to reduce the size of the paper to an area of 40 square inches. Find the width of the strip.

86. The sum of the areas of two circles is 100π square centimeters. The length of a radius of the larger circle is 2 centimeters more than the length of a radius of the smaller circle. Find the length of a radius of each circle.

87. The sum of the areas of two circles is 180π square inches. The length of a radius of the smaller circle is 6 inches less than the length of a radius of the larger circle. Find the length of a radius of each circle.

88. A strip of uniform width is shaded along both sides and both ends of a rectangular poster that is 18 inches by 14 inches. How wide is the strip if the unshaded portion of the poster has an area of 165 square inches?

Chapter 6 Summary

(6.1) The distributive property in the form $ab + ac = a(b + c)$ provides the basis for **factoring out a greatest common monomial or binomial factor**.

Rewriting an expression such as $ab + 3a + bc + 3c$ as $a(b + 3) + c(b + 3)$, and then factoring out the common binomial factor of $b + 3$ so that $a(b + 3) + c(b + 3)$ becomes $(b + 3)(a + c)$ is called **factoring by grouping**.

The property "$ab = 0$ if and only if $a = 0$ or $b = 0$" provides us with another technique for solving equations.

(6.2) The following factoring pattern is called the **difference of two squares**.

$$(a^2 - b^2) = (a - b)(a + b)$$

(6.3) The following multiplication pattern provides a technique for factoring trinomials of the form $x^2 + bx + c$.

$$(x + r)(x + s) = x^2 + rx + sx + rs$$
$$= x^2 + (r + s)x + rs.$$

sum of   product of
r and s r and s

(6.4) We presented two different techniques for factoring trinomials of the form $ax^2 + bx + c$. To review these techniques, return to Section 6.4 and study the examples.

(6.5) As a general guideline for **factoring completely**, always look for a greatest common monomial or binomial factor *first*, and then proceed with one or more of the following techniques.

1. Apply the difference-of-squares pattern.
2. Apply the perfect-square-trinomial pattern.
3. Factor a trinomial of the form $x^2 + bx + c$ into the product of two binomials.
4. Factor a trinomial of the form $ax^2 + bx + c$ into the product of two binomials.

Chapter 6 Review Problem Set

Factor completely. Indicate any that are not factorable using integers.

1. $x^2 - 9x + 14$
2. $3x^2 + 21x$
3. $9x^2 - 4$
4. $4x^2 + 8x - 5$
5. $25x^2 - 60x + 36$
6. $n^3 + 13n^2 + 40n$
7. $y^2 + 11y - 12$
8. $3xy^2 + 6x^2y$
9. $x^4 - 1$
10. $18n^2 + 9n - 5$
11. $x^2 + 7x + 24$
12. $4x^2 - 3x - 7$
13. $3n^2 + 3n - 90$
14. $x^3 - xy^2$
15. $2x^2 + 3xy - 2y^2$
16. $4n^2 - 6n - 40$
17. $5x + 5y + ax + ay$
18. $21t^2 - 5t - 4$
19. $2x^3 - 2x$
20. $3x^3 - 108x$
21. $16x^2 + 40x + 25$
22. $xy - 3x - 2y + 6$
23. $15x^2 - 7xy - 2y^2$
24. $6n^4 - 5n^3 + n^2$

Solve each of the following equations.

25. $x^2 + 4x - 12 = 0$
26. $x^2 = 11x$
27. $2x^2 + 3x - 20 = 0$
28. $9n^2 + 21n - 8 = 0$
29. $6n^2 = 24$
30. $16y^2 + 40y + 25 = 0$
31. $t^3 - t = 0$
32. $28x^2 + 71x + 18 = 0$
33. $x^2 + 3x - 28 = 0$
34. $(x - 2)(x + 2) = 21$
35. $5n^2 + 27n = 18$
36. $4n^2 + 10n = 14$
37. $2x^3 - 8x = 0$
38. $x^2 - 20x + 96 = 0$
39. $4t^2 + 17t - 15 = 0$
40. $3(x + 2) - x(x + 2) = 0$
41. $(2x - 5)(3x + 7) = 0$
42. $(x + 4)(x - 1) = 50$
43. $-7n - 2n^2 = -15$
44. $-23x + 6x^2 = -20$

Set up an equation and solve each of the following problems.

45. The larger of two numbers is one less than twice the smaller number. The difference of their squares is 33. Find the numbers.

46. The length of a rectangle is 2 centimeters less than 5 times the width of the rectangle. The area of the rectangle is 16 square centimeters. Find the length and width of the rectangle.

47. The combined area of two squares is 104 square inches. Each side of the larger square is five times as long as a side of the smaller square. Find the size of each square.

48. The longer leg of a right triangle is one unit less than twice the length of the shorter leg. The hypotenuse is one unit more than twice the length of the shorter leg. Find the lengths of the three sides of the triangle.

49. The product of two numbers is 26 and one of the numbers is one larger than six times the other number. Find the numbers.

50. Find three consecutive positive odd whole numbers such that the sum of the squares of the two smaller numbers is 9 more than the square of the largest number.

51. The number of books per shelf in a bookcase is 1 less than nine times the number of shelves. If the bookcase contains 140 books, find the number of shelves.

52. The combined area of a square and a rectangle is 225 square yards. The length of the rectangle is 8 times the width of the rectangle and the length of a side of the square is the same as the width of the rectangle. Find the dimensions of the square and the rectangle.

53. Find two consecutive integers such that the sum of their squares is 613.

54. If numerically the volume of a cube equals the total surface area of the cube, find the length of an edge of the cube.

55. The combined area of two circles is 53π square meters. The length of a radius of the larger circle is 1 meter more than three times the length of a radius of the smaller circle. Find the length of a radius of each circle.

56. The product of two consecutive odd whole numbers is one less than five times their sum. Find the integers.

57. Sandy has a photograph that is 14 centimeters long and 8 centimeters wide. She wants to reduce the length and width by the same amount so that the area is decreased by 40 square centimeters. By what amount should she reduce the length and width?

58. A strip of uniform width is plowed along both sides and both ends of a garden that is 120 feet long and 90 feet wide. How wide is the strip if the garden is one-half plowed?

Cumulative Review Problem Set

For Problems 1–6, evaluate each of the numerical expressions.

1. $(-3)^4$

2. -2^5

3. $\left(\dfrac{2}{3}\right)^{-1}$

4. $\dfrac{1}{4^{-2}}$

5. $\left(\dfrac{1}{2} - \dfrac{1}{3}\right)^{-2}$

6. $2^0 + 2^{-1} + 2^{-2}$

For Problems 7–12, evaluate each of the algebraic expressions for the given values of the variables.

7. $\dfrac{2x + 3y}{x - y}$ for $x = \dfrac{1}{2}$ and $y = -\dfrac{1}{3}$

8. $\dfrac{2}{5}n - \dfrac{1}{3}n - n + \dfrac{1}{2}n$ for $n = -\dfrac{3}{4}$

9. $\dfrac{3a - 2b - 4a + 7b}{-a - 3a + b - 2b}$ for $a = -1$ and $b = -\dfrac{1}{3}$

10. $-2(x - 4) + 3(2x - 1) - (3x - 2)$ for $x = -2$

11. $(x^2 + 2x - 4) - (x^2 - x - 2) + (2x^2 - 3x - 1)$ for $x = -1$

12. $2(n^2 - 3n - 1) - (n^2 + n + 4) - 3(2n - 1)$ for $n = 3$

For Problems 13–25, find the indicated products.

13. $(3x^2y^3)(-5xy^4)$

14. $(-6ab^4)(-2b^3)$

15. $(-2x^2y^5)^3$

16. $-3xy(2x - 5y)$

17. $(5x - 2)(3x - 1)$

18. $(7x - 1)(3x + 4)$

19. $(-x - 2)(2x + 3)$

20. $(7 - 2y)(7 + 2y)$

21. $(x - 2)(3x^2 - x - 4)$

22. $(2x - 5)(x^2 + x - 4)$

23. $(2n + 3)^3$

24. $(1 - 2n)^3$

25. $(x^2 - 2x + 6)(2x^2 + 5x - 6)$

For Problems 26–30, perform the indicated divisions.

26. $\dfrac{-52x^3y^4}{13xy^2}$

27. $\dfrac{-126a^3b^5}{-9a^2b^3}$

28. $\dfrac{56xy^2 - 64x^3y - 72x^4y^4}{8xy}$

29. $(2x^3 + 2x^2 - 19x - 21) \div (x + 3)$

30. $(3x^3 + 17x^2 + 6x - 4) \div (3x - 1)$

For Problems 31–34, simplify each expression and express answers using positive exponents only.

31. $(-2x^3)(3x^{-4})$

32. $\dfrac{4x^{-2}}{2x^{-1}}$

33. $(3x^{-1}y^{-2})^{-1}$

34. $(xy^2z^{-1})^{-2}$

For Problems 35–37, use scientific notation and the properties of exponents to help evaluate each numerical expression.

35. $(.00003)(4000)$

36. $(.0002)(.003)^2$

37. $\dfrac{.00034}{.0000017}$

For Problems 38–43, factor each expression completely. Indicate any that are not factorable using integers.

38. $7x^2 - 28$

39. $2ac - ad + 2bc - bd$

40. $6x^2 + x - 35$

41. $3x^2 + 12x - 36$

42. $2x^2 - x - 4$

43. $16x^4 - 1$

For Problems 44–55, solve each of the equations.

44. $-3(x - 1) + 2(x + 3) = -4$

45. $\dfrac{3n + 1}{5} + \dfrac{n - 2}{3} = \dfrac{2}{15}$

46. $.06x + .08(1500 - x) = 110$

47. $|-2x - 3| = 6$

48. $2x^2 -- 3x = 0$

49. $5n^2 = 125$

50. $x^2 + 2x - 48 = 0$

51. $14x^2 - 3x - 2 = 0$

52. $4x^3 - x^2 - 3x = 0$

53. $(n - 1)(n + 3) = 21$

54. $(5n - 3)(3n + 10) = 0$

55. $x(x + 2) - 4(x + 2) = 0$

For Problems 56–60, solve each of the inequalities.

56. $-5x + 3 > -4x + 5$

57. $\dfrac{3x}{4} - \dfrac{x}{2} \le \dfrac{5x}{6} - 1$

58. $.08(700 - x) + .11x \ge 65$

59. $|3x - 2| > 10$

60. $|2x + 7| < 7$

For Problems 61–70, set up an equation and solve each problem.

61. The sum of 4 and three times a certain number is the same as the sum of the number and 10. Find the number.

62. Fifteen percent of some number is 6. Find the number.

63. Lou has 18 coins consisting of dimes and quarters. If the total value of the coins is $3.30, how many coins of each denomination does he have?

64. A sum of $1500 is invested, part of it at 8% interest and the remainder at 9%. If the total interest amounts to $128, find the amount invested at each rate.

65. How many gallons of water must be added to 15 gallons of a 12% salt solution to change it to a 10% salt solution?

66. Two airplanes leave Atlanta at the same time and fly in opposite directions. If one travels at 400 miles per hour and the other at 450 miles per hour, how long will it take them to be 2975 miles apart?

67. The length of a rectangle is one meter more than twice its width. If the perimeter of the rectangle is 44 meters, find the length and width.

68. The length of a rectangle is two feet less than twice the width. If the area of the rectangle is 144 square feet, find its length and width.

69. The combined area of two circles is 65π square inches. The length of a radius of the larger circle is 3 inches more than the length of a radius of the smaller circle. Find the length of a radius of each circle.

70. One leg of a right triangle is 4 centimeters longer than the other leg. The hypotenuse is 8 centimeters longer than the shorter leg. Find the lengths of the three sides of the right triangle.

7 Algebraic Fractions

In Chapter 2 our study of common fractions naturally led to some work with simple algebraic fractions. Then in Chapters 5 and 6 the basic operations pertaining to polynomials were discussed. Now we can use some ideas about polynomials—specifically the factoring techniques—to expand our study of algebraic fractions.

7.1 Simplifying Algebraic Fractions

If the numerator and denominator of a fraction are polynomials, then the fraction is called an **algebric fraction** or a **rational expression**. The following are examples of algebraic fractions.

$$\frac{4}{x-2}; \quad \frac{x^2 + 2x - 4}{x^2 - 9}; \quad \frac{y + x^2}{xy - 3}; \quad \frac{x^3 + 2x^2 - 3x - 4}{x^2 - 2x - 6}.$$

Because division by zero must be avoided, no values can be assigned to variables that create a denominator of zero. Thus, the fraction $\dfrac{4}{x-2}$ is meaningful for all real number values of x except for $x = 2$. Rather than making a restriction for each individual fraction, we will simply assume that all denominators represent nonzero real numbers.

Recall that in our previous work the **fundamental principle of fractions** $\left(\dfrac{ak}{bk} = \dfrac{a}{b}\right)$ provided the basis for expressing fractions in reduced (or simplified) form as the next examples illustrate.

$$\frac{18}{24} = \frac{3 \cdot \cancel{6}}{4 \cdot \cancel{6}} = \frac{3}{4}; \qquad \frac{-42xy}{77y} = -\frac{2 \cdot 3 \cdot \cancel{7} \cdot x \cdot \cancel{y}}{\cancel{7} \cdot 11 \cdot \cancel{y}} = -\frac{6x}{11};$$

$$\frac{15x}{25x} = \frac{3 \cdot \cancel{5} \cdot \cancel{x}}{5 \cdot \cancel{5} \cdot \cancel{x}} = \frac{3}{5}; \qquad \frac{28x^2y^2}{-63x^2y^3} = \frac{4 \cdot \cancel{7} \cdot \cancel{x^2} \cdot \cancel{y^2}}{9 \cdot \cancel{7} \cdot \cancel{x^2} \cdot \underset{y}{\cancel{y^3}}} = -\frac{4}{9y}.$$

The factoring techniques from Chapter 6 can be used to factor numerators and/or denominators so that the fundamental principle of fractions can be applied. Several examples should clarify this process.

EXAMPLE 1 Simplify $\dfrac{x^2 + 6x}{x^2 - 36}$.

Solution $$\frac{x^2 + 6x}{x^2 - 36} = \frac{x(\cancel{x + 6})}{(x - 6)(\cancel{x + 6})} = \frac{x}{x - 6}. \qquad ■$$

EXAMPLE 2 Simplify $\dfrac{a + 2}{a^2 + 4a + 4}$.

Solution $$\frac{a + 2}{a^2 + 4a + 4} = \frac{1(\cancel{a + 2})}{(a + 2)(\cancel{a + 2})} = \frac{1}{a + 2}. \qquad ■$$

EXAMPLE 3   Simplify $\dfrac{x^2 + 4x - 21}{2x^2 + 15x + 7}$.

Solution $\dfrac{x^2 + 4x - 21}{2x^2 + 15x + 7} = \dfrac{(x - 3)(x + 7)}{(2x + 1)(x + 7)} = \dfrac{x - 3}{2x + 1}$.   ■

EXAMPLE 4   Simplify $\dfrac{a^2b + ab^2}{ab + b^2}$.

Solution $\dfrac{a^2b + ab^2}{ab + b^2} = \dfrac{ab(a + b)}{b(a + b)} = a$.   ■

EXAMPLE 5   Simplify $\dfrac{4x^3y - 36xy}{2x^2 - 4x - 30}$.

Solution $\dfrac{4x^3y - 36xy}{2x^2 - 4x - 30} = \dfrac{4xy(x^2 - 9)}{2(x^2 - 2x - 15)}$

$= \dfrac{\overset{2}{4xy}(x + 3)(x - 3)}{2(x - 5)(x + 3)} = \dfrac{2xy(x - 3)}{x - 5}$.   ■

Notice that in Example 5 we left the numerator of the final fraction in factored form. This is often done if polynomials other than monomials are involved. Either $\dfrac{2xy(x - 3)}{x - 5}$ or $\dfrac{2x^2y - 6xy}{x - 5}$ is an acceptable answer.

Remember that the quotient of any nonzero real number and its opposite is -1. For example, $\dfrac{7}{-7} = -1$ and $\dfrac{-9}{9} = -1$. Likewise, the indicated quotient of any polynomial and its opposite is equal to -1. For example,

$\dfrac{x}{-x} = -1$ because x and $-x$ are opposites;

$\dfrac{x - y}{y - x} = -1$ because $x - y$ and $y - x$ are opposites;

$\dfrac{a^2 - 9}{9 - a^2} = -1$ because $a^2 - 9$ and $9 - a^2$ are opposites.

The final examples of this section illustrate the use of this idea when simplifying algebraic fractions.

EXAMPLE 6   Simplify $\dfrac{14 - 7n}{n - 2}$.

Solution

$$\frac{14 - 7n}{n - 2} = \frac{7(2 - n)}{n - 2}$$

$$= 7(-1) \qquad \left(\frac{2 - n}{n - 2} = -1\right)$$

$$= -7.$$

EXAMPLE 7 Simplify $\dfrac{x^2 + 4x - 21}{15 - 2x - x^2}$.

Solution

$$\frac{x^2 + 4x - 21}{15 - 2x - x^2} = \frac{(x + 7)(x - 3)}{(5 + x)(3 - x)}$$

$$= \left(\frac{x + 7}{x + 5}\right)(-1) \qquad \left(\frac{x - 3}{3 - x} = -1\right)$$

$$= -\frac{x + 7}{x + 5} \quad \text{or} \quad \frac{-x - 7}{x + 5}.$$

■

Problem Set 7.1

For Problems 1–60, simplify each algebraic fraction.

1. $\dfrac{6x}{14y}$

2. $\dfrac{8y}{18x}$

3. $\dfrac{9xy}{24x}$

4. $\dfrac{12y}{20xy}$

5. $\dfrac{-15x^2y}{25x}$

6. $\dfrac{16x^3y^2}{-28x^2y}$

7. $\dfrac{-36x^4y^3}{-48x^6y^2}$

8. $\dfrac{-18x^3y}{-36xy^3}$

9. $\dfrac{12a^2b^5}{-54a^2b^3}$

10. $\dfrac{-24a^3b^3}{39a^5b^2}$

11. $\dfrac{32xy^2z^3}{72yz^4}$

12. $\dfrac{27x^2y^3z^4}{45x^3y^3z}$

13. $\dfrac{xy}{x^2 - 2x}$

14. $\dfrac{x^2 + 5x}{xy}$

15. $\dfrac{8x + 12y}{12}$

16. $\dfrac{8}{12x - 16y}$

17. $\dfrac{x^2 + 2x}{x^2 - 7x}$

18. $\dfrac{x^2 - 6x}{2x^2 + 6x}$

19. $\dfrac{7 - x}{x - 7}$

20. $\dfrac{x - 9}{9 - x}$

21. $\dfrac{15 - 3n}{n - 5}$

22. $\dfrac{2n^2 - 8n}{4 - n}$

23. $\dfrac{4x^3 - 4x}{1 - x^2}$

24. $\dfrac{9 - x^2}{3x^3 - 27x}$

25. $\dfrac{x^2 - 1}{3x^2 - 3x}$

26. $\dfrac{5x^2 + 25x}{x^2 - 25}$

27. $\dfrac{x^2 + xy}{x^2}$

28. $\dfrac{x^3}{x^3 - x^2y}$

29. $\dfrac{6x^3 - 15x^2y}{6x^2 + 24xy}$

30. $\dfrac{6x^2 + 42xy}{16x^3 - 8x^2y}$

31. $\dfrac{n^2 + 2n}{n^2 + 3n + 2}$

32. $\dfrac{n^2 + 9n + 18}{n^2 + 6n}$

33. $\dfrac{2n^2 + 5n - 3}{n^2 - 9}$

34. $\dfrac{3n^2 - 10n - 8}{n^2 - 16}$

35. $\dfrac{2x^2 + 17x + 35}{3x^2 + 19x + 20}$

36. $\dfrac{5x^2 - 32x + 12}{4x^2 - 27x + 18}$

37. $\dfrac{9(x - 1)^2}{12(x - 1)^3}$

38. $\dfrac{18(x + 2)^3}{16(x + 2)^2}$

39. $\dfrac{7x^2 + 61x - 18}{7x^2 + 19x - 6}$

40. $\dfrac{8x^2 - 51x + 18}{8x^2 + 29x - 12}$

41. $\dfrac{10a^2 + a - 3}{15a^2 + 4a - 3}$

42. $\dfrac{6a^2 - 11a - 10}{8a^2 - 22a + 5}$

43. $\dfrac{x^2 + 2xy - 3y^2}{2x^2 - xy - y^2}$

44. $\dfrac{x^2 - 3xy + 2y^2}{x^2 - 4y^2}$

45. $\dfrac{x^2 - 9}{-x^2 - 3x}$

46. $\dfrac{-x^2 - 2x}{x^2 - 4}$

47. $\dfrac{n^2 + 14n + 49}{8n + 56}$

48. $\dfrac{6n - 60}{n^2 - 20n + 100}$

49. $\dfrac{4n^2 - 12n + 9}{2n^2 - n - 3}$

50. $\dfrac{9n^2 + 30n + 25}{3n^2 - n - 10}$

51. $\dfrac{y^2 - 6y - 72}{y^2 - 8y - 84}$

52. $\dfrac{y^2 + 20y + 96}{y^2 + 23y + 120}$

53. $\dfrac{1 - x^2}{x - x^2}$

54. $\dfrac{2x + x^2}{4 - x^2}$

55. $\dfrac{6 - x - 2x^2}{12 + 7x - 10x^2}$

56. $\dfrac{15 + x - 2x^2}{21 - 10x + x^2}$

57. $\dfrac{x^2 + 7x - 18}{12 - 4x - x^2}$

58. $\dfrac{3x - 21}{28 - 4x}$

59. $\dfrac{5x - 40}{80 - 10x}$

60. $\dfrac{x^2 - x - 12}{8 + 2x - x^2}$

■ Miscellaneous Problems

For Problems 61–64, simplify each fraction. You will need to use factoring by grouping.

61. $\dfrac{xy - 3x + 2y - 6}{xy + 5x + 2y + 10}$

62. $\dfrac{xy + 4x - y - 4}{xy + 4x - 4y - 16}$

63. $\dfrac{xy - 6x + y - 6}{xy - 6x + 5y - 30}$

64. $\dfrac{xy - 7x - 5y + 35}{xy - 9x - 5y + 45}$

The link between positive and negative exponents $\left(a^{-n} = \dfrac{1}{a^n} \right)$ along with the property,

$\dfrac{a^n}{a^m} = a^{n-m}$, can also be used when reducing fractions. Consider the following example.

$$\dfrac{x^3}{x^7} = x^{3-7} = x^{-4} = \dfrac{1}{x^4}.$$

For Problems 65–70, use this approach to help express each fraction in reduced form. Express all answers with positive exponents only.

65. $\dfrac{x^3}{x^9}$

66. $\dfrac{x^4}{x^8}$

67. $\dfrac{x^4 y^3}{x^7 y^5}$

68. $\dfrac{x^5 y^2}{x^6 y^3}$

69. $\dfrac{28a^2 b^3}{-7a^5 b^3}$

70. $\dfrac{-44a^3 b^4}{4a^3 b^6}$

7.2 Multiplying and Dividing Algebraic Fractions

The earlier treatment of multiplication and division of rational numbers in $\dfrac{a}{b}$ form extends to algebraic fractions in general. Therefore, to multiply algebraic fractions we *multiply the numerators, multiply the denominators, and express the product in simplified form.* The following examples illustrate this idea.

1. $\dfrac{2x}{3y} \cdot \dfrac{5y}{4x} = \dfrac{\cancel{2} \cdot 5 \cdot \cancel{x} \cdot \cancel{y}}{3 \cdot \underset{2}{\cancel{4}} \cdot \cancel{x} \cdot \cancel{y}} = \dfrac{5}{6}.$ Notice that we used the commutative property of multiplication to rearrange factors in a more convenient form for recognizing common factors of the numerator and denominator.

2. $\dfrac{4a}{6b} \cdot \dfrac{8b}{12a^2} = \dfrac{\overset{4}{\cancel{4}} \cdot \overset{}{\cancel{8}} \cdot \cancel{a} \cdot \cancel{b}}{\underset{3}{\cancel{6}} \cdot \underset{3}{\cancel{12}} \cdot \underset{a}{a^2} \cdot \cancel{b}} = \dfrac{4}{9a}.$

3. $\dfrac{-9x^2}{15xy} \cdot \dfrac{5y^2}{7x^2y^3} = -\dfrac{\overset{3}{\cancel{9}} \cdot \cancel{5} \cdot x^2 \cdot y^2}{\underset{\cancel{5}}{\cancel{15}} \cdot 7 \cdot \underset{x}{x^3} \cdot \underset{y^2}{y^4}} = -\dfrac{3}{7xy^2}.$

When multiplying algebraic fractions, the numerators and/or denominators sometimes need to be factored so that common factors can be recognized. Consider the following examples.

EXAMPLE 1 Multiply and simplify $\dfrac{x}{x^2-9} \cdot \dfrac{x+3}{y}.$

Solution $\dfrac{x}{x^2-9} \cdot \dfrac{x+3}{y} = \dfrac{x(\cancel{x+3})}{(\cancel{x+3})(x-3)(y)} = \dfrac{x}{y(x-3)}.$

$\dfrac{x}{xy-3y}$ is also an acceptable answer.   ■

EXAMPLE 2 Multiply and simplify $\dfrac{x}{x^2+2x} \cdot \dfrac{x^2+10x+16}{5}.$

Solution $\dfrac{x}{x^2+2x} \cdot \dfrac{x^2+10x+16}{5} = \dfrac{\cancel{x}(\cancel{x+2})(x+8)}{\cancel{x}(\cancel{x+2})(5)} = \dfrac{x+8}{5}.$ ■

EXAMPLE 3 Multiply and simplify $\dfrac{a^2-3a}{a+5} \cdot \dfrac{a^2+3a-10}{a^2-5a+6}.$

Solution

$$\frac{a^2 - 3a}{a + 5} \cdot \frac{a^2 + 3a - 10}{a^2 - 5a + 6} = \frac{a(a-3)(a+5)(a-2)}{(a+5)(a-2)(a-3)} = a.$$ ■

EXAMPLE 4 Multiply and simplify $\dfrac{6n^2 + 7n - 3}{n + 1} \cdot \dfrac{n^2 - 1}{2n^2 + 3n}$.

Solution

$$\frac{6n^2 + 7n - 3}{n + 1} \cdot \frac{n^2 - 1}{2n^2 + 3n} = \frac{(2n+3)(3n - 1)(n+1)(n - 1)}{(n+1)(n)(2n+3)} = \frac{(3n - 1)(n - 1)}{n}.$$ ■

■ Dividing Algebraic Fractions

Recall that to divide two rational numbers in $\dfrac{a}{b}$ form we *invert the divisor and multiply.*

Symbolically, this is expressed as $\dfrac{a}{b} \div \dfrac{c}{d} = \dfrac{a}{b} \cdot \dfrac{d}{c}$. Furthermore, the numbers $\dfrac{c}{d}$ and $\dfrac{d}{c}$ are called reciprocals of each other because their product is one. Thus, division can also be described as *to divide by a fraction, multiply by its reciprocal.* Division of algebraic fractions is defined in the same way using the same vocabulary.

1. $\dfrac{4x}{7y} \div \dfrac{6x^2}{14y^2} = \dfrac{4x}{7y} \cdot \dfrac{14y^2}{6x^2} = \dfrac{\overset{2}{4} \cdot \overset{2}{14} \cdot x \cdot \overset{y}{y^2}}{\underset{3}{7} \cdot \underset{3}{6} \cdot \underset{x}{x^2} \cdot y} = \dfrac{4y}{3x}$.

2. $\dfrac{-8ab}{9b} \div \dfrac{18a^3}{15a^2b} = \dfrac{-8ab}{9b} \cdot \dfrac{15a^2b}{18a^3} = -\dfrac{\overset{4}{8} \cdot \overset{5}{15} \cdot a^3 \cdot \overset{b}{b^2}}{\underset{3}{9} \cdot \underset{9}{18} \cdot a^3 \cdot b} = -\dfrac{20b}{27}$.

3. $\dfrac{x^2y^3}{4ab} \div \dfrac{5xy^2}{-9a^2b} = \dfrac{x^2y^3}{4ab} \cdot \dfrac{-9a^2b}{5xy^2} = -\dfrac{9 \cdot \overset{x}{x^2} \cdot \overset{y}{y^3} \cdot \overset{a}{a^2} \cdot b}{4 \cdot 5 \cdot a \cdot b \cdot x \cdot y^2} = -\dfrac{9axy}{20}$.

The key idea when dividing fractions is to *first* convert to an equivalent multiplication problem and then to proceed to factor numerator and denominator completely and look for common factors.

EXAMPLE 5 Divide and simplify $\dfrac{x^2 - 4x}{xy} \div \dfrac{x^2 - 16}{y^3 + y^2}$.

Solution

$$\frac{x^2 - 4x}{xy} \div \frac{x^2 - 16}{y^3 + y^2} = \frac{x^2 - 4x}{xy} \cdot \frac{y^3 + y^2}{x^2 - 16} = \frac{x(x-4)(y^2)(y + 1)}{xy(x + 4)(x-4)} = \frac{y(y + 1)}{x + 4}.$$ ■

EXAMPLE 6 Divide and simplify $\dfrac{a^2 + 3a - 18}{a^2 + 4} \div \dfrac{1}{3a^2 + 12}$.

Solution

$$\frac{a^2 + 3a - 18}{a^2 + 4} \div \frac{1}{3a^2 + 12} = \frac{a^2 + 3a - 18}{a^2 + 4} \cdot \frac{3a^2 + 12}{1}$$

$$= \frac{(a + 6)(a - 3)(3)(a^2 + 4)}{a^2 + 4} = 3(a + 6)(a - 3). \quad ■$$

EXAMPLE 7 Divide and simplify $\dfrac{2n^2 - 7n - 4}{6n^2 + 7n + 2} \div (n - 4)$.

Solution

$$\frac{2n^2 - 7n - 4}{6n^2 + 7n + 2} \div (n - 4) = \frac{2n^2 - 7n - 4}{6n^2 + 7n + 2} \cdot \frac{1}{n - 4}$$

$$= \frac{(2n + 1)(n - 4)}{(2n + 1)(3n + 2)(n - 4)}$$

$$= \frac{1}{3n + 2}. \quad ■$$

In a problem such as Example 7, it may be helpful to write the divisor with a denominator of 1. Thus, $n - 4$ can be written as $\dfrac{n - 4}{1}$; its reciprocal then is obviously $\dfrac{1}{n - 4}$.

Problem Set 7.2

For Problems 1–40, perform the indicated multiplications and divisions and express answers in simplest form.

1. $\dfrac{5}{9} \cdot \dfrac{3}{10}$

2. $\dfrac{7}{8} \cdot \dfrac{12}{14}$

3. $\left(-\dfrac{3}{4}\right)\left(\dfrac{6}{7}\right)$

4. $\left(\dfrac{5}{6}\right)\left(-\dfrac{4}{15}\right)$

5. $\left(\dfrac{17}{9}\right) \div \left(-\dfrac{19}{9}\right)$

6. $\left(-\dfrac{15}{7}\right) \div \left(\dfrac{13}{14}\right)$

7. $\dfrac{8xy}{12y} \cdot \dfrac{6x}{14y}$

8. $\dfrac{9x}{15y} \cdot \dfrac{20xy}{18x}$

9. $\left(-\dfrac{5n^2}{18n}\right)\left(\dfrac{27n}{25}\right)$

10. $\left(\dfrac{4ab}{10}\right)\left(-\dfrac{30a}{22b}\right)$

11. $\dfrac{3a^2}{7} \div \dfrac{6a}{28}$

12. $\dfrac{4x}{11y} \div \dfrac{12x}{33}$

13. $\dfrac{18a^2b^2}{-27a} \div \dfrac{-9a}{5b}$

14. $\dfrac{24ab^2}{25b} \div \dfrac{-12ab}{15a^2}$

15. $24x^3 \div \dfrac{16x}{y}$

16. $14xy^2 \div \dfrac{7y}{9}$

17. $\dfrac{1}{15ab^3} \div \dfrac{-1}{12a}$

18. $\dfrac{-2}{7a^2b^3} \div \dfrac{1}{9ab^4}$

19. $\dfrac{18rs}{34} \div 9r$

20. $\dfrac{8rs}{3} \div 6s$

21. $\dfrac{y}{x + y} \cdot \dfrac{x^2 - y^2}{xy}$

22. $\dfrac{x^2 - 9}{6} \cdot \dfrac{8}{x - 3}$

23. $\dfrac{2x^2 + xy}{xy} \cdot \dfrac{y}{10x + 5y}$

24. $\dfrac{x^2 + y^2}{x - y} \cdot \dfrac{x^2 - xy}{3}$

25. $\dfrac{6ab}{4ab + 4b^2} \div \dfrac{7a - 7b}{a^2 - b^2}$

26. $\dfrac{4ab}{2a^2 - 2ab} \div \dfrac{ab + b}{3a - 3b}$

27. $\dfrac{x^2 + 11x + 30}{x^2 + 4} \cdot \dfrac{5x^2 + 20}{x^2 + 14x + 45}$

28. $\dfrac{x^2 + 15x + 54}{x^2 + 2} \cdot \dfrac{3x^2 + 6}{x^2 + 10x + 9}$

29. $\dfrac{2x^2 - 3xy + y^2}{4x^2y} \div \dfrac{x^2 - y^2}{6x^2y^2}$

30. $\dfrac{2x^2 + xy - y^2}{x^2y} \div \dfrac{5x^2 + 4xy - y^2}{y}$

31. $\dfrac{a + a^2}{15a^2 + 11a + 2} \cdot \dfrac{1 - a}{1 - a^2}$

32. $\dfrac{2a^2 - 11a - 21}{3a^2 + a} \cdot \dfrac{3a^2 - 11a - 4}{2a^2 - 5a - 12}$

33. $\dfrac{2x^2 - 2xy}{x^2 + 4x - 32} \cdot \dfrac{x^2 - 16}{5xy - 5y^2}$

34. $\dfrac{x^3 + 3x^2}{x^2 + 4x + 4} \cdot \dfrac{x^2 - 5x - 14}{x^2 + 3x}$

35. $\dfrac{2x^2 - xy - 3y^2}{(x + y)^2} \div \dfrac{4x^2 - 12xy + 9y^2}{10x - 15y}$

36. $\dfrac{x^2 + 4xy + 4y^2}{x^2} \div \dfrac{x^2 - 4y^2}{x^2 - 2xy}$

37. $\dfrac{(3t - 1)^2}{45t - 15} \div \dfrac{12t^2 + 5t - 3}{20t + 5}$

38. $\dfrac{5t^2 - 3t - 2}{(t - 1)^2} \div \dfrac{5t^2 + 32t + 12}{4t^2 - 3t - 1}$

39. $\dfrac{n^3 - n}{n^2 + 7n + 6} \cdot \dfrac{4n + 24}{n^2 - n}$

40. $\dfrac{2x^2 - 6x - 36}{x^2 + 2x - 48} \cdot \dfrac{x^2 + 5x - 24}{2x^2 - 18}$

For Problems 41–46, perform the indicated operations and express the answers in simplest form. Remember that multiplications and divisions are done in the order that they appear from left to right.

41. $\dfrac{6}{9y} \div \dfrac{30x}{12y^2} \cdot \dfrac{5xy}{4}$

42. $\dfrac{5xy^2}{12y} \cdot \dfrac{18x^2}{15y} \div \dfrac{3}{2xy}$

43. $\dfrac{8x^2}{xy - xy^2} \cdot \dfrac{x - 1}{8x^2 - 8y^2} \div \dfrac{xy}{x + y}$

44. $\dfrac{5x - 20}{x^2 - 9} \cdot \dfrac{x + 3}{x - 4} \div \dfrac{15}{x - 3}$

45. $\dfrac{x^2 + 9x + 18}{x^2 + 3x} \cdot \dfrac{x^2 + 5x}{x^2 - 25} \div \dfrac{x^2 + 8x}{x^2 + 3x - 40}$

46. $\dfrac{4x}{3x + 6y} \cdot \dfrac{5xy}{x^2 - 4} \div \dfrac{10}{x^2 + 4x + 4}$

7.3 Adding and Subtracting Algebraic Fractions

In Chapter 2 addition and subtraction of rational numbers in $\dfrac{a}{b}$ form were defined as follows.

Definition 7.1

If a, b, and c are integers and b is not zero, then

$$\frac{a}{b} + \frac{c}{b} = \frac{a+c}{b}. \qquad \text{(Addition)}$$

$$\frac{a}{b} - \frac{c}{b} = \frac{a-c}{b}. \qquad \text{(Subtraction)}$$

This definition extends to algebraic fractions in general. Thus, if the denominators of two algebraic fractions are the same, then the fractions can be added or subtracted by adding or subtracting the numerators and placing the result over the common denominator.

$$\frac{5}{x} + \frac{7}{x} = \frac{5+7}{x} = \frac{12}{x};$$

$$\frac{8}{xy} - \frac{3}{xy} = \frac{8-3}{xy} = \frac{5}{xy};$$

$$\frac{14}{2x+1} + \frac{15}{2x+1} = \frac{14+15}{2x+1} = \frac{29}{2x+1};$$

$$\frac{3}{a-1} - \frac{4}{a-1} = \frac{3-4}{a-1} = \frac{-1}{a-1} \quad \text{or} \quad -\frac{1}{a-1}.$$

In the next examples notice how our previous work with simplifying polynomials is again put to use.

$$\frac{x+3}{4} + \frac{2x-3}{4} = \frac{(x+3)+(2x-3)}{4} = \frac{3x}{4};$$

$$\frac{x+5}{7} - \frac{x+2}{7} = \frac{(x+5)-(x+2)}{7} = \frac{x+5-x-2}{7} = \frac{3}{7};$$

$$\frac{3x+1}{xy} + \frac{2x+3}{xy} = \frac{(3x+1)+(2x+3)}{xy} = \frac{5x+4}{xy};$$

$$\frac{2(3n+1)}{n} - \frac{3(n-1)}{n} = \frac{2(3n+1)-3(n-1)}{n}$$

$$= \frac{6n+2-3n+3}{n}$$

$$= \frac{3n+5}{n}.$$

It may be that the fraction that results from adding or subtracting two fractions needs to be simplified.

$$\frac{4x-3}{8} + \frac{2x+3}{8} = \frac{(4x-3)+(2x+3)}{8} = \frac{6x}{8} = \frac{3x}{4};$$

$$\frac{3n-1}{12} - \frac{n-5}{12} = \frac{(3n-1)-(n-5)}{12} = \frac{3n-1-n+5}{12}$$

$$= \frac{2n+4}{12}$$

$$= \frac{2(n+2)}{12} = \frac{n+2}{6};$$

$$\frac{-2x+3}{x^2-4} + \frac{3x-1}{x^2-4} = \frac{(-2x+3)+(3x-1)}{x^2-4} = \frac{x+2}{x^2-4}$$

$$= \frac{\cancel{x+2}}{\cancel{(x+2)}(x-2)}$$

$$= \frac{1}{x-2}.$$

Recall that to add or subtract rational numbers with different denominators we first change to equivalent fractions having a common denominator. In fact, we found that by using the least common denominator (LCD) our work was made easier. Let's carefully review the process because it will also work with algebraic fractions in general.

EXAMPLE 1 Add $\frac{3}{5} + \frac{1}{4}$.

Solution By inspection we see that the LCD is 20. Thus, both fractions can be changed to equivalent fractions having a denominator of 20.

$$\frac{3}{5} + \frac{1}{4} = \frac{3}{5}\left(\frac{4}{4}\right) + \frac{1}{4}\left(\frac{5}{5}\right) = \frac{12}{20} + \frac{5}{20} = \frac{17}{20}.$$
$$\qquad\quad \underset{\substack{\uparrow \\ \text{form} \\ \text{of } 1}}{} \quad \underset{\substack{\uparrow \\ \text{form} \\ \text{of } 1}}{} \qquad\qquad\qquad\qquad\qquad ■$$

EXAMPLE 2 Subtract $\frac{5}{18} - \frac{7}{24}$.

Solution If the LCD cannot be found by inspection, then the prime factorization forms can be used.

$$18 = 2 \cdot 3 \cdot 3$$
$$24 = 2 \cdot 2 \cdot 2 \cdot 3$$

$\Big)$ $\quad$ LCD $= 2 \cdot 2 \cdot 2 \cdot 3 \cdot 3 = 72.$

$$\frac{5}{18} - \frac{7}{24} = \frac{5}{18}\left(\frac{4}{4}\right) - \frac{7}{24}\left(\frac{3}{3}\right) = \frac{20}{72} - \frac{21}{72} = -\frac{1}{72}.$$

■

Now let's consider adding and subtracting algebraic fractions with different denominators.

EXAMPLE 3 Add $\dfrac{x-2}{4} + \dfrac{3x+1}{3}$.

Solution By inspection we see that the LCD is 12.

$$\frac{x-2}{4} + \frac{3x+1}{3} = \left(\frac{x-2}{4}\right)\left(\frac{3}{3}\right) + \left(\frac{3x+1}{3}\right)\left(\frac{4}{4}\right)$$

$$= \frac{3(x-2)}{12} + \frac{4(3x+1)}{12}$$

$$= \frac{3(x-2) + 4(3x+1)}{12}$$

$$= \frac{3x - 6 + 12x + 4}{12} = \frac{15x - 2}{12}.$$

■

EXAMPLE 4 Subtract $\dfrac{n-2}{2} - \dfrac{n-6}{6}$.

Solution By inspection, the LCD is 6.

$$\frac{n-2}{2} - \frac{n-6}{6} = \left(\frac{n-2}{2}\right)\left(\frac{3}{3}\right) - \frac{n-6}{6}$$

$$= \frac{3(n-2)}{6} - \frac{(n-6)}{6}$$

$$= \frac{3(n-2) - (n-6)}{6}$$

$$= \frac{3n - 6 - n + 6}{6}$$

$$= \frac{2n}{6}$$

$$= \frac{n}{3}. \qquad \text{Don't forget to simplify!}$$

■

Having the denominators contain variables does not create any serious difficulties; our approach remains basically the same.

EXAMPLE 5 Add $\dfrac{3}{4x} + \dfrac{7}{3x}$.

Solution   By inspection the LCD is $12x$.

$$\frac{3}{4x} + \frac{7}{3x} = \frac{3}{4x}\left(\frac{3}{3}\right) + \frac{7}{3x}\left(\frac{4}{4}\right) = \frac{9}{12x} + \frac{28}{12x} = \frac{37}{12x}.$$ ■

EXAMPLE 6 Subtract $\dfrac{11}{12x} - \dfrac{5}{14x}$.

Solution
$$\left.\begin{array}{l} 12x = 2 \cdot 2 \cdot 3 \cdot x \\[4pt] 14x = 2 \cdot 7 \cdot x \end{array}\right) \quad \text{LCD} = 2 \cdot 2 \cdot 3 \cdot 7 \cdot x = 84x.$$

$$\frac{11}{12x} - \frac{5}{14x} = \frac{11}{12x}\left(\frac{7}{7}\right) - \frac{5}{14x}\left(\frac{6}{6}\right)$$

$$= \frac{77}{84x} - \frac{30}{84x}$$

$$= \frac{47}{84x}.$$ ■

EXAMPLE 7 Add $\dfrac{2}{y} + \dfrac{4}{y-2}$.

Solution   By inspection the LCD is $y(y-2)$.

$$\frac{2}{y} + \frac{4}{y-2} = \frac{2}{y}\left(\frac{y-2}{y-2}\right) + \frac{4}{y-2}\left(\frac{y}{y}\right)$$

$$\underset{\substack{\uparrow \\ \text{form} \\ \text{of }1}}{} \qquad \underset{\substack{\uparrow \\ \text{form} \\ \text{of }1}}{}$$

$$= \frac{2(y-2)}{y(y-2)} + \frac{4y}{y(y-2)}$$

$$= \frac{2(y-2) + 4y}{y(y-2)}$$

$$= \frac{2y - 4 + 4y}{y(y-2)} = \frac{6y-4}{y(y-2)}.$$ ■

Notice the final result in Example 7. The numerator, $6y - 4$, can be factored into $2(3y - 2)$. However, since this produces no common factors with the denominator, the fraction cannot be simplified. Thus, the final answer can be left as $\dfrac{6y - 4}{y(y - 2)}$, or it would also be acceptable to express it as $\dfrac{2(3y - 2)}{y(y - 2)}$.

EXAMPLE 8 Subtract $\dfrac{4}{x + 2} - \dfrac{7}{x + 3}$.

Solution By inspection the LCD is $(x + 2)(x + 3)$.

$$\frac{4}{x + 2} - \frac{7}{x + 3} = \left(\frac{4}{x + 2}\right)\left(\frac{x + 3}{x + 3}\right) - \left(\frac{7}{x + 3}\right)\left(\frac{x + 2}{x + 2}\right)$$

$$= \frac{4(x + 3)}{(x + 2)(x + 3)} - \frac{7(x + 2)}{(x + 3)(x + 2)}$$

$$= \frac{4(x + 3) - 7(x + 2)}{(x + 2)(x + 3)}$$

$$= \frac{4x + 12 - 7x - 14}{(x + 2)(x + 3)}$$

$$= \frac{-3x - 2}{(x + 2)(x + 3)}. \qquad ■$$

Problem Set 7.3

For Problems 1–34, add or subtract as indicated. Be sure to express answers in simplest form.

1. $\dfrac{5}{x} + \dfrac{12}{x}$

2. $\dfrac{17}{x} - \dfrac{13}{x}$

3. $\dfrac{7}{3x} - \dfrac{5}{3x}$

4. $\dfrac{4}{5x} + \dfrac{3}{5x}$

5. $\dfrac{7}{2n} + \dfrac{1}{2n}$

6. $\dfrac{5}{3n} + \dfrac{4}{3n}$

7. $\dfrac{9}{4x^2} - \dfrac{13}{4x^2}$

8. $\dfrac{12}{5x^2} - \dfrac{22}{5x^2}$

9. $\dfrac{x + 1}{x} + \dfrac{3}{x}$

10. $\dfrac{x - 2}{x} + \dfrac{4}{x}$

11. $\dfrac{3}{x - 1} - \dfrac{6}{x - 1}$

12. $\dfrac{8}{x + 4} - \dfrac{10}{x + 4}$

13. $\dfrac{x + 1}{x} - \dfrac{1}{x}$

14. $\dfrac{2x + 3}{x} - \dfrac{3}{x}$

15. $\dfrac{3t - 1}{4} + \dfrac{2t + 3}{4}$

16. $\dfrac{4t - 1}{7} + \dfrac{8t - 5}{7}$

17. $\dfrac{7a + 2}{3} - \dfrac{4a - 6}{3}$

18. $\dfrac{9a - 1}{6} - \dfrac{4a - 2}{6}$

19. $\dfrac{4n+3}{8} + \dfrac{6n+5}{8}$

20. $\dfrac{2n-5}{10} + \dfrac{6n-1}{10}$

21. $\dfrac{3n-7}{6} - \dfrac{9n-1}{6}$

22. $\dfrac{2n-6}{5} - \dfrac{7n-1}{5}$

23. $\dfrac{5x-2}{7x} - \dfrac{8x+3}{7x}$

24. $\dfrac{4x+1}{3x} - \dfrac{2x+5}{3x}$

25. $\dfrac{3(x+2)}{4x} + \dfrac{6(x-1)}{4x}$

26. $\dfrac{4(x-3)}{5x} + \dfrac{2(x+6)}{5x}$

27. $\dfrac{6(n-1)}{3n} + \dfrac{3(n+2)}{3n}$

28. $\dfrac{2(n-4)}{3n} + \dfrac{4(n+2)}{3n}$

29. $\dfrac{2(3x-4)}{7x^2} - \dfrac{7x-8}{7x^2}$

30. $\dfrac{3(4x-3)}{8x^2} - \dfrac{11x-9}{8x^2}$

31. $\dfrac{a^2}{a+2} - \dfrac{4}{a+2}$

32. $\dfrac{n^2}{n-4} - \dfrac{16}{n-4}$

33. $\dfrac{3x}{(x-6)^2} - \dfrac{18}{(x-6)^2}$

34. $\dfrac{x^2+5x}{(x+1)^2} + \dfrac{4}{(x+1)^2}$

For Problems 35–80, add or subtract as indicated and express answers in simplest form.

35. $\dfrac{3x}{8} + \dfrac{5x}{4}$

36. $\dfrac{5x}{3} + \dfrac{2x}{9}$

37. $\dfrac{7n}{12} - \dfrac{4n}{3}$

38. $\dfrac{n}{6} - \dfrac{7n}{12}$

39. $\dfrac{y}{6} + \dfrac{3y}{4}$

40. $\dfrac{3y}{4} + \dfrac{7y}{5}$

41. $\dfrac{8x}{3} - \dfrac{3x}{7}$

42. $\dfrac{5y}{6} - \dfrac{3y}{8}$

43. $\dfrac{2x}{6} + \dfrac{3x}{5}$

44. $\dfrac{6x}{9} + \dfrac{7x}{12}$

45. $\dfrac{7n}{8} - \dfrac{3n}{9}$

46. $\dfrac{8n}{10} - \dfrac{7n}{15}$

47. $\dfrac{x+3}{5} + \dfrac{x-4}{2}$

48. $\dfrac{x-2}{5} + \dfrac{x+1}{6}$

49. $\dfrac{x-6}{9} + \dfrac{x+2}{3}$

50. $\dfrac{x-2}{4} + \dfrac{x+4}{8}$

51. $\dfrac{3n-1}{3} + \dfrac{2n+5}{4}$

52. $\dfrac{2n+3}{4} + \dfrac{4n-1}{7}$

53. $\dfrac{4n-3}{6} - \dfrac{3n+5}{18}$

54. $\dfrac{5n-2}{12} - \dfrac{4n+7}{6}$

55. $\dfrac{3x}{4} + \dfrac{x}{6} - \dfrac{5x}{8}$

56. $\dfrac{5x}{2} - \dfrac{3x}{4} - \dfrac{7x}{6}$

57. $\dfrac{x}{5} - \dfrac{3}{10} - \dfrac{7x}{12}$

58. $\dfrac{4x}{3} + \dfrac{5}{9} - \dfrac{11x}{6}$

59. $\dfrac{5}{8x} + \dfrac{1}{6x}$

60. $\dfrac{7}{8x} + \dfrac{5}{12x}$

61. $\dfrac{5}{6y} - \dfrac{7}{9y}$

62. $\dfrac{11}{9y} - \dfrac{8}{15y}$

63. $\dfrac{5}{12x} - \dfrac{11}{16x^2}$

64. $\dfrac{4}{9x} - \dfrac{7}{6x^2}$

65. $\dfrac{3}{2x} - \dfrac{2}{3x} + \dfrac{5}{4x}$

66. $\dfrac{3}{4x} - \dfrac{5}{6x} + \dfrac{10}{9x}$

67. $\dfrac{3}{x-5} + \dfrac{7}{x}$

68. $\dfrac{4}{x-8} + \dfrac{9}{x}$

69. $\dfrac{2}{n-1} - \dfrac{3}{n}$

70. $\dfrac{5}{n+3} - \dfrac{7}{n}$

71. $\dfrac{4}{n} - \dfrac{6}{n+4}$

72. $\dfrac{8}{n} - \dfrac{3}{n-9}$

73. $\dfrac{6}{x} - \dfrac{12}{2x+1}$

74. $\dfrac{2}{x} - \dfrac{6}{3x - 2}$ **75.** $\dfrac{4}{x + 4} + \dfrac{6}{x - 3}$ **76.** $\dfrac{7}{x - 2} + \dfrac{8}{x + 1}$

77. $\dfrac{3}{x - 2} - \dfrac{9}{x + 1}$ **78.** $\dfrac{5}{x - 1} - \dfrac{4}{x + 6}$ **79.** $\dfrac{3}{2x - 1} - \dfrac{4}{3x + 1}$

80. $\dfrac{6}{3x - 4} - \dfrac{4}{2x + 3}$

■ Miscellaneous Problems

Consider the addition problem $\dfrac{9}{x - 2} + \dfrac{4}{2 - x}$. Notice that the denominators, $x - 2$ and $2 - x$, are opposites of each other; that is, $-1(2 - x) = (x - 2)$. In such cases, adding the fractions can be accomplished as follows.

$$\frac{9}{x - 2} + \frac{4}{2 - x} = \frac{9}{x - 2} + \frac{4}{2 - x}\left(\frac{-1}{-1}\right)$$

$$\underset{\substack{\uparrow \\ \text{form} \\ \text{of } 1}}{}$$

$$= \frac{9}{x - 2} + \frac{-4}{x - 2}$$

$$= \frac{9 + (-4)}{x - 2}$$

$$= \frac{5}{x - 2}.$$

For Problems 81–86, use this approach to help with the additions and subtractions.

81. $\dfrac{7}{x - 1} - \dfrac{2}{1 - x}$ **82.** $\dfrac{5}{x - 3} + \dfrac{1}{3 - x}$

83. $\dfrac{x}{x - 4} + \dfrac{4}{4 - x}$ **84.** $\dfrac{-4}{a - 1} + \dfrac{2}{1 - a}$

85. $\dfrac{1}{x^2 - 9} + \dfrac{2}{x + 3} - \dfrac{3}{3 - x}$ **86.** $\dfrac{n}{2n - 1} - \dfrac{3}{1 - 2n}$

7.4 More on Addition and Subtraction of Algebraic Fractions

Success in adding and subtracting algebraic fractions depends upon both a good understanding of the process and the ability to perform the various computations. Regarding the computational work, you should adopt a carefully organized format showing as many steps as you need in order to minimize the chances of making careless errors. Don't be eager to find a shortcut for doing these kinds of problems.

Study the following examples very carefully. Notice the same basic procedure for each problem of (1) finding the LCD; (2) changing each fraction to an equivalent fraction having the LCD as its denominator; (3) adding or subtracting numerators and placing this result over the LCD; and (4) looking for possibilities to simplify the resulting fraction.

EXAMPLE 1 Add $\dfrac{3}{x^2 + 2x} + \dfrac{5}{x}$.

Solution

1st denominator: $x^2 + 2x = x(x + 2)$ $\left.\vphantom{\begin{matrix}a\\b\end{matrix}}\right\rangle$ $\longrightarrow$ LCD is $x(x + 2)$.
2nd denominator: x

$$\frac{3}{x^2 + 2x} + \frac{5}{x} = \frac{3}{x(x + 2)} + \frac{5}{x}\left(\frac{x + 2}{x + 2}\right)$$

              ↑   ↑
       This fraction   form
       has the LCD   of 1
       as its
       denominator.

$$= \frac{3}{x(x + 2)} + \frac{5(x + 2)}{x(x + 2)}$$

$$= \frac{3 + 5(x + 2)}{x(x + 2)}$$

$$= \frac{3 + 5x + 10}{x(x + 2)}$$

$$= \frac{5x + 13}{x(x + 2)}.$$ ■

EXAMPLE 2 Subtract $\dfrac{4}{x^2 - 4} - \dfrac{1}{x - 2}$.

Solution

$\begin{matrix} x^2 - 4 = (x + 2)(x - 2) \\ x - 2 = x - 2 \end{matrix}$ $\left.\vphantom{\begin{matrix}a\\b\end{matrix}}\right\rangle$ $\longrightarrow$ LCD is $(x + 2)(x - 2)$

$$\frac{4}{x^2 - 4} - \frac{1}{x - 2} = \frac{4}{(x + 2)(x - 2)} - \left(\frac{1}{x - 2}\right)\left(\frac{x + 2}{x + 2}\right)$$

$$= \frac{4}{(x + 2)(x - 2)} - \frac{1(x + 2)}{(x + 2)(x - 2)}$$

$$= \frac{4 - 1(x + 2)}{(x + 2)(x - 2)}$$

$$= \frac{4 - x - 2}{(x + 2)(x - 2)}$$

$$= \frac{-x + 2}{(x + 2)(x - 2)}$$

$$= \frac{-1(x - 2)}{(x + 2)(x - 2)} \longrightarrow \text{Note the changing of} \\ -x + 2 \text{ to } -1(x - 2).$$

$$= -\frac{1}{x + 2}. \qquad ■$$

EXAMPLE 3 Add $\dfrac{2}{a^2 - 9} + \dfrac{3}{a^2 + 5a + 6}$.

Solution

$$\left. \begin{array}{l} a^2 - 9 = (a + 3)(a - 3) \\ a^2 + 5a + 6 = (a + 3)(a + 2) \end{array} \right\} \longrightarrow \text{LCD is } (a + 3)(a - 3)(a + 2).$$

$$\frac{2}{a^2 - 9} + \frac{3}{a^2 + 5a + 6}$$

$$= \left(\frac{2}{(a + 3)(a - 3)} \right) \left(\frac{a + 2}{a + 2} \right) + \left(\frac{3}{(a + 3)(a + 2)} \right) \left(\frac{a - 3}{a - 3} \right)$$

$$\begin{array}{cc} \uparrow & \uparrow \\ \text{form} & \text{form} \\ \text{of 1} & \text{of 1} \end{array}$$

$$= \frac{2(a + 2)}{(a + 3)(a - 3)(a + 2)} + \frac{3(a - 3)}{(a + 3)(a - 3)(a + 2)}$$

$$= \frac{2(a + 2) + 3(a - 3)}{(a + 3)(a - 3)(a + 2)}$$

$$= \frac{2a + 4 + 3a - 9}{(a + 3)(a - 3)(a + 2)}$$

$$= \frac{5a - 5}{(a + 3)(a - 3)(a + 2)} \quad \text{or} \quad \frac{5(a - 1)}{(a + 3)(a - 3)(a + 2)}. \qquad ■$$

EXAMPLE 4 Perform the indicated operations.

$$\frac{2x}{x^2 - y^2} + \frac{3}{x + y} - \frac{2}{x - y}.$$

Solution

$$x^2 - y^2 = (x + y)(x - y)$$
$$x + y = x + y \qquad \left.\right\} \longrightarrow \text{LCD is } (x + y)(x - y).$$
$$x - y = x - y$$

$$\frac{2x}{x^2 - y^2} + \frac{3}{x + y} - \frac{2}{x - y}$$

$$= \frac{2x}{(x + y)(x - y)} + \left(\frac{3}{x + y}\right)\left(\frac{x - y}{x - y}\right) - \left(\frac{2}{x - y}\right)\left(\frac{x + y}{x + y}\right)$$

$$= \frac{2x}{(x + y)(x - y)} + \frac{3(x - y)}{(x + y)(x - y)} - \frac{2(x + y)}{(x + y)(x - y)}$$

$$= \frac{2x + 3(x - y) - 2(x + y)}{(x + y)(x - y)}$$

$$= \frac{2x + 3x - 3y - 2x - 2y}{(x + y)(x - y)}$$

$$= \frac{3x - 5y}{(x + y)(x - y)}. \qquad\qquad ■$$

■ Complex Fractions

Fractional forms that contain fractions in the numerators and/or denominators are called **complex fractions**. The following are examples of complex fractions.

$$\frac{\frac{2}{3}}{\frac{4}{5}}; \quad \frac{\frac{1}{x}}{\frac{3}{y}}; \quad \frac{\frac{1}{2} + \frac{1}{3}}{\frac{5}{6} - \frac{1}{4}}; \quad \frac{\frac{2}{x} + \frac{3}{y}}{\frac{5}{x} - \frac{1}{y^2}}.$$

It is often necessary to *simplify* a complex fraction, that is, to express it as a simple fraction. Let's illustrate this process with the four previous examples.

EXAMPLE 5 Simplify $\dfrac{\frac{2}{3}}{\frac{4}{5}}$.

Solution

This type of problem creates no difficulty since it is merely a division problem. Thus,

$$\frac{\frac{2}{3}}{\frac{4}{5}} = \frac{2}{3} \div \frac{4}{5} = \frac{\overset{1}{\cancel{2}}}{3} \cdot \frac{5}{\underset{2}{\cancel{4}}} = \frac{5}{6}.$$

EXAMPLE 6 Simplify $\dfrac{\dfrac{1}{x}}{\dfrac{3}{y}}$.

Solution $\dfrac{\dfrac{1}{x}}{\dfrac{3}{y}} = \dfrac{1}{x} \div \dfrac{3}{y} = \dfrac{1}{x} \cdot \dfrac{y}{3} = \dfrac{y}{3x}.$

EXAMPLE 7 Simplify $\dfrac{\dfrac{1}{2} + \dfrac{1}{3}}{\dfrac{5}{6} - \dfrac{1}{4}}$.

Solution A $\dfrac{\dfrac{1}{2} + \dfrac{1}{3}}{\dfrac{5}{6} - \dfrac{1}{4}} = \dfrac{\dfrac{3}{6} + \dfrac{2}{6}}{\dfrac{10}{12} - \dfrac{3}{12}} = \dfrac{\dfrac{5}{6}}{\dfrac{7}{12}} = \dfrac{5}{\overset{1}{6}} \cdot \dfrac{\overset{2}{12}}{7} = \dfrac{10}{7}.$

↖ Invert divisor and multiply.

Solution B The least common multiple of all four denominators (2, 3, 6, and 4) is 12. We shall multiply the entire complex fraction by a form of 1, namely, $\dfrac{12}{12}$.

$$\dfrac{\dfrac{1}{2} + \dfrac{1}{3}}{\dfrac{5}{6} - \dfrac{1}{4}} = \left(\dfrac{12}{12}\right)\left(\dfrac{\dfrac{1}{2} + \dfrac{1}{3}}{\dfrac{5}{6} - \dfrac{1}{4}}\right)$$

$$= \dfrac{12\left(\dfrac{1}{2} + \dfrac{1}{3}\right)}{12\left(\dfrac{5}{6} - \dfrac{1}{4}\right)}$$

$$= \dfrac{6 + 4}{10 - 3} = \dfrac{10}{7}.$$ ■

EXAMPLE 8 Simplify $\dfrac{\dfrac{2}{x} + \dfrac{3}{y}}{\dfrac{5}{x} - \dfrac{1}{y^2}}$

Solution A

$$\frac{\dfrac{2}{x}+\dfrac{3}{y}}{\dfrac{5}{x}-\dfrac{1}{y^2}}=\frac{\dfrac{2}{x}\left(\dfrac{y}{y}\right)+\dfrac{3}{y}\left(\dfrac{x}{x}\right)}{\dfrac{5}{x}\left(\dfrac{y^2}{y^2}\right)-\dfrac{1}{y^2}\left(\dfrac{x}{x}\right)}=\frac{\dfrac{2y}{xy}+\dfrac{3x}{xy}}{\dfrac{5y^2}{xy^2}-\dfrac{x}{xy^2}}$$

$$=\frac{\dfrac{2y+3x}{xy}}{\dfrac{5y^2-x}{xy^2}}$$

$$=\frac{2y+3x}{\cancel{xy}}\cdot\frac{\overset{y}{\cancel{xy^2}}}{5y^2-x}$$

$$=\frac{y(2y+3x)}{5y^2-x}.$$ ■

Solution B   The least common multiple of all four denominators (x, y, x, and y^2) is xy^2. We shall multiply the entire complex fraction by a form of 1, namely, $\dfrac{xy^2}{xy^2}$.

$$\frac{\dfrac{2}{x}+\dfrac{3}{y}}{\dfrac{5}{x}-\dfrac{1}{y^2}}=\left(\frac{xy^2}{xy^2}\right)\frac{\left(\dfrac{2}{x}+\dfrac{3}{y}\right)}{\left(\dfrac{5}{x}-\dfrac{1}{y^2}\right)}$$

$$=\frac{xy^2\left(\dfrac{2}{x}+\dfrac{3}{y}\right)}{xy^2\left(\dfrac{5}{x}-\dfrac{1}{y^2}\right)}$$

$$=\frac{2y^2+3xy}{5y^2-x}\quad\text{or}\cdot\quad\frac{y(2y+3x)}{5y^2-x}.$$ ■

Certainly either approach (Solution A or Solution B) will work for problems such as Examples 7 and 8. You should examine Solution B for each of the examples rather carefully. This approach works very effectively with algebraic complex fractions where the LCD of all the denominators of the simple fractions is easy to find. The two methods for simplifying a complex fraction can be summarized as follows.

1. Simplify the numerator and denominator of the fraction separately. Then divide the simplified numerator by the simplied denominator.

2. Multiply the numerator and denominator of the complex fraction by the least common multiple of all of the denominators appearing in the complex fraction.

Problem Set 7.4

For Problems 1–40, perform the indicated operations and express answers in simplest form.

1. $\dfrac{4}{x^2 - 4x} + \dfrac{3}{x}$

2. $\dfrac{3}{x^2 + 2x} + \dfrac{7}{x}$

3. $\dfrac{7}{x^2 + 2x} - \dfrac{5}{x}$

4. $\dfrac{9}{x^2 - 5x} - \dfrac{1}{x}$

5. $\dfrac{8}{n} - \dfrac{2}{n^2 - 6n}$

6. $\dfrac{6}{n} - \dfrac{4}{n^2 + 6n}$

7. $\dfrac{4}{n^2 + n} - \dfrac{4}{n}$

8. $\dfrac{8}{n^2 - 2n} + \dfrac{4}{n}$

9. $\dfrac{7}{2x} - \dfrac{x}{x^2 - x}$

10. $\dfrac{3x}{x^2 + 2x} + \dfrac{4}{5x}$

11. $\dfrac{3}{x^2 - 16} + \dfrac{5}{x + 4}$

12. $\dfrac{6}{x^2 - 9} + \dfrac{9}{x - 3}$

13. $\dfrac{8x}{x^2 - 1} - \dfrac{4}{x - 1}$

14. $\dfrac{6x}{x^2 - 4} - \dfrac{3}{x + 2}$

15. $\dfrac{4}{a^2 - 2a} + \dfrac{7}{a^2 + 2a}$

16. $\dfrac{3}{a^2 + 4a} + \dfrac{5}{a^2 - 4a}$

17. $\dfrac{1}{x^2 - 6x} - \dfrac{1}{x^2 + 6x}$

18. $\dfrac{3}{x^2 + 5x} - \dfrac{4}{x^2 - 5x}$

19. $\dfrac{n}{n^2 - 16} - \dfrac{2}{3n + 12}$

20. $\dfrac{n}{n^2 - 25} - \dfrac{2}{3n - 15}$

21. $\dfrac{5x}{6x + 4} + \dfrac{2x}{9x + 6}$

22. $\dfrac{7x}{3x - 12} + \dfrac{3x}{4x - 16}$

23. $\dfrac{x - 1}{5x + 5} - \dfrac{x - 4}{3x + 3}$

24. $\dfrac{2x + 1}{6x + 12} - \dfrac{3x - 4}{8x + 16}$

25. $\dfrac{2}{x^2 + 7x + 12} + \dfrac{3}{x^2 - 9}$

26. $\dfrac{x}{x^2 - 1} + \dfrac{3}{x^2 + 5x + 4}$

27. $\dfrac{x}{x^2 + 6x + 8} - \dfrac{5}{x^2 - 3x - 10}$

28. $\dfrac{x}{x^2 - x - 30} - \dfrac{7}{x^2 - 7x + 6}$

29. $\dfrac{a}{ab + b^2} - \dfrac{b}{a^2 + ab}$

30. $\dfrac{2x}{xy + y^2} - \dfrac{2y}{x^2 + xy}$

31. $\dfrac{3}{x - 5} - \dfrac{4}{x^2 - 25} + \dfrac{5}{x + 5}$

32. $\dfrac{2}{x + 1} + \dfrac{3}{x^2 - 1} - \dfrac{5}{x - 1}$

33. $\dfrac{10}{x^2 - 2x} + \dfrac{8}{x^2 + 2x} - \dfrac{3}{x^2 - 4}$

34. $\dfrac{1}{x^2 + 7x} - \dfrac{2}{x^2 - 7x} - \dfrac{5}{x^2 - 49}$

35. $\dfrac{3x}{x^2 + 7x + 10} - \dfrac{2}{x + 2} + \dfrac{3}{x + 5}$

36. $\dfrac{4}{x - 3} - \dfrac{3}{x - 6} + \dfrac{x}{x^2 - 9x + 18}$

37. $\dfrac{5x}{3x^2 + 7x - 20} - \dfrac{1}{3x - 5} - \dfrac{2}{x + 4}$

38. $\dfrac{4x}{6x^2 + 7x + 2} - \dfrac{2}{2x + 1} - \dfrac{4}{3x + 2}$

39. $\dfrac{2}{x + 4} - \dfrac{1}{x - 3} + \dfrac{2x + 1}{x^2 + x - 12}$

40. $\dfrac{3}{x - 5} - \dfrac{4}{x + 7} + \dfrac{3x - 27}{x^2 + 2x - 35}$

For Problems 41–60, simplify each of the complex fractions.

41. $\dfrac{\dfrac{1}{2} - \dfrac{3}{4}}{\dfrac{1}{6} + \dfrac{1}{3}}$

42. $\dfrac{\dfrac{3}{8} + \dfrac{1}{4}}{\dfrac{1}{2} + \dfrac{3}{16}}$

43. $\dfrac{\dfrac{2}{9} + \dfrac{1}{3}}{\dfrac{5}{6} - \dfrac{2}{3}}$

44. $\dfrac{\dfrac{7}{8} - \dfrac{1}{3}}{\dfrac{1}{6} + \dfrac{3}{4}}$

45. $\dfrac{3 - \dfrac{2}{3}}{2 + \dfrac{1}{4}}$

46. $\dfrac{4 + \dfrac{3}{5}}{\dfrac{1}{3} - 2}$

47. $\dfrac{\dfrac{3}{x}}{\dfrac{9}{y}}$

48. $\dfrac{\dfrac{-6}{a}}{\dfrac{8}{b}}$

49. $\dfrac{\dfrac{2}{x} + \dfrac{3}{y}}{\dfrac{5}{x} - \dfrac{1}{y}}$

50. $\dfrac{\dfrac{3}{x^2} - \dfrac{2}{x}}{\dfrac{4}{x} - \dfrac{7}{x^2}}$

51. $\dfrac{\dfrac{1}{y} - \dfrac{4}{x^2}}{\dfrac{7}{x} - \dfrac{3}{y}}$

52. $\dfrac{\dfrac{4}{ab} + \dfrac{2}{b}}{\dfrac{8}{a} + \dfrac{1}{b}}$

53. $\dfrac{\dfrac{6}{x} + 2}{\dfrac{3}{x} + 4}$

54. $\dfrac{1 - \dfrac{6}{y}}{3 - \dfrac{2}{y}}$

55. $\dfrac{\dfrac{3}{2x^2} - \dfrac{4}{x}}{\dfrac{5}{3x} + \dfrac{7}{x^2}}$

56. $\dfrac{\dfrac{4}{3x} + \dfrac{5}{x^2}}{\dfrac{7}{4x} - \dfrac{9}{x}}$

57. $\dfrac{\dfrac{x + 2}{4}}{\dfrac{1}{x} + \dfrac{3}{2}}$

58. $\dfrac{\dfrac{3}{x + 1} + 2}{-4 + \dfrac{2}{x + 1}}$

59. $\dfrac{\dfrac{1}{x - 1} - 2}{\dfrac{3}{x - 1} + 4}$

60. $\dfrac{\dfrac{3}{x - 2} + \dfrac{2}{x + 2}}{\dfrac{4}{x + 2} - \dfrac{5}{x - 2}}$

For Problems 61–71, answer each question with an algebraic fraction.

61. If jogging at a constant rate Joan can complete a race in 40 minutes, how much of the course has she completed at the end of m minutes?

62. If Kent can mow the entire lawn in m minutes, what fractional part of the lawn has he mowed at the end of 20 minutes?

63. If Sandy drove k kilometers at a rate of r kilometers per hour, how long did it take her to make the trip?

64. If Roy traveled m miles in h hours, what was his rate in miles per hour?

65. If l liters of gasoline cost d dollars, what is the price per liter?

66. If p pounds of candy cost c cents, what is the price per pound?

67. The product of two numbers is 34 and one of the numbers is n. What is the other number?

68. If a cold water faucet, when opened, can fill a tank in 3 hours, how much of the tank is filled at the end of h hours?

69. If the area of a rectangle is 47 square inches and the length is l inches, what is the width of the rectangle?

70. If the area of a rectangle is 56 square centimeters and the width is w centimeters, what is the length of the rectangle?

71. If the area of a triangle is 48 square feet and the length of one side is b feet, what is the length of the altitude to that side?

■ Miscellaneous Problems

For Problems 72–75, simplify each complex fraction.

72. $1 - \dfrac{n}{1 - \dfrac{1}{n}}$

73. $2 - \dfrac{3n}{1 + \dfrac{4}{n}}$

74. $\dfrac{3x}{4 - \dfrac{2}{x}} - 1$

75. $\dfrac{5x}{3 + \dfrac{1}{x}} + 2$

7.5 Fractional Equations

We shall consider fractional equations of two basic types in this text. One type has only constants as denominators and the other type has some variables in the denominators. In Chapter 3, we considered fractional equations involving only constants in the denominators. Let's review our approach to such equations since we will be using that same basic technique to solve any fractional equation.

EXAMPLE 1 Solve $\dfrac{x - 2}{3} + \dfrac{x + 1}{4} = \dfrac{1}{6}$.

Solution

$$\frac{x - 2}{3} + \frac{x + 1}{4} = \frac{1}{6}.$$

$$12\left(\frac{x - 2}{3} + \frac{x + 1}{4}\right) = 12\left(\frac{1}{6}\right) \qquad \text{Multiply both sides by 12, which is the LCD of all three denominators.}$$

$$12\left(\frac{x - 2}{3}\right) + 12\left(\frac{x + 1}{4}\right) = 12\left(\frac{1}{6}\right)$$

$$4(x - 2) + 3(x + 1) = 2$$

$$4x - 8 + 3x + 3 = 2$$

$$7x - 5 = 2$$

$$7x = 7$$

$$x = 1.$$

The solution set is $\{1\}$. (Check it!) ■

If an equation contains a variable in one or more denominators, then we proceed in essentially the same way *except we must avoid any value of the variable that makes a denominator zero.* Consider the following example.

EXAMPLE 2 Solve $\dfrac{3}{x} + \dfrac{1}{2} = \dfrac{5}{x}$.

Solution   First, we need to realize that x *cannot equal zero*. Then, we can proceed in the usual way.

$$\frac{3}{x} + \frac{1}{2} = \frac{5}{x}$$

$$2x\left(\frac{3}{x} + \frac{1}{2}\right) = 2x\left(\frac{5}{x}\right)$$ Multiply both sides by $2x$, which is the LCD of all denominators.

$$6 + x = 10$$

$$x = 4.$$

Check: $\dfrac{3}{x} + \dfrac{1}{2} = \dfrac{5}{x}$ becomes $\dfrac{3}{4} + \dfrac{1}{2} \overset{?}{=} \dfrac{5}{4}$ when $x = 4$.

$$\frac{3}{4} + \frac{2}{4} \overset{?}{=} \frac{5}{4}$$

$$\frac{5}{4} = \frac{5}{4}.$$

The solution set is $\{4\}$.   ■

EXAMPLE 3 Solve $\dfrac{5}{x+2} = \dfrac{2}{x-1}$.

Solution   Since neither denominator can be zero, we know that $x \neq -2$ and $x \neq 1$.

$$\frac{5}{x+2} = \frac{2}{x-1}$$

$$(x+2)(x-1)\left(\frac{5}{x+2}\right) = (x+2)(x-1)\left(\frac{2}{x-1}\right)$$ Multiply both sides by $(x+2)(x-1)$ which is the LCD.

$$5(x-1) = 2(x+2)$$

$$5x - 5 = 2x + 4$$

$$3x = 9$$

$$x = 3.$$

Since the only restrictions are $x \neq -2$ and $x \neq 1$, the solution set is $\{3\}$. (Check it!)   ■

EXAMPLE 4 Solve $\dfrac{2}{x-2} + 2 = \dfrac{x}{x-2}$.

Solution   Since no denominators can be zero, $x \neq 2$.

$$\frac{2}{x-2} + 2 = \frac{x}{x-2}$$

$$(x-2)\left(\frac{2}{x-2} + 2\right) = (x-2)\left(\frac{x}{x-2}\right) \qquad \text{Multiply by } x-2, \text{ which is the LCD.}$$

$$2 + 2(x-2) = x$$

$$2 + 2x - 4 = x$$

$$2x - 2 = x$$

$$x = 2.$$

Since 2 cannot be used because it will produce a denominator of zero, there is no solution to the given equation. The solution set is $\emptyset$. ■

Example 4 illustrates the importance of recognizing the restrictions that must be placed on possible values of a variable. We will always indicate such restrictions at the beginning of our solution.

EXAMPLE 5   Solve $\dfrac{125 - n}{n} = 4 + \dfrac{10}{n}$.

Solution

$$\frac{125 - n}{n} = 4 + \frac{10}{n}, \qquad n \neq 0 \qquad \text{Note the necessary restriction.}$$

$$n\left(\frac{125 - n}{n}\right) = n\left(4 + \frac{10}{n}\right) \qquad \text{Multiply both sides by } n.$$

$$125 - n = 4n + 10$$

$$115 = 5n$$

$$23 = n.$$

Since the only restriction is $n \neq 0$, the solution set is $\{23\}$. ■

You guessed it! We are now ready to solve more word problems, specifically those that translate into fractional equations.

PROBLEM 1   One number is 10 larger than another number. The indicated quotient of the smaller number divided by the larger number reduces to $\dfrac{3}{5}$. Find the numbers.

Solution   Let n represent the smaller number. Then $n + 10$ represents the larger number. The second sentence in the statement of the problem translates into the following equation.

$$\frac{n}{n+10} = \frac{3}{5}, \qquad n \neq = -10$$

$$5n = 3(n+10) \qquad \text{Cross products are equal.}$$
$$5n = 3n + 30$$
$$2n = 30$$
$$n = 15.$$

If n is 15, then $n + 10$ is 25. Thus, the numbers are 15 and 25. To check, consider the quotient of the smaller number divided by the larger number.

$$\frac{15}{25} = \frac{3 \cdot \cancel{5}}{5 \cdot \cancel{5}} = \frac{3}{5} \qquad\qquad ■$$

Before tackling the next word problem, let's review a basic relationship pertaining to division. We can always check a division problem by "multiplying the divisor times the quotient and adding the remainder." This can be expressed as

$$\text{dividend} = (\text{divisor})(\text{quotient}) + \text{remainder}.$$

Sometimes the remainder is expressed as a fractional part of the divisor. The above relationship then becomes

$$\frac{\text{dividend}}{\text{divisor}} = \text{quotient} + \frac{\text{remainder}}{\text{divisor}}$$

PROBLEM 2 The sum of two numbers is 57. If the larger is divided by the smaller, the quotient is 10 and the remainder is 2. Find the numbers.

Solution A Let n represent the smaller number. Then $57 - n$ represents the larger number. The relationship

$$\frac{\text{dividend}}{\text{divisor}} = \text{quotient} + \frac{\text{remainder}}{\text{divisor}}$$

can be used as a guideline.

$$\frac{57 - n}{n} = 10 + \frac{2}{n}, \qquad n \neq 0$$

$$n\left(\frac{57 - n}{n}\right) = n\left(10 + \frac{2}{n}\right)$$

$$57 - n = 10n + 2$$
$$55 = 11n$$
$$5 = n.$$

If n is 5, then $57 - n$ is 52. Thus, the numbers are 5 and 52. (Do they check?)

Solution B   Let n represent the larger number. Then $57 - n$ represents the smaller number.

$$\frac{n}{57 - n} = 10 + \frac{2}{57 - n}, \qquad n \neq 57$$

$$(57 - n)\left(\frac{n}{57 - n}\right) = (57 - n)\left(10 + \frac{2}{57 - n}\right)$$

$$n = 10(57 - n) + 2$$

$$n = 570 - 10n + 2$$

$$11n = 572$$

$$n = 52.$$

If n is 52 then $57 - n$ is 5. Thus, the numbers are 5 and 52. ■

Certainly, Solutions A and B are both acceptable ways of solving the problem. However, notice that the equation generated in Solution A is probably a little easier to solve than the one in Solution B. Sometimes it pays to look ahead as you declare what the variable is to represent. You may be able to make the computational work a little easier by a wise choice of what the variable represents.

In Chapter 4 we solved some uniform motion problems in which the formula $d = rt$ played an important role. Let's consider another one of those problems, keeping in mind that the formula $d = rt$ can also be written as $\dfrac{d}{r} = t$ or $\dfrac{d}{t} = r$.

PROBLEM 3   Wendy rides her bicycle 30 miles in the same time that it takes Kim to ride her bicycle 20 miles. If Wendy rides 5 miles per hour faster than Kim, find the rate of each.

Solution   Let r represent Kim's rate; then $r + 5$ represents Wendy's rate. The fact that their times are equal can be used as a guideline.

time of Kim   equals   time of Wendy
↓   ↓   ↓

$$\frac{\text{distance Kim rides}}{\text{rate Kim rides}} = \frac{\text{distance Wendy rides}}{\text{rate Wendy rides}}$$

$$\frac{20}{r} = \frac{30}{r + 5}$$

$$20(r + 5) = 30r$$

$$20r + 100 = 30r$$

$$100 = 10r$$

$$10 = r$$

Therefore, Kim rides at 10 miles per hour and Wendy rides at $10 + 5 = 15$ miles per hour. ■

Problem Set 7.5

For Problems 1–40, solve each of the equations.

1. $\dfrac{x}{2} + \dfrac{x}{3} = 10$

2. $\dfrac{x}{8} - \dfrac{x}{6} = -1$

3. $\dfrac{x}{6} - \dfrac{4x}{3} = \dfrac{1}{9}$

4. $\dfrac{3x}{4} + \dfrac{x}{5} = \dfrac{3}{10}$

5. $\dfrac{n}{2} + \dfrac{n-1}{6} = \dfrac{5}{2}$

6. $\dfrac{n+2}{7} + \dfrac{n}{3} = \dfrac{12}{7}$

7. $\dfrac{t-3}{4} + \dfrac{t+1}{9} = -1$

8. $\dfrac{t-2}{4} - \dfrac{t+3}{7} = 1$

9. $\dfrac{2x+3}{3} + \dfrac{3x-4}{4} = \dfrac{17}{4}$

10. $\dfrac{3x-1}{4} + \dfrac{2x-3}{5} = -2$

11. $\dfrac{x-4}{8} - \dfrac{x+5}{4} = 3$

12. $\dfrac{x+6}{9} - \dfrac{x-2}{5} = \dfrac{7}{15}$

13. $\dfrac{3x+2}{5} - \dfrac{2x-1}{6} = \dfrac{2}{15}$

14. $\dfrac{4x-1}{3} - \dfrac{2x+5}{8} = \dfrac{1}{6}$

15. $\dfrac{1}{x} + \dfrac{2}{3} = \dfrac{7}{6}$

16. $\dfrac{2}{x} + \dfrac{1}{4} = \dfrac{13}{20}$

17. $\dfrac{5}{3n} - \dfrac{1}{9} = \dfrac{1}{n}$

18. $\dfrac{9}{n} - \dfrac{1}{4} = \dfrac{7}{n}$

19. $\dfrac{1}{2x} + 3 = \dfrac{4}{3x}$

20. $\dfrac{2}{3x} + 1 = \dfrac{5}{4x}$

21. $\dfrac{4}{5t} - 1 = \dfrac{3}{2t}$

22. $\dfrac{1}{6t} - 2 = \dfrac{7}{8t}$

23. $\dfrac{-5}{4h} + \dfrac{7}{6h} = \dfrac{1}{4}$

24. $\dfrac{3}{h} + \dfrac{5}{2h} = 1$

25. $\dfrac{90-n}{n} = 10 + \dfrac{2}{n}$

26. $\dfrac{51-n}{n} = 7 + \dfrac{3}{n}$

27. $\dfrac{n}{49-n} = 3 + \dfrac{1}{49-n}$

28. $\dfrac{n}{57-n} = 10 + \dfrac{2}{57-n}$

29. $\dfrac{x}{x+3} - 2 = \dfrac{-3}{x+3}$

30. $\dfrac{4}{x-2} = \dfrac{5}{x+6}$

31. $\dfrac{7}{x+3} = \dfrac{5}{x-9}$)LCD

32. $\dfrac{x}{x-4} - 2 = \dfrac{4}{x-4}$

33. $\dfrac{x}{x+2} + 3 = \dfrac{1}{x+2}$

34. $\dfrac{x}{x-5} - 4 = \dfrac{3}{x-5}$

35. $-1 - \dfrac{5}{x-2} = \dfrac{3}{x-2}$

36. $\dfrac{x-1}{x} - 2 = \dfrac{3}{2}$

37. $1 + \dfrac{n+1}{2n} = \dfrac{3}{4}$

38. $\dfrac{3}{n-1} + 4 = \dfrac{2}{n-1}$

39. $\dfrac{h}{2} - \dfrac{h}{4} + \dfrac{h}{3} = 1$

40. $\dfrac{h}{4} + \dfrac{h}{5} - \dfrac{h}{6} = 1$

For Problems 41–52, set up an equation and solve each problem.

41. The numerator of a fraction is 8 less than the denominator. The fraction in its simplest form is $\dfrac{5}{6}$. Find the fraction.

42. One number is 12 larger than another number. The indicated quotient of the smaller number divided by the larger reduces to $\dfrac{2}{3}$. Find the numbers.

43. What number must be added to the numerator and denominator of $\frac{2}{5}$ to produce a fraction equivalent to $\frac{4}{5}$?

44. What number must be subtracted from the numerator and denominator of $\frac{29}{31}$ to produce a fraction equivalent to $\frac{11}{12}$?

45. The sum of two numbers is 65. If the larger number is divided by the smaller, the quotient is 8 and the remainder is 2. Find the numbers.

46. One number is 57 larger than another number. If the larger number is divided by the smaller, the quotient is 5 and the remainder is 5. Find the numbers.

47. The denominator of a fraction is 4 less than the numerator. If the numerator is increased by 6 and the denominator is doubled, the resulting fraction is equivalent to 1. Find the original fraction.

48. One number is 74 smaller than another number. If the larger number is divided by the smaller number, the quotient is 6 and the remainder is 4. Find the numbers.

49. It took Heidi 3 hours and 20 minutes more to ride her bicycle 125 miles than it took Abby to ride 75 miles. If they both rode at the same rate, find this rate.

50. Two trains left a depot traveling in opposite directions at the same rate. One train traveled 338 miles in 2 hours more time than it took the other train to travel 234 miles. Find the rate of the trains.

51. Kent drives his Mazda 270 miles in the same time that Dave drives his Datsun 250 miles. If Kent averages 4 miles per hour faster than Dave, find their rates.

52. An airplane travels 2050 miles in the same time that a car travels 260 miles. If the rate of the plane is 358 miles per hour greater than the rate of the car, find the rate of each.

■ Miscellaneous Problems

For Problems 53–56, solve each of the equations.

53. $\frac{3}{2n} + \frac{1}{n} = \frac{5}{3n}$

54. $\frac{1}{2n} + \frac{4}{n} = \frac{9}{2n}$

55. $\frac{n+1}{2} + \frac{n}{3} = \frac{1}{2}$

56. $\frac{1}{n+2} + \frac{2}{n+3} = \frac{3n+7}{(n+2)(n+3)}$

7.6 More Fractional Equations

Let's begin this section by considering a few more fractional equations. We will continue to solve them using the same basic techniques as in the previous section.

EXAMPLE 1 Solve $\dfrac{10}{8x-2} - \dfrac{6}{4x-1} = \dfrac{1}{9}$.

Solution

$$\dfrac{10}{8x-2} - \dfrac{6}{4x-1} = \dfrac{1}{9}, \qquad x \ne \dfrac{1}{4} \qquad \text{Do you see why } x$$

cannot equal $\dfrac{1}{4}$?

$$\dfrac{10}{2(4x-1)} - \dfrac{6}{4x-1} = \dfrac{1}{9} \qquad \text{Factor the first denominator.}$$

$$18(4x-1)\left(\dfrac{10}{2(4x-1)} - \dfrac{6}{4x-1}\right) = 18(4x-1)\left(\dfrac{1}{9}\right) \qquad \begin{array}{l}\text{Multiply both sides}\\ \text{by } 18(4x-1),\\ \text{which is the LCD.}\end{array}$$

$$9(10) - 18(6) = 2(4x-1)$$
$$90 - 108 = 8x - 2$$
$$-18 = 8x - 2$$
$$-16 = 8x$$
$$-2 = x. \qquad \text{Be sure that the solution } -2 \text{ checks!}$$

The solution set is $\{-2\}$. ■

Remark In the second step of the solution for Example 1 you may choose to reduce $\dfrac{10}{2(4x-1)}$ to $\dfrac{5}{4x-1}$. Then the left side, $\dfrac{5}{4x-1} - \dfrac{6}{4x-1}$, simplifies to $\dfrac{-1}{4x-1}$. This forms the proportion $\dfrac{-1}{4x-1} = \dfrac{1}{9}$ that can be easily solved using the cross product idea.

EXAMPLE 2 Solve $\dfrac{2n}{n+3} + \dfrac{5n}{n^2-9} = 2$.

Solution

$$\dfrac{2n}{n+3} + \dfrac{5n}{n^2-9} = 2, \qquad n \ne -3 \text{ and } n \ne 3$$

$$\dfrac{2n}{n+3} + \dfrac{5n}{(n+3)(n-3)} = 2$$

$$(n+3)(n-3)\left(\dfrac{2n}{n+3} + \dfrac{5n}{(n+3)(n-3)}\right) = (n+3)(n-3)(2)$$

$$2n(n-3) + 5n = 2(n^2-9)$$
$$2n^2 - 6n + 5n = 2n^2 - 18$$
$$-6n + 5n = -18 \qquad \text{Add } -2n^2 \text{ to both sides.}$$
$$-n = -18$$
$$n = 18.$$

The solution set is $\{18\}$. ■

EXAMPLE 3 Solve $n + \dfrac{1}{n} = \dfrac{10}{3}$.

Solution
$$n + \frac{1}{n} = \frac{10}{3}, \qquad n \neq 0$$

$$3n\left(n + \frac{1}{n}\right) = 3n\left(\frac{10}{3}\right)$$

$$3n^2 + 3 = 10n$$

$$3n^2 - 10n + 3 = 0$$

$$(3n - 1)(n - 3) = 0$$

Remember when we were using the factoring techniques to help solve equations of this type in Chapter 5!

$$3n - 1 = 0 \quad \text{or} \quad n - 3 = 0$$

$$3n = 1 \quad \text{or} \qquad n = 3$$

$$n = \frac{1}{3} \quad \text{or} \qquad n = 3.$$

The solution set is $\left\{\dfrac{1}{3}, 3\right\}$.   ■

■ Problem Solving

Recall that $\dfrac{2}{3}$ and $\dfrac{3}{2}$ are called multiplicative inverses, or *reciprocals*, of each other because their product is one. In general, the reciprocal of any nonzero real number n is the number $\dfrac{1}{n}$. Let's use this idea to solve a problem.

PROBLEM 1 The sum of a number and its reciprocal is $\dfrac{26}{5}$. Find the number.

Solution Let n represent the number. Then $\dfrac{1}{n}$ represents its reciprocal.

$$\text{number} + \text{its reciprocal} = \frac{26}{5}$$
$$\downarrow \qquad\qquad \downarrow \qquad\qquad \downarrow$$
$$n \quad + \quad \frac{1}{n} \quad = \frac{26}{5}, \qquad n \neq 0.$$

$$5n\left(n + \frac{1}{n}\right) = 5n\left(\frac{26}{5}\right)$$

$$5n^2 + 5 = 26n$$

$$5n^2 - 26n + 5 = 0$$
$$(5n - 1)(n - 5) = 0$$
$$5n - 1 = 0 \quad \text{or} \quad n - 5 = 0$$
$$5n = 1 \quad \text{or} \quad n = 5$$
$$n = \frac{1}{5} \quad \text{or} \quad n = 5.$$

If the number is $\frac{1}{5}$, its reciprocal is $\frac{1}{\frac{1}{5}} = 5$. If the number is 5, its reciprocal is $\frac{1}{5}$. ■

Now let's consider another uniform motion problem, one that presents a slight variation of those studied in the previous section. Again keep in mind that the distance-rate-time relationships are constantly being used in these problems.

PROBLEM 2   To travel 60 miles, it takes Sue, riding a moped, 2 hours less than it takes Ann, riding a bicycle, to travel 50 miles. Sue travels 10 miles per hour faster than Ann. Find the times and rates of both girls.

Solution   Let t represent Ann's time; then $t - 2$ represents Sue's time. We can record the information from the problem in the following table.

	Distance	Time	Rate $\left(r = \dfrac{d}{t} \right)$
Ann	50	t	$\dfrac{50}{t}$
Sue	60	$t - 2$	$\dfrac{60}{t - 2}$

The fact that Sue travels 10 miles per hour faster than Ann can be used as a guideline to set up an equation.

rate of Sue   equals   rate of Ann plus 10
$$\frac{60}{t - 2} \quad = \quad \frac{50}{t} + 10.$$

Solving this equation yields

$$\frac{60}{t - 2} = \frac{50}{t} + 10$$

$$t(t-2)\left(\frac{60}{t-2}\right) = t(t-2)\left(\frac{50}{t}+10\right)$$

$$60t = 50(t-2) + 10t(t-2)$$

$$60t = 50t - 100 + 10t^2 - 20t$$

$$0 = 10t^2 - 30t - 100$$

$$0 = t^2 - 3t - 10$$

$$0 = (t-5)(t+2)$$

$$t - 5 = 0 \quad \text{or} \quad t + 2 = 0$$

$$t = 5 \quad \text{or} \quad t = -2.$$

The negative solution must be disregarded, so Ann's time is 5 hours and Sue's time is $5 - 2 = 3$ hours. Ann's rate is $\frac{50}{5} = 10$ miles per hour and Sue's rate is $\frac{60}{3} = 20$ miles per hour. (Be sure that all of these results check back into the original problem!) ■

There is another class of problems commonly referred to as "work" problems, or sometimes as "rate-time" problems. For example, if a certain machine produces 120 items in 10 minutes, then we say that it is producing at a rate of $\frac{120}{10} = 12$ items per minute. Likewise, if a person can do a certain job in 5 hours, then he(she) is working at a rate of $\frac{1}{5}$ of the job per hour. In general, if Q is the quantity of something done in t units of time, then the rate, r, is given by $r = \frac{Q}{t}$. The rate is stated in terms of *so much quantity per unit of time*. The uniform motion problems discussed earlier are a special kind of rate-time problems where the "quantity" is distance. Furthermore, the use of tables to organize information, as we illustrated with the uniform motion problems, is also a convenient aid for some rate-time problems. Let's consider some problems.

PROBLEM 3 Printing press A can produce 35 fliers per minute and press B can produce 50 fliers per minute. Suppose that 2225 fliers are produced by first using A by itself for 15 minutes, and then using both A and B until the job is done. How long would the B printing press need to be used?

Solution Let m represent the number of minutes that the B printing press is used. Then $m + 15$ represents the number of minutes that the A press is used. The information in the problem can be organized as follows.

	Quantity	Time	Rate
A	$35(m + 15)$	$m + 15$	35
B	$50m$	m	50

Since the total quantity (total number of fliers) is 2225 fliers, we can set up and solve the following equation.

$$35(m + 15) + 50m = 2225$$
$$35m + 525 + 50m = 2225$$
$$85m = 1700$$
$$m = 20.$$

Therefore, the B printing press must be used for 20 minutes. ■

PROBLEM 4   Bill can mow a lawn in 45 minutes and Jennifer can mow the same lawn in 30 minutes. How long would it take the two of them working together to mow the lawn?

Remark   Before you look at the solution of this problem, *estimate* the answer. Remember that Jennifer can mow the lawn by herself in 30 minutes.

Solution   Bill's rate is $\dfrac{1}{45}$ of the lawn per minute and Jennifer's rate is $\dfrac{1}{30}$ of the lawn per minute. If we let m represent the number of minutes that they work together, then $\dfrac{1}{m}$ represents the rate when working together. Therefore, since the sum of the individual rates must equal the rate working together, we can set up and solve the following equation.

$$\frac{1}{30} + \frac{1}{45} = \frac{1}{m}, \qquad m \neq 0$$
$$90m\left(\frac{1}{30} + \frac{1}{45}\right) = 90m\left(\frac{1}{m}\right)$$
$$3m + 2m = 90$$
$$5m = 90$$
$$m = 18.$$

It should take them 18 minutes to mow the lawn when working together. ■

PROBLEM 5   It takes Amy twice as long to deliver papers as it does Nancy. How long would it take each girl by herself if they can deliver the papers together in 40 minutes?

Solution   Let m represent the number of minutes that it takes Nancy by herself; then $2m$ represents Amy's time by herself. Therefore, Nancy's rate is $\dfrac{1}{m}$ and Amy's rate is $\dfrac{1}{2m}$. Since the combined rate is $\dfrac{1}{40}$, we can set up and solve the following equation.

$$\frac{1}{m} + \frac{1}{2m} = \frac{1}{40}, \qquad m \neq 0$$

$$40\left(\frac{1}{m} + \frac{1}{2m}\right) = 40m\left(\frac{1}{40}\right)$$

$$40 + 20 = m$$

$$60 = m.$$

Therefore, Nancy can deliver the papers by herself in 60 minutes and Amy can deliver them by herself in 2(60) = 120 minutes. ■

One final example of this section illustrates another approach that some people find meaningful for work problems. A basic idea used in this approach pertains to representing the fractional parts of a job. For example, if a person can do a certain job in 7 hours, then at the end of 3 hours he(she) has finished $\frac{3}{7}$ of the job. (Again a constant rate of work is being assumed.) At the end of 5 hours, $\frac{5}{7}$ of the job has been done, and, in general, at the end of h hours, $\frac{h}{7}$ of the job has been completed. Let's use this idea to solve a work problem.

PROBLEM 6 Together Al and Larry can paint a house in $3\frac{3}{5}$ days. Al can paint the house alone in 6 days. How long would it take Larry to paint the house by himself?

Solution Let n represent the number of days that it would take Larry by himself. Now we can use the following guideline to set up an equation. (Notice that "1" represents the one job to be done, which is painting the house.)

fractional part of job done by Al when working together		fractional part of job done by Larry when working together		total job
↓		↓		↓

$$\frac{3\frac{3}{5}}{6} + \frac{3\frac{3}{5}}{n} = 1, \qquad n \neq 0.$$

$$\frac{\frac{18}{5}}{6} + \frac{\frac{18}{5}}{n} = 1 \qquad \text{Remember our work with complex fractions!}$$

$$\frac{18}{30} + \frac{18}{5n} = 1$$

$$\frac{3}{5} + \frac{18}{5n} = 1$$

$$5n\left(\frac{3}{5} + \frac{18}{5n}\right) = 5n(1) \qquad \text{Multiply both sides by } 5n.$$

$$3n + 18 = 5n$$

$$18 = 2n$$

$$9 = n.$$

It would take Larry 9 days to paint the house by himself.   ■

Problem Set 7.6

For Problems 1–32, solve each equation.

1. $\dfrac{4}{x} + \dfrac{7}{6} = \dfrac{1}{x} + \dfrac{2}{3x}$

2. $\dfrac{2}{3x} - \dfrac{9}{x} = -\dfrac{25}{9}$

3. $\dfrac{3}{2x + 2} + \dfrac{4}{x + 1} = \dfrac{11}{12}$

4. $\dfrac{5}{2x - 6} + \dfrac{1}{x - 3} = \dfrac{7}{2}$

5. $\dfrac{5}{2n - 10} - \dfrac{3}{n - 5} = 1$

6. $\dfrac{7}{3x + 6} - \dfrac{2}{x + 2} = 2$

7. $\dfrac{3}{2t} - \dfrac{5}{t} = \dfrac{7}{5t} + 1$

8. $\dfrac{2}{3t} + \dfrac{3}{4t} = 1 - \dfrac{5}{2t}$

9. $\dfrac{x}{x - 2} + \dfrac{4}{x + 2} = 1$

10. $\dfrac{2x}{x + 1} - \dfrac{3}{x - 1} = 2$

11. $\dfrac{x}{x - 4} - \dfrac{2x}{x + 4} = -1$

12. $\dfrac{2x}{x + 2} + \dfrac{x}{x - 2} = 3$

13. $\dfrac{3n}{n + 3} - \dfrac{n}{n - 3} = 2$

14. $\dfrac{4n}{n - 5} - \dfrac{2n}{n + 5} = 2$

15. $\dfrac{3}{t^2 - 4} + \dfrac{5}{t + 2} = \dfrac{2}{t - 2}$

16. $\dfrac{t}{2t - 8} + \dfrac{16}{t^2 - 16} = \dfrac{1}{2}$

17. $\dfrac{4}{x - 1} - \dfrac{2x - 3}{x^2 - 1} = \dfrac{6}{x + 1}$

18. $\dfrac{3x - 1}{x^2 - 9} + \dfrac{4}{x + 3} = \dfrac{5}{x - 3}$

19. $8 + \dfrac{5}{y^2 + 2y} = \dfrac{3}{y + 2}$

20. $2 + \dfrac{4}{y - 1} = \dfrac{4}{y^2 - y}$

21. $n + \dfrac{1}{n} = \dfrac{17}{4}$

22. $n + \dfrac{3}{n} = 4$

23. $\dfrac{15}{4n} + \dfrac{15}{4(n + 4)} = 1$

24. $\dfrac{10}{7x} + \dfrac{10}{7(x + 3)} = 1$

25. $x - \dfrac{5x}{x - 2} = \dfrac{-10}{x - 2}$

26. $\dfrac{x + 1}{x - 3} - \dfrac{3}{x} = \dfrac{12}{x^2 - 3x}$

27. $\dfrac{t}{4t - 4} + \dfrac{5}{t^2 - 1} = \dfrac{1}{4}$

28. $\dfrac{x}{3x - 6} + \dfrac{4}{x^2 - 4} = \dfrac{1}{3}$

29. $\dfrac{3}{n - 5} + \dfrac{4}{n + 7} = \dfrac{2n + 11}{n^2 + 2n - 35}$

30. $\dfrac{2}{n + 3} + \dfrac{3}{n - 4} = \dfrac{2n - 1}{n^2 - n - 12}$

31. $\dfrac{a}{a + 2} + \dfrac{3}{a + 4} = \dfrac{14}{a^2 + 6a + 8}$

32. $3 + \dfrac{6}{t - 3} = \dfrac{6}{t^2 - 3t}$

For Problems 33–50, set up an equation and solve the problem.

33. The sum of a number and twice its reciprocal is $\dfrac{9}{2}$. Find the number.

34. The sum of a number and three times its reciprocal is 4. Find the number.

35. A number is $\dfrac{21}{10}$ larger than its reciprocal. Find the number.

36. The reciprocal of a number subtracted from the number yields $\dfrac{5}{6}$. Find the number.

37. Celia rides her bicycle 60 miles in 2 hours less than it takes Tom to ride his bicycle 85 miles. If Celia rides 3 miles per hour faster than Tom, find their rates.

38. To travel 300 miles, it takes a freight train 2 hours more than it does an express train to travel 280 miles. The rate of the express is 20 miles per hour greater than the rate of the freight. Find the rates of both trains.

39. Jeff rides his bicycle out into the country 40 miles. On the way back, he takes a different route that is 2 miles longer and it takes him an hour longer to return. If his rate on the way out into the country is 4 miles per hour faster than his rate back, find both rates.

40. Rita jogs for 8 miles and then walks an additional 12 miles. She jogs at a rate twice her walking rate and the entire distance of 20 miles was covered in 4 hours. Find the rate she jogs and the rate that she walks.

41. A water tank can be filled by an inlet pipe in 5 minutes. A drain pipe will empty the tank in 6 minutes. If by mistake the drain is left open as the tank is being filled, how long will it take before the tank overflows?

42. Mark can do a job in 10 minutes. Phil can do the same job in 15 minutes. If they work together, how long will it take them to complete the job?

43. It takes Barry twice as long to deliver papers as it does Mike. How long would it take each if they can deliver the papers together in 40 minutes?

44. Working together Cindy and Sharon can address envelopes in 12 minutes. Cindy could do the addressing by herself in 20 minutes. How long would it take Sharon to address the envelopes by herself?

45. It takes Pat 12 hours to complete a task. After he had been working for 3 hours, he was joined by his brother Mike and together they finished the job in 5 hours. How long would it take Mike to do the job by himself?

46. Working together Pam and Laura can complete a job in $1\dfrac{1}{2}$ hours. When working alone it

takes Laura 4 hours longer than Pam to do the job. How long does it take each of them working alone?

47. A computer room has two card readers. Card reader A reads at the rate of 600 cards per minute and reader B at the rate of 850 cards per minute. Suppose 9400 cards have been read by using card reader A by itself for 6 minutes, and then using both of them until the job is done. How long would card reader B need to be used?

48. It takes two pipes 3 hours to fill a water tank. Pipe B can fill the tank alone in 8 hours more than it takes pipe A to fill the tank alone. How long would it take each pipe to fill the tank by itself?

49. Amelia can type 600 words in 5 minutes less than it takes Paul to type 600 words. If Amelia types at a rate 20 words per minute faster than Paul, find the rate of each.

50. A computer company markets two card readers. Card reader B can read 8000 cards in 2 minutes less time than it takes card reader A to read 7500 cards. If the rate of card reader A is 250 cards per minute less than the rate of reader B, find each rate.

■ Miscellaneous Problems

For Problems 51–53, solve each equation.

51. $\dfrac{3x-1}{x^2-9} + \dfrac{4}{x+3} = \dfrac{7}{x-3}$

52. $\dfrac{x-2}{x^2-1} + \dfrac{3}{x+1} = \dfrac{-5}{x-1}$

53. $\dfrac{7x-12}{x^2-16} - \dfrac{5}{x+4} = \dfrac{2}{x-4}$

Chapter 7 Summary

(7.1) The **fundamental principle of fractions** $\left(\dfrac{ak}{bk} = \dfrac{a}{b}\right)$ provides the basis for simplifying algebraic fractions.

(7.2) To **multiply** algebraic fractions, *multiply the numerators, multiply the denominators, and express the product in simplified form.*

To **divide** algebraic fractions, *invert the divisor and multiply.*

(7.3) **Addition** and **subtraction** of algebraic fractions are based on the following.

$$\frac{a}{b} + \frac{c}{b} = \frac{a+c}{b}. \qquad \text{(Addition)}$$

$$\frac{a}{b} - \frac{c}{b} = \frac{a-c}{b}. \qquad \text{(Subtraction)}$$

(7.4) The following procedure can be followed when adding and subtracting fractions.

1. Find the least common denominator.

2. Change each fraction to an equivalent fraction having the LCD as its denominator.

3. Add or subtract the numerators and place this result over the LCD.

4. Look for possibilities to simplify the final fraction.

Fractional forms that contain fractions in the numerators and/or denominators are called **complex fractions**.

(7.5) and (7.6) To solve a fractional equation it is often convenient to begin by multiplying both sides of the equation by the LCD, which *clears the equation of all fractions.*

Chapter 7 Review Problem Set

Simplify each of the following algebraic fractions.

1. $\dfrac{56x^2y}{72xy^3}$

2. $\dfrac{x^2 - 9x}{x^2 - 6x - 27}$

3. $\dfrac{3n^2 - n - 10}{n^2 - 4}$

4. $\dfrac{16a^2 + 24a + 9}{20a^2 + 7a - 6}$

Perform the indicated operations and express answers in simplest form.

5. $\dfrac{7x^2y^2}{12y^3} \cdot \dfrac{18y}{28x}$

6. $\dfrac{x^2y}{x^2 + 2x} \cdot \dfrac{x^2 - x - 6}{y}$

7. $\dfrac{n^2 - 2n - 24}{n^2 + 11n + 28} \div \dfrac{n^3 - 6n^2}{n^2 - 49}$

8. $\dfrac{4a^2 + 4a + 1}{(a + 6)^2} \div \dfrac{6a^2 - 5a - 4}{3a^2 + 14a - 24}$

9. $\dfrac{3x + 4}{5} + \dfrac{2x - 7}{4}$

10. $\dfrac{7}{3x} + \dfrac{5}{4x} - \dfrac{2}{8x^2}$

11. $\dfrac{7}{n} + \dfrac{3}{n - 1}$

12. $\dfrac{2}{a - 4} - \dfrac{3}{a - 2}$

13. $\dfrac{2x}{x^2 - 3x} - \dfrac{3}{4x}$

14. $\dfrac{2}{x^2 + 7x + 10} + \dfrac{3}{x^2 - 25}$

15. $\dfrac{5x}{x^2 - 4x - 21} - \dfrac{3}{x - 7} + \dfrac{4}{x + 3}$

Solve each of the following equations.

16. $\dfrac{2x - 1}{3} + \dfrac{3x - 2}{4} = \dfrac{5}{6}$

17. $\dfrac{5}{3x} - 2 = \dfrac{7}{2x} + \dfrac{1}{5x}$

18. $\dfrac{67 - x}{x} = 6 + \dfrac{4}{x}$

19. $\dfrac{5}{2n + 3} = \dfrac{6}{3n - 2}$

20. $\dfrac{x}{x-3} + \dfrac{5}{x+3} = 1$

21. $n + \dfrac{1}{n} = 2$

22. $\dfrac{n-1}{n^2 + 8n - 9} - \dfrac{n}{n+9} = 4$

23. $\dfrac{6}{7x} - \dfrac{1}{6} = \dfrac{5}{6x}$

24. $n + \dfrac{1}{n} = \dfrac{5}{2}$

25. $\dfrac{n}{5} = \dfrac{10}{n-5}$

26. $\dfrac{-1}{2x-5} + \dfrac{2x-4}{4x^2 - 25} = \dfrac{5}{6x+15}$

27. $1 + \dfrac{1}{n-1} = \dfrac{1}{n^2 - n}$

Set up an equation and solve each of the following problems.

28. The sum of two numbers is 75. If the larger number is divided by the smaller, the quotient is 9 and the remainder is 5. Find the numbers.

29. It takes Nancy three times as long to complete a task as it does Becky. How long would it take each of them to complete the task if working together they can do it in 2 hours?

30. The sum of a number and twice its reciprocal is 3. Find the number.

31. The denominator of a fraction is twice the numerator. If 4 is added to the numerator and 18 to the denominator, a fraction that is equivalent to $\dfrac{4}{9}$ is produced. Find the original fraction.

32. Lanette can ride her moped 44 miles in the same time that Todd rides his bicycle 30 miles. If Lanette rides 7 miles per hour faster than Todd, find their rates.

33. Jim rode his bicycle 36 miles in 4 hours. For the first 20 miles he rode at a constant rate and then for the last 16 miles he reduced his rate by 2 miles per hour. Find his rate for the last 16 miles.

34. An inlet pipe can fill a tank in 10 minutes. A drain can empty the tank in 12 minutes. If the tank is empty and both the inlet pipe and the drain are open, how long will it take before the tank overflows?

35. Sue can type 1000 words in 5 minutes more time than it takes Corinne to type 840 words. If Sue's rate is 6 words per minute less than Corinne's rate, find their typing speeds.

Cumulative Review Problem Set

For Problems 1–8, evaluate each of the algebraic expressions for the given values of the variables. You may want to first simplify the expression or change its form by factoring.

1. $3x - 2xy - 7x + 5xy$ for $x = \dfrac{1}{2}$ and $y = 3$

2. $7(a-b) - 3(a-b) - (a-b)$ for $a = -3$ and $b = -5$

3. $\dfrac{xy + yz}{y}$ for $x = \dfrac{2}{3}, y = \dfrac{5}{6}$, and $z = \dfrac{3}{4}$

4. $ab + b^2$ for $a = 0.4$ and $b = 0.6$

5. $x^2 - y^2$ for $x = -6$ and $y = 4$

6. $x^2 + 5x - 36$ for $x = -9$

7. $\dfrac{x^2 + 2x}{x^2 + 5x + 6}$ for $x = -6$

8. $\dfrac{x^2 + 3x - 10}{x^2 - 9x + 14}$ for $x = 4$

For Problems 9–16, evaluate each of the expressions.

9. 3^{-3}

10. $\left(\dfrac{2}{3}\right)^{-1}$

11. $\left(\dfrac{1}{2} + \dfrac{1}{3}\right)^0$

12. $\left(\dfrac{1}{3} + \dfrac{1}{4}\right)^{-1}$

13. -4^{-2}

14. $\left(\dfrac{3}{4}\right)^{-2}$

15. $\dfrac{1}{\left(\dfrac{2}{5}\right)^{-2}}$

16. $(-3)^{-3}$

For Problems 17–32, perform the indicated operations and express answers in simplest form.

17. $\dfrac{7}{5x} + \dfrac{2}{x} - \dfrac{3}{2x}$

18. $\dfrac{4x}{5y} \div \dfrac{12x^2}{10y^2}$

19. $\dfrac{4}{x - 6} + \dfrac{3}{x + 4}$

20. $\dfrac{2}{x^2 - 4x} - \dfrac{3}{x^2}$

21. $\dfrac{x^2 - 8x}{x^2 - x - 56} \cdot \dfrac{x^2 - 49}{3xy}$

22. $\dfrac{5}{x^2 - x - 12} - \dfrac{3}{x - 4}$

23. $(-5x^2y)(7x^3y^4)$

24. $(9ab^3)^2$

25. $(-3n^2)(5n^2 + 6n - 2)$

26. $(5x - 1)(3x + 4)$

27. $(2x + 5)^2$

28. $(x + 2)(2x^2 - 3x - 1)$

29. $(x^2 - x - 1)(x^2 + 2x - 3)$

30. $(-2x - 1)(3x - 7)$

31. $\dfrac{24x^2y^3 - 48x^4y^5}{8xy^2}$

32. $(28x^2 - 19x - 20) \div (4x - 5)$

For Problems 33–42, factor completely.

33. $3x^3 + 15x^2 + 27x$

34. $x^2 - 100$

35. $5x^2 - 22x + 8$

36. $8x^2 - 22x - 63$

37. $n^2 + 25n + 144$

38. $nx + ny - 2x - 2y$

39. $3x^3 - 3x$

40. $2x^3 - 6x^2 - 108x$

41. $36x^2 - 60x + 25$

42. $3x^2 - 5xy - 2y^2$

For Problems 43–58, solve each of the equations.

43. $3(x - 2) - 2(x + 6) = -2(x + 1)$

44. $x^2 = -11x$

45. $0.2x - 3(x - 0.4) = 1$

46. $|2x - 1| = 7$

47. $\dfrac{3n - 1}{4} = \dfrac{5n + 2}{7}$

48. $5n^2 - 5 = 0$

49. $x^2 + 5x - 6 = 0$

50. $n + \dfrac{4}{n} = 4$

51. $\dfrac{2x + 1}{2} + \dfrac{3x - 4}{3} = 1$

52. $2(x - 1) - x(x - 1) = 0$

53. $\dfrac{3}{2x} - 1 = \dfrac{5}{3x} + 2$

54. $6t^2 + 19t - 7 = 0$

55. $(2x - 1)(x - 8) = 0$

56. $(x + 1)(x + 6) = 24$

57. $\dfrac{x}{x - 2} - \dfrac{7}{x + 1} = 1$

58. $\dfrac{1}{n} - \dfrac{2}{n - 1} = \dfrac{3}{n}$

8 Coordinate Geometry and Linear Systems

On a real number line a one-to-one correspondence is established between the set of real numbers and the points on a line. To each point on a line there corresponds a real number and to each real number there corresponds a point on the line. The real number line was used in Section 4.3 to graph the solutions for inequalities of **one variable**. The idea of associating points with numbers can be extended so that a correspondence is established between pairs of real numbers and points in a plane. This will allow us to graph the solutions of equations containing **two variables**.

8.1 Cartesian Coordinate System

Consider two number lines (one vertical and one horizontal) perpendicular to each other at the point associated with zero on both lines (Figure 8.1). These number lines are referred to as the horizontal and vertical **axes** or together as the **coordinate axes**. They partition the plane into four parts called **quadrants**. The quadrants are numbered counterclockwise from I to IV as indicated in Figure 8.1. The point of intersection of the two axes is called the **origin**.

It is now possible to set up a one-to-one correspondence between ordered pairs of real numbers and the points in a plane. To each ordered pair of real numbers there corresponds a unique point in the plane and to each point there corresponds a unique ordered pair of real numbers. We have indicated a part of this correspondence in Figure 8.2. The ordered pair (3, 1) corresponds to point A and means that the point A

Figure 8.1

Figure 8.2

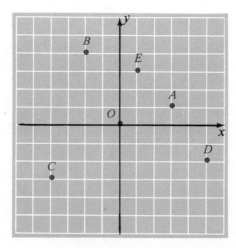

$A(3, 1)$
$B(-2, 4)$
$C(-4, -3)$
$D(5, -2)$
$O(0, 0)$
$E(1, 3)$

is located 3 units to the right of and one unit up from the origin. (The ordered pair $(0,0)$ corresponds to the origin.) The ordered pair $(-2,4)$ corresponds to point B and that means that point B is located 2 units to the left and 4 units up from the origin. Make sure that you agree with all of the other locations in Figure 8.2.

In general, the real numbers a and b in an ordered pair (a,b), associated with a point, are referred to as the **coordinates of the point**. The first number, a, called the **abscissa**, is the directed distance of the point from the vertical axis measured parallel to the horizontal axis. The second number, b, called the **ordinate**, is the directed distance of the point from the horizontal axis measured parallel to the vertical axis (Figure 8.3(a)). Thus, in the first quadrant all points have a positive abscissa and a positive ordinate. In the second quadrant all points have a negative abscissa and a positive ordinate. We have indicated the sign situations for all four quadrants in Figure 8.3(b). This system of associating points with ordered pairs of real numbers is called the **Cartesian coordinate system** or the **rectangular coordinate system**.

Figure 8.3(a)

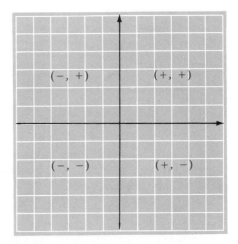

Figure 8.3(b)

Historically, the Cartesian coordinate system provided the basis for the development of a branch of mathematics called **analytic geometry**; at the present time it is often referred to as **coordinate geometry**. René Descartes, a French mathematician of the 17th century, was able to transform geometric problems into an algebraic setting and then use the tools of algebra to solve the problems.

Basically, there are two kinds of problems in analytic geometry.

1. Given an algebraic equation, find its geometric graph.
2. Given a set of conditions pertaining to a geometric figure, find its algebraic equation.

We will consider a few problems of type 1 at this time.

Let's begin by considering the solutions for the equation $y = x + 3$. A *solution* of an equation in two variables is an ordered pair of real numbers that satisfies the equation. When using the variables x and y, we agree that the first number of an ordered pair is a value for x and the second number is a value for y. We see that $(1, 4)$ is a solution for $y = x + 3$ because when x is replaced by 1 and y by 4, the true numerical statement $4 = 1 + 3$ is obtained. Likewise, $(-1, 2)$ is a solution for $y = x + 3$ because $2 = -1 + 3$ is a true statement. Infinitely many pairs of real numbers that satisfy $y = x + 3$ can be found by arbitrarily choosing values for x and for each value of x chosen, determining a corresponding value for y. Let's use a table to record some of the solutions for $y = x + 3$.

Choose x	Determine y from $y = x + 3$	Solution for $y = x + 3$
0	3	$(0, 3)$
1	4	$(1, 4)$
3	6	$(3, 6)$
5	8	$(5, 8)$
-1	2	$(-1, 2)$
-3	0	$(-3, 0)$
-5	-2	$(-5, -2)$

Now we can graph the ordered pairs as points on a Cartesian coordinate system using the horizontal axis as the **x-axis** and the vertical axis as the **y-axis** (Figure 8.4(a)). The straight line containing the points in Figure 8.4(b) is called the **graph of the equation** $y = x + 3$.

Figure 8.4(a)

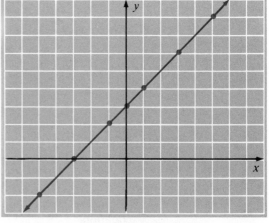

Figure 8.4(b)

The following examples further illustrate the process of graphing equations.

EXAMPLE 1 Graph $y = x^2$.

Solution First we set up a table of some of the solutions.

x	y	Solutions (x, y)
0	0	$(0, 0)$
1	1	$(1, 1)$
2	4	$(2, 4)$
3	9	$(3, 9)$
-1	1	$(-1, 1)$
-2	4	$(-2, 4)$
-3	9	$(-3, 9)$

Then we plot the points associated with the solutions (Figure 8.5(a)). Finally, we connect the points with a smooth curve (Figure 8.5(b)). The curve in Figure 8.5(b) is called a **parabola**, which is not important to us in this course, but does become a major topic of study in other mathematics courses.)

Figure 8.5 (a)

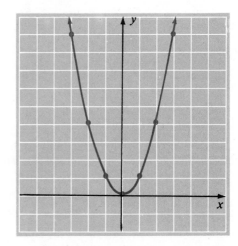

Figure 8.5 (b) ■

How many solutions do we need to have in a table of values? There is no definite answer to this question other than *a sufficient number so that the graph of the equation is determined*. In other words, we need to plot points until we can tell the nature of the curve.

EXAMPLE 2 Graph $2x + 3y = 6$.

Solution   First, let's change the form of the equation to make it easier to find solutions. We can either solve for x in terms of y or for y in terms of x. Doing the latter, we obtain:

$$2x + 3y = 6$$
$$3y = 6 - 2x$$
$$y = \frac{6 - 2x}{3}.$$

Now a table of values can be set up. Plotting these points and connecting them produces Figure 8.6.

Figure 8.6

x	y
0	2
3	0
6	-2
-3	4
-6	6

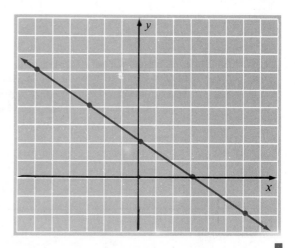

Look carefully at the table of values in Example 2. Notice that values of x were chosen so that integers were obtained for y. This is not necessary but does make things easier from a computational viewpoint. Also, plotting points associated with pairs of integers is more exact than getting involved with fractions.

To graph an equation in two variables x and y, keep in mind the following steps.

1. Solve the equation for y in terms of x or for x in terms of y if it is not already in such a form.

2. Set up a table of ordered pairs that satisfy the equation.

3. Plot the points associated with the ordered pairs.

4. Connect the points with a smooth curve.

Let's conclude this section with two more examples illustrating Step 1.

EXAMPLE 3   Solve $4x + 9y = 12$ for y.

Solution

$$4x + 9y = 12$$

$$9y = 12 - 4x \qquad \text{Subtracted } 4x \text{ from both sides.}$$

$$y = \frac{12 - 4x}{9}. \qquad \text{Divided both sides by 9.} \qquad \blacksquare$$

EXAMPLE 4   Solve $4x - 5y = 6$ for y.

Solution

$$4x - 5y = 6$$

$$-5y = 6 - 4x \qquad \text{Subtracted } 4x \text{ from both sides.}$$

$$y = \frac{6 - 4x}{-5} \qquad \text{Divided both sides by } -5.$$

$$y = \frac{4x - 6}{5}. \qquad \frac{6 - 4x}{-5} \text{ can be changed to } \frac{-6 + 4x}{5} \text{ by multiplying numerator and denominator by } -1. \qquad \blacksquare$$

Problem Set 8.1

For Problems 1–10, solve the given equation for the variable indicated.

1. $3x + 7y = 13$ for y **2.** $5x + 9y = 17$ for y **3.** $x - 3y = 9$ for x

4. $2x - 7y = 5$ for x **5.** $-x + 5y = 14$ for y **6.** $-2x - y = 9$ for y

7. $-3x + y = 7$ for x **8.** $-x - y = 9$ for x **9.** $-2x + 3y = -5$ for y

10. $3x - 4y = -7$ for y

For Problems 11–34, graph each of the equations.

11. $y = x + 1$ **12.** $y = x + 4$ **13.** $y = x - 2$ **14.** $y = -x - 1$

15. $y = (x - 2)^2$ **16.** $y = (x + 1)^2$ **17.** $y = x^2 - 2$ **18.** $y = x^2 + 1$

19. $y = \dfrac{1}{2}x + 3$ **20.** $y = \dfrac{1}{2}x - 2$ **21.** $x + 2y = 4$ **22.** $x + 3y = 6$

23. $2x - 5y = 10$ **24.** $5x - 2y = 10$ **25.** $y = x^3$ **26.** $y = x^4$

27. $y = -x^2$ **28.** $y = -x^3$ **29.** $y = x$ **30.** $y = -x$

31. $y = -3x + 2$ **32.** $3x - y = 4$ **33.** $y = 2x^2$ **34.** $y = -3x^2$

8.2   Linear Equations in Two Variables

The following table summarizes some of our results from graphing equations in the previous section and accompanying problem set.

Equation	Type of graph produced
$y = x + 3$	straight line
$y = x^2$	parabola
$2x + 3y = 6$	straight line
$y = -3x + 2$	straight line
$y = x^2 - 2$	parabola
$y = (x - 2)^2$	parabola
$5x - 2y = 10$	straight line
$3x - y = 4$	straight line
$y = x^3$	no name will be given at this time, but not a straight line
$y = x$	straight line
$y = \dfrac{1}{2}x + 3$	straight line

In the above table pay special attention to the equations that produced a straight line graph. They are called **linear equations in two variables**. In general, any equation of the form $Ax + By = C$, where A, B, and C are constants (A and B not both zero) and x and y are variables, is a linear equation in two variables and its graph is a straight line.

Two points of clarification about the previous description of a linear equation in two variables should be made. First, the choice of x and y for variables is arbitrary. Any two letters could be used to represent the variables. An equation such as $3m + 2n = 7$ can be considered a linear equation in two variables. So that we are not constantly changing the labeling of the coordinate axes when graphing equations, however, it is much easier to use the same two variables in all equations. Thus, we will go along with convention and use x and y as our variables. Secondly, the statement "any equation of the form $Ax + By = C$" technically means "any equation of the form $Ax + By = C$ or *equivalent* to the form." For example, the equation $y = x + 3$, which has a straight line graph, is equivalent to $-x + y = 3$.

The knowledge that any equation of the form $Ax + By = C$ produces a straight line graph, along with the fact that two points determine a straight line, makes graphing linear equations in two variables a simple process. We merely find two solutions, plot the corresponding points, and connect the points with a straight line. It is probably wise to find a third point as a check point. Let's consider an example.

EXAMPLE 1 Graph $2x - 3y = 6$.

Solution

Let $x = 0$, then $2(0) - 3y = 6$

$$-3y = 6$$
$$y = -2.$$

Thus, $(0, -2)$ is a solution.

Let $y = 0$, then $2x - 3(0) = 6$

$$2x = 6$$
$$x = 3.$$

Thus, $(3, 0)$ is a solution.

Let $x = -3$, then $2(-3) - 3y = 6$

$$-6 - 3y = 6$$
$$-3y = 12$$
$$y = -4.$$

Thus, $(-3, -4)$ is a solution.

Plotting the points associated with these three solutions and connecting them with a straight line produces the graph of $2x - 3y = 6$ in Figure 8.7.

Figure 8.7

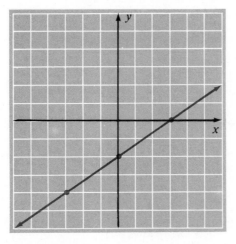

Let us briefly review our approach to Example 1. Notice that we did not begin our solution by either solving for y in terms of x or for x in terms of y. The reason for this is that we know the graph is a straight line and therefore there is no need for an extensive table of values. Thus, no real benefit is achieved by changing the form of the original equation. The first two solutions indicate where the line intersects the coordinate axes. The ordinate of the point $(0, -2)$ is called the **y-intercept** and the abscissa of the point $(3, 0)$ the **x-intercept** of this graph. That is, the graph of the equation $2x - 3y = 6$ has a y-intercept of -2 and an x-intercept of 3. In general, the intercepts are often easy to find. You can let $x = 0$ and solve for y to find the y-intercept and let $y = 0$ and solve for x to find the x-intercept. The third solution, $(-3, -4)$, served as a check point. If it had not been on the line determined by the two intercepts, then we would have known that an error had been committed.

EXAMPLE 2 Graph $x + 2y = 4$.

Solution Without showing all of our work, the following table indicates the intercepts and a check point.

x	y	
0	2	intercepts
4	0	
2	1	check points

Plotting the points $(0, 2)$, $(4, 0)$, and $(2, 1)$, and connecting them with a straight line produces the graph in Figure 8.8.

Figure 8.8

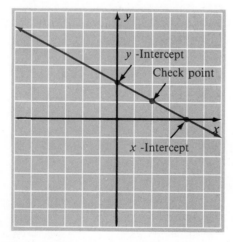

EXAMPLE 3 Graph $2x + 3y = 7$.

Solution The intercepts and a check point are given in the following table.

Finding intercepts may involve fractions, but the computation is usually rather easy. The points from the table are plotted and the graph of $2x + 3y = 7$ is shown in Figure 8.9.

Figure 8.9

x	y	
0	$\dfrac{7}{3}$	intercepts
$\dfrac{7}{2}$	0	
2	1	check point

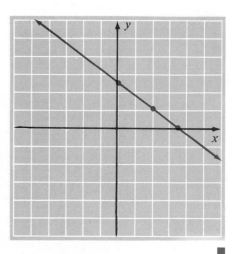

EXAMPLE 4 Graph $y = 2x$.

Solution Notice that $(0,0)$ is a solution; thus, this line intersects both axes at the origin. Since both the x-intercept and y-intercept are determined by the origin, $(0,0)$, another point is necessary to graph the line. Then a third point should be found as a check point. These results are summarized in the following table.

The graph of $y = 2x$ is shown in Figure 8.10.

Figure 8.10

x	y	
0	0	intercept
2	4	additional point
-1	-2	check point

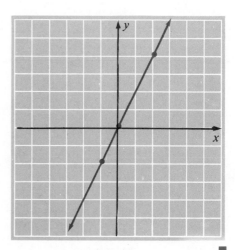

EXAMPLE 5 Graph $x = 3$.

Solution Since we are considering linear equations in *two variables*, the equation $x = 3$ is equivalent to $x + 0(y) = 3$. Now we can see that any value of y can be used, but the x-value must always be 3. Therefore, some of the solutions are $(3, 0)$, $(3, 1)$, $(3, 2)$, $(3, -1)$, and $(3, -2)$. The graph of all of the solutions is the vertical line indicated in Figure 8.11.

Figure 8.11

Problem Set 8.2

Use the techniques of this section to aid in sketching a graph for each of the following equations.

1. $x + y = 2$ 2. $x + y = 4$ 3. $x - y = 3$

4. $x - y = 1$ 5. $x - y = -4$ 6. $-x + y = 5$

7. $x + 2y = 2$ 8. $x + 3y = 5$ 9. $3x - y = 6$

10. $2x - y = -4$ 11. $3x - 2y = 6$ 12. $2x - 3y = 4$

13. $x - y = 0$ 14. $x + y = 0$ 15. $y = 3x$

16. $y = -2x$ 17. $x = -2$ 18. $y = 3$

19. $y = 0$ 20. $x = 0$ 21. $y = -2x - 1$

22. $y = 3x - 4$ 23. $y = \dfrac{1}{2}x + 1$ 24. $y = \dfrac{2}{3}x - 2$

25. $y = -\dfrac{1}{3}x - 2$ 26. $y = -\dfrac{3}{4}x - 1$ 27. $4x + 5y = -10$

28. $3x + 5y = -9$ 29. $-2x + y = -4$ 30. $-3x + y = -5$

31. $3x - 4y = 7$ 32. $4x - 3y = 10$ 33. $y + 4x = 0$

34. $y - 5x = 0$ 35. $x = 2y$ 36. $x = -3y$

■ Miscellaneous Problems

From our previous work with absolute value we know that $|x + y| = 2$ is equivalent to $x + y = 2$ or $x + y = -2$. Therefore, the graph of $|x + y| = 2$ consists of the two lines $x + y = 2$ and $x + y = -2$. Graph each of the following.

37. $|x + y| = 1$ **38.** $|x - y| = 2$ **39.** $|2x + y| = 4$ **40.** $|3x - y| = 6$

8.3 The Slope of a Line

In Figure 8.12 note that the line associated with $4x - y = 4$ is "steeper" than the line associated with $2x - 3y = 6$. Mathematically, the concept of **slope** is used to discuss the "steepness" of lines. The slope of a line is the ratio of the vertical change compared to the horizontal change as we move from one point on a line to another point. This is indicated in Figure 8.13 using the points P_1 and P_2.

Figure 8.12

Figure 8.13

    A precise definition for slope can be given by considering the coordinates of the points P_1, P_2, and Q in Figure 8.14. Since P_1 and P_2 represent any two points on the line, we will assign the coordinates (x_1, y_1) to P_1 and (x_2, y_2) to P_2. The point Q is the same distance from the y-axis as P_2 and the same distance from the x-axis as P_1. Thus, we assign the coordinates (x_2, y_1) to Q (Figure 8.14). It should now be apparent that the vertical change is $y_2 - y_1$ and the horizontal change is $x_2 - x_1$. Thus, the following definition for slope is given.

Figure 8.14

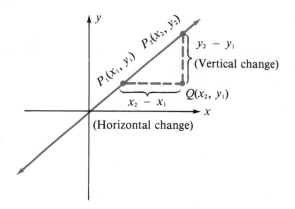

Definition 8.1

If points P_1 and P_2 with coordinates (x_1, y_1) and (x_2, y_2) respectively, are any two different points on a line, then the slope of the line (denoted by m) is

$$m = \frac{y_2 - y_1}{x_2 - x_1}, \qquad x_1 \neq x_2.$$

Using Definition 8.1, the slope of a line can be easily determined if the coordinates of two points on the line are known.

EXAMPLE 1   Find the slope of the line determined by each of the following pairs of points.

(a) $(2, 1)$ and $(4, 6)$ (b) $(3, 2)$ and $(-4, 5)$ (c) $(-4, -3)$ and $(-1, -3)$

Solution   (a) Let $(2, 1)$ be P_1 and $(4, 6)$ be P_2.

$$m = \frac{y_2 - y_1}{x_2 - x_1} = \frac{6 - 1}{4 - 2} = \frac{5}{2}.$$

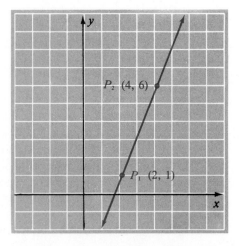

Figure 8.15

(b) Let $(3, 2)$ be P_1 and $(-4, 5)$ be P_2.

$$m = \frac{y_2 - y_1}{x_2 - x_1} = \frac{5 - 2}{-4 - 3}$$

$$= \frac{3}{-7} = -\frac{3}{7}.$$

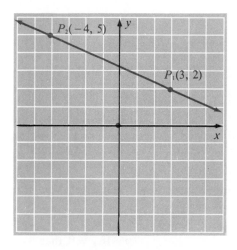

Figure 8.16

(c) Let $(-4, -3)$ be P_1 and $(-1, -3)$ be P_2.

$$m = \frac{y_2 - y_1}{x_2 - x_1} = \frac{-3 - (-3)}{-1 - (-4)}$$

$$= \frac{0}{3} = 0.$$

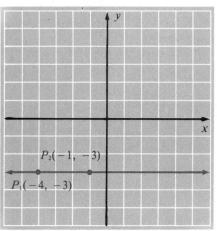

Figure 8.17

The designation of P_1 and P_2 for such problems is arbitrary and will not affect the value of the slope. For example, in part (a) of Example 1 suppose that we let $(4, 6)$ be P_1 and $(2, 1)$ be P_2. Then we obtain

$$m = \frac{y_2 - y_1}{x_2 - x_2} = \frac{1 - 6}{2 - 4} = \frac{-5}{-2} = \frac{5}{2}.$$

The various parts of Example 1 also illustrate the three basic possibilities for slope; that is, the slope of a line can be *positive*, *negative*, or *zero*. A line having a positive slope rises as we move from left to right as is the case in part (a). A line having a negative slope falls as we move from left to right, as in part (b). A horizontal line, as in part (c), has a slope of 0. Finally, we need to realize that the concept of slope is undefined for vertical lines. This is due to the fact that for any vertical line the change in x as we move from one point to another is zero. Thus, the ratio $\dfrac{y_2 - y_1}{x_2 - x_1}$ will have a denominator of zero and be undefined. So in Definition 8.1 the restriction $x_1 \neq x_2$ is made.

EXAMPLE 2   Find the slope of the line determined by the equation $3x + 4y = 12$.

Solution   Since any two points on the line can be used to determine the slope of the line, let's find the intercepts.

$$\text{If } x = 0, \text{ then } 3(0) + 4y = 12$$
$$4y = 12$$
$$y = 3.$$
$$\text{If } y = 0, \text{ then } 3x + 4(0) = 12$$
$$3x = 12$$
$$x = 4.$$

Using $(0, 3)$ as P_1 and $(4, 0)$ as P_2, we obtain

$$m = \frac{y_2 - y_1}{x_2 - x_1} = \frac{0 - 3}{4 - 0} = \frac{-3}{4} = -\frac{3}{4}. \qquad ■$$

One final idea pertaining to the concept of slope needs to be emphasized. The slope of a line is a **ratio** of vertical change compared to horizontal change. A slope of $\dfrac{3}{4}$ means that for every 3 units of vertical change there must be a corresponding 4 units of horizontal change. So starting at some point on the line, we could move to other points on the line as follows.

$$\frac{3}{4} = \frac{6}{8}: \qquad \text{by moving 6 units } \textit{up} \text{ and 8 units to the } \textit{right};$$

$$\frac{3}{4} = \frac{15}{20}: \qquad \text{by moving 15 units } \textit{up} \text{ and 20 units to the } \textit{right};$$

$$\frac{3}{4} = \frac{\frac{3}{2}}{2}: \qquad \text{by moving } 1\frac{1}{2} \text{ units } \textit{up} \text{ and 2 units to the } \textit{right};$$

$$\frac{3}{4} = \frac{-3}{-4}: \qquad \text{by moving 3 units } \textit{down} \text{ and 4 units to the } \textit{left}.$$

Likewise, a slope of $-\dfrac{5}{6}$ indicates that starting at some point on the line we could move to other points on the line as follows.

$$-\frac{5}{6} = \frac{-5}{6}:$$ by moving 5 units *down* and 6 units to the *right*;

$$-\frac{5}{6} = \frac{5}{-6}$$ by moving 5 units *up* and 6 units to the *left*;

$$-\frac{5}{6} = \frac{-10}{12}:$$ by moving 10 units *down* and 12 units to the *right*;

$$-\frac{5}{6} = \frac{15}{-18}:$$ by moving 15 units *up* and 18 units to the *left*.

Problem Set 8.3

For Problems 1–20, find the slope of the line determined by each pair of points.

1. $(7, 5), (3, 2)$ **2.** $(9, 10), (6, 2)$ **3.** $(-1, 3), (-6, -4)$

4. $(-2, 5), (-7, -1)$ **5.** $(2, 8), (7, 2)$ **6.** $(3, 9), (8, 4)$

7. $(-2, 5), (1, -5)$ **8.** $(-3, 4), (2, -6)$ **9.** $(4, -1), (-4, -7)$

10. $(5, -3), (-5, -9)$ **11.** $(3, -4), (2, -4)$ **12.** $(-3, -6), (5, -6)$

13. $(-6, -1), (-2, -7)$ **14.** $(-8, -3), (-2, -11)$ **15.** $(-2, 4), (-2, -6)$

16. $(-4, -5), (-4, 9)$ **17.** $(-1, 10), (-9, 2)$ **18.** $(-2, 12), (-10, 2)$

19. $(a, b), (c, d)$ **20.** $(a, 0), (0, b)$

21. Find y if the line through $(7, 8)$ and $(2, y)$ has a slope of $\dfrac{4}{5}$.

22. Find y if the line through $(12, 14)$ and $(3, y)$ has a slope of $\dfrac{4}{3}$.

23. Find x if the line through $(-2, -4)$ and $(x, 2)$ has a slope of $-\dfrac{3}{2}$.

24. Find x if the line through $(6, -4)$ and $(x, 6)$ has a slope of $-\dfrac{5}{4}$.

For Problems 25–32, you are given one point on a line and the slope of the line. Find the coordinates of three other points on the line.

25. $(3, 2), m = \dfrac{2}{3}$ **26.** $(4, 1), m = \dfrac{5}{6}$ **27.** $(-2, -4), m = \dfrac{1}{2}$

28. $(-6, -2), m = \dfrac{2}{5}$ **29.** $(-3, 4), m = -\dfrac{3}{4}$ **30.** $(-2, 6), m = -\dfrac{3}{7}$

31. $(4, -5), m = -2$ **32.** $(6, -2), m = 4$

For Problems 33–40, sketch the line determined by each pair of points and decide whether the slope of the line is *positive*, *negative*, or *zero*.

33. $(2, 8), (7, 1)$ **34.** $(1, -2), (7, -8)$ **35.** $(-1, 3), (-6, -2)$

36. $(7, 3), (4, -6)$ **37.** $(-2, 4), (6, 4)$ **38.** $(-3, -4), (5, -4)$

39. $(-3, 5) (2, -7)$ **40.** $(-1, -1), (1, -9)$

For Problems 41–60, find the coordinates of two points on the given line and then use those coordinates to find the slope of the line.

41. $3x + 2y = 6$ **42.** $4x + 3y = 12$ **43.** $5x - 4y = 20$

44. $7x - 3y = 21$ **45.** $x + 5y = 6$ **46.** $2x + y = 4$

47. $2x - y = -7$ **48.** $x - 4y = -6$ **49.** $y = 3$

50. $x = 6$ **51.** $-2x + 5y = 9$ **52.** $-3x - 7y = 10$

53. $6x - 5y = -30$ **54.** $7x - 6y = -42$ **55.** $y = -3x - 1$

56. $y = -2x + 5$ **57.** $y = 4x$ **58.** $y = 6x$

59. $y = \dfrac{2}{3} x - \dfrac{1}{2}$ **60.** $y = -\dfrac{3}{4} x + \dfrac{1}{5}$

8.4   Writing Equations of Lines

As we stated in Section 8.1, basically there are two types of problems in analytic or coordinate geometry.

1. Given an algebraic equation, find its geometric graph.
2. Given a set of conditions pertaining to a geometric figure, determine its algebraic equation.

Problems of type 1 have been a part of the first two sections of this chapter. At this time we want to consider a few problems of type 2 dealing with straight lines. In other words, given certain facts about a line, we need to be able to write its algebraic equation.

EXAMPLE 1   Find the equation of the line having a slope of $\dfrac{3}{4}$ and containing the point $(1, 2)$.

Solution   First, let's draw the line as indicated in Figure 8.18. Since the slope is $\dfrac{3}{4}$, a second point can be determined by moving 3 units up and 4 units to the right of the given point $(1, 2)$. (The point $(5, 5)$ merely helps to draw the line; it will not be used in analyzing the problem.) Now let's choose a point (x, y) that represents all points on the line other than the given point $(1, 2)$. The slope determined by $(1, 2)$ and (x, y) is $\dfrac{3}{4}$. Thus,

Figure 8.18

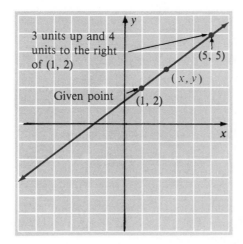

$$\frac{y - 2}{x - 1} = \frac{3}{4}$$

$$3(x - 1) = 4(y - 2)$$

$$3x - 3 = 4y - 8$$

$$3x - 4y = -5.$$

■

EXAMPLE 2   Find the equation of the line containing $(3, 4)$ and $(-2, 5)$.

Solution   First, let's draw the line determined by the two given points (Figure 8.19).

Figure 8.19

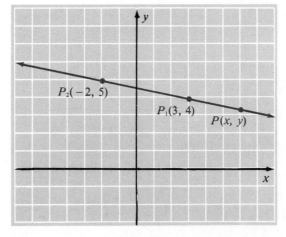

Since two points are known, the slope can be found.

$$m = \frac{y_2 - y_1}{x_2 - x_1} = \frac{5 - 4}{-2 - 3} = \frac{1}{-5} = -\frac{1}{5}.$$

Now we can use the same approach as in Example 1. Form an equation using a variable point (x, y), one of the two given points, (we chose P_1), and the slope of $-\frac{1}{5}$.

$$\frac{y - 4}{x - 3} = \frac{1}{-5} \qquad -\frac{1}{5} = \frac{1}{-5}$$

$$x - 3 = -5y + 20$$
$$x + 5y = 23. \qquad\qquad ■$$

EXAMPLE 3 Find the equation of the line that has a slope of $\frac{1}{4}$ and a y-intercept of 2.

Solution A y-intercept of 2 means that the point $(0, 2)$ is on the line. Since the slope is $\frac{1}{4}$, another point can be found by moving 1 unit up and 4 units to the right of $(0, 2)$. The line is drawn in Figure 8.20. Choosing a variable point (x, y) we can proceed as in the previous examples.

$$\frac{y - 2}{x - 0} = \frac{1}{4}$$

$$x = 4y - 8$$
$$x - 4y = -8.$$

Figure 8.20

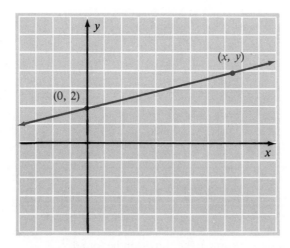

Perhaps it would be helpful for you to pause for a moment and look back over Examples 1, 2, and 3. Notice that the same basic approach is used in all three examples; that is, we chose a variable point (x, y) and used it, along with another known point, to determine the equation of the line. We should also recognize that the approach taken in these examples can be generalized to produce some special forms of straight lines.

■ Point-Slope Form

EXAMPLE 4 Find the equation of the line having a slope of m and containing the point (x_1, y_1).

Solution Choosing (x, y) to represent any other point on the line (Figure 8.21) the slope of the line is given by

$$m = \frac{y - y_1}{x - x_1}$$

from which we obtain

$$y - y_1 = m(x - x_1).$$

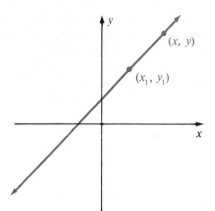

Figure 8.21

The equation

$$\boxed{y - y_1 = m(x - x_1)}$$

is referred to as the **point-slope form** of the equation of a straight line. Instead of the approach used in Example 1, we could use the point-slope form to write the equation of a line with a given slope and containing a given point, as the next example illustrates.

EXAMPLE 5 Write the equation of the line having a slope of $\frac{3}{5}$ and containing the point $(2, -4)$.

Solution Substituting $\frac{3}{5}$ for m, 2 for x_1, and -4 for y_1 in the point-slope form, we obtain

$$y - y_1 = m(x - x_1)$$

$$y - (-4) = \frac{3}{5}(x - 2)$$

$$y + 4 = \frac{3}{5}(x - 2)$$

$$5(y + 4) = 3(x - 2)$$
$$5y + 20 = 3x - 6$$
$$26 = 3x - 5y.$$

■

■ Slope-Intercept Form

Now consider the equation of a line having a slope of m and a y-intercept of b (Figure 8.22).

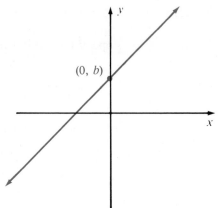

Figure 8.22

A y-intercept of b means that the line contains the point $(0, b)$; therefore, we can use the point-slope form as follows.

$$y - y_1 = m(x - x_1)$$
$$y - b = m(x - 0) \qquad y_1 = b \quad \text{and} \quad x_1 = 0$$
$$y - b = mx$$
$$y = mx + b$$

The equation
$$\boxed{y = mx + b}$$

is called the **slope-intercept form** of the equation of a straight line. It can be used for two primary purposes, as the next two examples illustrate.

EXAMPLE 6 Find the equation of the line that has a slope of $\frac{1}{4}$ and a y-intercept of 2.

Solution   This is a restatement of Example 3, but this time we will use the slope-intercept form of a line ($y = mx + b$) to write its equation. From the statement of the problem we know that $m = \dfrac{1}{4}$ and $b = 2$. Thus, substituting these values for m and b in $y = mx + b$, we obtain

$$y = mx + b$$

$$y = \frac{1}{4}x + 2$$

$$4y = x + 8$$

$$x - 4y = -8. \qquad \text{(Same result as in Example 3.)} \qquad ■$$

EXAMPLE 7 Find the slope of the line whose equation is $2x + 3y = 4$.

Solution   In Section 8.3 we solved this type of problem by determining the coordinates of two points on the line and then used the slope formula. Now we have another approach for such a problem. Let's change the given equation to slope-intercept form; that is to say, let's solve the equation for y in terms of x.

$$2x + 3y = 4$$

$$3y = -2x + 4$$

$$y = -\frac{2}{3}x + \frac{4}{3}.$$

Comparing this result to $y = mx + b$, we see that $m = -\dfrac{2}{3}$ and $b = \dfrac{4}{3}$. ■

In general, **if the equation of a nonvertical line is written in slope-intercept form, the coefficient of x is the slope of the line and the constant term is the y-intercept.** (Remember that the concept of slope is not defined for a vertical line.) Let's use that idea in a few more examples.

EXAMPLE 8 Find the slope and y-intercept of each of the following lines.
(a) $5x - 4y = 12$ (b) $-y = 3x - 4$ (c) $y = 2$

Solution   (a) Changing $5x - 4y = 12$ to slope-intercept form, we obtain

$$5x - 4y = 12$$

$$-4y = -5x + 12$$

$$4y = 5x - 12$$

$$y = \frac{5}{4}x - 3.$$

The slope of the line is $\dfrac{5}{4}$ (the coefficient of x) and the y-intercept is -3 (the constant term).

(b) Multiplying both sides of the given equation by -1 will change it to slope-intercept form.

$$-y = 3x - 4;$$
$$y = -3x + 4.$$

The slope of the line is -3 and the y-intercept is 4.

(c) The equation $y = 2$ can be written as

$$y = 0(x) + 2.$$

The slope of the line is 0 and the y-intercept is 2. (A line with a slope of 0 is horizontal.) ■

Problem Set 8.4

For Problems 1–12, find the equation of the line that contains the given point and has the given slope. Express equations in the form $Ax + By = C$, where A, B, and C are integers.

1. $(2, 3)$, $m = \dfrac{2}{3}$

2. $(5, 2)$, $m = \dfrac{3}{7}$

3. $(-3, -5)$, $m = \dfrac{1}{2}$

4. $(5, -6)$, $m = \dfrac{3}{5}$

5. $(-4, 8)$, $m = -\dfrac{1}{3}$

6. $(-2, -4)$, $m = -\dfrac{5}{6}$

7. $(3, -7)$, $m = 0$

8. $(-3, 9)$, $m = 0$

9. $(0, 0)$, $m = -\dfrac{4}{9}$

10. $(0, 0)$, $m = \dfrac{5}{11}$

11. $(-6, -2)$, $m = 3$

12. $(2, -10)$, $m = -2$

For Problems 13–22, find the equation of the line that contains the two given points. Express equations in the form $Ax + By = C$, where A, B, and C are integers.

13. $(2, 3)$ and $(7, 10)$

14. $(1, 4)$ and $(9, 10)$

15. $(3, -2)$ and $(-1, 4)$

16. $(-2, 8)$ and $(4, -2)$

17. $(-1, -2)$ and $(-6, -7)$

18. $(-8, -7)$ and $(-3, -1)$

19. $(0, 0)$ and $(-3, -5)$

20. $(5, -8)$ and $(0, 0)$

21. $(0, 4)$ and $(7, 0)$

22. $(-2, 0)$ and $(0, -9)$

For Problems 23–32, find the equation of the line with the given slope and y-intercept. Express equations in the form $Ax + By = C$, where A, B, and C are integers.

23. $m = \dfrac{3}{5}$ and $b = 2$

24. $m = \dfrac{5}{9}$ and $b = 4$

25. $m = 2$ and $b = -1$

26. $m = 4$ and $b = -3$

27. $m = -\dfrac{1}{6}$ and $b = -4$

28. $m = -\dfrac{5}{7}$ and $b = -1$

29. $m = -1$ and $b = \dfrac{5}{2}$

30. $m = -2$ and $b = \dfrac{7}{3}$

31. $m = -\dfrac{5}{9}$ and $b = -\dfrac{1}{2}$

32. $m = -\dfrac{7}{12}$ and $b = -\dfrac{2}{3}$

For Problems 33–44, determine the slope and y-intercept of the line represented by the given equation.

33. $y = -2x - 5$

34. $y = \dfrac{2}{3}x + 4$

35. $3x - 5y = 15$

36. $7x + 5y = 35$

37. $-4x + 9y = 18$

38. $-6x + 7y = -14$

39. $-y = -\dfrac{3}{4}x + 4$

40. $5x - 2y = 0$

41. $-2x - 11y = 11$

42. $-y = \dfrac{2}{3}x + \dfrac{11}{2}$

43. $9x + 7y = 0$

44. $-5x - 13y = 26$

■ Miscellaneous Problems

The following two properties pertain to parallel and perpendicular lines.

1. Two nonvertical lines are parallel if and only if their slopes are equal.
2. Two nonvertical lines are perpendicular if and only if the product of their slopes is -1.

For Problems 45–50, determine whether each pair of lines are (a) parallel, (b) perpendicular, or (c) intersecting lines that are not perpendicular.

45. $5x - 2y = 6$
$2x + 5y = 9$

46. $2x - y = 9$
$6x - 3y = 5$

47. $4x - 3y = 12$
$3x - 4y = 12$

48. $9x + 2y = 18$
$2x - 9y = 13$

49. $x - 3y = 7$
$5x - 15y = 9$

50. $7x - 2y = 14$
$7x + 2y = 5$

51. Write the equation of the line that contains the point $(4, 3)$ and is parallel to the line $2x - 3y = 6$.

52. Write the equation of the line that contains the point $(-1, 3)$ and is perpendicular to the line $3x - y = 4$.

8.5 Solving Linear Systems by Graphing

Suppose that we graph $x - 2y = 4$ and $x + 2y = 8$ on the same set of axes, as indicated in Figure 8.23. The ordered pair, $(6, 1)$, which is associated with the point

Figure 8.23

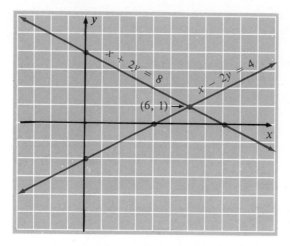

of intersection of the two lines, satisfies both equations. That is to say, $(6, 1)$ is a solution for $x - 2y = 4$ *and* for $x + 2y = 8$. To check this, we can substitute 6 for x and 1 for y in both equations.

$x - 2y = 4$ becomes $6 - 2(1) = 4$;

$x + 2y = 8$ becomes $6 + 2(1) = 8$.

Thus, we say that $\{(6, 1)\}$ is the solution set of the system

$$\begin{pmatrix} x - 2y = 4 \\ x + 2y = 8 \end{pmatrix}.$$

Two or more linear equations in two variables considered together are called a **system of linear equations**. The following are systems of linear equations.

$$\begin{pmatrix} x - 2y = 4 \\ x + 2y = 8 \end{pmatrix}; \qquad \begin{pmatrix} 5x - 3y = 9 \\ 3x + 7y = 12 \end{pmatrix}; \qquad \begin{pmatrix} 4x - y = 5 \\ 2x + y = 9 \\ 7x - 2y = 13 \end{pmatrix}.$$

To **solve a system of linear equations** means to find all of the ordered pairs that are solutions of all of the equations in the system. In this book we shall consider only systems of *two* linear equations in two variables. There are several techniques for solving systems of linear equations. We will use three of them in this chapter: a graphing method in this section and two other methods in the following sections.

To solve a system of linear equations by **graphing**, you proceed as we did in the opening discussion of this section. The equations are graphed on the same set of axes and then the ordered pairs associated with any points of intersection are the solutions to the system. Let's consider another example.

EXAMPLE 1 Solve the system $\begin{pmatrix} x + \ y = \ \ 5 \\ x - 2y = -4 \end{pmatrix}$.

Solution   Let's find the intercepts and a check point for each of the lines.

$x + y = 5$

x	y	
0	5	intercepts
5	0	
2	3	check point

$x - 2y = -4$

x	y	
0	2	intercepts
-4	0	
-2	1	check point

Figure 8.24 shows the graphs of the two equations.

Figure 8.24

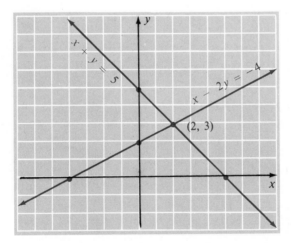

It appears that $(2, 3)$ is the solution of the system. To check we can substitute 2 for x and 3 for y in both equations.

$x + y$ becomes $2 + 3 = 5$;   (a true statement)

$x - 2y = -4$ becomes $2 - 2(3) = -4$.   (a true statement)

Therefore, $\{(2, 3)\}$ is the solution set.   ■

It should be evident that solving systems of equations by graphing requires accurate graphs. In fact, unless the solutions are integers it is really quite difficult to obtain exact solutions from a graph. For this reason the systems in this section have integral solutions. Furthermore, checking a solution takes on additional significance when using the graphing approach. By checking you can be absolutely sure that you are *reading* the correct solution from the graph.

Figure 8.25 shows the three possible cases for the graph of a system of two linear equations in two variables.

Case I.   The graphs of the two equations are two lines intersecting in one point. There is *one solution* and the system is called a **consistent system**.

Case II.   The graphs of the two equations are parallel lines. There is *no solution* and the system is called an **inconsistent system**.

Case III.   The graphs of the two equations are the same line. There *are infinitely many* solutions to the system. Any pair of real numbers that satisfies one of the equations will also satisfy the other equation, and we say the equations are **dependent**.

Figure 8.25

Case I

Case II

Case III

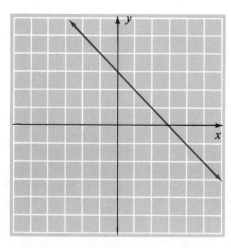

Thus, as we solve a system of two linear equations in two variables we know what to expect. The system will have no solutions, one ordered pair as a solution, or infinitely many ordered pairs as solutions. Most of the systems that we will be working with in this text will have one solution.

An example of Case I was given in Example 1 (Figure 8.24). The next two examples illustrate the other cases.

EXAMPLE 2 Solve the system $\begin{pmatrix} 2x + 3y = 6 \\ 2x + 3y = 12 \end{pmatrix}$.

Solution

$2x + 3y = 6$

x	y
0	2
3	0
-3	4

$2x + 3y = 12$

x	y
0	4
6	0
3	2

Figure 8.26 shows the graph of the system. Since the lines are parallel, there is *no solution* to the system. The solution set is $\varnothing$.

Figure 8.26

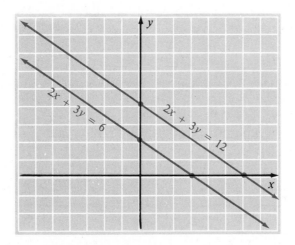

EXAMPLE 3 Solve the system $\begin{pmatrix} x + y = 3 \\ 2x + 2y = 6 \end{pmatrix}$.

Solution

$x + y = 3$

x	y
0	3
3	0
1	2

$2x + 2y = 6$

x	y
0	3
3	0
1	2

Figure 8.27 shows the graph of this system. Since the graph of both equations is the same line, there are *infinitely many solutions* to the system. Any ordered pair of real numbers that satisfies one equation will also satisfy the other equation.

Figure 8.27

Problem Set 8.5

For Problems 1–10, decide whether the given ordered pair is a solution of the given system of equations.

1. $\begin{pmatrix} 5x + y = 9 \\ 3x - 2y = 4 \end{pmatrix}$ $(1, 4)$

2. $\begin{pmatrix} 3x - 2y = 8 \\ 2x + y = 3 \end{pmatrix}$ $(2, -1)$

3. $\begin{pmatrix} x - 3y = 17 \\ 2x + 5y = -21 \end{pmatrix}$ $(2, -5)$

4. $\begin{pmatrix} 2x + 7y = -5 \\ 4x - y = 6 \end{pmatrix}$ $(1, -1)$

5. $\begin{pmatrix} y = 2x \\ 3x - 4y = 5 \end{pmatrix}$ $(-1, -2)$

6. $\begin{pmatrix} -x + 2y = 8 \\ x - 2y = 8 \end{pmatrix}$ $(-2, 3)$

7. $\begin{pmatrix} 6x - 5y = 5 \\ 3x + 4y = -4 \end{pmatrix}$ $(0, -1)$

8. $\begin{pmatrix} y = 3x - 1 \\ 5x - 2y = -1 \end{pmatrix}$ $(3, 8)$

9. $\begin{pmatrix} -3x - y = 4 \\ -2x + 3y = -23 \end{pmatrix}$ $(4, -5)$

10. $\begin{pmatrix} 5x + 4y = -15 \\ 2x - 7y = -6 \end{pmatrix}$ $(-3, 0)$

For Problems 11–30, use the graphing method to solve each system.

11. $\begin{pmatrix} x + y = 1 \\ x - y = 3 \end{pmatrix}$

12. $\begin{pmatrix} x - y = 2 \\ x + y = -4 \end{pmatrix}$

13. $\begin{pmatrix} x + 2y = 4 \\ 2x - y = 3 \end{pmatrix}$

14. $\begin{pmatrix} 2x - y = -8 \\ x + y = 2 \end{pmatrix}$

15. $\begin{pmatrix} x + 3y = 6 \\ x + 3y = 3 \end{pmatrix}$

16. $\begin{pmatrix} y = -2x \\ y - 3x = 0 \end{pmatrix}$

17. $\begin{pmatrix} x + y = 0 \\ x - y = 0 \end{pmatrix}$ **18.** $\begin{pmatrix} 3x - y = 3 \\ 3x - y = -3 \end{pmatrix}$ **19.** $\begin{pmatrix} 3x - 2y = 5 \\ 2x + 5y = -3 \end{pmatrix}$

20. $\begin{pmatrix} 2x + 3y = 1 \\ 4x - 3y = -7 \end{pmatrix}$ **21.** $\begin{pmatrix} y = -2x + 3 \\ 6x + 3y = 9 \end{pmatrix}$ **22.** $\begin{pmatrix} y = 2x + 5 \\ x + 3y = -6 \end{pmatrix}$

23. $\begin{pmatrix} y = 5x - 2 \\ 4x + 3y = 13 \end{pmatrix}$ **24.** $\begin{pmatrix} y = x - 2 \\ 2x - 2y = 4 \end{pmatrix}$ **25.** $\begin{pmatrix} y = 4 - 2x \\ y = 7 - 3x \end{pmatrix}$

26. $\begin{pmatrix} y = 3x + 4 \\ y = 5x + 8 \end{pmatrix}$ **27.** $\begin{pmatrix} y = 2x \\ 3x - 2y = -2 \end{pmatrix}$ **28.** $\begin{pmatrix} y = 3x \\ 4x - 3y = 5 \end{pmatrix}$

29. $\begin{pmatrix} 7x - 2y = -8 \\ x = -2 \end{pmatrix}$ **30.** $\begin{pmatrix} 3x + 8y = -1 \\ y = -2 \end{pmatrix}$

8.6 The Addition Method

We have used the addition property of equality (if $a = b$, then $a + c = b + c$) to help solve equations containing one variable. An extension of the addition property forms the basis for another method of solving systems of linear equations. Property 8.1 states that two equations can be added and the resulting equation will be equivalent to the original ones. Let's use this property to help solve a system of equations.

Property 8.1

For all real numbers a, b, c, and d, if $a = b$, and $c = d$, then

$$a + c = b + d.$$

EXAMPLE 1 Solve $\begin{pmatrix} x + y = 12 \\ x - y = 2 \end{pmatrix}$.

Solution

$$\begin{array}{r} x + y = 12 \\ x - y = 2 \\ \hline 2x = 14 \end{array}$$ This is the result of adding the two equations.

Solving this new equation we obtain

$$2x = 14$$
$$\boxed{x = 7.}$$

Now we can substitute the value of 7 for x in one of the original equations. Thus,

$$x + y = 12$$
$$7 + y = 12$$
$$\boxed{y = 5.}$$

To check we see that

$$x + y = 12 \quad \text{and} \quad x - y = 2;$$
$$7 + 5 = 12 \quad \text{and} \quad 7 - 5 = 2.$$

Thus, the solution set of the system is $\{(7, 5)\}$. ■

Note in Example 1 that by adding the two original equations, a simple equation containing only one variable was obtained. Adding equations to eliminate a variable is the key idea behind the **addition method** for solving systems of linear equations. The next example further illustrates this point.

EXAMPLE 2 Solve $\begin{pmatrix} 2x + 3y = -26 \\ 4x - 3y = 2 \end{pmatrix}$

Solution

$$2x + 3y = -26$$
$$\underline{4x - 3y = 2}$$
$$6x = -24 \qquad \text{Added the two equations.}$$
$$\boxed{x = -4.}$$

Substitute -4 for x in one of the two original equations.

$$2x + 3y = -26$$
$$2(-4) + 3y = -26$$
$$-8 + 3y = -26$$
$$3y = -18$$
$$\boxed{y = -6.}$$

The solution set is $\{(-4, -6)\}$. ■

It may be necessary to change the form of one, or perhaps both, of the original equations before adding them. The next example illustrates this idea.

EXAMPLE 3 Solve $\begin{pmatrix} y = x - 21 \\ x + y = -3 \end{pmatrix}$.

Solution $y = x - 21$ Subtract x from both sides $-x + y = -21$
                $x + y = -3$ Leave alone. $\underline{\quad x + y = -3 \quad}$
                                             $2y = -24$
                                             $\boxed{y = -12.}$

Substitute -12 for y in one of the original equations.

$$x + y = -3$$
$$x + (-12) = -3$$
$$\boxed{x = 9.}$$

The solution set is $\{(9, -12)\}$.   ■

Frequently, the multiplication property of equality needs to be applied first so that adding the equations will eliminate a variable, as the next example illustrates.

EXAMPLE 4 Solve $\begin{pmatrix} 2x + 5y = 29 \\ 3x - y = 1 \end{pmatrix}$.

Solution   Notice that adding the equations as they are would not eliminate a variable. However, we should observe that multiplying the bottom equation by 5 and then adding this newly formed, but equivalent, equation to the top equation will eliminate the y's.

$2x + 5y = 29$ Leave alone. $2x + 5y = 29$
$3x - y = 1$ Multiply both sides by 5. $\underline{15x - 5y = 5}$
                                           $17x = 34$
                                           $\boxed{x = 2}$

Substitute 2 for x in one of the original equations.

$$3x - y = 1$$
$$3(2) - y = 1$$
$$6 - y = 1$$
$$-y = -5$$
$$\boxed{y = 5.}$$

The solution set is $\{(2, 5)\}$.   ■

Notice in these problems that after finding the value of one of the variables, this number is substituted into one of the *original* equations to find the value of the other variable. It doesn't matter which of the two original equations you use, so pick the one easiest to solve.

Sometimes the multiplication property needs to be applied to both equations. Let's look at an example of this type.

EXAMPLE 5 Solve $\begin{pmatrix} 2x + 3y = & 4 \\ 9x - 2y = & -13 \end{pmatrix}$.

Solution A

$$2x + 3y = 4 \qquad \underline{\text{Multiply both sides by 9.}} \qquad 18x + 27y = 36$$
$$9x - 2y = -13 \qquad \underline{\text{Multiply both sides by } -2.} \qquad \underline{-18x + 4y = 26}$$
$$31y = 62$$
$$\boxed{y = 2.}$$

Substitute 2 for y in one of the original equations.

$$2x + 3y = 4$$
$$2x + 3(2) = 4$$
$$2x + 6 = 4$$
$$2x = -2$$
$$x = -1.$$

The solution set is $\{(-1, 2)\}$.

Solution B

$$2x + 3y = 4 \qquad \underline{\text{Multiply both sides by 2.}} \qquad 4x + 6y = 8$$
$$9x - 2y = -13 \qquad \underline{\text{Multiply both sides by 3.}} \qquad \underline{27x - 6y = -39}$$
$$31x = -31$$
$$\boxed{x = -1.}$$

Substitute -1 for x in one of the original equations.

$$2x + 3y = 4$$
$$2(-1) + 3y = 4$$
$$-2 + 3y = 4$$
$$3y = 6$$
$$\boxed{y = 2.}$$

The solution set is $\{(-1, 2)\}$. ■

Look back over Solutions A and B for Example 5 carefully. Especially notice the first steps where the multiplication property of equality is applied to the two equations. In Solution A we multiplied by numbers so that adding the resulting equations eliminated the x variable. In Solution B we multiplied so that the y variable was eliminated when we added the resulting equations. Either approach will work; pick the one that involves the easiest computation.

EXAMPLE 6 Solve $\begin{pmatrix} 3x - 4y = 7 \\ 5x + 3y = 9 \end{pmatrix}$.

$$3x - 4y = 7 \quad \underrightarrow{\text{Multiply both sides by 3.}} \quad 9x - 12y = 21$$
$$5x + 3y = 9 \quad \underrightarrow{\text{Multiply both sides by 4.}} \quad \underline{20x + 12y = 36}$$
$$29x \qquad\quad = 57$$
$$x = \frac{57}{29}.$$

Since substituting $\dfrac{57}{29}$ for x into one of the original equations will produce some messy calculations, let's solve for y by eliminating the x's.

$$3x - 4y = 7 \quad \underrightarrow{\text{Multiply both sides by } -5.} \quad -15x + 20y = -35$$
$$5x + 3y = 9 \quad \underrightarrow{\text{Multiply both sides by 3.}} \quad \underline{15x + 9y = \quad 27}$$
$$29y = \quad -8$$
$$y = -\frac{8}{29}.$$

The solution set is $\left\{ \left(\dfrac{57}{29}, -\dfrac{8}{29} \right) \right\}$.   ■

■ Problem Solving

Many word problems that we solved earlier in this text using one equation in one variable can also be solved using a system of two linear equations in two variables. In fact, many times you may find that it seems quite natural to use two variables. It may also seem more meaningful at times to use variables other than x and y. Let's consider some examples.

PROBLEM 1 The cost of 3 tennis balls and 2 golf balls is $7. Furthermore, the cost of 6 tennis balls and 3 golf balls is $12. Find the cost of 1 tennis ball and the cost of 1 golf ball.

Solution Let's use t to represent the cost of one tennis ball and g the cost of one golf ball. The problem translates into the following system of equations.

$$3t + 2g = 7 \qquad \text{The cost of 3 tennis balls and 2 golf balls is \$7.}$$
$$6t + 3g = 12 \qquad \text{The cost of 6 tennis balls and 3 golf balls is \$12.}$$

Solving this system by the addition method, we obtain

$$3t + 2g = 7 \quad \underrightarrow{\text{Multiply both sides by } -2.} \quad -6t - 4g = -14$$
$$6t + 3g = 12 \quad \underrightarrow{\text{Leave alone.}} \qquad \underline{6t + 3g = \quad 12}$$
$$-g = -2$$
$$\boxed{g = 2.}$$

Substitute 2 for g in one of the original equations.

$$3t + 2g = 7$$
$$3t + 2(2) = 7$$
$$3t + 4 = 7$$
$$3t = 3$$
$$\boxed{t = 1.}$$

The cost of a tennis ball is $1 and the cost of a golf ball is $2.   ■

PROBLEM 2   Niki invested $500, part of it at 9% interest and the rest at 10%. Her total interest earned for a year was $48. How much did she invest at each rate?

Solution   Let x represent the amount invested at 9%. Let y represent the amount invested at 10%. The problem translates into the following system.

$$x + y = 500 \qquad \text{Niki invested \$500.}$$
$$0.09x + 0.10y = 48. \qquad \text{Her total interest earned for a year was \$48.}$$

Multiplying the second equation by 100 produces $9x + 10y = 4800$. Then, we have the following equivalent system to solve.

$$\begin{pmatrix} x + y = 500 \\ 9x + 10y = 4800. \end{pmatrix}$$

Using the addition method, we can proceed as follows.

$$x + y = 500 \quad \underrightarrow{\text{Multiply both sides by } -9.} \quad -9x - 9y = -4500$$
$$9x + 10y = 4800 \quad \underrightarrow{\text{Leave alone.}} \qquad \underline{9x + 10y = \quad 4800}$$
$$\boxed{y = \quad 300.}$$

Substituting 300 for y in $x + y = 500$ yields

$$x + y = 500$$
$$x + 300 = 500$$
$$\boxed{x = 200.}$$

Therefore, we know that $200 was invested at 9% interest and $300 at 10%.   ■

Before you tackle the word problems in this next problem set, it might be helpful to review the problem solving suggestions offered in Section 4.2. Those suggestions continue to apply here except that we now have the flexibility of using two equations and two unknowns. Don't forget that to check a word problem you need to see if your answers satisfy the conditions stated in the problem.

Problem Set 8.6

For Problems 1–24, solve each system using the addition method.

1. $\begin{pmatrix} x + y = 14 \\ x - y = -2 \end{pmatrix}$

2. $\begin{pmatrix} x - y = -14 \\ x + y = 6 \end{pmatrix}$

3. $\begin{pmatrix} x + 4y = -21 \\ 3x - 4y = 1 \end{pmatrix}$

4. $\begin{pmatrix} -3x + 2y = -21 \\ 3x - 7y = 36 \end{pmatrix}$

5. $\begin{pmatrix} y = 6 - x \\ x - y = -18 \end{pmatrix}$

6. $\begin{pmatrix} x + y = -10 \\ x = y + 6 \end{pmatrix}$

7. $\begin{pmatrix} 5x + y = 23 \\ 3x - 2y = 19 \end{pmatrix}$

8. $\begin{pmatrix} 4x - 5y = -36 \\ x + 2y = 30 \end{pmatrix}$

9. $\begin{pmatrix} x + 2y = 5 \\ 3x - 2y = 6 \end{pmatrix}$

10. $\begin{pmatrix} 2x - y = 3 \\ -2x + 5y = 7 \end{pmatrix}$

11. $\begin{pmatrix} y = -x \\ 2x - y = -2 \end{pmatrix}$

12. $\begin{pmatrix} 3x = 2y \\ 8x + 20y = 19 \end{pmatrix}$

13. $\begin{pmatrix} 4x + 5y = 9 \\ 5x - 6y = -50 \end{pmatrix}$

14. $\begin{pmatrix} 2x + 7y = 46 \\ 3x - 4y = -18 \end{pmatrix}$

15. $\begin{pmatrix} 9x - 7y = 29 \\ 5x - 3y = 17 \end{pmatrix}$

16. $\begin{pmatrix} 7x + 5y = -6 \\ 4x + 3y = -4 \end{pmatrix}$

17. $\begin{pmatrix} 6x + 5y = -6 \\ 8x - 3y = 21 \end{pmatrix}$

18. $\begin{pmatrix} 3x - 8y = -23 \\ 7x + 4y = -48 \end{pmatrix}$

19. $\begin{pmatrix} 2x - 7y = -1 \\ 9x + 4y = -2 \end{pmatrix}$

20. $\begin{pmatrix} 6x - 2y = -3 \\ 5x - 9y = 1 \end{pmatrix}$

21. $\begin{pmatrix} x + y = 750 \\ .07x + .08y = 57.5 \end{pmatrix}$

22. $\begin{pmatrix} x + y = 700 \\ .06x + .09y = 54 \end{pmatrix}$

23. $\begin{pmatrix} .09x + .11y = 31 \\ y = x + 100 \end{pmatrix}$

24. $\begin{pmatrix} .08x + .1y = 56 \\ y = 2x \end{pmatrix}$

For Problems 25–40, solve each problem by setting up and solving a system of two linear equations in two variables.

25. The sum of two numbers is 30 and their difference is 12. Find the numbers.

26. The sum of two numbers is 20. If twice the smaller is subtracted from the larger, the result is 2. Find the numbers.

27. The difference of two numbers is 7. If three times the smaller is subtracted from twice the larger, the result is 6. Find the numbers.

28. The difference of two numbers is 17. If the larger is increased by three times the smaller, the result is 37. Find the numbers.

29. One number is twice another number. The sum of three times the smaller and five times the larger is 78. Find the numbers.

30. One number is three times another number. The sum of four times the smaller and seven times the larger is 175. Find the numbers.

31. Three lemons and 2 apples cost $1.05. Two lemons and 3 apples cost $1.20. Find the cost of 1 lemon and also the cost of 1 apple.

32. Betty bought 30 stamps for $6. Some of them were 17-cent stamps and the rest were 22-cent stamps. How many of each kind did she buy?

33. Larry has $1.45 in change in his pocket, consisting of dimes and quarters. If he has a total of 10 coins, how many of each kind does he have?

34. Cindy has $0.90 in change in her purse, consisting of nickels and dimes. If she has a total of 13 coins, how many of each kind does she have in her purse?

35. A library buys a total of 35 books costing $462. Some of the books cost $12 per book and the rest cost $14 per book. How many books of each price did they buy?

36. Some punch containing 10% grapefruit juice is mixed with some punch containing 5% grapefruit juice to produce 5 quarts of punch that is 8% grapefruit juice. How many quarts of 10% and 5% grapefruit juice must be used?

37. A 10% salt solution is to be mixed with a 15% salt solution to produce 10 gallons of a 13% salt solution. How many gallons of the 10% solution and how many gallons of the 15% solution will be needed?

38. The income from a student production was $21,000. The price of a student ticket was $3 and nonstudent tickets were $5 each. Five thousand people attended the production. How many tickets of each kind were sold?

39. Sidney invested $1300, part of it at 10% and the rest at 12%. If his total yearly interest was $146, how much did he invest at each rate?

40. Heather invested $1100, part of it at 8% and the rest at 9%. Her total interest earned for a year was $95. How much did she invest at each rate?

■   Miscellaneous Problems

For Problems 41–44, solve each system using the method of addition.

41. $\left(\begin{array}{l} \frac{1}{2}x - \frac{1}{3}y = -2 \\ \frac{3}{2}x + \frac{2}{3}y = \ \ 34 \end{array} \right)$

42. $\left(\begin{array}{l} x - 2y = 0 \\ 3x + 5y = 0 \end{array} \right)$

43. $\left(\begin{array}{l} x - 4y = 6 \\ 2x - 8y = 3 \end{array} \right)$

44. $\left(\begin{array}{l} y = 1 - 2x \\ 6x + 3y = 3 \end{array} \right)$

8.7   The Substitution Method

There is a third method of solving systems of equations called the **substitution method**. Like the elimination method, it produces exact solutions and can be used on any system of linear equations; however, we will find that some systems lend themselves more to the substitution method than others. Let's consider a few examples to illustrate the use of the substitution method.

EXAMPLE 1 Solve $\begin{pmatrix} y = x + 10 \\ x + y = 14 \end{pmatrix}$.

Solution Since the first equation states that y equals $x + 10$, we can substitute $x + 10$ for y in the second equation.

$$x + y = 14 \quad \underline{\text{Substitute} \quad x + 10 \text{ for } y.} \quad x + (x + 10) = 14$$

Now we have an equation with one variable that can be solved in the usual way.

$$x + (x + 10) = 14$$
$$2x + 10 = 14$$
$$2x = 4$$
$$\boxed{x = 2.}$$

Substituting 2 for x in one of the original equations we can find the value of y.

$$y = x + 10$$
$$y = 2 + 10$$
$$\boxed{y = 12.}$$

The solution set is $\{(2, 12)\}$. ■

EXAMPLE 2 Solve $\begin{pmatrix} 3x + 5y = -7 \\ x = 2y + 5 \end{pmatrix}$.

Solution Since the second equation states that x equals $2y + 5$ we can substitute $2y + 5$ for x in the first equation.

$$3x + 5y = -7 \quad \underline{\text{Substitute } 2y + 5 \text{ for } x.} \quad 3(2y + 5) + 5y = -7$$

Solving this equation we obtain

$$3(2y + 5) + 5y = -7$$
$$6y + 15 + 5y = -7$$
$$11y + 15 = -7$$
$$11y = -22$$
$$\boxed{y = -2.}$$

Substituting -2 for y in one of the two original equations produces

$$x = 2y + 5$$
$$x = 2(-2) + 5$$
$$x = -4 + 5$$
$$\boxed{x = 1.}$$

The solution set is $\{(1, -2)\}$. ■

Note that the key idea behind the substitution method is also the elimination of a variable, but the elimination is done by a substitution rather than by addition of the equations. The substitution method is especially convenient to use if at least one of the equations is of the form y *equals* or x *equals*. In Example 1, the first equation is of the form y *equals* and in Example 2 the second equation is of the form x *equals*.
Let's consider another example using the substitution method.

EXAMPLE 3 Solve $\begin{pmatrix} 2x + 3y = -30 \\ y = \dfrac{2}{3}x - 6 \end{pmatrix}$.

Solution The second equation allows us to substitute $\dfrac{2}{3}x - 6$ for y in the first equation.

$$2x + 3y = -30 \quad \xrightarrow{\text{Substitute } \frac{2}{3}x - 6 \text{ for } y.} \quad 2x + 3\left(\frac{2}{3}x - 6\right) = -30$$

Solving this equation produces

$$2x + 3\left(\frac{2}{3}x - 6\right) = -30$$
$$2x + 2x - 18 = -30$$
$$4x - 18 = -30$$
$$4x = -12$$
$$\boxed{x = -3.}$$

Now we can substitute -3 for x in one of the original equations.

$$y = \frac{2}{3}x - 6$$

$$y = \frac{2}{3}(-3) - 6$$

$$y = -2 - 6$$
$$\boxed{y = -8.}$$

The solution set is $\{(-3, -8)\}$. ■

It may be necessary to change the form of one of the equations before a substitution can be made. The following examples illustrate this point.

EXAMPLE 4 Solve $\begin{pmatrix} 4x - 5y = 55 \\ x + y = -2 \end{pmatrix}$.

Solution The form of the second equation can easily be changed to make it ready for the substitution method.

$$x + y = -2$$
$$y = -2 - x. \qquad \text{Added } -x \text{ to both sides.}$$

Now, $-2 - x$ can be substituted for y in the first equation.

$$4x - 5y = 55 \quad \underline{\text{Substitute } -2 - x \text{ for } y.} \quad 4x - 5(-2 - x) = 55$$

Solving this equation, we obtain

$$4x - 5(-2 - x) = 55$$
$$4x + 10 + 5x = 55$$
$$9x + 10 = 55$$
$$9x = 45$$
$$\boxed{x = 5.}$$

Substituting 5 for x in one of the *original* equations produces

$$x + y = -2$$
$$5 + y = -2$$
$$\boxed{y = -7.}$$

The solution set is $\{(5, -7)\}$. ■

In Example 4, we could have started by changing the form of the first equation to make it ready for substitution. However, you should be able to *look ahead* and see that this would produce a fractional form to substitute. We were able to avoid any messy calculations with fractions by changing the form of the second equation instead of the first. Sometimes, when using the substitution method, you cannot avoid the use of fractional forms. The next example is a case in point.

EXAMPLE 5 Solve $\left(\begin{array}{l} 3x + 2y = 8 \\ 2x - 3y = -38 \end{array} \right)$.

Solution Looking ahead we see that changing the form of either equation will produce a fractional form. Therefore, let's merely pick the first equation and solve for y.

$$3x + 2y = 8$$
$$2y = 8 - 3x \qquad \text{Added } -3x \text{ to both sides.}$$
$$y = \frac{8 - 3x}{2}. \qquad \text{Multiplied both sides by } \frac{1}{2}.$$

Now we can substitute $\dfrac{8 - 3x}{2}$ for y in the second equation and determine the value of x.

$$2x - 3y = -38$$
$$2x - 3\left(\frac{8 - 3x}{2} \right) = -38$$
$$2x - \frac{24 - 9x}{2} = -38$$
$$4x - 24 + 9x = -76 \qquad \text{Multiplied both sides by 2.}$$
$$13x - 24 = -76$$
$$13x = -52$$
$$\boxed{x = -4.}$$

Substituting -4 for x in one of the original equations, we obtain

$$3x + 2y = 8$$
$$3(-4) + 2y = 8$$
$$-12 + 2y = 8$$
$$2y = 20$$
$$\boxed{y = 10.}$$

The solution set is $\{(-4, 10)\}$. ■

■ Which Method to Use

We have now studied three methods of solving systems of linear equations—the *graphing method*, the *addition method*, and the *substitution method*. As we indicated earlier, the graphing method is quite restrictive and works well only if the solutions are integers or if only approximate answers are needed. Both the elimination and the substitution methods can be used to obtain exact solutions for any system of linear equations in two variables. The method chosen may depend upon the original form of the equations. Let's consider two examples to illustrate this point.

EXAMPLE 6 Solve $\begin{pmatrix} 7x - 5y = -52 \\ y = 3x - 4 \end{pmatrix}$.

Solution Since the second equation indicates that $3x - 4$ can be substituted for y, this system lends itself to the substitution method.

$$7x - 5y = -52 \quad \underrightarrow{\text{Substitute } 3x - 4 \text{ for } y.} \quad 7x - 5(3x - 4) = -52$$

Solving this equation, we obtain

$$7x - 5(3x - 4) = -52$$
$$7x - 15x + 20 = -52$$
$$-8x + 20 = -52$$
$$-8x = -72$$
$$\boxed{x = 9.}$$

Substituting 9 for x in one of the original equations produces

$$y = 3x - 4$$
$$y = 3(9) - 4$$
$$y = 27 - 4$$
$$\boxed{y = 23.}$$

The solution set is $\{(9, 23)\}$. ■

EXAMPLE 7 Solve $\begin{pmatrix} 10x + 7y = 19 \\ 2x - 6y = -11 \end{pmatrix}$.

Solution Because changing the form of either of the two equations in preparation for the substitution method would produce a fractional form, we are probably better off to use the addition method. Furthermore, we should notice that the coefficients of x lend themselves to the addition method.

$$
\begin{array}{lll}
10x + 7y = 19 & \underrightarrow{\text{Leave alone.}} & 10x + 7y = 19 \\
2x - 6y = -11 & \underrightarrow{\text{Multiply by } -5.} & -10x + 30y = 55 \\
& & \overline{37y = 74} \\
& & \boxed{y = 2.}
\end{array}
$$

Substituting 2 for y in the first equation of the given system produces

$$10x + 7y = 19$$
$$10x + 7(2) = 19$$
$$10x + 14 = 19$$
$$10x = 5$$
$$x = \frac{5}{10} = \frac{1}{2}.$$

The solution set is $\left\{\left(\frac{1}{2}, 2\right)\right\}$. ■

■ Problem Solving

Let's conclude this section with three more word problems.

PROBLEM 1 The length of a rectangle is 1 centimeter less than 3 times the width. The perimeter of the rectangle is 94 centimeters. Find the length and width of the rectangle.

Solution Let w represent the width of the rectangle. Let l represent the length of the rectangle.

The problem translates into the following system of equations.

$l = 3w - 1;$ The length of a rectangle is 1 centimeter less than 3 times the width.

$2l + 2w = 94.$ The perimeter of the rectangle is 94 centimeters.

Multiplying both sides of the second equation by one-half produces the equivalent equation $l + w = 47$; so we have the following system to solve.

$$\begin{pmatrix} l = 3w - 1 \\ l + w = 47 \end{pmatrix}.$$

The first equation indicates we can substitute $3w - 1$ for l in the second equation.

$$l + w = 47 \quad \xrightarrow{\text{Substitute } 3w - 1 \text{ for } l.} \quad 3w - 1 + w = 47$$

Solving this equation yields

$$3w - 1 + w = 47$$
$$4w - 1 = 47$$
$$4w = 48$$
$$\boxed{w = 12.}$$

Substituting 12 for w in one of the original equations produces

$$l = 3w - 1$$
$$l = 3(12) - 1$$
$$l = 36 - 1$$
$$\boxed{l = 35.}$$

The rectangle is 12 centimeters wide and 35 centimeters long. ■

PROBLEM 2 The sum of the digits of a two-digit number is 7. The tens digit is 1 more than twice the units digit. Find the number.

Solution Let t represent the tens digit. Let u represent the units digit.

$$t + u = 7; \qquad \text{The sum of the digits is 7.}$$
$$t = 2u + 1. \qquad \text{The tens digit is 1 more than twice the units digit.}$$

The second equation indicates that $2u + 1$ can be substituted for t in the first equation.

$$t + u = 7$$
$$(2u + 1) + u = 7$$
$$3u + 1 = 7$$
$$3u = 6$$
$$\boxed{u = 2.}$$

Substituting 2 for u in one of the original equations yields

$$t = 2u + 1$$
$$t = 2(2) + 1$$
$$\boxed{t = 5.}$$

The tens digit is 5 and the units digit is 2; the number is 52. ■

The two-variable expression $10t + u$ can be used to represent any two-digit number. The t represents the tens digit and u the units digit. For example, if $t = 4$ and $u = 7$, then $10t + u$ becomes $10(4) + 7$ or 47. Now let's consider the following problem.

PROBLEM 3   A two-digit number is 7 times its units digit. The sum of the digits is 8. Find the number.

Solution   Let t represent the tens digit. Let u represent the units digit.

$10t + u = 7u;$ A two-digit number is 7 times its units digit.

$t + u = 8.$ The sum of the digits is 8.

The first equation can be simplified to $5t - 3u = 0$; thus we have the following system to solve.

$$\begin{pmatrix} 5t - 3u = 0 \\ t + \ u = 8. \end{pmatrix}$$

Using the addition method, we proceed as follows.

$5t - 3u = 0$ Leave alone. $\longrightarrow$ $5t - 3u = \ \ 0$

$t + u = 8$ Multiply both sides by 3. $\longrightarrow$ $3t + 3u = 24$

$\overline{}$

$8t = 24$

$\boxed{t = 3.}$

Substituting 3 for t in $t + u = 8$ yields

$t + u = 8$

$3 + u = 8$

$\boxed{u = 5.}$

The tens digit is 3 and the units digit is 5; the number is 35.   ■

Problem Set 8.7

For Problems 1–26, solve each system using the substitution method.

1. $\begin{pmatrix} y = 2x - 1 \\ x + y = 14 \end{pmatrix}$

2. $\begin{pmatrix} y = 3x + 4 \\ x + y = 52 \end{pmatrix}$

3. $\begin{pmatrix} x - y = -14 \\ y = -3x - 2 \end{pmatrix}$

4. $\begin{pmatrix} x - y = -23 \\ y = -2x + 5 \end{pmatrix}$

5. $\begin{pmatrix} 4x - 3y = -6 \\ y = -2x + 7 \end{pmatrix}$

6. $\begin{pmatrix} 8x - y = -8 \\ y = 4x + 5 \end{pmatrix}$

7. $\begin{pmatrix} x + y = 1 \\ 3x + 6y = 7 \end{pmatrix}$

8. $\begin{pmatrix} 2x - 4y = -9 \\ x + y = 3 \end{pmatrix}$

9. $\begin{pmatrix} 2x - y = 12 \\ x = \dfrac{3}{4}y \end{pmatrix}$

10. $\begin{pmatrix} 4x - 5y = 6 \\ y = \dfrac{2}{3}x \end{pmatrix}$

11. $\begin{pmatrix} y = \dfrac{3}{2}x \\ 6x - 5y = 15 \end{pmatrix}$

12. $\begin{pmatrix} x = \dfrac{3}{5}y \\ 4x - 3y = 12 \end{pmatrix}$

13. $\begin{pmatrix} 4y - 1 = x \\ x = 3y + 10 \end{pmatrix}$

14. $\begin{pmatrix} y = 5x - 2 \\ y = 2x + 7 \end{pmatrix}$

15. $\begin{pmatrix} 7x + 2y = -2 \\ 6x + 5y = 18 \end{pmatrix}$

16. $\begin{pmatrix} 2x + 3y = 31 \\ 3x + 5y = -20 \end{pmatrix}$

17. $\begin{pmatrix} 8x - 3y = -9 \\ x + 5y = -71 \end{pmatrix}$

18. $\begin{pmatrix} 9x - 2y = -18 \\ 4x - y = -7 \end{pmatrix}$

19. $\begin{pmatrix} 4x - 6y = 1 \\ 2x + 3y = 4 \end{pmatrix}$

20. $\begin{pmatrix} 2x + 3y = 1 \\ 4x - 6y = -1 \end{pmatrix}$

21. $\begin{pmatrix} 5x + 7y = 3 \\ 3x - 2y = 0 \end{pmatrix}$

22. $\begin{pmatrix} 7x - 3y = 4 \\ 2x + 5y = 0 \end{pmatrix}$

23. $\begin{pmatrix} .05x + .07y = 33 \\ y = x + 300 \end{pmatrix}$

24. $\begin{pmatrix} .06x + .08y = 15 \\ y = x + 100 \end{pmatrix}$

25. $\begin{pmatrix} x + y = 13 \\ .05x + .1y = 1.15 \end{pmatrix}$

26. $\begin{pmatrix} x + y = 17 \\ .1x + .25y = 3.2 \end{pmatrix}$

For Problems 27–46, solve each system using either the addition or the substitution method, whichever seems most appropriate to you.

27. $\begin{pmatrix} 5x - 4y = 14 \\ 7x + 3y = -32 \end{pmatrix}$

28. $\begin{pmatrix} 2x + 3y = 13 \\ 3x - 5y = -28 \end{pmatrix}$

29. $\begin{pmatrix} 2x + 9y = 6 \\ y = -x \end{pmatrix}$

30. $\begin{pmatrix} y = 3x + 2 \\ 4x - 3y = -21 \end{pmatrix}$

31. $\begin{pmatrix} x + y = 22 \\ .6x + .5y = 12 \end{pmatrix}$

32. $\begin{pmatrix} x + y = 13 \\ .4x + .5y = 6 \end{pmatrix}$

33. $\begin{pmatrix} 4x - y = 0 \\ 7x + 2y = 9 \end{pmatrix}$

34. $\begin{pmatrix} x - 2y = 0 \\ 5x = 8y + 12 \end{pmatrix}$

35. $\begin{pmatrix} 2x + y = 1 \\ 6x - 7y = -57 \end{pmatrix}$

36. $\begin{pmatrix} 7x - 9y = 11 \\ x - 3y = 3 \end{pmatrix}$

37. $\begin{pmatrix} 6x - y = -1 \\ 10x + 2y = 13 \end{pmatrix}$

38. $\begin{pmatrix} x + 4y = -5 \\ 6x - 5y = -1 \end{pmatrix}$

39. $\begin{pmatrix} 4x - 9y = -97 \\ x + y = 5 \end{pmatrix}$

40. $\begin{pmatrix} 4x = 5y - 5 \\ 8x - 3y = 11 \end{pmatrix}$

41. $\begin{pmatrix} 3x - 8y = -5 \\ x = 2y \end{pmatrix}$

42. $\begin{pmatrix} x + y = -3 \\ 5x + 6y = -22 \end{pmatrix}$

43. $\begin{pmatrix} 5y - 2x = -4 \\ 10y = 3x + 4 \end{pmatrix}$

44. $\begin{pmatrix} 3x - 7 = 2y + 15 \\ 3x = 6y + 18 \end{pmatrix}$

45. $\begin{pmatrix} x = -y - 1 \\ 6x - 5y = 4 \end{pmatrix}$

46. $\begin{pmatrix} 4x + 7y = 1 \\ y = 2x + 3 \end{pmatrix}$

For Problems 47–60, solve each problem by setting up and solving a system of two linear equations in two variables. Use either the addition or the substitution method to solve the systems.

47. The sum of two numbers is 46 and the difference of the numbers is 22. Find the numbers.

48. The sum of two numbers is 52. The larger number is two more than four times the smaller number. Find the numbers.

49. A motel rents double rooms at $28 per day and single rooms at $19 per day. If a total of 50 rooms was rented one day for $1265, how many of each kind were rented?

50. The total receipts from a concert amounted to $2600. Student tickets were sold at $4 each and nonstudent tickets at $6 each. The number of student tickets sold was 5 times the number of nonstudent tickets sold. How many student tickets and how many nonstudent tickets were sold?

51. The sum of the digits of a two-digit number is 9. The two-digit number is 9 times its units digit. Find the number.

52. The sum of the digits of a two-digit number is 13. If the digits are reversed, the new number is 9 larger than the original number. Find the number.

53. Larry has a number of dimes and quarters totaling $12.05. The number of quarters is 5 more than twice the number of dimes. How many coins of each kind does he have?

54. Sue has 3 times as many nickels as pennies. Together her pennies and nickels have a value of $4.80. How many pennies and how many nickels does she have?

55. The sum of the digits of a two-digit number is 12. The tens digit is 3 times as large as the units digit. Find the number.

56. The units digit of a two-digit number is 1 more than 4 times the tens digit. The sum of the digits is 11. Find the number.

57. Tina invested some money at 8% interest and some money at 9%. She invested $250 more at 9% than she invested at 8%. Her total yearly interest from the two investments was $48. How much did Tina invest at each rate?

58. One solution contains 40% alcohol and a second solution contains 70% alcohol. How many liters of each solution should be mixed to make 30 liters that contain 50% alcohol?

59. One solution contains 30% alcohol and a second solution contains 70% alcohol. How many liters of each solution should be mixed to make 10 liters containing 40% alcohol?

60. The length of a rectangle is 3 meters less than 4 times the width. The perimeter of the rectangle is 74 meters. Find the length and width of the rectangle.

■ **Miscellaneous Problems**

For Problems 61–64, use the substitution method to solve each system.

61. $\left(\begin{array}{l} y = \dfrac{2}{3}x - \dfrac{1}{2} \\ \dfrac{3}{2}x - y = -\dfrac{1}{3} \end{array} \right)$

62. $\left(\begin{array}{l} 5x - 3y = 0 \\ 4x + 7y = 0 \end{array} \right)$

63. $\left(\begin{array}{l} y = 2x - 1 \\ 6x - 3y = 3 \end{array} \right)$

64. $\left(\begin{array}{l} x = -4y + 5 \\ 2x + 8y = -1 \end{array} \right)$

Chapter 8 Summary

(8.1) The **Cartesian** or **rectangular coordinate system** involves a one-to-one correspondence between ordered pairs of real numbers and the points of a plane. The system provides the basis for a study of coordinate geometry—a link between algebra and geometry dealing with two basic kinds of problems.

1. Given an algebraic equation, find its geometric graph.
2. Given a set of conditions pertaining to a geometric figure, find its algebraic equation.

One graphing technique is to plot a sufficient number of points until the graph of the equation is determined.

(8.2) Any equation of the form $Ax + By = C$, where A, B, and C are constants (A and B not both zero) and x and y are variables, is a **linear equation in two variables** and its graph is a **straight line**.

To **graph** a linear equation we can find two solutions (the intercepts are usually easy to determine), plot the corresponding points, and connect the points with a straight line.

(8.3) If points P_1 and P_2 with coordinates (x_1, y_1) and (x_2, y_2), respectively, are any two points on a line, then the **slope** of the line (denoted by m) is given by

$$m = \frac{y_2 - y_1}{x_2 - x_1}, \qquad x_1 \neq x_2.$$

The slope of a line is a **ratio** of vertical change compared to horizontal change. The slope of a line can be negative, positive, or zero. The concept of slope is not defined for vertical lines.

(8.4) You should review Examples 1, 2, and 3 of this section to pull together a general approach to writing equations of lines given certain conditions.

The equation

$$y - y_1 = m(x - x_1)$$

is called the **point-slope form** of the equation of a straight line.

The equation

$$y = mx + b$$

is called the **slope-intercept form** of the equation of a straight line. If the equation of a nonvertical line is written in slope-intercept form, then the coefficient of x is the slope of the line and the constant term is the y-intercept.

(8.5) Solving a system of two linear equations by graphing produces one of the following three possibilities.

1. The graphs of the two equations are two intersecting lines indicating **one solution** for the system called a **consistent system**.
2. The graphs of the two equations are two parallel lines indicating **no solution** for the system called an **inconsistent system**.

3. The graphs of the two equations are the same line indicating **infinitely many solutions** for the system. The equations are referred to as a set of **dependent** equations.

(8.6) The **addition method** for solving a system of two linear equations relies on the property "if $a = b$ and $c = d$, then $a + c = b + d$." Equations are added **to eliminate a variable**.

(8.7) The **substitution method** for solving a system of two linear equations involves solving one of the two equations for one variable and **substituting** this expression into the other equation.

To review the problem of choosing a method for solving a particular system of two linear equations, return to Section 8.7 and study Examples 6 and 7 one more time.

Chapter 8 Review Problem Set

For Problems 1–6, graph each of the equations.

1. $2x - 5y = 10$

2. $y = -\dfrac{1}{3}x + 1$

3. $y = 2x^2 + 1$

4. $y = -x^2 - 2$

5. $2x - 3y = 0$

6. $y = -x^3$

7. Find the slope of the line determined by $(3, -4)$ and $(-2, 5)$.

8. Find the slope of the line $5x - 6y = 30$.

9. Write the equation of the line that has a slope of $-\dfrac{5}{7}$ and contains the point $(2, -3)$.

10. Write the equation of the line containing $(2, 5)$ and $(-1, -3)$.

11. Write the equation of the line having a slope of $\dfrac{2}{9}$ and a y-intercept of -1.

12. Write the equation of the line containing $(2, 4)$ and which is perpendicular to the x-axis.

13. Solve the system $\begin{pmatrix} 2x + y = 4 \\ x - y = 5 \end{pmatrix}$ by using the graphing method.

For Problems 14–25, solve each system by using either the addition method or the substitution method.

14. $\begin{pmatrix} 2x - y = 1 \\ 3x - 2y = -5 \end{pmatrix}$

15. $\begin{pmatrix} 2x + 5y = 7 \\ x = -3y + 1 \end{pmatrix}$

16. $\begin{pmatrix} 3x + 2y = 7 \\ 4x - 5y = 3 \end{pmatrix}$

17. $\begin{pmatrix} 9x + 2y = 140 \\ x + 5y = 135 \end{pmatrix}$

18. $\begin{pmatrix} \dfrac{1}{2}x + \dfrac{1}{4}y = -5 \\ \dfrac{2}{3}x - \dfrac{1}{2}y = 0 \end{pmatrix}$

19. $\begin{pmatrix} x + y = 1000 \\ 0.07x + 0.09y = 82 \end{pmatrix}$

20. $\begin{pmatrix} y = 5x + 2 \\ 10x - 2y = 1 \end{pmatrix}$ **21.** $\begin{pmatrix} 5x - 7y = 9 \\ y = 3x - 2 \end{pmatrix}$

22. $\begin{pmatrix} 10t + u = 6u \\ t + u = 12 \end{pmatrix}$ **23.** $\begin{pmatrix} t = 2u \\ 10t + u - 36 = 10u + t \end{pmatrix}$

24. $\begin{pmatrix} u = 2t + 1 \\ 10t + u + 10u + t = 110 \end{pmatrix}$ **25.** $\begin{pmatrix} y = -\dfrac{2}{3}x \\ \dfrac{1}{3}x - y = -9 \end{pmatrix}$

Solve each of the following problems by setting up and solving a system of two linear equations in two variables.

26. The sum of two numbers is 113. The larger number is 1 less than twice the smaller number. Find the numbers.

27. Mark invested a certain amount of money at 9% annual interest and $50 more than amount at 11%. He received $55.50 in interest. How much did he invest at each rate?

28. Cindy has 43 coins consisting of nickels and dimes. The total value of the coins is $3.40. How many coins of each kind does she have?

29. The length of a rectangle is 1 inch more than three times the width. If the perimeter of the rectangle is 50 inches, find the length and width.

30. The sum of the digits of a two-digit number is 13 and the units digit is one more than twice the tens digit. Find the number.

31. The sum of the digits of a two-digit number is 9. The number formed by reversing the digits is 27 larger than the original number. Find the original number.

32. Two angles are complementary and one of them is 6° less than twice the other one. Find the measure of each angle.

33. Two angles are supplementary and the larger angle is 20° less than three times the smaller angle. Find the measure of each angle.

34. Four cheeseburgers and 5 milkshakes cost a total of $8.35. Two milkshakes cost $.35 more than one cheeseburger. Find the cost of a cheeseburger and also find the cost of a milkshake.

35. Three cans of prune juice and 2 cans of tomato juice cost $3.85. On the other hand, 2 cans of prune juice and 3 cans of tomato juice cost $3.55. Find the cost per can of each.

9 Square Roots and Radicals

In Section 2.3, $\sqrt{2}$ and $\sqrt{3}$ were used as examples of irrational numbers. Irrational numbers, in decimal form, are nonrepeating decimals. For example $\sqrt{2} = 1.414213562373\ldots$, where the three dots indicate that the decimal expansion continues indefinitely. It was further stated that "more work will be done with the irrationals in Chapter 9." The time has come for us to extend our skills relative to the set of irrational numbers.

9.1 Square Roots and Radicals

To **square a number** means to raise it to the second power—that is, to use the number as a factor twice.

$$3^2 = 3 \cdot 3 = 9; \qquad \text{(read "three squared equals nine")}$$
$$7^2 = 7 \cdot 7 = 49;$$
$$(-3)^2 = (-3)(-3) = 9;$$
$$(-7)^2 = (-7)(-7) = 49.$$

A **square root of a number** is one of its two equal factors. Thus, 3 is a square root of 9 because $3 \cdot 3 = 9$. Likewise, -3 is also a square root of 9 because $(-3)(-3) = 9$. The concept of square root is defined as follows.

Definition 9.1

a is a square root of b if $a^2 = b$.

The following generalizations are a direct consequence of Definition 9.1.

1. Every positive real number has two square roots; one is positive and the other is negative. They are opposites of each other.

2. Negative real numbers have no real number square roots. (This follows from Definition 9.1 since any nonzero real number is positive when squared.)

3. The square root of 0 is 0.

The symbol $\sqrt{}$, called a **radical sign**, is used to indicate the nonnegative square root. The number under the radical sign is called the **radicand**. The entire expression, such as $\sqrt{49}$, is called a **radical**.

$\sqrt{49} = 7$ ($\sqrt{49}$ indicates the **nonnegative** or **principal** square root of 49.)

$-\sqrt{49} = -7$ ($-\sqrt{49}$ indicates the negative square root of 49.)

$\sqrt{0} = 0$ (Zero has only one square root. Technically, we could write $-\sqrt{0} = -0 = 0$.)

$\sqrt{-4}$ is not a real number.

$-\sqrt{-4}$ is not a real number.

If a is a nonnegative integer that is the square of an integer, then $\sqrt{a}$ and $-\sqrt{a}$ are rational numbers. For example, $\sqrt{1}$, $\sqrt{4}$, and $\sqrt{25}$ are the rational numbers 1, 2, and 5, respectively. The numbers 1, 4, and 25 are called perfect squares because each represents the square of some integer.

If a is a positive integer that is *not* the square of an integer, then $\sqrt{a}$ and $-\sqrt{a}$ are irrational numbers. For example, $\sqrt{7}$, $-\sqrt{7}$, $\sqrt{23}$, $-\sqrt{31}$, $\sqrt{52}$, and $-\sqrt{75}$ are irrational numbers.

It is often necessary to find a rational approximation for an irrational number. Rational approximations for square roots that are irrational numbers are commonly found by using a table or a calculator. Table 9.1 contains the squares and square roots (to the nearest one-thousandth) of all whole numbers from 1 through 100, inclusive. From the table we can find rational approximations as follows.

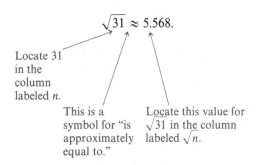

$$\sqrt{31} \approx 5.568.$$

Locate 31 in the column labeled n.

This is a symbol for "is approximately equal to."

Locate this value for $\sqrt{31}$ in the column labeled $\sqrt{n}$.

Be sure that you agree with the following values taken from Table 9.1.

$\sqrt{14} \approx 3.742;$

$\sqrt{46} \approx 6.782;$

$\sqrt{65} \approx 8.062;$

$\sqrt{81} = 9.$ (Note the equality symbol. $\sqrt{81}$ is exactly the rational number 9. The three zeroes are given in the table only to be consistent with the other entries.)

Table 9.1   Squares and Approximate Square Roots

n	n^2	$\sqrt{n}$	n	n^2	$\sqrt{n}$
1	1	1.000	51	2601	7.141
2	4	1.414	52	2704	7.211
3	9	1.732	53	2809	7.280
4	16	2.000	54	2916	7.348
5	25	2.236	55	3025	7.416
6	36	2.449	56	3136	7.483
7	49	2.646	57	3249	7.550
8	64	2.828	58	3364	7.616
9	81	3.000	59	3481	7.681

Table 9.1 (continued)

n	n^2	$\sqrt{n}$	n	n^2	$\sqrt{n}$
10	100	3.162	60	3600	7.746
11	121	3.317	61	3721	7.810
12	144	3.464	62	3844	7.874
13	169	3.606	63	3969	7.937
14	196	3.742	64	4096	8.000
15	225	3.873	65	4225	8.062
16	256	4.000	66	4356	8.124
17	289	4.123	67	4489	8.185
18	324	4.243	68	4624	8.246
19	361	4.359	69	4761	8.307
20	400	4.472	70	4900	8.367
21	441	4.583	71	5041	8.426
22	484	4.690	72	5184	8.485
23	529	4.796	73	5329	8.544
24	576	4.899	74	5476	8.602
25	625	5.000	75	5625	8.660
26	676	5.099	76	5776	8.718
27	729	5.196	77	5929	8.775
28	784	5.292	78	6084	8.832
29	841	5.385	79	6241	8.888
30	900	5.477	80	6400	8.944
31	961	5.568	81	6561	9.000
32	1024	5.657	82	6724	9.055
33	1089	5.745	83	6889	9.110
34	1156	5.831	84	7056	9.165
35	1225	5.916	85	7225	9.220
36	1296	6.000	86	7396	9.274
37	1369	6.083	87	7569	9.327
38	1444	6.164	88	7744	9.381
39	1521	6.245	89	7921	9.434
40	1600	6.325	90	8100	9.487
41	1681	6.403	91	8281	9.539
42	1764	6.481	92	8464	9.592
43	1849	6.557	93	8649	9.644
44	1936	6.633	94	8836	9.695
45	2025	6.708	95	9025	9.747
46	2116	6.782	96	9216	9.798
47	2209	6.856	97	9409	9.849
48	2304	6.928	98	9604	9.899
49	2401	7.000	99	9801	9.950
50	2500	7.071	100	10000	10.000

The column labeled n^2 in Table 9.1 contains the squares of the whole numbers from 1 through 100, inclusive. For example, from the table we obtain $25^2 = 625$; $37^2 = 1369$; $62^2 = 3844$; and $89^2 = 7921$. Since $25^2 = 625$, we can state that $\sqrt{625} = 25$. Thus, the column labeled n^2 also provides us with additional square root facts.

$$\sqrt{1936} = 44.$$

Located 1936   Locate this value
in the column   for $\sqrt{1936}$ in the
labeled n^2.   column labeled n.

Now suppose that we want to find a value for $\sqrt{500}$. Scanning the column labeled n^2, we see that 500 does not appear; however, it is between two values given in the table. We can reason as follows.

$$22^2 = 484;$$
$$\leftarrow 500$$
$$23^2 = 529.$$

Since 500 is closer to 484 than to 529, we approximate $\sqrt{500}$ to be closer to 22 than to 23. Thus, we write $\sqrt{500} \approx 22$. This is merely a whole number approximation, but for some purposes it may be sufficiently precise.

Let's look at another example of this type.

EXAMPLE 1   Find a whole number approximation for $\sqrt{7836}$.

Solution   From $88^2 = 7744;$
Table $\leftarrow 7836$
9.1 $89^2 = 7921.$

7836	7921
7744	7836
92	85

Since 7836 is closer to 7921 than to 7744, we obtain

$$\sqrt{7836} \approx 89.$$ ■

Recall our use of the distributive property as the basis for combining similar terms. For example,

$$3x + 2x = (3 + 2)x = 5x; \qquad 7y - 4y = (7 - 4)y = 3y;$$
$$9a^2 + 5a^2 = (9 + 5)a^2 = 14a^2.$$

In a like manner, expressions containing radicals can often be simplified by using the distributive property as follows.

$$5\sqrt{2} + 7\sqrt{2} = (5 + 7)\sqrt{2} = 12\sqrt{2};$$
$$8\sqrt{5} - 2\sqrt{5} = (8 - 2)\sqrt{5} = 6\sqrt{5};$$
$$4\sqrt{7} + 6\sqrt{7} + 3\sqrt{11} - \sqrt{11} = (4 + 6)\sqrt{7} + (3 - 1)\sqrt{11} = 10\sqrt{7} + 2\sqrt{11}.$$

Note that **to add or subtract square roots they must have the same radicand**. Thus, an expression such as $7\sqrt{3} + 2\sqrt{5}$ cannot be simplified. Also note the form used to indicate multiplication when a radical is involved. For example, $5 \cdot \sqrt{2}$ is written as $5\sqrt{2}$.

The process of finding a rational approximation for an expression containing radicals can be made easier by first simplifying the expression. Consider the following examples.

EXAMPLE 2   Find a rational approximation, to the nearest one tenth, for $5\sqrt{2} + 4\sqrt{2}$.

Solution   First, let's simplify the given expression.

$$5\sqrt{2} + 4\sqrt{2} = 9\sqrt{2}.$$

Now from Table 9.1 we find that $\sqrt{2} = 1.41$, to the nearest hundredth. (We expressed the value for $\sqrt{2}$ to the nearest hundredth because the final result is to be rounded to the nearest tenth.) Therefore,

$$9\sqrt{2} = 9(1.41) = 12.69 = 12.7, \quad \text{to the nearest tenth.} \qquad ■$$

EXAMPLE 3   Find a rational approximation, to the nearest tenth, for

$$6\sqrt{3} + \sqrt{3} + 7\sqrt{5} - 3\sqrt{5}.$$

Solution   First let's simplify the given expression.

$$6\sqrt{3} + \sqrt{3} + 7\sqrt{5} - 3\sqrt{5} = 7\sqrt{3} + 4\sqrt{5}.$$

Then from Table 9.1 we obtain

$$\sqrt{3} = 1.73 \text{ and } \sqrt{5} = 2.24, \quad \text{to the nearest hundredth.}$$

Therefore,

$$7\sqrt{3} + 4\sqrt{5} = 7(1.73) + 4(2.24)$$
$$= 12.11 + 8.96$$
$$= 21.07$$
$$= 21.1, \quad \text{to the nearest tenth.} \qquad ■$$

As mentioned earlier, calculators provide another source for finding approximate values for square roots. Many calculators now have a special square root button. If you have access to a calculator, you may want to check some of our entries in Table 9.1. Your calculator will give more than three decimal places, but by rounding to the nearest one thousandth you should be able to obtain results that agree with the table.

The process of simplifying radical expressions—as we did in Examples 2 and 3—before approximating the square roots is helpful even when using a calculator.

Problem Set 9.1

For Problems 1–12, evaluate each radical.

1. $\sqrt{49}$	**2.** $\sqrt{100}$	**3.** $-\sqrt{64}$	**4.** $-\sqrt{36}$
5. $\sqrt{121}$	**6.** $\sqrt{144}$	**7.** $\sqrt{3600}$	**8.** $\sqrt{2500}$
9. $-\sqrt{1600}$	**10.** $-\sqrt{900}$	**11.** $\sqrt{6400}$	**12.** $\sqrt{400}$

For Problems 13–20, use Table 9.1 to find a rational approximation, to the nearest thousandth, for each radical.

13. $\sqrt{19}$	**14.** $\sqrt{33}$	**15.** $\sqrt{50}$	**16.** $\sqrt{66}$
17. $\sqrt{75}$	**18.** $\sqrt{90}$	**19.** $\sqrt{94}$	**20.** $\sqrt{98}$

For Problems 21–28, use Table 9.1 to find an exact value for each radical.

21. $\sqrt{196}$	**22.** $\sqrt{484}$	**23.** $\sqrt{5329}$	**24.** $\sqrt{6241}$
25. $\sqrt{841}$	**26.** $\sqrt{441}$	**27.** $\sqrt{7396}$	**28.** $\sqrt{8649}$

For Problems 29–40, use Table 9.1 to help find a whole number approximation for each of the radicals.

29. $\sqrt{40}$	**30.** $\sqrt{130}$	**31.** $\sqrt{4325}$	**32.** $\sqrt{7500}$
33. $\sqrt{1175}$	**34.** $\sqrt{1700}$	**35.** $\sqrt{9501}$	**36.** $\sqrt{8050}$
37. $\sqrt{6614}$	**38.** $\sqrt{5825}$	**39.** $\sqrt{3400}$	**40.** $\sqrt{2250}$

For Problems 41–52, simplify each expression by using the distributive property.

41. $7\sqrt{2} + 14\sqrt{2}$	**42.** $9\sqrt{3} + 4\sqrt{3}$
43. $17\sqrt{7} - 9\sqrt{7}$	**44.** $19\sqrt{5} - 8\sqrt{5}$
45. $6\sqrt{3} - 15\sqrt{3}$	**46.** $7\sqrt{6} - 21\sqrt{6}$
47. $9\sqrt{5} + 3\sqrt{5} - 6\sqrt{5}$	**48.** $8\sqrt{7} + 13\sqrt{7} - 9\sqrt{7}$
49. $8\sqrt{2} - 4\sqrt{3} - 9\sqrt{2} + 6\sqrt{3}$	**50.** $7\sqrt{5} - 9\sqrt{6} + 14\sqrt{5} - 2\sqrt{6}$

51. $6\sqrt{7} + 5\sqrt{10} - 8\sqrt{10} - 4\sqrt{7} - 11\sqrt{7} + \sqrt{10}$

52. $\sqrt{3} - \sqrt{5} + 4\sqrt{5} - 3\sqrt{3} - 9\sqrt{5} - 16\sqrt{3}$

For Problems 53–66, find a rational approximation, to the nearest one tenth, for each of the radical expressions. Remember to simplify the expression first, if possible, and then to use Table 9.1 to find the necessary approximations.

53. $9\sqrt{3} + \sqrt{3}$

54. $6\sqrt{2} + 14\sqrt{2}$

55. $9\sqrt{5} - 3\sqrt{5}$

56. $18\sqrt{6} - 12\sqrt{6}$

57. $14\sqrt{2} - 15\sqrt{2}$

58. $7\sqrt{3} - 12\sqrt{3}$

59. $8\sqrt{7} - 4\sqrt{7} + 6\sqrt{7}$

60. $9\sqrt{5} - 2\sqrt{5} + \sqrt{5}$

61. $4\sqrt{3} - 2\sqrt{2}$

62. $3\sqrt{2} + \sqrt{3} - \sqrt{5}$

63. $9\sqrt{6} - 3\sqrt{5} + 2\sqrt{6} - 7\sqrt{5} - \sqrt{6}$

64. $8\sqrt{7} - 2\sqrt{10} + 4\sqrt{7} - 3\sqrt{10} - 7\sqrt{7} + 4\sqrt{10}$

65. $4\sqrt{11} - 5\sqrt{11} - 7\sqrt{11} + 2\sqrt{11} - 3\sqrt{11}$

66. $14\sqrt{13} - 17\sqrt{13} + 3\sqrt{13} - 4\sqrt{13} - 5\sqrt{13}$

■ Miscellaneous Problems

For Problems 67–72, evaluate each radical.

67. $\sqrt{\dfrac{4}{9}}$

68. $\sqrt{\dfrac{25}{16}}$

69. $\sqrt{.04}$

70. $\sqrt{.64}$

71. $\sqrt{.0009}$

72. $\sqrt{.0016}$

9.2 Simplifying Radicals

Note the following facts pertaining to square roots.

$$\sqrt{4 \cdot 9} = \sqrt{36} = 6$$

and

$$\sqrt{4}\sqrt{9} = 2 \cdot 3 = 6.$$

Thus, we observe $\sqrt{4 \cdot 9} = \sqrt{4}\sqrt{9}$. Such facts illustrate a general property.

Property 9.1

For any nonnegative real numbers a and b,

$$\sqrt{ab} = \sqrt{a}\sqrt{b}.$$

In other words, we say that **the square root of a product is equal to the product of the square roots**.

Property 9.1 and the definition of square root provide the basis for changing radicals to **simplest radical form**. Let's use a few examples to illustrate the meaning of simplest radical form.

1. $\sqrt{8} = \sqrt{4 \cdot 2} = \sqrt{4}\sqrt{2} = 2\sqrt{2}.$

        4 is a $\sqrt{4} = 2$
        perfect
        square

2. $\sqrt{45} = \sqrt{9 \cdot 5} = \sqrt{9}\sqrt{5} = 3\sqrt{5}.$

        9 is a $\sqrt{9} = 3$
        perfect
        square

3. $\sqrt{48} = \sqrt{16 \cdot 3} = \sqrt{16}\sqrt{3} = 4\sqrt{3}.$

        16 is a $\sqrt{16} = 4$
        perfect
        square

The first step in each example is to express the radicand of the given radical as the product of two factors, at least one of which is a perfect square other than 1. Also observe the radicands of the final radicals. In each case, the radicand *cannot* be expressed as the product of two factors at least one of which is a perfect square other than 1. We say that the final radicals, $2\sqrt{2}$, $3\sqrt{5}$, and $4\sqrt{3}$, are in simplest radical form.

You may vary the steps somewhat in changing to simplest radical form, but the final result should be the same. Consider another sequence of steps to change $\sqrt{48}$ to simplest form.

$$\sqrt{48} = \sqrt{4 \cdot 12} = \sqrt{4}\sqrt{12} = 2\sqrt{12} = 2\sqrt{4 \cdot 3} = 2\sqrt{4}\sqrt{3} = 2 \cdot 2 \cdot \sqrt{3} = 4\sqrt{3}.$$

    4 is a   This is   4 is a   same result
    perfect   not in   perfect   as in example (3)
    square   simplest square
                form.

Another variation of the technique for changing radicals to simplest form is to prime factor the radicand and then to look for perfect squares in exponential form. Let's use this idea and redo the previous examples.

4. $\sqrt{8} = \sqrt{2 \cdot 2 \cdot 2} = \sqrt{2^2 \cdot 2} = \sqrt{2^2}\sqrt{2} = 2\sqrt{2}.$

        prime 2^2 is a
        factors   perfect
        of 8   square

5. $\sqrt{45} = \sqrt{3 \cdot 3 \cdot 5} = \sqrt{3^2 \cdot 5} = \sqrt{3^2}\sqrt{5} = 3\sqrt{5}.$

       ↑   ↑

     prime 3^2 is a
     factors   perfect
     of 45   square

6. $\sqrt{48} = \sqrt{2 \cdot 2 \cdot 2 \cdot 2 \cdot 3} = \sqrt{2^4 \cdot 3} = \sqrt{2^4}\sqrt{3} = 2^2\sqrt{3} = 4\sqrt{3}.$

       ↑   ↑   ↑

     prime factors 2^4 is a $\sqrt{2^4} = 2^2$
     of 48   perfect   since
                      square $2^2 \cdot 2^2 = 2^4$

The following examples further illustrate the process of changing to simplest radical form. Only the major steps are shown, so be sure that you can fill in the details.

7. $\sqrt{56} = \sqrt{4}\sqrt{14} = 2\sqrt{14}.$

8. $\sqrt{75} = \sqrt{25}\sqrt{3} = 5\sqrt{3}.$

9. $\sqrt{108} = \sqrt{2 \cdot 2 \cdot 3 \cdot 3 \cdot 3} = \sqrt{2^2 \cdot 3^2}\sqrt{3} = 6\sqrt{3}.$

10. $5\sqrt{12} = 5\sqrt{4}\sqrt{3} = 5 \cdot 2 \cdot \sqrt{3} = 10\sqrt{3}.$

The following property is used when changing radicals involving variables in the radicand to simplest radical form.

Property 9.2

If a is any nonnegative real number, then

$$\sqrt{a^2} = a.$$

The next few examples illustrate the use of Property 9.2 along with our previous knowledge of radicals.

11. $\sqrt{x^2 y} = \sqrt{x^2}\sqrt{y} = x\sqrt{y}.$

12. $\sqrt{4x^3} = \sqrt{4x^2}\sqrt{x} = 2x\sqrt{x}.$

           ↑

     $4x^2$ is a
     perfect square
     since
     $(2x)(2x) = 4x^2$

13. $\sqrt{8xy^3} = \sqrt{4y^2}\sqrt{2xy} = 2y\sqrt{2xy}.$

14. $\sqrt{27x^5 y^3} = \sqrt{9x^4 y^2}\sqrt{3xy} = 3x^2 y\sqrt{3xy}.$

                        ↑

               $\sqrt{9x^4 y^2} = 3x^2 y$

If the numerical coefficient of the radicand is quite large, you may want to look at it in prime factored form. The next example illustrates this idea.

$$
\begin{aligned}
\textbf{15.} \quad \sqrt{180a^6b^3} &= \sqrt{2 \cdot 2 \cdot 3 \cdot 3 \cdot 5 \cdot a^6 \cdot b^2 \cdot b} \\
&= \sqrt{36a^6b^2}\sqrt{5b} \\
&= 6a^3b\sqrt{5b}.
\end{aligned}
$$

Simplifying expressions containing radicals often requires that you first change the radicals to simplest form and then apply the distributive property.

EXAMPLE 1 Simplify $5\sqrt{8} + 3\sqrt{2}$.

Solution

$$
\begin{aligned}
5\sqrt{8} + 3\sqrt{2} &= 5\sqrt{4}\sqrt{2} + 3\sqrt{2} \\
&= 5 \cdot 2 \cdot \sqrt{2} + 3\sqrt{2} \\
&= 10\sqrt{2} + 3\sqrt{2} \\
&= (10 + 3)\sqrt{2} \\
&= 13\sqrt{2}.
\end{aligned}
$$

■

EXAMPLE 2 Simplify $2\sqrt{27} - 5\sqrt{48} + 4\sqrt{3}$.

Solution

$$
\begin{aligned}
2\sqrt{27} - 5\sqrt{48} + 4\sqrt{3} &= 2\sqrt{9}\sqrt{3} - 5\sqrt{16}\sqrt{3} + 4\sqrt{3} \\
&= 2 \cdot 3 \cdot \sqrt{3} - 5 \cdot 4 \cdot \sqrt{3} + 4\sqrt{3} \\
&= 6\sqrt{3} - 20\sqrt{3} + 4\sqrt{3} \\
&= (6 - 20 + 4)\sqrt{3} \\
&= -10\sqrt{3}.
\end{aligned}
$$

■

EXAMPLE 3 Simplify $\dfrac{1}{4}\sqrt{45} + \dfrac{1}{3}\sqrt{20}$.

Solution

$$
\begin{aligned}
\frac{1}{4}\sqrt{45} + \frac{1}{3}\sqrt{20} &= \frac{1}{4}\sqrt{9}\sqrt{5} + \frac{1}{3}\sqrt{4}\sqrt{5} \\[2mm]
&= \frac{1}{4} \cdot 3 \cdot \sqrt{5} + \frac{1}{3} \cdot 2 \cdot \sqrt{5} \\[2mm]
&= \frac{3}{4}\sqrt{5} + \frac{2}{3}\sqrt{5} \\[2mm]
&= \left(\frac{3}{4} + \frac{2}{3}\right)\sqrt{5} = \left(\frac{9}{12} + \frac{8}{12}\right)\sqrt{5} \\[2mm]
&= \frac{17}{12}\sqrt{5}.
\end{aligned}
$$

■

Problem Set 9.2

For Problems 1–30, change each radical to simplest radical form.

1. $\sqrt{24}$	**2.** $\sqrt{54}$	**3.** $\sqrt{18}$	**4.** $\sqrt{50}$
5. $\sqrt{27}$	**6.** $\sqrt{12}$	**7.** $\sqrt{40}$	**8.** $\sqrt{90}$
9. $\sqrt{28}$	**10.** $\sqrt{63}$	**11.** $\sqrt{80}$	**12.** $\sqrt{125}$
13. $\sqrt{117}$	**14.** $\sqrt{126}$	**15.** $4\sqrt{72}$	**16.** $8\sqrt{98}$
17. $3\sqrt{75}$	**18.** $5\sqrt{108}$	**19.** $-5\sqrt{20}$	**20.** $-6\sqrt{45}$
21. $-8\sqrt{96}$	**22.** $-4\sqrt{54}$	**23.** $\frac{3}{2}\sqrt{8}$	**24.** $\frac{5}{2}\sqrt{32}$
25. $\frac{3}{4}\sqrt{12}$	**26.** $\frac{4}{5}\sqrt{27}$	**27.** $-\frac{2}{3}\sqrt{45}$	**28.** $-\frac{3}{5}\sqrt{125}$
29. $-\sqrt{150}$	**30.** $-\sqrt{112}$		

For Problems 31–52, express each radical in simplest radical form. All variables represent nonnegative real numbers.

31. $\sqrt{x^2y^3}$	**32.** $\sqrt{xy^4}$	**33.** $\sqrt{2x^2y}$	**34.** $\sqrt{3x^2y^2}$
35. $\sqrt{8x^2}$	**36.** $\sqrt{24x^3}$	**37.** $\sqrt{27a^3b}$	**38.** $\sqrt{45a^2b^4}$
39. $\sqrt{144ab^5}$	**40.** $\sqrt{121a^3b^6}$	**41.** $\sqrt{63x^4y^2}$	**42.** $\sqrt{28x^3y}$
43. $3\sqrt{48x^2}$	**44.** $5\sqrt{12x^2y^2}$	**45.** $-6\sqrt{72x^7}$	**46.** $-8\sqrt{80y^9}$
47. $\frac{2}{9}\sqrt{54xy}$	**48.** $\frac{4}{3}\sqrt{20xy}$	**49.** $\frac{3}{4}\sqrt{52x^5y^6}$	**50.** $\frac{5}{4}\sqrt{56x^5y^7}$
51. $-\frac{2}{3}\sqrt{169a^8}$	**52.** $-\frac{2}{7}\sqrt{196a^{10}}$		

For Problems 53–66, simplify each expression.

53. $7\sqrt{32} + 5\sqrt{2}$	**54.** $6\sqrt{48} + 5\sqrt{3}$
55. $4\sqrt{45} - 9\sqrt{5}$	**56.** $7\sqrt{24} - 12\sqrt{6}$
57. $3\sqrt{50} + 4\sqrt{18}$	**58.** $6\sqrt{98} + 3\sqrt{32}$
59. $4\sqrt{63} - 7\sqrt{28}$	**60.** $2\sqrt{40} - 7\sqrt{90}$
61. $5\sqrt{12} + 3\sqrt{27} - 2\sqrt{75}$	**62.** $4\sqrt{18} - 6\sqrt{50} - 3\sqrt{72}$
63. $\frac{1}{2}\sqrt{20} + \frac{2}{3}\sqrt{45} - \frac{1}{4}\sqrt{80}$	**64.** $\frac{1}{3}\sqrt{12} - \frac{3}{2}\sqrt{48} + \frac{3}{4}\sqrt{108}$
65. $3\sqrt{8} - 5\sqrt{20} - 7\sqrt{18} - 9\sqrt{125}$	**66.** $5\sqrt{27} - 3\sqrt{24} + 8\sqrt{54} - 7\sqrt{75}$

■ Miscellaneous Problems

67. A rational approximation, to the nearest tenth, for $\sqrt{125}$ can be found by first changing it to simplest radical form and then using Table 9.1 on Page 338–339.

$$\sqrt{125} = \sqrt{25}\sqrt{5} = 5\sqrt{5}.$$

From Table 9.1, we obtain $\sqrt{5} = 2.24$, to the nearest hundredth. Therefore,

$$\sqrt{125} = 5\sqrt{5} = 5(2.24) = 11.20 = 11.2, \quad \text{to the nearest tenth.}$$

Find a rational approximation, to the nearest tenth, for each of the following.

(a) $\sqrt{128}$ (b) $\sqrt{147}$ (c) $\sqrt{150}$

(d) $\sqrt{160}$ (e) $\sqrt{135}$ (f) $\sqrt{200}$

68. (a) Is $\sqrt{x^2} = x$ true for $x = 5$? For $x = -5$?

     (b) Is $\sqrt{x^2} = |x|$ true for $x = 5$? For $x = -5$?

9.3 More on Simplifying Radicals

Another property of square roots is motivated by the following examples.

$$\sqrt{\frac{36}{9}} = \sqrt{4} = 2$$

and

$$\frac{\sqrt{36}}{\sqrt{9}} = \frac{6}{3} = 2.$$

Thus, we see that $\sqrt{\dfrac{36}{9}} = \dfrac{\sqrt{36}}{\sqrt{9}}$. The following general property can be stated.

Property 9.3

For any nonnegative real numbers a and b, b not zero,

$$\sqrt{\frac{a}{b}} = \frac{\sqrt{a}}{\sqrt{b}}.$$

Property 9.3 states that **the square root of the quotient is equal to the quotient of the square roots.**

To evaluate a radical such as $\sqrt{\dfrac{25}{4}}$, for which the numerator and denominator of the fractional radicand are perfect squares, you may use Property 9.3 or merely rely on the definition of square root.

$$\sqrt{\frac{25}{4}} = \frac{\sqrt{25}}{\sqrt{4}} = \frac{5}{2}$$

or

$$\sqrt{\frac{25}{4}} = \frac{5}{2} \quad \text{because} \quad \frac{5}{2} \cdot \frac{5}{2} = \frac{25}{4}.$$

Sometimes it is easier to do the indicated division first and then to find the square root, as the next example illustrates.

$$\sqrt{\frac{324}{9}} = \sqrt{36} = 6.$$

The next three examples present situations where only the denominator of the radicand is a perfect square.

EXAMPLE 1 Simplify $\sqrt{\dfrac{13}{4}}$.

Solution $$\sqrt{\frac{13}{4}} = \frac{\sqrt{13}}{\sqrt{4}} = \frac{\sqrt{13}}{2}. \qquad\qquad ■$$

EXAMPLE 2 Simplify $\sqrt{\dfrac{28}{9}}$.

Solution $$\sqrt{\frac{28}{9}} = \frac{\sqrt{28}}{\sqrt{9}} = \frac{\sqrt{28}}{3} = \frac{\sqrt{4}\sqrt{7}}{3} = \frac{2\sqrt{7}}{3}. \qquad\qquad ■$$

Don't stop here.
The radical in
the numerator
can be simplified.

EXAMPLE 3 Simplify $\dfrac{\sqrt{12}}{\sqrt{16}}$.

Solution A $$\frac{\sqrt{12}}{\sqrt{16}} = \frac{\sqrt{12}}{4} = \frac{\sqrt{4}\sqrt{3}}{4} = \frac{2\sqrt{3}}{4} = \frac{\sqrt{3}}{2}.$$

Solution B $$\frac{\sqrt{12}}{\sqrt{16}} = \sqrt{\frac{12}{16}} = \sqrt{\frac{3}{4}} = \frac{\sqrt{3}}{\sqrt{4}} = \frac{\sqrt{3}}{2}. \qquad\qquad ■$$

Reduce the
fraction.

The two approaches to Example 3 again illustrate the need to *think first* and then *push the pencil*. You may find one approach easier than another.

Now let's consider an example where neither the numerator nor the denominator of the radicand is a perfect square.

EXAMPLE 4 Simplify $\sqrt{\dfrac{2}{3}}$.

Solution A

$$\sqrt{\frac{2}{3}} = \frac{\sqrt{2}}{\sqrt{3}} = \frac{\sqrt{2}}{\sqrt{3}} \cdot \overset{\text{form of } 1}{\frac{\sqrt{3}}{\sqrt{3}}} = \frac{\sqrt{6}}{3}.$$

Solution B

$$\sqrt{\frac{2}{3}} = \sqrt{\frac{2}{3} \cdot \frac{3}{3}} = \sqrt{\frac{6}{9}} = \frac{\sqrt{6}}{\sqrt{9}} = \frac{\sqrt{6}}{3}.$$ ■

The process used to simplify the radical in Example 4 is referred to as **rationalizing the denominator**. Notice that the denominator becomes a rational number. The process of rationalizing the denominator can be accomplished in more than one way, as illustrated in the next example.

EXAMPLE 5 Simplify $\dfrac{\sqrt{5}}{\sqrt{8}}$.

Solution A

$$\frac{\sqrt{5}}{\sqrt{8}} = \frac{\sqrt{5}}{\sqrt{8}} \cdot \frac{\sqrt{8}}{\sqrt{8}} = \frac{\sqrt{40}}{8} = \frac{\sqrt{4}\sqrt{10}}{8} = \frac{2\sqrt{10}}{8} = \frac{\sqrt{10}}{4}.$$

Solution B

$$\frac{\sqrt{5}}{\sqrt{8}} = \frac{\sqrt{5}}{\sqrt{8}} \cdot \frac{\sqrt{2}}{\sqrt{2}} = \frac{\sqrt{10}}{\sqrt{16}} = \frac{\sqrt{10}}{4}.$$

Solution C

$$\frac{\sqrt{5}}{\sqrt{8}} = \frac{\sqrt{5}}{\sqrt{4}\sqrt{2}} = \frac{\sqrt{5}}{2\sqrt{2}} = \frac{\sqrt{5}}{2\sqrt{2}} \cdot \frac{\sqrt{2}}{\sqrt{2}} = \frac{\sqrt{10}}{4}$$ ■

Before considering more examples, let's summarize some ideas pertaining to the simplifying of radicals. An algebraic expression containing square root radicals is said to be in **simplest radical form** if the following conditions are satisfied.

1. No fraction appears within a radical sign. $\left(\sqrt{\dfrac{2}{3}} \text{ violates this condition.}\right)$

2. No radical appears in the denominators. $\left(\dfrac{5}{\sqrt{8}} \text{ violates this condition.}\right)$

3. No radicand when expressed in prime factored form contains a factor raised to a power greater than 1. ($\sqrt{8} = \sqrt{2^3}$ violates this condition.)

Condition 3 can also be stated as "no radicand contains a factor that is a perfect square other than 1." Again $\sqrt{8}$ violates this condition since $\sqrt{8} = \sqrt{4 \cdot 2}$ and 4 is a perfect square.

Study the following examples and check our final radicals according to the three conditions previously mentioned.

EXAMPLE 6 Simplify each of the following.

(a) $\dfrac{3}{\sqrt{x}}$ (b) $\sqrt{\dfrac{2x}{3y}}$ (c) $\dfrac{3\sqrt{5}}{\sqrt{6}}$ (d) $\sqrt{\dfrac{4x^2}{9y}}$

Solution

(a) $\dfrac{3}{\sqrt{x}} = \dfrac{3}{\sqrt{x}} \cdot \dfrac{\sqrt{x}}{\sqrt{x}} = \dfrac{3\sqrt{x}}{x}$.

(b) $\sqrt{\dfrac{2x}{3y}} = \dfrac{\sqrt{2x}}{\sqrt{3y}} = \dfrac{\sqrt{2x}}{\sqrt{3y}} \cdot \dfrac{\sqrt{3y}}{\sqrt{3y}} = \dfrac{\sqrt{6xy}}{3y}$.

(c) $\dfrac{3\sqrt{5}}{\sqrt{6}} = \dfrac{3\sqrt{5}}{\sqrt{6}} \cdot \dfrac{\sqrt{6}}{\sqrt{6}} = \dfrac{3\sqrt{30}}{6} = \dfrac{\sqrt{30}}{2}$.

(d) $\sqrt{\dfrac{4x^2}{9y}} = \dfrac{\sqrt{4x^2}}{\sqrt{9y}} = \dfrac{2x}{\sqrt{9}\sqrt{y}} = \dfrac{2x}{3\sqrt{y}} = \dfrac{2x}{3\sqrt{y}} \cdot \dfrac{\sqrt{y}}{\sqrt{y}} = \dfrac{2x\sqrt{y}}{3y}$. ■

Let's return again to the idea of simplifying expressions containing radicals. Sometimes it may appear as if no simplifying can be done; however, after the individual radicals have been changed to simplest form then the distributive property may apply. Consider the following examples.

EXAMPLE 7 Simplify $5\sqrt{2} + \dfrac{3}{\sqrt{2}}$.

Solution

$$5\sqrt{2} + \dfrac{3}{\sqrt{2}} = 5\sqrt{2} + \dfrac{3}{\sqrt{2}} \cdot \dfrac{\sqrt{2}}{\sqrt{2}} = 5\sqrt{2} + \dfrac{3\sqrt{2}}{2}$$

$$= \left(5 + \dfrac{3}{2}\right)\sqrt{2} = \left(\dfrac{10}{2} + \dfrac{3}{2}\right)\sqrt{2}$$

$$= \dfrac{13}{2}\sqrt{2} \quad \text{or} \quad \dfrac{13\sqrt{2}}{2}. \quad ■$$

EXAMPLE 8 Simplify $\sqrt{\dfrac{3}{2}} + \sqrt{24}$.

Solution

$$\sqrt{\frac{3}{2}} + \sqrt{24} = \frac{\sqrt{3}}{\sqrt{2}} + \sqrt{24}$$

$$= \frac{\sqrt{3}}{\sqrt{2}} \cdot \frac{\sqrt{2}}{\sqrt{2}} + \sqrt{4}\sqrt{6}$$

$$= \frac{\sqrt{6}}{2} + 2\sqrt{6}$$

$$= \left(\frac{1}{2} + 2\right)\sqrt{6}$$

$$= \left(\frac{1}{2} + \frac{4}{2}\right)\sqrt{6}$$

$$= \frac{5}{2}\sqrt{6}.$$
■

EXAMPLE 9 Simplify $3\sqrt{\dfrac{7}{4}} - 14\sqrt{\dfrac{1}{7}} + 5\sqrt{28}$.

Solution

$$3\sqrt{\frac{7}{4}} - 14\sqrt{\frac{1}{7}} + 5\sqrt{28} = \frac{3\sqrt{7}}{\sqrt{4}} - \frac{14\sqrt{1}}{\sqrt{7}} + 5\sqrt{4}\sqrt{7}$$

$$= \frac{3\sqrt{7}}{2} - \frac{14}{\sqrt{7}} \cdot \frac{\sqrt{7}}{\sqrt{7}} + 10\sqrt{7}$$

$$= \frac{3\sqrt{7}}{2} - \frac{14\sqrt{7}}{7} + 10\sqrt{7}$$

$$= \frac{3\sqrt{7}}{2} - 2\sqrt{7} + 10\sqrt{7}$$

$$= \left(\frac{3}{2} - 2 + 10\right)\sqrt{7}$$

$$= \left(\frac{3}{2} - \frac{4}{2} + \frac{20}{2}\right)\sqrt{7}$$

$$= \frac{19}{2}\sqrt{7}.$$
■

Problem Set 9.3

For Problems 1–10, evaluate each radical.

1. $\sqrt{\dfrac{16}{25}}$ 2. $\sqrt{\dfrac{4}{49}}$ 3. $-\sqrt{\dfrac{81}{9}}$ 4. $-\sqrt{\dfrac{64}{16}}$ 5. $\sqrt{\dfrac{1}{64}}$

6. $\sqrt{\dfrac{100}{121}}$ 7. $\sqrt{\dfrac{169}{144}}$ 8. $-\sqrt{\dfrac{196}{144}}$ 9. $-\sqrt{\dfrac{25}{256}}$ 10. $-\sqrt{\dfrac{289}{225}}$

For Problems 11–40, change each radical to simplest radical form.

11. $\sqrt{\dfrac{19}{25}}$ 12. $\sqrt{\dfrac{17}{4}}$ 13. $\sqrt{\dfrac{8}{49}}$ 14. $\sqrt{\dfrac{24}{25}}$ 15. $\dfrac{\sqrt{28}}{\sqrt{9}}$

16. $\dfrac{\sqrt{45}}{\sqrt{16}}$ 17. $\dfrac{\sqrt{12}}{\sqrt{36}}$ 18. $\dfrac{\sqrt{20}}{\sqrt{64}}$ 19. $\sqrt{\dfrac{3}{2}}$ 20. $\sqrt{\dfrac{2}{5}}$

21. $\sqrt{\dfrac{5}{8}}$ 22. $\sqrt{\dfrac{7}{12}}$ 23. $\dfrac{\sqrt{56}}{\sqrt{8}}$ 24. $\dfrac{\sqrt{55}}{\sqrt{11}}$ 25. $\dfrac{\sqrt{63}}{\sqrt{7}}$

26. $\dfrac{\sqrt{96}}{\sqrt{6}}$ 27. $\dfrac{\sqrt{5}}{\sqrt{18}}$ 28. $\dfrac{\sqrt{3}}{\sqrt{32}}$ 29. $\dfrac{\sqrt{4}}{\sqrt{27}}$ 30. $\dfrac{\sqrt{9}}{\sqrt{48}}$

31. $\sqrt{\dfrac{1}{24}}$ 32. $\sqrt{\dfrac{1}{12}}$ 33. $\dfrac{2\sqrt{3}}{\sqrt{5}}$ 34. $\dfrac{3\sqrt{2}}{\sqrt{6}}$ 35. $\dfrac{4\sqrt{2}}{3\sqrt{3}}$

36. $\dfrac{2\sqrt{5}}{7\sqrt{8}}$ 37. $\dfrac{3\sqrt{7}}{4\sqrt{12}}$ 38. $\dfrac{6\sqrt{12}}{5\sqrt{24}}$ 39. $\sqrt{4\dfrac{1}{9}}$ 40. $\sqrt{3\dfrac{1}{4}}$

For Problems 41–60, change each radical to simplest radical form. All variables represent positive real numbers.

41. $\dfrac{3}{\sqrt{x}}$ 42. $\dfrac{2}{\sqrt{xy}}$ 43. $\dfrac{5}{\sqrt{2x}}$ 44. $\dfrac{7}{\sqrt{3y}}$ 45. $\sqrt{\dfrac{3}{x}}$

46. $\sqrt{\dfrac{8}{x}}$ 47. $\sqrt{\dfrac{12}{x^2}}$ 48. $\sqrt{\dfrac{27}{4y^2}}$ 49. $\dfrac{\sqrt{2x}}{\sqrt{5y}}$ 50. $\dfrac{\sqrt{3y}}{\sqrt{32x}}$

51. $\dfrac{\sqrt{5x}}{\sqrt{27y}}$ 52. $\dfrac{\sqrt{3x^2}}{\sqrt{5y^2}}$ 53. $\dfrac{\sqrt{2x^3}}{\sqrt{8y}}$ 54. $\dfrac{\sqrt{5x^2}}{\sqrt{45y^3}}$ 55. $\sqrt{\dfrac{9}{x^3}}$

56. $\sqrt{\dfrac{25}{y^5}}$ 57. $\dfrac{4}{\sqrt{x^7}}$ 58. $\dfrac{14}{\sqrt{x^5}}$ 59. $\dfrac{3\sqrt{x}}{2\sqrt{y^3}}$ 60. $\dfrac{5\sqrt{x}}{7\sqrt{xy}}$

For Problems 61–72, simplify each expression.

61. $7\sqrt{3} + \sqrt{\dfrac{1}{3}}$ 62. $-3\sqrt{2} + \sqrt{\dfrac{1}{2}}$ 63. $4\sqrt{10} - \sqrt{\dfrac{2}{5}}$

64. $8\sqrt{5} - 3\sqrt{\dfrac{1}{5}}$ **65.** $-2\sqrt{5} - 5\sqrt{\dfrac{1}{5}}$ **66.** $6\sqrt{7} + 4\sqrt{\dfrac{1}{7}}$

67. $-3\sqrt{6} - \dfrac{5\sqrt{2}}{\sqrt{3}}$ **68.** $4\sqrt{8} - \dfrac{6}{\sqrt{2}}$ **69.** $4\sqrt{12} + \dfrac{3}{\sqrt{3}} - 5\sqrt{27}$

70. $3\sqrt{8} - 4\sqrt{18} - \dfrac{6}{\sqrt{2}}$ **71.** $\dfrac{9\sqrt{5}}{\sqrt{3}} - 6\sqrt{60} + \dfrac{10\sqrt{3}}{\sqrt{5}}$ **72.** $-2\sqrt{3} - 3\sqrt{48} + \dfrac{3}{\sqrt{3}}$

■ Miscellaneous Problems

73. (a) Use 1.414 as a rational approximation for $\sqrt{2}$ and find the value of $\dfrac{1}{\sqrt{2}}$ to the nearest one-thousandth.

    (b) Change $\dfrac{1}{\sqrt{2}}$ to simplest radical form.

    (c) Use 1.414 as a rational approximation for $\sqrt{2}$ and find the value of $\dfrac{\sqrt{2}}{2}$ to the nearest one-thousandth.

    (d) Use 1.414 as a rational approximation for $\sqrt{2}$ and find the value of each of the following to the nearest one-thousandth.

        (i) $\dfrac{3}{\sqrt{2}}$ (ii) $\dfrac{5}{3\sqrt{2}}$ (iii) $\dfrac{2}{4\sqrt{2}}$ (iv) $\sqrt{8} - \dfrac{2}{\sqrt{2}}$

9.4 Products and Quotients Involving Radicals

Property 9.1 ($\sqrt{ab} = \sqrt{a}\sqrt{b}$) is used to multiply square roots and in some cases to simplify the resulting radical. The following examples illustrate several types of multiplication problems involving radicals.

EXAMPLE 1 Multiply and simplify where possible.

    (a) $\sqrt{3}\sqrt{12}$ **(b)** $\sqrt{3}\sqrt{15}$ **(c)** $\sqrt{7}\sqrt{8}$ **(d)** $\sqrt{5}\sqrt{11}$ **(e)** $(3\sqrt{2})(4\sqrt{3})$

Solution   **(a)** $\sqrt{3}\sqrt{12} = \sqrt{36} = 6.$

    (b) $\sqrt{3}\sqrt{15} = \sqrt{45} = \sqrt{9}\sqrt{5} = 3\sqrt{5}.$

    (c) $\sqrt{7}\sqrt{8} = \sqrt{56} = \sqrt{4}\sqrt{14} = 2\sqrt{14}.$

    (d) $\sqrt{5}\sqrt{11} = \sqrt{55}.$

    (e) $(3\sqrt{2})(4\sqrt{3}) = 3 \cdot 4 \cdot \sqrt{2} \cdot \sqrt{3} = 12\sqrt{6}.$ ■

Recall the use of the distributive property when finding the product of a monomial and a polynomial. For example, $2x(3x + 4) = 2x(3x) + 2x(4) = 6x^2 + 8x$. In a similar manner, the distributive property and Property 9.1 provide the basis for finding certain special products involving radicals. The following examples illustrate this idea.

EXAMPLE 2 Multiply and simplify where possible.

(a) $\sqrt{2}(\sqrt{3} + \sqrt{5})$ (b) $\sqrt{3}(\sqrt{12} - \sqrt{6})$ (c) $\sqrt{8}(\sqrt{2} - 3)$ (d) $\sqrt{x}(\sqrt{x} + \sqrt{y})$

Solution

(a) $\sqrt{2}(\sqrt{3} + \sqrt{5}) = \sqrt{2}\sqrt{3} + \sqrt{2}\sqrt{5} = \sqrt{6} + \sqrt{10}$.

(b) $\sqrt{3}(\sqrt{12} - \sqrt{6}) = \sqrt{3}\sqrt{12} - \sqrt{3}\sqrt{6}$
$$= \sqrt{36} - \sqrt{18}$$
$$= 6 - \sqrt{9}\sqrt{2}$$
$$= 6 - 3\sqrt{2}.$$

(c) $\sqrt{8}(\sqrt{2} - 3) = \sqrt{8}\sqrt{2} - \sqrt{8}(3)$
$$= \sqrt{16} - 3\sqrt{8}$$
$$= 4 - 3\sqrt{4}\sqrt{2}$$
$$= 4 - 6\sqrt{2}.$$

(d) $\sqrt{x}(\sqrt{x} + \sqrt{y}) = \sqrt{x}\sqrt{x} + \sqrt{x}\sqrt{y}$
$$= x + \sqrt{xy}. \qquad\blacksquare$$

The distributive property plays a central role in finding the product of two binomials. For example, $(x + 2)(x + 3) = x(x + 3) + 2(x + 3) = x^2 + 3x + 2x + 6 = x^2 + 5x + 6$. Finding the product of two binomial expressions involving radicals can be handled in a similar fashion.

EXAMPLE 3 Multiply and simplify.

(a) $(\sqrt{3} + \sqrt{5})(\sqrt{2} + \sqrt{6})$ (b) $(\sqrt{7} - 3)(\sqrt{7} + 6)$

(c) $(\sqrt{8} + \sqrt{5})(\sqrt{8} - \sqrt{5})$ (d) $(\sqrt{x} - 4)(\sqrt{x} + 4)$

Solution

(a) $(\sqrt{3} + \sqrt{5})(\sqrt{2} + \sqrt{6}) = \sqrt{3}(\sqrt{2} + \sqrt{6}) + \sqrt{5}(\sqrt{2} + \sqrt{6})$
$$= \sqrt{3}\sqrt{2} + \sqrt{3}\sqrt{6} + \sqrt{5}\sqrt{2} + \sqrt{5}\sqrt{6}$$
$$= \sqrt{6} + \sqrt{18} + \sqrt{10} + \sqrt{30}$$
$$= \sqrt{6} + 3\sqrt{2} + \sqrt{10} + \sqrt{30}.$$

(b) $(\sqrt{7} - 3)(\sqrt{7} + 6) = \sqrt{7}(\sqrt{7} + 6) - 3(\sqrt{7} + 6)$
$$= \sqrt{7}\sqrt{7} + 6\sqrt{7} - 3\sqrt{7} - 18$$
$$= 7 + 6\sqrt{7} - 3\sqrt{7} - 18$$
$$= -11 + 3\sqrt{7}.$$

(c) $(\sqrt{8} + \sqrt{5})(\sqrt{8} - \sqrt{5}) = \sqrt{8}(\sqrt{8} - \sqrt{5}) + \sqrt{5}(\sqrt{8} - \sqrt{5})$

$$= \sqrt{8}\sqrt{8} - \sqrt{8}\sqrt{5} + \sqrt{5}\sqrt{8} - \sqrt{5}\sqrt{5}$$

$$= 8 - \sqrt{40} + \sqrt{40} - 5$$

$$= 3.$$

(d) $(\sqrt{x} - 4)(\sqrt{x} + 4) = \sqrt{x}(\sqrt{x} + 4) - 4(\sqrt{x} + 4)$

$$= \sqrt{x}\sqrt{x} + 4\sqrt{x} - 4\sqrt{x} - 16$$

$$= x - 16.$$ ■

Note parts (c) and (d) of Example 3. In each case the final product is in rational form. This suggests the way to rationalize the denominator of an expression containing a binomial denominator with radicals. Consider the following example.

EXAMPLE 4 Simplify $\dfrac{4}{\sqrt{5} + \sqrt{2}}$.

Solution
$$\frac{4}{\sqrt{5} + \sqrt{2}} = \frac{4}{\sqrt{5} + \sqrt{2}} \cdot \frac{\sqrt{5} - \sqrt{2}}{\sqrt{5} - \sqrt{2}} \qquad \frac{\sqrt{5} - \sqrt{2}}{\sqrt{5} - \sqrt{2}} \text{ is merely a form of 1.}$$

$$= \frac{4(\sqrt{5} - \sqrt{2})}{(\sqrt{5} + \sqrt{2})(\sqrt{5} - \sqrt{2})}$$

$$= \frac{4(\sqrt{5} - \sqrt{2})}{5 - 2}$$

$$= \frac{4(\sqrt{5} - \sqrt{2})}{3} \quad \text{or} \quad \frac{4\sqrt{5} - 4\sqrt{2}}{3}.$$

Either answer
is acceptable. ■

The next four examples further illustrate the process of rationalizing and simplifying expressions containing binomial denominators.

EXAMPLE 5 Rationalize the denominator and simplify $\dfrac{\sqrt{3}}{\sqrt{6} - 2}$.

Solution
$$\frac{\sqrt{3}}{\sqrt{6} - 2} = \frac{\sqrt{3}}{\sqrt{6} - 2} \cdot \frac{\sqrt{6} + 2}{\sqrt{6} + 2}$$

$$= \frac{\sqrt{3}(\sqrt{6} + 2)}{(\sqrt{6} - 2)(\sqrt{6} + 2)}$$

$$= \frac{\sqrt{18} + 2\sqrt{3}}{6 - 4}$$

$$= \frac{3\sqrt{2} + 2\sqrt{3}}{2}. \qquad (\sqrt{18} = \sqrt{9}\sqrt{2} = 3\sqrt{2})$$

■

EXAMPLE 6 Rationalize the denominator and simplify $\dfrac{14}{2\sqrt{3} + \sqrt{5}}$.

Solution

$$\frac{14}{2\sqrt{3} + \sqrt{5}} = \frac{14}{2\sqrt{3} + \sqrt{5}} \cdot \frac{2\sqrt{3} - \sqrt{5}}{2\sqrt{3} - \sqrt{5}}$$

$$= \frac{14(2\sqrt{3} - \sqrt{5})}{(2\sqrt{3} + \sqrt{5})(2\sqrt{3} - \sqrt{5})}$$

$$= \frac{14(2\sqrt{3} - \sqrt{5})}{12 - 5}$$

$$= \frac{14(2\sqrt{3} - \sqrt{5})}{7}$$

$$= 2(2\sqrt{3} - \sqrt{5}) \quad \text{or} \quad 4\sqrt{3} - 2\sqrt{5}.$$

■

EXAMPLE 7 Rationalize the denominator and simplify $\dfrac{\sqrt{x} + 2}{\sqrt{x} - 3}$.

Solution

$$\frac{\sqrt{x} + 2}{\sqrt{x} - 3} = \frac{\sqrt{x} + 2}{\sqrt{x} - 3} \cdot \frac{\sqrt{x} + 3}{\sqrt{x} + 3}$$

$$= \frac{(\sqrt{x} + 2)(\sqrt{x} + 3)}{(\sqrt{x} - 3)(\sqrt{x} + 3)}$$

$$= \frac{x + 2\sqrt{x} + 3\sqrt{x} + 6}{x - 9}$$

$$= \frac{x + 5\sqrt{x} + 6}{x - 9}.$$

■

EXAMPLE 8 Rationalize the denominator and simplify $\dfrac{3 + \sqrt{2}}{\sqrt{2} - 6}$.

Solution

$$\frac{3 + \sqrt{2}}{\sqrt{2} - 6} = \frac{3 + \sqrt{2}}{\sqrt{2} - 6} \cdot \frac{\sqrt{2} + 6}{\sqrt{2} + 6}$$

$$= \frac{(3 + \sqrt{2})(\sqrt{2} + 6)}{(\sqrt{2} - 6)(\sqrt{2} + 6)}$$

$$= \frac{3\sqrt{2} + 18 + 2 + 6\sqrt{2}}{2 - 36}$$

$$= \frac{9\sqrt{2} + 20}{-34}$$

$$= -\frac{9\sqrt{2} + 20}{34}. \qquad \left(\frac{a}{-b} = -\frac{a}{b}\right)$$ ■

Problem Set 9.4

For Problems 1–20, multiply and simplify where possible.

1. $\sqrt{7}\sqrt{5}$ 2. $\sqrt{3}\sqrt{10}$ 3. $\sqrt{6}\sqrt{8}$

4. $\sqrt{6}\sqrt{12}$ 5. $\sqrt{5}\sqrt{10}$ 6. $\sqrt{2}\sqrt{12}$

7. $\sqrt{3}\sqrt{27}$ 8. $\sqrt{2}\sqrt{32}$ 9. $\sqrt{8}\sqrt{12}$

10. $\sqrt{12}\sqrt{20}$ 11. $(3\sqrt{3})(5\sqrt{7})$ 12. $(4\sqrt{2})(7\sqrt{5})$

13. $(-2\sqrt{2})(3\sqrt{7})$ 14. $(5\sqrt{3})(-6\sqrt{10})$ 15. $(3\sqrt{6})(4\sqrt{6})$

16. $(2\sqrt{7})(5\sqrt{7})$ 17. $(5\sqrt{2})(4\sqrt{12})$ 18. $(3\sqrt{2})(2\sqrt{27})$

19. $(4\sqrt{3})(2\sqrt{15})$ 20. $(7\sqrt{7})(2\sqrt{14})$

For Problems 21–42, find the products by applying the distributive property. Express answers in simplest radical form.

21. $\sqrt{2}(\sqrt{3} + \sqrt{5})$ 22. $\sqrt{3}(\sqrt{5} + \sqrt{7})$

23. $\sqrt{6}(\sqrt{2} - 5)$ 24. $\sqrt{8}(\sqrt{3} - 4)$

25. $\sqrt{3}(\sqrt{6} + \sqrt{7})$ 26. $\sqrt{2}(\sqrt{8} + \sqrt{10})$

27. $\sqrt{12}(\sqrt{6} - \sqrt{8})$ 28. $\sqrt{15}(\sqrt{3} - \sqrt{5})$

29. $4\sqrt{3}(\sqrt{2} - 2\sqrt{5})$ 30. $3\sqrt{2}(3\sqrt{5} - 2\sqrt{7})$

31. $(\sqrt{2} + 6)(\sqrt{2} + 9)$ 32. $(\sqrt{3} + 4)(\sqrt{3} + 7)$

33. $(\sqrt{6} - 5)(\sqrt{6} + 3)$ 34. $(\sqrt{7} - 6)(\sqrt{7} + 1)$

35. $(\sqrt{3} + \sqrt{6})(\sqrt{6} + \sqrt{8})$ 36. $(\sqrt{2} + \sqrt{6})(\sqrt{8} + \sqrt{5})$

37. $(5 + \sqrt{10})(5 - \sqrt{10})$ 38. $(\sqrt{11} - 2)(\sqrt{11} + 2)$

39. $(3\sqrt{2} - \sqrt{3})(3\sqrt{2} + \sqrt{3})$

40. $(5\sqrt{6} - \sqrt{2})(5\sqrt{6} + \sqrt{2})$

41. $(5\sqrt{3} + 2\sqrt{6})(5\sqrt{3} - 2\sqrt{6})$

42. $(4\sqrt{5} + 5\sqrt{7})(4\sqrt{5} - 5\sqrt{7})$

For Problems 43–54, find each product and express answers in simplest radical form. All variables represent nonnegative real numbers.

43. $\sqrt{xy}\sqrt{x}$

44. $\sqrt{x^2 y}\sqrt{y}$

45. $\sqrt{3x}\sqrt{6y}$

46. $\sqrt{2a}\sqrt{6b}$

47. $(4\sqrt{a})(3\sqrt{ab})$

48. $(5\sqrt{a})(6\sqrt{a})$

49. $\sqrt{2x}(\sqrt{3x} - \sqrt{6y})$

50. $\sqrt{6x}(\sqrt{2x} - \sqrt{4y})$

51. $(\sqrt{x} + 5)(\sqrt{x} - 3)$

52. $(3 + \sqrt{x})(7 - \sqrt{x})$

53. $(\sqrt{x} + 7)(\sqrt{x} - 7)$

54. $(\sqrt{x} + 9)(\sqrt{x} - 9)$

For Problems 55–70, rationalize the denominators and simplify. All variables represent positive real numbers.

55. $\dfrac{3}{\sqrt{2} + 4}$

56. $\dfrac{5}{\sqrt{3} + 7}$

57. $\dfrac{8}{\sqrt{6} - 2}$

58. $\dfrac{10}{3 - \sqrt{7}}$

59. $\dfrac{2}{\sqrt{5} + \sqrt{3}}$

60. $\dfrac{3}{\sqrt{6} + \sqrt{5}}$

61. $\dfrac{10}{2 - 3\sqrt{3}}$

62. $\dfrac{5}{3\sqrt{2} - 4}$

63. $\dfrac{4}{\sqrt{x} - 2}$

64. $\dfrac{7}{\sqrt{x} + 4}$

65. $\dfrac{\sqrt{x}}{\sqrt{x} + 3}$

66. $\dfrac{\sqrt{y}}{\sqrt{y} - 6}$

67. $\dfrac{\sqrt{a} + 2}{\sqrt{a} - 5}$

68. $\dfrac{\sqrt{a} - 3}{\sqrt{a} + 1}$

69. $\dfrac{2 + \sqrt{3}}{3 - \sqrt{2}}$

70. $\dfrac{3 - \sqrt{5}}{4 + \sqrt{8}}$

9.5 Solving Radical Equations

Equations containing radicals with variables in the radicand are called **radical equations**. The following are examples of radical equations involving one variable.

$$\sqrt{x} = 3; \qquad \sqrt{2x + 1} = 5; \qquad \sqrt{3x + 4} = -4;$$

$$\sqrt{5s - 2} = \sqrt{2s + 19}; \qquad \sqrt{2y - 4} = y - 2; \qquad \sqrt{x + 6} = x.$$

In order to solve such equations we need the following property of equality.

Property 9.4

For real numbers a and b, if $a = b$, then

$$a^2 = b^2.$$

Property 9.4 states that we can **square both sides of an equation**. However, squaring both sides of an equation sometimes produces results that do not satisfy the original equation. Let's consider two examples to illustrate the point.

EXAMPLE 1   Solve $\sqrt{x} = 3$.

Solution

$$\sqrt{x} = 3$$
$$(\sqrt{x})^2 = 3^2 \qquad \text{Square both sides.}$$
$$x = 9.$$

Since $\sqrt{9} = 3$, the solution set is $\{9\}$.   ■

EXAMPLE 2   Solve $\sqrt{x} = -3$.

Solution

$$\sqrt{x} = -3$$
$$(\sqrt{x})^2 = (-3)^2 \qquad \text{Square both sides.}$$
$$x = 9.$$

Since $\sqrt{9} \neq -3$, 9 is not a solution and the solution set is $\varnothing$.   ■

In general, squaring both sides of an equation will produce an equation that has all of the solutions of the original equation, but it may also have some extra solutions that will not satisfy the original equation. (Such extra solutions are called **extraneous solutions** or **roots**.) Therefore, when using the "squaring" property (Property 9.4) you *must* check each potential solution in the original equation.

Let's consider some examples to illustrate different situations that arise when solving radical equations.

EXAMPLE 3   Solve $\sqrt{2x + 1} = 5$.

Solution

$$\sqrt{2x + 1} = 5$$
$$(\sqrt{2x + 1})^2 = 5^2 \qquad \text{Square both sides.}$$
$$2x + 1 = 25$$
$$2x = 24$$
$$x = 12.$$

Check: $\sqrt{2x + 1} = 5$
$$\sqrt{2(12) + 1} \overset{?}{=} 5$$
$$\sqrt{24 + 1} \overset{?}{=} 5$$
$$\sqrt{25} \overset{?}{=} 5$$
$$5 = 5.$$

The solution set is $\{12\}$.   ■

EXAMPLE 4 Solve $\sqrt{3x + 4} = -4$.

Solution

$$\sqrt{3x + 4} = -4$$
$$(\sqrt{3x + 4})^2 = (-4)^2 \qquad \text{Square both sides.}$$
$$3x + 4 = 16$$
$$3x = 12$$
$$x = 4.$$

Check:
$$\sqrt{3x + 4} = -4$$
$$\sqrt{3(4) + 4} \overset{?}{=} -4$$
$$\sqrt{16} \overset{?}{=} -4$$
$$4 \neq -4.$$

Since 4 does not check, (4 is an extraneous root) the equation $\sqrt{3x + 4} = -4$ has no real number solutions. The solution set is $\varnothing$. ■

EXAMPLE 5 Solve $\sqrt{5s - 2} = \sqrt{2s + 19}$.

Solution

$$\sqrt{5s - 2} = \sqrt{2s + 19}$$
$$(\sqrt{5s - 2})^2 = (\sqrt{2s + 19})^2 \qquad \text{Square both sides.}$$
$$5s - 2 = 2s + 19$$
$$3s = 21$$
$$s = 7.$$

Check:
$$\sqrt{5s - 2} = \sqrt{2s + 19}$$
$$\sqrt{5(7) - 2} \overset{?}{=} \sqrt{2(7) + 19}$$
$$\sqrt{33} = \sqrt{33}$$

The solution set is $\{7\}$. ■

EXAMPLE 6 Solve $\sqrt{2y - 4} = y - 2$.

Solution

$$\sqrt{2y - 4} = y - 2$$
$$(\sqrt{2y - 4})^2 = (y - 2)^2 \qquad \text{Square both sides.}$$
$$2y - 4 = y^2 - 4y + 4$$
$$0 = y^2 - 6y + 8$$
$$0 = (y - 4)(y - 2) \qquad \text{Factor right side.}$$
$$y - 4 = 0 \quad \text{or} \quad y - 2 = 0 \qquad \text{Remember the property:}$$
$$\qquad\qquad\qquad\qquad\qquad\qquad ab = 0 \text{ if and only if}$$
$$\qquad\qquad\qquad\qquad\qquad\qquad a = 0 \text{ or } b = 0.$$

$$y = 4 \quad \text{or} \quad y = 2$$

Check: $\sqrt{2y-4} = y - 2$ $\sqrt{2y-4} = y - 2$

$\sqrt{2(4)-4} \overset{?}{=} 4 - 2$ or $\sqrt{2(2)-4} \overset{?}{=} 2 - 2$

$\sqrt{4} \overset{?}{=} 2$ $\sqrt{0} \overset{?}{=} 0$

$2 = 2$ $0 = 0$

The solution set is $\{2, 4\}$.   ■

EXAMPLE 7 Solve $\sqrt{x+6} = x$.

Solution $\sqrt{x+6} = x$

$\sqrt{x} = x - 6$ Added -6 to both sides so that the term with the radical sign is alone on one side of the equation.

$(\sqrt{x})^2 = (x-6)^2$

$x = x^2 - 12x + 36$

$0 = x^2 - 13x + 36$

$0 = (x-4)(x-9)$

$x - 4 = 0$ or $x - 9 = 0$ Apply: $ab = 0$ if and only if $a = 0$ or $b = 0$.

$x = 4$ or $x = 9$

Check: $\sqrt{x+6} = x$ $\sqrt{x+6} = x$

$\sqrt{4} + 6 \overset{?}{=} 4$ or $\sqrt{9} + 6 \overset{?}{=} 9$

$2 + 6 \overset{?}{=} 4$ $3 + 6 \overset{?}{=} 9$

$8 \neq 4$ $9 = 9$

The solution set is $\{9\}$.   ■

Note in Example 7 that we changed the form of the original equation, $\sqrt{x+6} = x$, to $\sqrt{x} = x - 6$ before squaring both sides. Squaring both sides of $\sqrt{x+6} = x$ produces $x + 12\sqrt{x} + 36 = x^2$, a more complex equation that still contains a radical. So again, it pays to think ahead a few steps before carrying out the details of the problem.

Problem Set 9.5

Solve each of the following equations. *Be sure* to check all potential solutions in the original equation.

1. $\sqrt{x} = 7$ **2.** $\sqrt{x} = 12$ **3.** $\sqrt{2x} = 6$

4. $\sqrt{3x} = 9$ **5.** $\sqrt{3x} = -6$ **6.** $\sqrt{2x} = -8$

7. $\sqrt{4x} = 3$

8. $\sqrt{5x} = 4$

9. $3\sqrt{x} = 2$

10. $4\sqrt{x} = 3$

11. $\sqrt{2n - 3} = 5$

12. $\sqrt{3n + 1} = 7$

13. $\sqrt{5y + 2} = -1$

14. $\sqrt{4n - 3} - 4 = 0$

15. $\sqrt{6x - 5} - 3 = 0$

16. $\sqrt{5x + 3} = -4$

17. $5\sqrt{x} = 30$

18. $6\sqrt{x} = 42$

19. $\sqrt{3a - 2} = \sqrt{2a + 4}$

20. $\sqrt{4a + 3} = \sqrt{5a - 4}$

21. $\sqrt{7x - 3} = \sqrt{4x + 3}$

22. $\sqrt{8x - 6} = \sqrt{4x + 11}$

23. $2\sqrt{y + 1} = 5$

24. $3\sqrt{y - 2} = 4$

25. $\sqrt{x + 3} = x + 3$

26. $\sqrt{x + 7} = x + 7$

27. $\sqrt{-2x + 28} = x - 2$

28. $\sqrt{-2x} = x + 4$

29. $\sqrt{3n - 4} = \sqrt{n}$

30. $\sqrt{5n - 1} = \sqrt{2n}$

31. $\sqrt{3x} = x - 6$

32. $2\sqrt{x} = x - 3$

33. $4\sqrt{x} + 5 = x$

34. $\sqrt{-x - 6} = x$

35. $\sqrt{x^2 + 27} = x + 3$

36. $\sqrt{x^2 - 35} = x - 5$

37. $\sqrt{x^2 + 2x + 3} = x + 2$

38. $\sqrt{x^2 + x + 4} = x + 3$

39. $\sqrt{8x} - 2 = x$

40. $\sqrt{2x - 4} = x - 6$

■ **Miscellaneous Problems**

Solve each of the following equations.

41. $\sqrt{x + 1} = 5 - \sqrt{x - 4}$

42. $\sqrt{x - 2} = \sqrt{x + 7} - 1$

43. $\sqrt{2n - 1} + \sqrt{n - 5} = 3$

44. $\sqrt{2n + 1} - \sqrt{n - 3} = 2$

Chapter 9 Summary

(9.1) The number a is a square root of b if $a^2 = b$. The symbol, $\sqrt{}$, is called a **radical sign** and is used to indicate the **nonnegative** or **principal square root**. The number under the radical sign is called the **radicand**. An entire expression, such as $\sqrt{37}$, is called a **radical**.

You should be able to use Table 9.1 to find rational approximations for irrational numbers.

The distributive property provides the basis for simplifying a radical expression such as $5\sqrt{2} + 7\sqrt{2}$ to $12\sqrt{2}$.

(9.2) The property "$\sqrt{ab} = \sqrt{a}\sqrt{b}$" provides the basis for changing radicals to **simplest radical form**.

The property "$\sqrt{a^2} = a$" is used when changing radicals involving variables in the radicand to simplest radical form.

(9.3) The property "$\sqrt{\dfrac{a}{b}} = \dfrac{\sqrt{a}}{\sqrt{b}}$" is used when **rationalizing the denominator**.

An algebraic expression containing square root radicals is said to be in **simplest radical form** if the following conditions are satisfied.

1. No fraction appears within a radical sign.

$$\left(\sqrt{\frac{3}{5}} \text{ violates this condition.}\right)$$

2. No radical appears in the denominators.

$$\left(\frac{3}{\sqrt{10}} \text{ violates this condition.}\right)$$

3. No radicand when expressed in prime factored form contains a factor raised to a power greater than 1.

$$(\sqrt{8} = \sqrt{2^3} \text{ violates this condition.})$$

(9.4) The property "$\sqrt{ab} = \sqrt{a}\sqrt{b}$" is used to multiply square roots and to simplify the resulting radical when appropriate.

(9.5) The property of equality "if $a = b$, then $a^2 = b^2$" is used to solve equations containing radicals.

In general, squaring both sides of an equation will produce an equation that has all of the solutions of the original equation, but it may also have some extra solutions that will not satisfy the original equation. Therefore, when using the "squaring" property, you must check each potential solution in the original equation.

Chapter 9 Review Problem Set

Evaluate each of the following without using a table or a calculator.

1. $\sqrt{64}$ 2. $-\sqrt{49}$ 3. $\sqrt{1600}$ 4. $\sqrt{\frac{81}{25}}$ 5. $-\sqrt{\frac{4}{9}}$ 6. $\sqrt{\frac{49}{36}}$

Change each of the following to simplest radical form.

7. $\sqrt{20}$ 8. $\sqrt{32}$ 9. $5\sqrt{8}$ 10. $\sqrt{80}$ 11. $-3\sqrt{24}$

12. $\frac{\sqrt{12}}{\sqrt{49}}$ 13. $\frac{\sqrt{36}}{\sqrt{7}}$ 14. $\sqrt{\frac{7}{8}}$ 15. $\sqrt{\frac{8}{24}}$ 16. $\frac{3\sqrt{2}}{\sqrt{5}}$

17. $\frac{4\sqrt{3}}{\sqrt{12}}$ 18. $\frac{5\sqrt{2}}{2\sqrt{3}}$ 19. $\frac{-3\sqrt{2}}{\sqrt{27}}$ 20. $\frac{4\sqrt{6}}{3\sqrt{12}}$

If $\sqrt{3} \approx 1.732$, find a rational approximation, to the nearest one-thousandth, for each of the following.

21. $\sqrt{27}$ **22.** $\dfrac{2}{\sqrt{3}}$ **23.** $3\sqrt{12} + \sqrt{48}$ **24.** $2\sqrt{27} - 2\sqrt{75}$

Change each of the following to simplest radical form. All variables represent positive real numbers.

25. $\sqrt{12a^2b^3}$ **26.** $\sqrt{50xy^4}$ **27.** $3x\sqrt{2x^5}$ **28.** $-4t\sqrt{8t^3}$

29. $\dfrac{4}{3}\sqrt{27xy^2}$ **30.** $\dfrac{3}{4}\sqrt{24x^3y^5}$ **31.** $\dfrac{\sqrt{2x}}{\sqrt{5y}}$ **32.** $\dfrac{\sqrt{72x}}{\sqrt{16y}}$

33. $\sqrt{\dfrac{4}{x}}$ **34.** $\sqrt{\dfrac{2x^3}{9}}$ **35.** $\dfrac{3\sqrt{x}}{4\sqrt{y^3}}$ **36.** $\dfrac{-2\sqrt{x^2y}}{5\sqrt{xy}}$

Find the following products and express answers in simplest radical form.

37. $(\sqrt{6})(\sqrt{12})$ **38.** $(2\sqrt{3})(3\sqrt{6})$ **39.** $(-5\sqrt{8})(2\sqrt{2})$
40. $(-3\sqrt{12})(-4\sqrt{7})$ **41.** $\sqrt{2}(\sqrt{8} + \sqrt{20})$ **42.** $3\sqrt{5}(\sqrt{8} - 2\sqrt{12})$
43. $(\sqrt{3} + \sqrt{5})(\sqrt{3} + \sqrt{7})$ **44.** $(2\sqrt{3} + 3\sqrt{2})(\sqrt{3} - 5\sqrt{2})$
45. $(\sqrt{6} + 2\sqrt{7})(3\sqrt{6} - \sqrt{7})$ **46.** $(3 + 2\sqrt{5})(4 - 3\sqrt{5})$

Rationalize the denominators and simplify.

47. $\dfrac{5}{\sqrt{7} - \sqrt{5}}$ **48.** $\dfrac{\sqrt{6}}{\sqrt{3} - \sqrt{2}}$ **49.** $\dfrac{2}{3\sqrt{2} - \sqrt{6}}$ **50.** $\dfrac{\sqrt{6}}{3\sqrt{7} + 2\sqrt{10}}$

Simplify each of the following radical expressions.

51. $2\sqrt{50} + 3\sqrt{72} - 5\sqrt{8}$ **52.** $\sqrt{8x} - 3\sqrt{18x}$ **53.** $5\sqrt{12} - 6\sqrt{27} - \sqrt{48}$

54. $3\sqrt{10} + \sqrt{\dfrac{2}{5}}$ **55.** $4\sqrt{20} - \dfrac{3}{\sqrt{5}} + \sqrt{45}$ **56.** $\sqrt{\dfrac{2}{3}} - 2\sqrt{54}$

Solve each of the following equations.

57. $\sqrt{5x + 6} = 6$ **58.** $\sqrt{6x + 1} = \sqrt{3x + 13}$ **59.** $3\sqrt{n} = n$
60. $\sqrt{y + 5} = y + 5$ **61.** $\sqrt{-3a + 10} = a - 2$ **62.** $3 - \sqrt{2x - 1} = 2$

10 Quadratic Equations

Solving equations has been a central theme throughout this text. Let's pause for a moment and reflect back upon the different types of equations that we have solved.

Type of equation	Examples
First-degree equations of one variable	$4x + 3 = 7x + 1$; $3(x - 6) = 9$; $\dfrac{x + 2}{3} + \dfrac{x + 3}{4} = 1$; $0.2y + 0.3 = 1.4$.
Second-degree equations of one variable *that are factorable*	$x^2 + 3x = 0$; $x^2 + 5x + 6 = 0$; $x^2 - 4 = 0$; $x^2 + 10x + 25 = 0$.
Fractional equations	$\dfrac{3}{x} + \dfrac{2}{x} = 4$; $\dfrac{5}{a - 2} = \dfrac{6}{a + 3}$; $\dfrac{2}{x^2 - 4} + \dfrac{5}{x + 2} = \dfrac{6}{x - 2}$.
Radical equations	$\sqrt{x} = 4$; $\sqrt{y + 2} = 3$; $\sqrt{a + 1} = \sqrt{2a - 7}$.
Systems of equations	$\begin{pmatrix} 2x + 3y = 4 \\ 5x - y = 7 \end{pmatrix}$; $\begin{pmatrix} 3a + 5b = 9 \\ 7a - 9b = 12 \end{pmatrix}$.

As indicated in the above table, we have learned how to solve second-degree equations of one variable, but only those that are factorable. In this chapter we will extend our work to include more general types of second-degree equations in one variable.

10.1 Quadratic Equations

A second-degree equation in one variable contains the variable with an exponent of two, but no higher power. Such equations are also called **quadratic equations**. The following are examples of quadratic equations.

$$x^2 = 25; \qquad y^2 + 6y = 0; \qquad x^2 + 7x - 4 = 0;$$
$$4y^2 + 2y - 1 = 0; \qquad 5x^2 + 2x - 1 = 2x^2 + 6x - 5.$$

A quadratic equation in the variable x can also be defined as any equation that can be written in the form $ax^2 + bx + c = 0$, where a, b, and c are real numbers and $a \neq 0$. The form $ax^2 + bx + c = 0$ is referred to as the **standard form** of a quadratic equation.

In Chapter 6 you solved quadratic equations (the term "quadratic" was not used at that time) by factoring and applying the property "$ab = 0$ if and only if $a = 0$ or $b = 0$." Let's review a few examples of that type.

EXAMPLE 1 Solve $x^2 - 13x = 0$.

Solution

$$x^2 - 13x = 0$$
$$x(x - 13) = 0 \qquad \text{Factor left side of equation.}$$
$$x = 0 \quad \text{or} \quad x - 13 = 0 \qquad \text{Apply: } ab = 0 \text{ if and only if } a = 0 \text{ or } b = 0.$$
$$x = 0 \quad \text{or} \qquad x = 13.$$

The solution set is $\{0, 13\}$. (Don't forget to check these solutions!) ■

EXAMPLE 2 Solve $y^2 = 49$.

Solution

$$y^2 = 49$$
$$y^2 - 49 = 0$$
$$(y + 7)(y - 7) = 0 \qquad \text{Factor left side.}$$
$$y + 7 = 0 \quad \text{or} \quad y - 7 = 0 \qquad \text{Apply: } ab = 0 \text{ if and only if } a = 0 \text{ or } b = 0.$$
$$y = -7 \quad \text{or} \qquad y = 7.$$

The solution set is $\{-7, 7\}$. ■

EXAMPLE 3 Solve $n^2 + 2n - 24 = 0$.

Solution

$$n^2 + 2n - 24 = 0$$
$$(n + 6)(n - 4) = 0 \qquad \text{Factor left side.}$$
$$n + 6 = 0 \quad \text{or} \quad n - 4 = 0 \qquad \text{Apply: } ab = 0 \text{ if and only if } a = 0 \text{ or } b = 0.$$
$$n = -6 \quad \text{or} \qquad n = 4.$$

The solution set is $\{-6, 4\}$. ■

EXAMPLE 4   Solve $x^2 + 6x + 9 = 0$.

Solution

$$x^2 + 6x + 9 = 0$$

$$(x + 3)(x + 3) = 0 \qquad \text{Factor left side.}$$

$$x + 3 = 0 \quad \text{or} \quad x + 3 = 0 \qquad \text{Apply: } ab = 0 \text{ if and only if } a = 0 \text{ or } b = 0$$

$$x = -3 \quad \text{or} \qquad x = -3.$$

The solution set is $\{-3\}$.   ■

Let's consider the equation $x^2 = a$, where x is the variable and a is any nonnegative real number. We can solve this equation as follows.

$$x^2 = a$$

$$x^2 = (\sqrt{a})^2 \qquad \qquad ((\sqrt{a})^2 = a)$$

$$x^2 - (\sqrt{a})^2 = 0$$

$$(x - \sqrt{a})(x + \sqrt{a}) = 0 \qquad \text{Factor left side.}$$

$$x - \sqrt{a} = 0 \quad \text{or} \quad x + \sqrt{a} = 0 \qquad \text{Apply: } ab = 0 \text{ if and only if } a = 0$$
$$\qquad \qquad \qquad \qquad \qquad \qquad \qquad \text{or } b = 0$$

$$x = \sqrt{a} \quad \text{or} \qquad x = -\sqrt{a}.$$

The solutions are $\sqrt{a}$ and $-\sqrt{a}$.

We shall consider this result as a general property and use it to solve certain types of quadratic equations.

Property 10.1

For any nonnegative real number a,

$$x^2 = a \quad \text{if and only if} \quad x = \sqrt{a} \quad \text{or} \quad x = -\sqrt{a}.$$

(The statement "$x = \sqrt{a}$ or $x = -\sqrt{a}$" can be written as $x = \pm\sqrt{a}$.)

Property 10.1 along with our knowledge of square roots makes it very easy to solve quadratic equations of the form $x^2 = a$.

EXAMPLE 5   Solve $x^2 = 81$.

Solution

$$x^2 = 81$$

$$x = \pm\sqrt{81} \qquad \text{Apply Property 10.1.}$$

$$x = \pm 9.$$

The solution set is $\{-9, 9\}$.   ■

EXAMPLE 6 Solve $x^2 = 8$.

Solution

$$x^2 = 8$$
$$x = \pm\sqrt{8}$$
$$x = \pm 2\sqrt{2}. \qquad (\sqrt{8} = \sqrt{4}\sqrt{2} = 2\sqrt{2})$$

The solution set is $\{-2\sqrt{2}, 2\sqrt{2}\}$. ■

EXAMPLE 7 Solve $5n^2 = 12$.

Solution

$$5n^2 = 12$$
$$n^2 = \frac{12}{5}$$
$$n = \pm\sqrt{\frac{12}{5}}$$
$$n = \pm\frac{2\sqrt{15}}{5}. \qquad \left(\sqrt{\frac{12}{5}} = \frac{\sqrt{12}}{\sqrt{5}} = \frac{\sqrt{12}}{\sqrt{5}} \cdot \frac{\sqrt{5}}{\sqrt{5}} = \frac{\sqrt{60}}{5} = \frac{2\sqrt{15}}{5}\right)$$

The solution set is $\left\{-\dfrac{2\sqrt{15}}{5}, \dfrac{2\sqrt{15}}{5}\right\}$. ■

EXAMPLE 8 Solve $(x - 2)^2 = 16$.

Solution

$$(x - 2)^2 = 16$$
$$x - 2 = \pm 4$$
$$x - 2 = 4 \quad \text{or} \quad x - 2 = -4$$
$$x = 6 \quad \text{or} \quad x = -2.$$

The solution set is $\{-2, 6\}$. ■

EXAMPLE 9 Solve $(x + 5)^2 = 27$.

Solution

$$(x + 5)^2 = 27$$
$$x + 5 = \pm\sqrt{27}$$
$$x + 5 = \pm 3\sqrt{3} \qquad (\sqrt{27} = \sqrt{9}\sqrt{3} = 3\sqrt{3})$$
$$x + 5 = 3\sqrt{3} \quad \text{or} \quad x + 5 = -3\sqrt{3}$$
$$x = -5 + 3\sqrt{3} \quad \text{or} \quad x = -5 - 3\sqrt{3}.$$

The solution set is $\{-5 - 3\sqrt{3}, -5 + 3\sqrt{3}\}$. ■

It may be necessary to do a bit of changing of form before Property 10.1 can be applied. The next example illustrates this point.

EXAMPLE 10   Solve $3(2x - 3)^2 + 8 = 44$.

Solution

$$3(2x - 3)^2 + 8 = 44$$
$$3(2x - 3)^2 = 36$$
$$(2x - 3)^2 = 12$$
$$2x - 3 = \pm\sqrt{12} \qquad \text{Apply Property 10.1}$$

$2x - 3 = \sqrt{12}$ or $2x - 3 = -\sqrt{12}$

$2x = 3 + \sqrt{12}$ or $2x = 3 - \sqrt{12}$

$$x = \frac{3 + \sqrt{12}}{2} \quad \text{or} \quad x = \frac{3 - \sqrt{12}}{2}$$

$$x = \frac{3 + 2\sqrt{3}}{2} \quad \text{or} \quad x = \frac{3 - 2\sqrt{3}}{2}. \qquad (\sqrt{12} = \sqrt{4}\sqrt{3} = 2\sqrt{3})$$

The solution set is $\left\{\dfrac{3 - 2\sqrt{3}}{2}, \dfrac{3 + 2\sqrt{3}}{2}\right\}$.   ■

It should be noted that quadratic equations of the form $x^2 = a$, where a is a *negative* number, have no real number solutions. For example, $x^2 = -4$ has no real number solutions since any real number squared is nonnegative. In a like manner an equation such as $(x + 3)^2 = -14$ has no real number solutions.

■ Using the Pythagorean Theorem

Our work with radicals, Property 10.1, and the Pythagorean Theorem merge to form a basis for solving a variety of problems pertaining to right triangles. First, let's restate the Pythagorean Theorem.

Pythagorean Theorem

If for a right triangle, a and b are the measures of the legs, and c is the measure of the hypotenuse, then

$$a^2 + b^2 = c^2.$$

(The *hypotenuse* is the side opposite the right angle and the *legs* are the other two sides.)

PROBLEM 1 Find c in the accompanying figure.

3 cm

4 cm

Solution Applying the Pythagorean Theorem we obtain

$$c^2 = a^2 + b^2$$
$$c^2 = 3^2 + 4^2$$
$$c^2 = 9 + 16$$
$$c^2 = 25$$
$$c = 5.$$

The length of c is 5 centimeters. ■

Remark: Don't forget that the equation $c^2 = 25$ does have two solutions, 5 and -5. However, since we are finding the lengths of line segments, the negative solutions need to be disregarded.

PROBLEM 2 Find a in the accompanying figure.

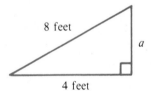

8 feet

a

4 feet

Solution
$$a^2 + 4^2 = 8^2$$
$$a^2 + 16 = 64$$
$$a^2 = 48$$
$$a = \sqrt{48} = \sqrt{16}\sqrt{3} = 4\sqrt{3} \text{ feet.}$$ ■

There are two special kinds of right triangles that are used frequently in later mathematics courses. An **isosceles right triangle** is a right triangle having both legs of the same length. Let's consider a problem involving an isosceles right triangle.

PROBLEM 3 Find the length of each leg of an isosceles right triangle having a hypotenuse with a measure of 8 meters.

Solution

Let's sketch an isosceles right triangle and let x represent the length of each leg. Then by applying the Pythagorean Theorem we obtain

$$x^2 + x^2 = 64$$
$$2x^2 = 64$$
$$x^2 = 32$$
$$x = \sqrt{32} = \sqrt{16}\sqrt{2} = 4\sqrt{2}.$$

Each leg is $4\sqrt{2}$ meters long. ■

Another special kind of right triangle is one that contains acute angles of $30°$ and $60°$. In such a right triangle, often referred to as a $30°$–$60°$ right triangle, *the side opposite the 30° angle is equal in length to one-half of the length of the hypotenuse*. This relationship, along with the Pythagorean Theorem, provides us with another problem solving technique.

PROBLEM 4 Suppose that the side opposite the $60°$ angle in a $30°$–$60°$ right triangle is 6 inches long. Find the length of the other leg and the length of the hypotenuse.

Solution

Let x represent the length of the side opposite the $30°$ angle. Then $2x$ represents the length of the hypotenuse, as illustrated in the accompanying sketch. Therefore, we can apply the Pythagorean Theorem as follows.

$$x^2 + 6^2 = (2x)^2$$
$$x^2 + 36 = 4x^2$$
$$36 = 3x^2$$
$$12 = x^2$$
$$\sqrt{12} = x$$
$$2\sqrt{3} = x.$$

The side opposite the $30°$ angle is $2\sqrt{3}$ inches long and the hypotenuse is $2(2\sqrt{3}) = 4\sqrt{3}$ inches long. ■

Problem Set 10.1

For Problems 1–18, solve each quadratic equation by factoring and applying the property: $ab = 0$ if and only if $a = 0$ or $b = 0$.

1. $x^2 + 15x = 0$ **2.** $x^2 - 11x = 0$ **3.** $n^2 = 12n$

4. $n^2 = -21n$ **5.** $3y^2 = 15y$ **6.** $8y^2 = -56y$

7. $x^2 - 9x + 8 = 0$ **8.** $x^2 + 16x + 48 = 0$ **9.** $x^2 - 5x - 14 = 0$

10. $x^2 - 5x - 36 = 0$ **11.** $n^2 + 5n - 6 = 0$ **12.** $n^2 + 3n - 28 = 0$

13. $6y^2 + 7y - 5 = 0$ **14.** $4y^2 - 21y - 18 = 0$ **15.** $30x^2 - 37x + 10 = 0$

16. $42x^2 + 67x + 21 = 0$ **17.** $4x^2 - 4x + 1 = 0$ **18.** $9x^2 + 12x + 4 = 0$

For Problems 19–56, use Property 10.1 to help solve each quadratic equation. Express irrational solutions in simplest radical form.

19. $x^2 = 64$ **20.** $x^2 = 169$ **21.** $x^2 = \dfrac{25}{9}$

22. $x^2 = \dfrac{4}{81}$ **23.** $4x^2 = 64$ **24.** $5x^2 = 500$

25. $n^2 = 14$ **26.** $n^2 = 22$ **27.** $n^2 + 16 = 0$

28. $n^2 = 24$ **29.** $y^2 = 32$ **30.** $y^2 + 25 = 0$

31. $3x^2 - 54 = 0$ **32.** $4x^2 - 108 = 0$ **33.** $2x^2 = 9$

34. $3x^2 = 16$ **35.** $8n^2 = 25$ **36.** $12n^2 = 49$

37. $(x - 1)^2 = 4$ **38.** $(x - 2)^2 = 9$ **39.** $(x + 3)^2 = 25$

40. $(x + 5)^2 = 36$ **41.** $(3x - 2)^2 = 49$ **42.** $(4x + 3)^2 = 1$

43. $(x + 6)^2 = 5$ **44.** $(x - 7)^2 = 6$ **45.** $(n - 1)^2 = 8$

46. $(n + 1)^2 = 12$ **47.** $(2n + 3)^2 = 20$ **48.** $(3n - 2)^2 = 28$

49. $(4x - 1)^2 = -2$ **50.** $(5x + 3)^2 - 32 = 0$ **51.** $(3x - 5)^2 - 40 = 0$

52. $(2x + 9)^2 + 6 = 0$ **53.** $2(7x - 1)^2 + 5 = 37$ **54.** $3(4x - 5)^2 - 50 = 25$

55. $2(x + 8)^2 - 9 = 91$ **56.** $2(x - 7)^2 - 7 = 101$

For Problems 57–62, a and b represent the lengths of the legs of a right triangle and c represents the length of the hypotenuse. Express answers in simplest radical form.

57. Find c if $a = 1$ inch and $b = 7$ inches.

58. Find c if $a = 2$ inches and $b = 6$ inches.

59. Find a if $c = 8$ meters and $b = 6$ meters.

60. Find a if $c = 11$ centimeters and $b = 7$ centimeters.

61. Find b if $c = 12$ feet and $a = 10$ feet.

62. Find b if $c = 10$ yards and $a = 4$ yards.

For Problems 63–68, use the accompanying figure. Express answers in simplest radical form.

63. If $c = 8$ inches, find a and b.

64. If $c = 6$ inches, find a and b.

65. If $a = 6$ feet, find b and c.

66. If $a = 5$ feet, find b and c.

67. If $b = 12$ meters, find a and c.

68. If $b = 5$ centimeters, find a and c.

For Problems 69–72, use the accompanying *isosceles right triangle*. **Express answers in simplest radical form.**

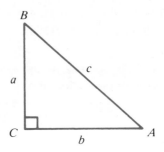

69. If $b = 10$ inches, find a and c.

70. If $a = 7$ inches, find b and c.

71. If $c = 9$ meters, find a and b.

72. If $c = 5$ meters, find a and b.

■ Miscellaneous Problems

73. Sometimes we simply need to determine whether a particular radical expression is positive or negative. For example, is $-6 + \sqrt{39}$ a positive or negative number? We can determine this by approximating a value for $\sqrt{39}$. Since $6^2 = 36$, we know that $\sqrt{39}$ is a little larger than 6. Therefore, $-6 + \sqrt{39}$ has to be positive.

Determine whether each of the following is positive or negative.

(**a**) $-8 + \sqrt{56}$ (**b**) $-7 + \sqrt{47}$ (**c**) $9 - \sqrt{77}$ (**d**) $12 - \sqrt{130}$

(**e**) $-6 + 5\sqrt{2}$ (**f**) $-10 + 6\sqrt{3}$ (**g**) $-13 + \sqrt{150}$ (**h**) $-14 + \sqrt{200}$

10.2 Completing the Square

Thus far we have solved quadratic equations by factoring or by applying Property 10.1 (if $x^2 = a$, then $x = \sqrt{a}$ or $x = -\sqrt{a}$). In this section we will consider another method called **completing the square** that will give us the power to solve *any* quadratic equation.

A factoring technique that we studied in Chapter 6 was based on recognizing *perfect trinomial squares*. In each of the following, the trinomial on the right side, which is the result of squaring a binomial on the left side, is **a perfect trinomial square.**

$$(x + 5)^2 = x^2 + 10x + 25;$$
$$(x + 7)^2 = x^2 + 14x + 49;$$
$$(x - 3)^2 = x^2 - 6x + 9;$$
$$(x - 6)^2 = x^2 - 12x + 36.$$

We need to pay attention to the following special relationship. In each of the previous trinomial squares the constant term is equal to the square of one-half of the coefficient of the x-term. For example,

$$x^2 + 10x + 25 \qquad \frac{1}{2}(10) = 5 \quad \text{and} \quad 5^2 = 25;$$

$$x^2 - 12x + 36 \qquad \frac{1}{2}(12) = 6 \quad \text{and} \quad 6^2 = 36.$$

This relationship allows us to form a perfect trinomial square by adding the proper constant term. For example, suppose that we want to form a perfect trinomial square from $x^2 + 8x$. Since $\frac{1}{2}(8) = 4$ and $4^2 = 16$, the perfect square $x^2 + 8x + 16$ can be formed.

Let's use the previous ideas to help solve some quadratic equations.

EXAMPLE 1 Solve $x^2 + 8x - 1 = 0$ by the method of completing the square.

Solution

$$x^2 + 8x - 1 = 0$$
$$x^2 + 8x = 1$$
$$x^2 + 8x + 16 = 1 + 16 \qquad \text{16 is added to the left side to form a perfect trinomial square and therefore 16 also has to be added to the right side.}$$

$$(x + 4)^2 = 17.$$

(Now we can proceed as we did with such equations in the last section.)

$$x + 4 = \pm\sqrt{17}$$
$$x + 4 = \sqrt{17} \qquad \text{or} \quad x + 4 = -\sqrt{17}$$
$$x = -4 + \sqrt{17} \quad \text{or} \qquad x = -4 - \sqrt{17}.$$

The solution set is $\{-4 - \sqrt{17}, -4 + \sqrt{17}\}$. ■

Observe that the method of completing the square to solve a quadratic equation is merely what the name implies. A perfect trinomial square is formed; then the equation can be changed to the necessary form for using the property: if $x^2 = a$, then $x = \sqrt{a}$ or $x = -\sqrt{a}$. Let's consider another example.

EXAMPLE 2 Solve $x^2 - 2x - 11 = 0$ by the method of completing the square.

Solution

$$x^2 - 2x - 11 = 0$$
$$x^2 - 2x = 11$$
$$x^2 - 2x + 1 = 11 + 1$$
$$(x - 1)^2 = 12$$

$$x - 1 = \pm\sqrt{12}$$
$$x - 1 = \pm 2\sqrt{3}$$
$$x - 1 = 2\sqrt{3} \quad \text{or} \quad x - 1 = -2\sqrt{3}$$
$$x = 1 + 2\sqrt{3} \quad \text{or} \quad x = 1 - 2\sqrt{3}.$$

The solution set is $\{1 - 2\sqrt{3}, 1 + 2\sqrt{3}\}$. ■

In the next example the coefficient of the x-term is odd, which means that taking one-half of it puts us in the realm of fractions. The use of common fractions rather than decimals makes our previous work with radicals applicable.

EXAMPLE 3 Solve $x^2 - 3x + 1 = 0$ by the method of completing the square.

Solution

$$x^2 - 3x + 1 = 0$$
$$x^2 - 3x = -1$$
$$x^2 - 3x + \frac{9}{4} = -1 + \frac{9}{4}$$
$$\left(x - \frac{3}{2}\right)^2 = \frac{5}{4}$$
$$x - \frac{3}{2} = \pm\sqrt{\frac{5}{4}}$$
$$x - \frac{3}{2} = \pm\frac{\sqrt{5}}{2}$$
$$x - \frac{3}{2} = \frac{\sqrt{5}}{2} \quad \text{or} \quad x - \frac{3}{2} = -\frac{\sqrt{5}}{2}$$
$$x = \frac{3}{2} + \frac{\sqrt{5}}{2} \quad \text{or} \quad x = \frac{3}{2} - \frac{\sqrt{5}}{2}$$
$$x = \frac{3 + \sqrt{5}}{2} \quad \text{or} \quad x = \frac{3 - \sqrt{5}}{2}.$$

The solution set is $\left\{\dfrac{3 - \sqrt{5}}{2}, \dfrac{3 + \sqrt{5}}{2}\right\}$. ■

The relationship for a perfect trinomial square that states that **the constant term is equal to the square of one-half of the coefficient of the x-term** holds only if the coefficient of x^2 is 1. Thus, a slight adjustment needs to be made when solving quadratic equations having a coefficient of x^2 other than 1. The next example shows how to make this adjustment.

EXAMPLE 4 Solve $2x^2 + 12x - 3 = 0$ by the method of completing the square.

Solution
$$2x^2 + 12x - 3 = 0$$
$$2x^2 + 12x = 3$$
$$x^2 + 6x = \frac{3}{2} \qquad \text{Multiply both sides by } \frac{1}{2}.$$
$$x^2 + 6x + 9 = \frac{3}{2} + 9$$
$$(x + 3)^2 = \frac{21}{2}$$
$$x + 3 = \pm\sqrt{\frac{21}{2}}$$
$$x + 3 = \pm\frac{\sqrt{42}}{2} \qquad \left(\sqrt{\frac{21}{2}} = \frac{\sqrt{21}}{\sqrt{2}} = \frac{\sqrt{21}}{\sqrt{2}} \cdot \frac{\sqrt{2}}{\sqrt{2}} = \frac{\sqrt{42}}{2}\right)$$

$$x + 3 = \frac{\sqrt{42}}{2} \qquad \text{or} \quad x + 3 = -\frac{\sqrt{42}}{2}$$
$$x = -3 + \frac{\sqrt{42}}{2} \quad \text{or} \qquad x = -3 - \frac{\sqrt{42}}{2}$$
$$x = \frac{-6 + \sqrt{42}}{2} \quad \text{or} \qquad x = \frac{-6 - \sqrt{42}}{2}.$$

The solution set is $\left\{\dfrac{-6 - \sqrt{42}}{2}, \dfrac{-6 + \sqrt{42}}{2}\right\}$. ■

As mentioned earlier, the method of completing the square can be used to solve *any* quadratic equation. To illustrate this point, let's use it to solve an equation that could also be solved by factoring.

EXAMPLE 5 Solve $x^2 + 2x - 8 = 0$ by the method of completing the square and by factoring.

Solution (by completing the square)
$$x^2 + 2x - 8 = 0$$
$$x^2 + 2x = 8$$
$$x^2 + 2x + 1 = 8 + 1$$
$$(x + 1)^2 = 9$$
$$(x + 1) = \pm 3$$
$$x + 1 = 3 \quad \text{or} \quad x + 1 = -3$$
$$x = 2 \quad \text{or} \qquad x = -4.$$

The solution set is $\{-4, 2\}$.

Solution (by factoring) $x^2 + 2x - 8 = 0$

$$(x + 4)(x - 2) = 0$$

$$x + 4 = 0 \quad \text{or} \quad x - 2 = 0$$

$$x = -4 \quad \text{or} \quad x = 2.$$

The solution set is $\{-4, 2\}$. ■

No claim is made that using the method of completing the square with an equation such as the one in Example 5 is easier than the factoring technique. However, it is important for you to recognize that the method of completing the square will work with any quadratic equation.

Our final example of this section illustrates the point that the method of completing the square will identify those quadratic equations that have no real number solutions.

EXAMPLE 6 Solve $x^2 + 10x + 30 = 0$ by the method of completing the square.

Solution

$$x^2 + 10x + 30 = 0$$

$$x^2 + 10x = -30$$

$$x^2 + 10x + 25 = -30 + 25$$

$$(x + 5)^2 = -5.$$

We can stop here and reason as follows: any value of x will yield a nonnegative value for $(x + 5)^2$; thus, it cannot equal -5. The original equation, $x^2 + 10x + 30 = 0$, has *no solutions in the set of real numbers*. The solution set is $\varnothing$. ■

Problem Set 10.2

For Problems 1–32, use the method of completing the square to help solve each quadratic equation.

1. $x^2 + 8x - 1 = 0$ 2. $x^2 - 4x - 1 = 0$ 3. $x^2 + 10x + 2 = 0$

4. $x^2 + 8x + 3 = 0$ 5. $x^2 - 4x - 4 = 0$ 6. $x^2 + 6x - 11 = 0$

7. $x^2 + 6x + 12 = 0$ 8. $n^2 - 10n = 7$ 9. $n^2 + 2n = 17$

10. $n^2 + 12n + 40 = 0$ 11. $x^2 + x - 3 = 0$ 12. $x^2 + 3x - 5 = 0$

13. $a^2 - 5a = 2$ 14. $a^2 - 7a = 4$ 15. $2x^2 + 8x - 3 = 0$

16. $2x^2 - 12x + 1 = 0$ 17. $3x^2 + 12x - 2 = 0$ 18. $3x^2 - 6x + 2 = 0$

19. $2t^2 - 4t + 1 = 0$ 20. $4t^2 + 8t + 5 = 0$ 21. $5n^2 + 10n + 6 = 0$

22. $2n^2 + 5n - 1 = 0$ 23. $-n^2 + 9n = 4$ 24. $-n^2 - 7n = 2$

25. $2x^2 + 3x - 1 = 0$ 26. $3x^2 - x - 3 = 0$ 27. $3x^2 + 2x - 2 = 0$

28. $9x = 3x^2 - 1$ **29.** $n(n + 2) = 168$ **30.** $n(n + 4) = 140$

31. $n(n - 4) = 165$ **32.** $n(n - 2) = 288$

For Problems 33–42, solve each quadratic equation by using (a) the factoring method, and (b) the method of completing the square.

33. $x^2 + 4x - 12 = 0$ **34.** $x^2 - 6x - 40 = 0$ **35.** $x^2 + 12x + 27 = 0$

36. $x^2 + 18x + 77 = 0$ **37.** $n^2 - 3n - 40 = 0$ **38.** $n^2 + 9n - 36 = 0$

39. $2n^2 - 9n + 4 = 0$ **40.** $6n^2 - 11n - 10 = 0$ **41.** $4n^2 + 4n - 15 = 0$

42. $4n^2 + 12n - 7 = 0$

■ Miscellaneous Problem

43. Use the method of completing the square to solve $ax^2 + bx + c = 0$ for x, where a, b, and c are real numbers and $a \neq 0$.

10.3 The Quadratic Formula

The method of completing the square can be used to solve *any* quadratic equation. The equation $ax^2 + bx + c = 0$, where a, b, and c are real numbers with $a \neq 0$, can be used to represent *any* quadratic equation. These two ideas merge to produce the **quadratic formula**, a formula that can be used to solve *any* quadratic equation. The merging is accomplished by using the method of completing the square to solve the equation $ax^2 + bx + c = 0$ as follows.

$$ax^2 + bx + c = 0$$

$$ax^2 + bx = -c$$

$$x^2 + \frac{b}{a}x = -\frac{c}{a} \qquad \text{Multiply both sides by } \frac{1}{a}.$$

$$x^2 + \frac{b}{a}x + \frac{b^2}{4a^2} = -\frac{c}{a} + \frac{b^2}{4a^2} \qquad \text{Complete the square by adding } \frac{b^2}{4a^2} \text{ to both sides.}$$

$$\left(x + \frac{b}{2a}\right)^2 = \frac{b^2 - 4ac}{4a^2}. \qquad \text{The right side is combined into a single term.}$$

If $b^2 - 4ac \geq 0$, then

$$x + \frac{b}{2a} = \pm\sqrt{\frac{b^2 - 4ac}{4a^2}}$$

$$x + \frac{b}{2a} = \frac{\pm\sqrt{b^2 - 4ac}}{2a}$$

$$x + \frac{b}{2a} = \frac{\sqrt{b^2 - 4ac}}{2a} \qquad \text{or} \qquad x + \frac{b}{2a} = \frac{-\sqrt{b^2 - 4ac}}{2a}$$

$$x = -\frac{b}{2a} + \frac{\sqrt{b^2 - 4ac}}{2a} \qquad \text{or} \qquad x = -\frac{b}{2a} - \frac{\sqrt{b^2 - 4ac}}{2a}$$

$$x = \frac{-b + \sqrt{b^2 - 4ac}}{2a} \qquad \text{or} \qquad x = \frac{-b - \sqrt{b^2 - 4ac}}{2a}.$$

The solutions are $\dfrac{-b + \sqrt{b^2 - 4ac}}{2a}$ and $\dfrac{-b - \sqrt{b^2 - 4ac}}{2a}$.

The **quadratic formula** is usually stated as follows.

$$x = \frac{-b \pm \sqrt{b^2 - 4ac}}{2a}$$

It can be used to solve any quadratic equation by expressing the equation in standard form and substituting the values for a, b, and c into the formula. Consider the following examples.

EXAMPLE 1 Solve $x^2 + 7x + 10 = 0$ by using the quadratic formula.

Solution The equation $x^2 + 7x + 10 = 0$ is in standard form, so $a = 1$, $b = 7$, and $c = 10$. Substituting these values into the formula we obtain

$$x = \frac{-b \pm \sqrt{b^2 - 4ac}}{2a}$$

$$x = \frac{-7 \pm \sqrt{7^2 - 4(1)(10)}}{2(1)}$$

$$x = \frac{-7 \pm \sqrt{9}}{2}$$

$$x = \frac{-7 \pm 3}{2}$$

$$x = \frac{-7 + 3}{2} \qquad \text{or} \qquad x = \frac{-7 - 3}{2}$$

$$x = -2 \qquad \text{or} \qquad x = -5.$$

The solution set is $\{-5, -2\}$. ■

EXAMPLE 2 Solve $x^2 - 3x = 1$ by using the quadratic formula.

Solution

$$x^2 - 3x = 1$$
$$x^2 - 3x - 1 = 0 \qquad\qquad\qquad \text{Change to standard form.}$$

$$x = \frac{-(-3) \pm \sqrt{(-3)^2 - 4(1)(-1)}}{2(1)}$$

$$x = \frac{3 \pm \sqrt{9 + 4}}{2}$$

$$x = \frac{3 \pm \sqrt{13}}{2}.$$

The solution set is $\left\{ \dfrac{3 - \sqrt{13}}{2}, \dfrac{3 + \sqrt{13}}{2} \right\}.$ ■

EXAMPLE 3 Solve $15n^2 - n - 2 = 0$ by using the quadratic formula.

Solution Although the variable x is commonly used in the statement of the quadratic formula, it should be evident that any variable could be used.

$$15n^2 - n - 2 = 0$$

$$n = \frac{-(-1) \pm \sqrt{(-1)^2 - 4(15)(-2)}}{2(15)}$$

$$n = \frac{1 \pm \sqrt{1 + 120}}{30}$$

$$n = \frac{1 \pm \sqrt{121}}{30}$$

$$n = \frac{1 \pm 11}{30}$$

$$n = \frac{1 + 11}{30} \quad \text{or} \quad n = \frac{1 - 11}{30}$$

$$n = \frac{12}{30} \quad \text{or} \quad n = \frac{-10}{30}$$

$$n = \frac{2}{5} \quad \text{or} \quad n = -\frac{1}{3}.$$

The solution set is $\left\{ -\dfrac{1}{3}, \dfrac{2}{5} \right\}.$ ■

EXAMPLE 4 Solve $t^2 - 5t - 84 = 0$ by using the quadratic formula.

Solution

$$t^2 - 5t - 84 = 0$$

$$t = \frac{-(-5) \pm \sqrt{(-5)^2 - 4(1)(-84)}}{2(1)}$$

$$t = \frac{5 \pm \sqrt{25 + 336}}{2}$$

$$t = \frac{5 \pm \sqrt{361}}{2}$$

$$t = \frac{5 \pm 19}{2}$$

$$t = \frac{5 + 19}{2} \quad \text{or} \quad t = \frac{5 - 19}{2}$$

$$t = \frac{24}{2} \quad \text{or} \quad t = \frac{-14}{2}$$

$$t = 12 \quad \text{or} \quad t = -7.$$

The solution set is $\{-7, 12\}$.   ■

Quadratic equations that have no real number solutions can be easily identified when using the quadratic formula. The final example of this section illustrates this point.

EXAMPLE 5 Solve $x^2 - 2x + 8 = 0$ by using the quadratic formula.

Solution

$$x^2 - 2x + 8 = 0$$

$$x = \frac{-(-2) \pm \sqrt{(-2)^2 - 4(1)(8)}}{2(1)}$$

$$x = \frac{2 \pm \sqrt{4 - 32}}{2}$$

$$x = \frac{2 \pm \sqrt{-28}}{2}.$$

Since $\sqrt{-28}$ is not a real number, we conclude that the given equation has no real number solutions. Its solution set is $\varnothing$. (More work will be done with this type of equation in Section 11.2.)   ■

Problem Set 10.3

Use the quadratic formula to solve each of the following quadratic equations.

1. $x^2 - 5x - 6 = 0$ 2. $x^2 + 3x - 4 = 0$ 3. $x^2 + 5x = 36$

4. $x^2 - 8x = -12$ 5. $n^2 - 2n - 5 = 0$ 6. $n^2 - 4n - 1 = 0$

7. $a^2 - 5a - 2 = 0$ 8. $a^2 + 3a + 1 = 0$ 9. $x^2 - 2x + 6 = 0$

10. $x^2 - 8x + 16 = 0$ 11. $y^2 + 4y + 2 = 0$ 12. $n^2 + 6n + 11 = 0$

13. $x^2 - 6x = 0$ 14. $x^2 + 8x = 0$ 15. $2x^2 = 7x$

16. $3x^2 = -10x$ 17. $n^2 - 34n + 288 = 0$ 18. $n^2 + 27n + 182 = 0$

19. $x^2 + 2x - 80 = 0$ 20. $x^2 - 15x + 54 = 0$ 21. $t^2 + 4t + 4 = 0$

22. $t^2 + 6t - 5 = 0$ 23. $6x^2 + x - 2 = 0$ 24. $4x^2 - x - 3 = 0$

25. $5x^2 + 3x - 2 = 0$ 26. $6x^2 - x - 2 = 0$ 27. $12x^2 + 19x = -5$

28. $2x^2 + 7x - 6 = 0$ 29. $2x^2 + 5x - 6 = 0$ 30. $2x^2 + 3x - 3 = 0$

31. $3x^2 + 4x - 1 = 0$ 32. $3x^2 + 2x - 4 = 0$ 33. $16x^2 + 24x + 9 = 0$

34. $9x^2 - 30x + 25 = 0$ 35. $4n^2 + 8n - 1 = 0$ 36. $4n^2 + 6n - 1 = 0$

37. $6n^2 + 9n + 1 = 0$ 38. $5n^2 + 8n + 1 = 0$ 39. $2y^2 - y - 4 = 0$

40. $3t^2 + 6t + 5 = 0$ 41. $4t^2 + 5t + 3 = 0$ 42. $5x^2 + x - 1 = 0$

43. $7x^2 + 5x - 4 = 0$ 44. $6x^2 + 2x - 3 = 0$ 45. $7 = 3x^2 - x$

46. $-2x^2 + 3x = -4$ 47. $n^2 + 23n = -126$ 48. $n^2 + 2n = 195$

10.4   Solving Quadratic Equations — Which Method?

Let's summarize the three basic methods of solving quadratic equations presented in this chapter by solving a specific quadratic equation using all three techniques. Consider the equation $x^2 + 4x - 12 = 0$.

Factoring Method

$x^2 + 4x - 12 = 0$

$(x + 6)(x - 2) = 0$

$x + 6 = 0$ or $x - 2 = 0$

$x = -6$ or $x = 2.$

The solution set is $\{-6, 2\}$.

Completing the Square Method

$x^2 + 4x - 12 = 0$

$x^2 + 4x = 12$

$$x^2 + 4x + 4 = 12 + 4$$
$$(x + 2)^2 = 16$$
$$x + 2 = \pm\sqrt{16}$$
$$x + 2 = 4 \quad \text{or} \quad x + 2 = -4$$
$$x = 2 \quad \text{or} \quad x = -6.$$

The solution set is $\{-6, 2\}$.

Quadratic Formula Method

$$x^2 + 4x - 12 = 0$$

$$x = \frac{-4 \pm \sqrt{4^2 - 4(1)(-12)}}{2(1)}$$

$$x = \frac{-4 \pm \sqrt{64}}{2}$$

$$x = \frac{-4 \pm 8}{2}$$

$$x = \frac{-4 + 8}{2} \quad \text{,or} \quad x = \frac{-4 - 8}{2}$$

$$x = 2 \qquad \text{or} \quad x = -6.$$

The solution set is $\{-6, 2\}$.

We also discussed the use of the property "$x^2 = a$ if and only if $x = \pm\sqrt{a}$," for certain types of quadratic equations. For example, $x^2 = 4$ can be solved easily by applying the property and obtaining $x = \sqrt{4}$ or $x = -\sqrt{4}$; thus, the solutions are 2 and -2.

Which method should be used to solve a particular quadratic equation? Let's consider some examples in which the different techniques are used. However, keep in mind that usually this is a decision *you* must make as the need arises. So become as familiar as you can with the strengths and weaknesses of each method.

EXAMPLE 1   Solve $2x^2 + 12x - 54 = 0$.

Solution   First, it is very helpful to recognize a factor of 2 in each of the terms on the left side.

$$2x^2 + 12x - 54 = 0$$

$$x^2 + 6x - 27 = 0. \qquad \text{Multiply both sides by } \frac{1}{2}.$$

Now we should recognize that the left side can be factored. Thus, we can proceed as follows.

$$(x + 9)(x - 3) = 0$$
$$x + 9 = 0 \quad \text{or} \quad x - 3 = 0$$
$$x = -9 \quad \text{or} \quad x = 3.$$

The solution set is $\{-9, 3\}$. ■

EXAMPLE 2 Solve $(4x + 3)^2 = 16$.

Solution The form of this equation lends itself to the use of the property "$x^2 = a$ if and only if $x \pm \sqrt{a}$."

$$(4x + 3)^2 = 16$$
$$4x + 3 = \pm\sqrt{16}$$
$$4x + 3 = 4 \quad \text{or} \quad 4x + 3 = -4$$
$$4x = 1 \quad \text{or} \quad 4x = -7$$
$$x = \frac{1}{4} \quad \text{or} \quad x = -\frac{7}{4}.$$

The solution set is $\left\{-\dfrac{7}{4}, \dfrac{1}{4}\right\}$. ■

EXAMPLE 3 Solve $n + \dfrac{1}{n} = 5$.

Solution First, we need to *clear the equation of fractions* by multiplying both sides by n.

$$n + \frac{1}{n} = 5, \quad n \neq 0$$

$$n\left(n + \frac{1}{n}\right) = 5(n)$$

$$n^2 + 1 = 5n.$$

Now, let's change it to standard form.

$$n^2 - 5n + 1 = 0.$$

Since the left side cannot be factored using integers, we must solve the equation by using either the method of completing the square or the quadratic formula.

Using the formula we obtain

$$n = \frac{-(-5) \pm \sqrt{(-5)^2 - 4(1)(1)}}{2(1)}$$

$$n = \frac{5 \pm \sqrt{21}}{2}.$$

The solution set is $\left\{\dfrac{5 - \sqrt{21}}{2}, \dfrac{5 + \sqrt{21}}{2}\right\}$.   ■

EXAMPLE 4   Solve $t^2 = \sqrt{2}\,t$.

Solution   A quadratic equation without a constant term can be easily solved by the factoring method.

$$t^2 = \sqrt{2}\,t$$
$$t^2 - \sqrt{2}\,t = 0$$
$$t(t - \sqrt{2}) = 0$$
$$t = 0 \quad \text{or} \quad t - \sqrt{2} = 0$$
$$t = 0 \quad \text{or} \quad t = \sqrt{2}.$$

The solution set is $\{0, \sqrt{2}\}$. (Check each of these solutions in the given equation.)   ■

EXAMPLE 5   Solve $x^2 - 28x + 192 = 0$.

Solution   Determining whether or not the left side is factorable presents a bit of a problem because of the size of the constant term. Therefore, let's not concern ourselves with trying to factor, but merely use the quadratic formula.

$$x^2 - 28x + 192 = 0$$

$$x = \frac{-(-28) \pm \sqrt{(-28)^2 - 4(1)(192)}}{2(1)}$$

$$x = \frac{28 \pm \sqrt{784 - 768}}{2}$$

$$x = \frac{28 \pm \sqrt{16}}{2}$$

$$x = \frac{28 + 4}{2} \quad \text{or} \quad x = \frac{28 - 4}{2}$$

$$x = 16 \quad \text{or} \quad x = 12.$$

The solution set is $\{12, 16\}$.   ■

EXAMPLE 6 Solve $x^2 + 12x = 17$.

Solution   The form of this equation and the fact that the coefficient of x is even makes the method of completing the square a reasonable approach.

$$x^2 + 12x = 17$$
$$x^2 + 12x + 36 = 17 + 36$$
$$(x + 6)^2 = 53$$
$$x + 6 = \pm\sqrt{53}$$
$$x = -6 \pm \sqrt{53}.$$

The solution set is $\{-6 - \sqrt{53}, -6 + \sqrt{53}\}$.   ■

Problem Set 10.4

Solve each of the following quadratic equations using the method that seems most appropriate to you.

1. $x^2 + 4x = 45$

2. $x^2 + 4x = 60$

3. $(5n + 6)^2 = 49$

4. $(3n - 1)^2 = 25$

5. $t^2 - t - 2 = 0$

6. $t^2 + 2t - 3 = 0$

7. $8x = 3x^2$

8. $5x^2 = 7x$

9. $9x^2 - 6x + 1 = 0$

10. $4x^2 + 36x + 81 = 0$

11. $5n^2 = \sqrt{8}\,n$

12. $\sqrt{3}\,n = 2n^2$

13. $n^2 - 14n = 19$

14. $n^2 - 10n = 14$

15. $5x^2 - 2x - 7 = 0$

16. $3x^2 - 4x - 2 = 0$

17. $15x^2 + 28x + 5 = 0$

18. $20y^2 - 7y - 6 = 0$

19. $x^2 - \sqrt{8}\,x - 7 = 0$

20. $x^2 + \sqrt{5}\,x - 5 = 0$

21. $y^2 + 5y = 84$

22. $y^2 + 7y = 60$

23. $2n = 3 + \dfrac{3}{n}$

24. $n + \dfrac{1}{n} = 7$

25. $3x^2 - 9x - 12 = 0$

26. $2x^2 + 10x - 28 = 0$

27. $2x^2 - 3x + 7 = 0$

28. $3x^2 - 2x + 5 = 0$

29. $n(n - 46) = -480$

30. $n(n + 42) = -432$

31. $n - \dfrac{3}{n} = -1$

32. $n - \dfrac{2}{n} = \dfrac{3}{4}$

33. $x + \dfrac{1}{x} = \dfrac{25}{12}$

34. $x + \dfrac{1}{x} = \dfrac{65}{8}$

35. $t^2 + 12t + 36 = 49$

36. $t^2 - 10t + 25 = 16$

37. $x^2 - 28x + 187 = 0$

38. $x^2 - 33x + 266 = 0$

39. $\dfrac{x^2}{3} - x = -\dfrac{1}{2}$

40. $\dfrac{2x - 1}{3} = \dfrac{5}{x + 2}$

41. $\dfrac{2}{x + 2} - \dfrac{1}{x} = 3$

42. $\dfrac{3}{x - 1} + \dfrac{2}{x} = \dfrac{3}{2}$

43. $\dfrac{2}{3n - 1} = \dfrac{n + 2}{6}$

44. $\dfrac{x^2}{2} = x + \dfrac{1}{4}$

45. $(n - 2)(n + 4) = 7$

46. $(n + 3)(n - 8) = -30$

10.5 Solving Problems Using Quadratic Equations

Remember the common thread of this text as indicated by the following diagram.

develop   use skills   use equations
skills   → to solve   → to solve word
          equations   problems

Now we should be ready to use our skills pertaining to systems of equations (Chapter 8) and quadratic equations to help solve additional types of word problems. Before considering such problems, let's review and update the problem solving suggestions given in Chapter 3.

Suggestions for Solving Word Problems

1. Read the problem carefully, making certain that you understand the meanings of all the words. Be especially alert for any technical terms used in the statement of the problem.

2. Read the problem a second time (perhaps even a third time) to get an overview of the situation being described and to determine the known facts, as well as what is to be found.

3. Sketch any figure, diagram, or chart that might be helpful in analyzing the problem.

*4. Choose *meaningful* variables to represent the unknown quantities. Use one or two variables, whichever seems easiest. The term "meaningful" refers to the choice of letters to use as variables. Choose letters that have some significance for the problem under consideration. For example, if the problem deals with the length and width of a rectangle, then *l* and *w* would seem like natural choices for the variables.

*5. Look for *guidelines* that can be used to help set up equations. A guideline might be a formula such as "area of a rectangular region equals length times width" or a statement of a relationship such as "the product of the two numbers is 98."

*6. (a) Form an equation containing the variable that translates the conditions of the guideline from English to algebra; or

    (b) form two equations containing the two variables that translate the guidelines from English to algebra.

*7. Solve the equation (system of equations) and use the solution (solutions) to determine all facts requested in the problem.

8. **Check all answers back into the original statement of the problem.**

The asterisks indicate suggestions that have been revised to include using systems of equations to solve problems. Keep these suggestions in mind as you study the examples and work the problems in this section.

PROBLEM 1 The length of a rectangular region is 2 centimeters longer than its width. The area of the region is 35 square centimeters. Find the length and width of the rectangle.

Solution Let l represent the length. Let w represent the width.

l

Area is 35 cm².

w

The area formula for a rectangle, $A = lw$, and the statement "the length of a rectangular region is 2 centimeters longer than its width" can be used as guidelines to form a system of equations.

$$\left(\begin{array}{l} lw = 35 \\ \ \ l = w + 2. \end{array} \right)$$

The second equation indicates that $w + 2$ can be substituted for l. Making this substitution in the first equation yields

$$(w + 2)(w) = 35.$$

Solving this quadratic equation by factoring, we obtain

$$
\begin{aligned}
w^2 + 2w &= 35 \\
w^2 + 2w - 35 &= 0 \\
(w + 7)(w - 5) &= 0 \\
w + 7 = 0 \quad &\text{or} \quad w - 5 = 0. \\
w = -7 \quad &\text{or} \quad w = 5
\end{aligned}
$$

The width of a rectangle cannot be a negative number, so the solution -7 is discarded. Thus, the width of the rectangle is 5 centimeters and the length $(w + 2)$ is 7 centimeters. ■

PROBLEM 2 Find two consecutive whole numbers whose product is 506.

Solution Let n represent the smaller whole number. Then $n + 1$ represents the next larger whole number. The phrase "whose product is 506" translates into the equation

$$n(n + 1) = 506.$$

Changing this quadratic equation to standard form produces

$$
\begin{aligned}
n^2 + n &= 506 \\
n^2 + n - 506 &= 0.
\end{aligned}
$$

Because of the size of the constant term, let's not try to factor; instead, use the quadratic formula.

$$n = \frac{-1 \pm \sqrt{1^2 - 4(1)(-506)}}{2(1)}$$

$$n = \frac{-1 \pm \sqrt{2025}}{2}$$

$$n = \frac{-1 \pm 45}{2} \qquad \text{From Table 9.1 on Page 339 we obtain } \sqrt{2025} = 45.$$

$$n = \frac{-1 + 45}{2} \quad \text{or} \quad n = \frac{-1 - 45}{2}$$

$$n = 22 \qquad \text{or} \quad n = -23.$$

Since we are looking for whole numbers, the solution -23 has to be discarded. Therefore, the whole numbers are 22 and 23. ■

PROBLEM 3   The perimeter of a rectangular lot is 100 meters and its area is 616 square meters. Find the length and width of the lot.

Solution   Let l represent the length. Let w represent the width.

Area is
616 m^2.
Perimeter is
100 m.

w

l

$$\begin{pmatrix} lw = 616 \\ 2l + 2w = 100. \end{pmatrix} \quad \longleftarrow \quad \text{Area is 616 m}^2.$$
$$\longleftarrow \quad \text{Perimeter is 100 m.}$$

Multiplying the second equation by $\dfrac{1}{2}$ produces $l + w = 50$, which can be changed to $l = 50 - w$. Substituting $50 - w$ for l in the first equation produces the quadratic equation

$$(50 - w)(w) = 616$$
$$50w - w^2 = 616$$
$$w^2 - 50w = -616.$$

Using the method of completing the square we obtain

$$w^2 - 50w + 625 = -616 + 625$$
$$(w - 25)^2 = 9$$
$$w - 25 = \pm 3$$
$$w - 25 = 3 \quad \text{or} \quad w - 25 = -3$$
$$w = 28 \quad \text{or} \quad w = 22.$$

If $w = 28$, then $l = 50 - w = 22$.
If $w = 22$, then $l = 50 - w = 28$.

The rectangle is 28 meters by 22 meters or 22 meters by 28 meters.   ■

PROBLEM 4 Find two numbers such that their sum is 2 and their product is -1.

Solution Let n represent one of the numbers. Let m represent the other number.

$$\begin{pmatrix} n + m = 2 \\ nm = -1 \end{pmatrix} \begin{matrix} \longleftarrow \text{ Their sum is 2.} \\ \longleftarrow \text{ Their product is } -1. \end{matrix}$$

The first equation can be changed to $m = 2 - n$; then $2 - n$ can be substituted for m in the second equation.

$$n(2 - n) = -1$$
$$2n - n^2 = -1$$
$$-n^2 + 2n + 1 = 0$$
$$n^2 - 2n - 1 = 0$$
$$n = \frac{-(-2) \pm \sqrt{(-2)^2 - 4(1)(-1)}}{2(1)}$$
$$n = \frac{2 \pm \sqrt{8}}{2} = \frac{2 \pm 2\sqrt{2}}{2} = 1 \pm \sqrt{2}.$$

If $n = 1 + \sqrt{2}$, then $m = 2 - (1 + \sqrt{2})$
$$= 2 - 1 - \sqrt{2}$$
$$= 1 - \sqrt{2}.$$

If $n = 1 - \sqrt{2}$, then $m = 2 - (1 - \sqrt{2})$
$$= 2 - 1 + \sqrt{2}$$
$$= 1 + \sqrt{2}.$$

The numbers are $1 + \sqrt{2}$ and $1 - \sqrt{2}$.   Perhaps you should check these numbers in the original statement of the problem!   ■

Finally, let's consider a uniform motion problem similar to those we solved in Chapter 7. However, we now have the flexibility of using two equations in two variables.

PROBLEM 5   Larry drove 156 miles in one hour more than it took Mike to drive 108 miles. Mike drove at an average rate of 2 miles per hour faster than Larry. How fast did each one travel?

Solution   We can represent the unknown rates and times in the following ways.

Let r represent Larry's rate.

Let t represent Larry's time.

Then $r + 2$ represents Mike's rate

and $t - 1$ represents Mike's time.

Since "distance equals rate times time," the following system can be set up.

$$\left(\begin{array}{l} rt = 156 \\ (r + 2)(t - 1) = 108. \end{array} \right)$$

Solving the first equation for r produces $r = \dfrac{156}{t}$. Substituting $\dfrac{156}{t}$ for r in the second equation and simplifying, we obtain

$$\left(\frac{156}{t} + 2 \right)(t - 1) = 108$$

$$156 - \frac{156}{t} + 2t - 2 = 108$$

$$2t - \frac{156}{t} + 154 = 108$$

$$2t - \frac{156}{t} + 46 = 0$$

$$2t^2 - 156 + 46t = 0$$

$$2t^2 + 46t - 156 = 0$$

$$t^2 + 23t - 78 = 0.$$

This quadratic equation can be solved by factoring.

$$(t + 26)(t - 3) = 0$$

$$t + 26 = 0 \qquad \text{or} \quad t - 3 = 0$$

$$t = -26 \quad \text{or} \qquad t = 3.$$

The negative solution must be disregarded. So, Mike's time is $3 - 1 = 2$ hours. Larry's rate is $\frac{156}{3} = 52$ miles per hour and Mike's rate is $\frac{108}{2} = 54$ miles per hour.　　　　　　　　　　　　　　　　　　　　　　　　　　　■

Problem Set 10.5

Solve each of the following problems.

1. Find two consecutive whole numbers whose product is 306.

2. Find two consecutive whole numbers whose product is 702.

3. The sum of two positive integers is 44 and their product is 475. Find the integers.

4. Two positive integers differ by 6. Their product is 616. Find the integers.

5. Find two numbers such that their sum is 6 and their product is 4.

6. Find two numbers such that their sum is 4 and their product is 1.

7. The sum of a number and its reciprocal is $\frac{3\sqrt{2}}{2}$. Find the number.

8. The sum of a number and its reciprocal is $\frac{73}{24}$. Find the number.

9. Each of three consecutive even whole numbers is squared. The three results are added and the sum is 596. Find the numbers.

10. Each of three consecutive whole numbers is squared. The three results are added and the sum is 245. Find the three whole numbers.

11. The sum of the square of a number and the square of one-half of the number is 80. Find the number.

12. The difference between the square of a positive number and the square of one-half the number is 243. Find the number.

13. Find the length and width of a rectangle if its length is 4 meters less than twice the width and the area of the rectangle is 96 square meters.

14. The length of a rectangular region is 4 centimeters longer than its width. The area of the region is 45 square centimeters. Find the length and width of the rectangle.

15. The perimeter of a rectangle is 80 centimeters and its area is 375 square centimeters. Find the length and width of the rectangle.

16. The perimeter of a rectangle is 132 yards and its area is 1080 square yards. Find the length and width of the rectangle.

17. The area of a tennis court is 2106 square feet. The length of the court is $\frac{26}{9}$ times the width. Find the length and width of a tennis court.

18. The area of a badminton court is 880 square feet. The length of the court is 2.2 times the width. Find the length and width of the court.

19. An auditorium contains 300 seats. There are 5 fewer rows than the number of seats per row. Find the number of rows and the number of seats per row.

20. Three hundred seventy-five trees were planted in an orchard. The number of trees per row was 10 more than the number of rows. How many rows of trees are in the orchard?

21. The area of a rectangular region is 63 square feet. If the length and width are each increased by 3 feet, the area is increased by 57 square feet. Find the length and width of the original rectangle.

22. The area of a circle is numerically equal to twice the circumference of the circle. Find the length of a radius of the circle.

23. The sum of the lengths of the two legs of a right triangle is 14 inches. If the length of the hypotenuse is 10 inches, find the length of each leg.

24. A page for a magazine contains 70 square inches of type. The height of a page is twice the width. If the margin around the type is to be 2 inches uniformly, what are the dimensions of the page?

25. A 5-by-7-inch picture is surrounded by a frame of uniform width. The area of the picture and frame together is 80 square inches. Find the width of the frame.

26. A rectangular piece of cardboard is 3 inches longer than it is wide. From each corner, a square piece 2 inches on a side is cut out. The flaps are then turned up to form an open box having a volume of 140 cubic inches. Find the length and width of the original piece of cardboard.

27. A class trip was to cost $3000. If there had been 10 more students, it would have cost each student $25 less. How many students took the trip?

28. Simon mowed some lawns and earned $40. It took him 3 hours longer than he anticipated and thus he earned $3 per hour less than he anticipated. How long did he expect it to take him?

29. A piece of wire 56 inches long is cut into two pieces and each piece is bent into the shape of a square. If the sum of the areas of the two squares is 100 square inches, find the length of each piece of wire.

30. By increasing the speed of a car by 10 miles per hour, it is possible to make a trip of 200 miles in 1 hour less time. What was the original speed for the trip?

31. On a 50-mile bicycle ride, Irene averaged 4 miles per hour faster for the first 36 miles than she did for last 14 miles. The entire trip of 50 miles took 3 hours. Find her rate for the first 36 miles.

32. One side of a triangle is 1 foot more than twice the length of the altitude to that side. If the area of the triangle is 18 square feet, find the length of a side and the length of the altitude to that side.

Chapter 10 Summary

(10.1)   A **quadratic equation** in the variable x is any equation that can be written in the form $ax^2 + bx + c = 0$, where a, b, and c are real numbers and $a \neq 0$.

Quadratic equations that are factorable using integers can be solved by factoring and applying the property "$ab = 0$ if and only if $a = 0$ or $b = 0$."

The property "$x^2 = a$ if and only if $x = \pm\sqrt{a}$" can be used to solve certain types of quadratic equations.

Many times equations of the form $x^2 = a$ are produced when working with the Pythagorean Theorem.

Don't forget: (1) in an isosceles right triangle, the lengths of the two legs are equal, and (2) in a $30° - 60°$ right triangle, the length of the leg opposite the $30°$ angle is one-half of the length of the hypotenuse.

(10.2)   You should be able to solve quadratic equations by the method of **completing the square**. To review this method, look back over the examples in Section 10.2.

(10.3)   The **quadratic formula** is usually stated as

$$x = \frac{-b \pm \sqrt{b^2 - 4ac}}{2a}.$$

It can be used to solve any quadratic equation after it has been written in the form $ax^2 + bx + c = 0$.

(10.4)   To review the strengths and weaknesses of the three basic methods for solving a quadratic equation (factoring, completing the square, the quadratic formula), go back over the examples in Section 10.4.

(10.5)   Our knowledge pertaining to systems of equations and quadratic equations provides us with a stronger basis for solving word problems.

Chapter 10 Review Problem Set

Solve each of the following quadratic equations.

1. $(2x + 7)^2 = 25$

2. $x^2 + 8x = -3$

3. $21x^2 - 13x + 2 = 0$

4. $x^2 = 17x$

5. $n - \dfrac{4}{n} = -3$

6. $n^2 - 26n + 165 = 0$

7. $3a^2 + 7a - 1 = 0$

8. $4x^2 - 4x + 1 = 0$

9. $5x^2 + 6x + 7 = 0$

10. $3x^2 + 18x + 15 = 0$

11. $3(x - 2)^2 - 2 = 4$

12. $x^2 + 4x - 14 = 0$

13. $y^2 = 45$

14. $x(x - 6) = 27$

15. $x^2 = x$

16. $n^2 - 4n - 3 = 6$

17. $n^2 - 44n + 480 = 0$

18. $\dfrac{x^2}{4} = x + 1$

19. $\dfrac{5x - 2}{3} = \dfrac{2}{x + 1}$

20. $\dfrac{-1}{3x - 1} = \dfrac{2x + 1}{-2}$

21. $\dfrac{5}{x - 3} + \dfrac{4}{x} = 6$

22. $\dfrac{1}{x + 2} - \dfrac{2}{x} = 3$

Set up an equation or a system of equations to help solve each of the following problems.

23. The perimeter of a rectangle is 42 inches and its area is 108 square inches. Find the length and width of the rectangle.

24. Find two consecutive whole numbers whose product is 342.

25. Each of three consecutive odd numbers is squared. The three results are added and the sum is 251. Find the numbers.

26. The combined area of two squares is 50 square meters. Each side of the larger square is three times as long as a side of the smaller square. Find the lengths of the sides of each square.

27. The difference of the lengths of the two legs of a right triangle is 2 yards. If the length of the hypotenuse is $2\sqrt{13}$ yards find the length of each leg.

28. Tony bought a number of shares of stock for a total of $720. A month later the value of the stock increased by $8 per share and he sold all but 20 shares and regained his original investment plus a profit of $80. How many shares did he sell and at what price per share?

29. A company has a rectangular parking lot 40 meters wide and 60 meters long. They plan to increase the area of the lot by 1100 square meters by adding a strip of equal width to one side and one end. Find the width of the strip to be added.

30. Jay traveled 225 miles in two hours less time than it took Jean to travel 336 miles. If Jay's rate was 3 miles per hour less than Jean's rate, find each rate.

31. The length of the hypotenuse of an isosceles right triangle is 12 inches. Find the length of each leg.

32. In a $30°-60°$ right triangle the side opposite the $60°$ angle is 8 centimeters long. Find the length of the hypotenuse.

Cumulative Review Problem Set

For Problems 1–6, evaluate each of the numerical expressions.

1. -2^7

2. $\left(\dfrac{1}{4}\right)^{-3}$

3. $\left(\dfrac{1}{3}-\dfrac{1}{4}\right)^{-2}$

4. $-\sqrt{64}$

5. $\sqrt{\dfrac{4}{9}}$

6. $3^0 + 3^{-1} + 3^{-2}$

For Problems 7–10, evaluate each algebraic expression for the given values of the variables.

7. $3(2x-1) - 4(2x+3) - (x+6)$ for $x = -4$

8. $(3x^2 - 4x - 6) - (3x^2 + 3x + 1)$ for $x = 6$

9. $2(a-b) - 3(2a+b) + 2(a-3b)$ for $a = -2$ and $b = 3$

10. $x^2 - 2xy + y^2$ for $x = 5$ and $y = -2$

For Problems 11–20, perform the indicated operations and express answers in simplest form.

11. $\dfrac{3}{4x} + \dfrac{5}{2x} - \dfrac{7}{x}$

12. $\dfrac{3}{x-2} - \dfrac{4}{x+3}$

13. $\dfrac{3x}{7y} \div \dfrac{6x}{35y^2}$

14. $\dfrac{x-2}{x^2+x-6} \cdot \dfrac{x^2+6x+9}{x^2-x-12}$

15. $\dfrac{7}{x^2+3x-18} - \dfrac{8}{x-3}$

16. $(-3xy)(-4y^2)(5x^3y)$

17. $(9x-2)(3x+4)$

18. $(-x-1)(5x+7)$

19. $(3x+1)(2x^2-x-4)$

20. $(10x^3-8x^2-17x-3) \div (5x+1)$

For Problems 21–26, factor each expression completely.

21. $12x^3+14x^2-40x$

22. $12x^2-27$

23. $xy+3x-2y-6$

24. $30+19x-5x^2$

25. $4x^4-4$

26. $21x^2+22x-8$

For Problems 27–38, change each radical expression to simplest radical form.

27. $4\sqrt{28}$

28. $-3\sqrt{48}$

29. $-\sqrt{45}$

30. $\sqrt{\dfrac{18}{49}}$

31. $\sqrt{\dfrac{36}{5}}$

32. $\dfrac{2\sqrt{3}}{3\sqrt{2}}$

33. $\dfrac{5\sqrt{8}}{6\sqrt{12}}$

34. $\sqrt{27x^3y^4}$

35. $\sqrt{72xy^5}$

36. $\dfrac{3\sqrt{x^3y}}{4\sqrt{xy}}$

37. $\dfrac{-2\sqrt{ab^2}}{5\sqrt{b}}$

38. $\dfrac{2}{3}\sqrt{40x^5}$

For Problems 39–43, find each product and express answers in simplest radical form.

39. $(3\sqrt{8})(4\sqrt{2})$

40. $(-2\sqrt{6})(5\sqrt{3})$

41. $6\sqrt{2}(9\sqrt{8}-3\sqrt{12})$

42. $(2\sqrt{3}+\sqrt{5})(\sqrt{3}-2\sqrt{5})$

43. $(3\sqrt{2}-\sqrt{7})(3\sqrt{2}+\sqrt{7})$

For Problems 44–46, rationalize the denominators and simplify.

44. $\dfrac{4}{\sqrt{3}+\sqrt{2}}$

45. $\dfrac{-6}{3\sqrt{5}-\sqrt{6}}$

46. $\dfrac{\sqrt{2}}{3\sqrt{3}+2\sqrt{2}}$

For Problems 47–50, simplify each of the radical expressions.

47. $3\sqrt{50}-7\sqrt{72}+4\sqrt{98}$

48. $4\sqrt{27}-2\sqrt{50}-6\sqrt{48}-3\sqrt{18}$

49. $\dfrac{2}{3}\sqrt{20}-\dfrac{3}{4}\sqrt{45}+\sqrt{80}$

50. $\dfrac{1}{2}\sqrt{75}-\dfrac{2}{3}\sqrt{27}+\dfrac{1}{4}\sqrt{12}$

For Problems 51–54, graph each of the equations.

51. $3x-6y=-6$

52. $y=-2x^2+1$

53. $y=-2x^3$

54. $y=-x$

For Problems 55–60, solve each of the problems.

55. Find the slope of the line determined by the points $(-3,6)$ and $(2,-4)$.

56. Find the slope of the line determined by the equation $4x-7y=12$.

57. Write the equation of the line that has a slope of $\dfrac{2}{3}$ and contains the point $(7, 2)$.

58. Write the equation of the line that contains the points $(-4, 1)$ and $(-1, -3)$.

59. Write the equation of the line having a slope of $-\dfrac{1}{4}$ and a y-intercept of -3.

60. Write the equation of the line that is perpendicular to the x-axis and contains the point $(4, -5)$.

For Problems 61–64, solve each of the systems by using either the substitution method or the addition method.

61. $\begin{pmatrix} y = 3x - 5 \\ 3x + 4y = -5 \end{pmatrix}$

62. $\begin{pmatrix} 4x - 3y = -20 \\ 3x + 5y = 14 \end{pmatrix}$

63. $\begin{pmatrix} \dfrac{1}{2}x - \dfrac{2}{3}y = -11 \\ \dfrac{1}{3}x + \dfrac{5}{6}y = 8 \end{pmatrix}$

64. $\begin{pmatrix} 2x + 7y = 22 \\ 4x - 5y = -13 \end{pmatrix}$

For Problems 65–86, solve each of the equations.

65. $-2(n - 1) + 4(2n - 3) = 4(n + 6)$

66. $\dfrac{4}{x - 1} = \dfrac{-1}{x + 6}$

67. $|2x - 6| = 3$

68. $\dfrac{t - 1}{3} - \dfrac{t + 2}{4} = -\dfrac{5}{12}$

69. $(3x - 1)^2 = 9$

70. $x^2 + 4x = 8$

71. $-7 - 2n - 6n = 7n - 5n + 12$

72. $\dfrac{n - 5}{2} = 3 - \dfrac{n + 4}{5}$

73. $.11x + .14(x + 400) = 181$

74. $(2t + 1)(t - 2) = 7$

75. $\dfrac{x}{60 - x} = 7 + \dfrac{4}{60 - x}$

76. $1 + \dfrac{x + 1}{2x} = \dfrac{3}{4}$

77. $\dfrac{4}{x - 2} - \dfrac{2x - 3}{x^2 - 4} = \dfrac{5}{x + 2}$

78. $n - \dfrac{5n}{n - 2} = \dfrac{-10}{n - 2}$

79. $\sqrt{3n} - 2 = 7$

80. $\sqrt{x + 4} = x + 4$

81. $4\sqrt{x} + 5 = x$

82. $(x + 1)^2 = 12$

83. $x^2 - 6x - 3 = 0$

84. $3n^2 - n - 3 = 0$

85. $2n^2 + 5n - 6 = 0$

86. $7 = 3t^2 - t$

For Problems 87–94, solve each of the inequalities.

87. $-3n - 4 \le 11$

88. $-5 > 3n - 4 - 7n$

89. $2(x - 2) + 3(x + 4) > 6$

90. $\dfrac{1}{2}n - \dfrac{2}{3}n < -1$

91. $\dfrac{x+1}{2} + \dfrac{x-2}{6} < \dfrac{3}{8}$

92. $\dfrac{x-3}{7} - \dfrac{x-2}{4} \le \dfrac{9}{14}$

93. $|3x - 2| < 7$

94. $|-2x + 4| > 2$

For Problems 95–110, set up an equation, an inequality, or a system of equations to help solve each problem.

95. If two angles are supplementary and the larger angle is 15 less than twice the smaller angle, find the measure of each angle.

96. The sum of two numbers is 50. If the larger number is 2 less than 3 times the smaller number, find the numbers.

97. Nick has 47 coins consisting of nickels, dimes, and quarters. The number of dimes is 1 more than twice the number of nickels and the number of quarters is 4 more than 3 times the number of nickels. Find the number of coins of each denomination.

98. If a home valued at $70,000 is assessed $1050 in real estate taxes, at the same rate how much are the taxes on a home assessed at $90,000?

99. A retailer has some skirts that cost her $30 each. She wants to sell them at a profit of 60% of the cost. What price should she charge for the skirts?

100. Rosa leaves a town traveling in her car at a rate of 45 miles per hour. One hour later, Polly leaves the same town traveling the same route at a rate of 55 miles per hour. How long will it take Polly to overtake Rosa?

101. How many milliliters of pure acid must be added to 100 milliliters of a 10% acid solution to obtain a 20% solution?

102. Andy has scores of 85, 90, and 86 on his first three algebra tests. What score must he make on the fourth test to have an average of 88 or better for the four tests?

103. The Cubs have won 70 games and lost 72 games. They have 20 more games to play. To win more than 50% of all their games, how many of the 20 games remaining must they win?

104. The area of a rectangle is twice the area of a square. If the rectangle is 16 inches long and the width of the rectangle is the same as the length of a side of the square, find the dimensions of both the rectangle and the square.

105. The length of a rectangle is one and one-half times its width. If the area of the rectangle is 54 square feet, find its length and width.

106. An apple orchard contains 112 trees. The number of trees per row is 6 more than the number of rows. Find the number of rows.

107. The sum of two numbers is 66. If the larger number is divided by the smaller, the quotient is 8 and the remainder is 3. Find the numbers.

108. Seth can do a job in 20 minutes. Butch can do the same job in 30 minutes. If they work together, how long will it take them to complete the job?

109. The perimeter of a rectangle is 36 meters and its area is 72 square meters. Find its length and width.

110. Doris traveled 180 miles in 2 hours less time than it took Ellen to travel 250 miles. Doris drove 10 miles per hour faster than Ellen. How fast did each one travel?

11 Additional Topics

In this chapter we will expand our knowledge of some topics discussed in earlier chapters and also consider some new topics that are studied in more detail in later mathematics courses. This should provide you with the opportunity to tie some ideas together and to get a glimpse of what lies ahead in intermediate algebra.

11.1 Complex Numbers

In Chapter 10 we found that some quadratic equations have no real number solutions. For example, the equation $x^2 = -4$ has no real number solutions because $2^2 = 4$ and $(-2)^2 = 4$. In this section we will consider a set of numbers that contains some numbers whose squares are negative real numbers. Then in the next section we shall see that this set of numbers, called the set of **complex numbers**, provides solutions for not only equations such as $x^2 = -4$, but also for any quadratic equation with real number coefficients in one variable.

Our work with complex numbers is based on the following definition.

Definition 11.1

The number i is such that

$$i = \sqrt{-1} \quad \text{and} \quad i^2 = -1.$$

The number i is not a real number and is often called the **imaginary unit**, but the number i^2 is the real number -1.

In Chapter 9 we used the property $\sqrt{a}\sqrt{b} = \sqrt{ab}$ to multiply radicals and to express radicals in simplest radical form. This property also holds if *only one of a or b* is negative. Thus, square root radicals containing negative numbers as radicands can be simplified as follows.

$$\sqrt{-4} = \sqrt{-1}\sqrt{4} = i(2); \quad \text{(usually written as } 2i\text{)}$$
$$\sqrt{-13} = \sqrt{-1}\sqrt{13} = i\sqrt{13};$$
$$\sqrt{-12} = \sqrt{-1}\sqrt{12} = i\sqrt{12} = i\sqrt{4}\sqrt{3} = 2i\sqrt{3};$$
$$\sqrt{-18} = \sqrt{-1}\sqrt{18} = i\sqrt{18} = i\sqrt{9}\sqrt{2} = 3i\sqrt{2}.$$

The imaginary unit i is used to define a complex number as follows.

Definition 11.2

A **complex number** is any number that can be expressed in the form

$a + bi$

where a and b are real numbers.

The form $a + bi$ is called the **standard form** of a complex number. The real number a is called the **real part** of the complex number and b is called the **imaginary part**. The following examples illustrate this terminology.

1. The number $3 + 4i$ is a complex number in standard form having a real part of 3 and an imaginary part of 4.
2. The number $-5 - 2i$ can be written in the standard form $-5 + (-2i)$ and is thus a complex number having a real part of -5 and an imaginary part of -2. (The form $-5 - 2i$ is often used, knowing that it means $-5 + (-2i)$.)
3. The number $-7i$ can be written in the standard form $0 + (-7i)$; thus it is a complex number having a real part of 0 and an imaginary part of -7.
4. The number 4 can be written in the standard form $4 + 0i$ and is thus a complex number having a real part of 4 and an imaginary part of 0.

Number 4 shows us that all real numbers can be considered complex numbers.

The commutative, associative, and distributive properties hold for all complex numbers and form the basis for manipulating with complexes. The following two statements describe addition and subtraction of complex numbers.

$$(a + bi) + (c + di) = (a + c) + (b + d)i; \qquad \text{(Addition)}$$
$$(a + bi) - (c + di) = (a - c) + (b - d)i. \qquad \text{(Subtraction)}$$

To add complex numbers, add their real parts and add their imaginary parts. To subtract complex numbers, subtract their real parts and subtract their imaginary parts. Consider the following examples.

1. $(3 + 5i) + (4 + 7i) = (3 + 4) + (5 + 7)i = 7 + 12i.$
2. $(-5 + 2i) + (7 - 8i) = (-5 + 7) + (2 + (-8))i$

$$= 2 + (-6i). \qquad \text{(usually written as } 2 - 6i)$$

3. $\left(\dfrac{1}{2} + \dfrac{3}{4}i\right) + \left(\dfrac{2}{3} + \dfrac{1}{5}i\right) = \left(\dfrac{1}{2} + \dfrac{2}{3}\right) + \left(\dfrac{3}{4} + \dfrac{1}{5}\right)i$

$$= \left(\dfrac{3}{6} + \dfrac{4}{6}\right) + \left(\dfrac{15}{20} + \dfrac{4}{20}\right)i$$

$$= \dfrac{7}{6} + \dfrac{19}{20}i.$$

4. $(8 + 9i) - (5 + 4i) = (8 - 5) + (9 - 4)i = 3 + 5i$
5. $(2 - 3i) - (5 + 9i) = (2 - 5) + (-3 - 9)i = -3 + (-12i)$

$$= -3 - 12i$$

6. $(-1 + 6i) - (-2 - i) = (-1 - (-2)) + (6 - (-1))i = 1 + 7i.$

■ Multiplying Complex Numbers

Since complex numbers have a *binomial form*, we find the product of two complex numbers in the same way as we find the product of two binomials. Consider the following examples.

EXAMPLE 1 Multiply $(3 + 2i)(4 + 5i)$.

Solution Multiplying each term of the first number times each term of the second number and simplifying yields

$$(3 + 2i)(4 + 5i) = 3(4) + 3(5i) + 2i(4) + 2i(5i)$$
$$= 12 + 15i + 8i + 10i^2$$
$$= 12 + (15 + 8)i + 10(-1) \qquad (i^2 = -1)$$
$$= 2 + 23i.$$

■

EXAMPLE 2 Multiply $(6 - 3i)(-5 + 2i)$.

Solution
$$(6 - 3i)(-5 + 2i) = 6(-5) + 6(2i) - (3i)(-5) - (3i)(2i)$$
$$= -30 + 12i + 15i - 6i^2$$
$$= -30 + 27i - 6(-1)$$
$$= -24 + 27i.$$

■

EXAMPLE 3 Find the indicated product $(5 - i)^2$.

Solution Remember that $(a - b)^2$ means $(a - b)(a - b)$.

$$(5 - i)^2 = (5 - i)(5 - i) = 5(5) - 5(i) - i(5) - (i)(-i)$$
$$= 25 - 5i - 5i + i^2$$
$$= 25 - 10i + (-1)$$
$$= 24 - 10i.$$

■

To find products such as $(4i)(3i)$ or $2i(4 + 6i)$, we could change each number to standard *binomial form* and then proceed as in the previous examples, but it is easier to handle them as follows.

$$(4i)(3i) = 4 \cdot 3 \cdot i \cdot i = 12i^2 = 12(-1) = -12;$$
$$2i(4 + 6i) = 2i(4) + 2i(6i)$$
$$= 8i + 12i^2$$
$$= 8i + 12(-1)$$
$$= -12 + 8i.$$

Problem Set 11.1

For Problems 1–12, write each radical in terms of i and simplify. For example, $\sqrt{-20} = i\sqrt{20} = i\sqrt{4}\sqrt{5} = 2i\sqrt{5}$.

1. $\sqrt{-64}$ **2.** $\sqrt{-81}$ **3.** $\sqrt{-\dfrac{25}{9}}$ **4.** $\sqrt{-\dfrac{49}{16}}$ **5.** $\sqrt{-11}$

6. $\sqrt{-17}$ **7.** $\sqrt{-50}$ **8.** $\sqrt{-32}$ **9.** $\sqrt{-48}$ **10.** $\sqrt{-45}$

11. $\sqrt{-54}$ **12.** $\sqrt{-28}$

For Problems 13–30, add or subtract the complex numbers as indicated.

13. $(3 + 8i) + (5 + 9i)$ **14.** $(7 + 10i) + (2 + 3i)$

15. $(7 - 6i) + (3 - 4i)$ **16.** $(8 - 2i) + (-7 + 3i)$

17. $(10 + 4i) - (6 + 2i)$ **18.** $(12 + 7i) - (5 + i)$

19. $(5 + 2i) - (7 + 8i)$ **20.** $(3 + i) - (7 + 4i)$

21. $(-2 - i) - (3 - 4i)$ **22.** $(-7 - 3i) - (8 - 9i)$

23. $(-4 - 7i) + (-8 - 9i)$ **24.** $(-1 - 2i) + (-6 - 6i)$

25. $(0 - 6i) + (-10 + 2i)$ **26.** $(4 - i) + (4 + i)$

27. $(-9 + 7i) - (-8 - 5i)$ **28.** $(-12 + 6i) - (-7 + 2i)$

29. $(-10 - 4i) - (10 + 4i)$ **30.** $(-8 - 2i) - (8 + 2i)$

For Problems 31–50, find each product and express it in the standard form of a complex number $(a + bi)$.

31. $(7i)(8i)$ **32.** $(-6i)(7i)$ **33.** $2i(6 + 3i)$

34. $3i(-4 + 9i)$ **35.** $-4i(-5 - 6i)$ **36.** $-5i(7 - 8i)$

37. $(2 + 3i)(5 + 4i)$ **38.** $(4 + 2i)(6 + 5i)$ **39.** $(7 - 3i)(8 + i)$

40. $(9 - 3i)(2 - 5i)$ **41.** $(-2 - 3i)(6 - 3i)$ **42.** $(-3 - 8i)(1 - i)$

43. $(-1 - 4i)(-2 - 7i)$ **44.** $(-3 - 7i)(-6 - 10i)$ **45.** $(4 + 5i)^2$

46. $(7 - 2i)^2$ **47.** $(5 - 6i)(5 + 6i)$ **48.** $(7 - 3i)(7 + 3i)$

49. $(-2 + i)(-2 - i)$ **50.** $(-5 - 8i)(-5 + 8i)$

11.2 Quadratic Equations: Complex Solutions

As stated in the previous section, the set of complex numbers provides solutions for all quadratic equations having real number coefficients. In other words, every quadratic equation of the form $ax^2 + bx + c = 0$, where a, b, and c are real numbers and $a \neq 0$, has a solution (or solutions) from the set of complex numbers.

To find solutions for quadratic equations we continue to use the techniques of factoring, completing the square, the quadratic formula, and the property "if $x^2 = a$, then $x = \pm\sqrt{a}$". Let's consider some examples.

EXAMPLE 1 Solve $x^2 = -4$.

Solution Using the property "if $x^2 = a$, then $x = \pm\sqrt{a}$" we can proceed as follows.

$$x^2 = -4$$
$$x = \pm\sqrt{-4}$$
$$x = \pm 2i. \qquad (\sqrt{-4} = \sqrt{-1}\sqrt{4} = 2i).$$

Check:

$x^2 = -4$	$x^2 = -4$
$(2i)^2 \overset{?}{=} -4$	$(-2i)^2 \overset{?}{=} -4$
$4i^2 \overset{?}{=} -4$	$4i^2 \overset{?}{=} -4$
$4(-1) \overset{?}{=} -4$	$4(-1) \overset{?}{=} -4$
$-4 = -4$	$-4 = -4$

The solution set is $\{-2i, 2i\}$. ■

EXAMPLE 2 Solve $(x - 2)^2 = -7$.

Solution
$$(x - 2)^2 = -7$$
$$x - 2 = \pm\sqrt{-7}$$
$$x - 2 = \pm i\sqrt{7} \qquad (\sqrt{-7} = \sqrt{-1}\sqrt{7} = i\sqrt{7})$$
$$x = 2 \pm i\sqrt{7}.$$

Check:

$(x - 2)^2 = -7$	$(x - 2)^2 = -7$
$(2 + i\sqrt{7} - 2)^2 \overset{?}{=} -7$	$(2 - i\sqrt{7} - 2)^2 \overset{?}{=} -7$
$(i\sqrt{7})^2 \overset{?}{=} -7$	$(-i\sqrt{7})^2 \overset{?}{=} -7$
$7i^2 \overset{?}{=} -7$	$7i^2 \overset{?}{=} -7$
$7(-1) \overset{?}{=} -7$	$7(-1) \overset{?}{=} -7$
$-7 = -7$	$-7 = -7$

The solution set is $\{2 - i\sqrt{7}, 2 + i\sqrt{7}\}$. ■

EXAMPLE 3 Solve $x^2 + 2x = -10$.

Solution The form of the equation lends itself to **completing the square** so let's proceed as follows.

$$x^2 + 2x = -10$$
$$x^2 + 2x + 1 = -10 + 1$$
$$(x + 1)^2 = -9$$
$$x + 1 = \pm 3i$$
$$x = -1 \pm 3i.$$

Check:

$$x^2 + 2x = -10 \qquad\qquad x^2 + 2x = -10$$
$$(-1 + 3i)^2 + 2(-1 + 3i) \overset{?}{=} -10 \qquad (-1 - 3i)^2 + 2(-1 - 3i) \overset{?}{=} -10$$
$$1 - 6i + 9i^2 - 2 + 6i \overset{?}{=} -10 \qquad 1 + 6i + 9i^2 - 2 - 6i \overset{?}{=} -10$$
$$-1 + 9i^2 \overset{?}{=} -10 \qquad\qquad -1 + 9i^2 \overset{?}{=} -10$$
$$-1 + 9(-1) \overset{?}{=} -10 \qquad\qquad -1 + 9(-1) \overset{?}{=} -10$$
$$-10 = -10 \qquad\qquad -10 = -10$$

The solution set is $\{-1 - 3i, -1 + 3i\}$. ■

EXAMPLE 4 Solve $x^2 - 2x + 2 = 0$.

Solution Using the quadratic formula, we obtain the following solutions.

$$x^2 - 2x + 2 = 0$$

$$x = \frac{2 \pm \sqrt{(-2)^2 - 4(1)(2)}}{2} \qquad \left(x = \frac{-b \pm \sqrt{b^2 - 4ac}}{2a} \right)$$

$$x = \frac{2 \pm \sqrt{4 - 8}}{2}$$

$$x = \frac{2 \pm \sqrt{-4}}{2}$$

$$x = \frac{2 \pm 2i}{2}$$

$$x = 1 \pm i.$$

Check:

$$x^2 - 2x + 2 = 0 \qquad\qquad x^2 - 2x + 2 = 0$$
$$(1 + i)^2 - 2(1 + i) + 2 \overset{?}{=} 0 \qquad (1 - i)^2 - 2(1 - i) + 2 \overset{?}{=} 0$$
$$1 + 2i + i^2 - 2 - 2i + 2 = 0 \qquad 1 - 2i + i^2 - 2 + 2i \overset{?}{=} 0$$
$$1 + i^2 \overset{?}{=} 0 \qquad\qquad 1 + i^2 \overset{?}{=} 0$$
$$1 - 1 \overset{?}{=} 0 \qquad\qquad 1 - 1 \overset{?}{=} 0$$
$$0 = 0 \qquad\qquad 0 = 0$$

The solution set is $\{1 - i, 1 + i\}$. ■

EXAMPLE 5 Solve $x^2 + 3x - 10 = 0$.

Solution We can factor $x^2 + 3x - 10$ and proceed as follows.

$$x^2 + 3x - 10 = 0$$
$$(x + 5)(x - 2) = 0$$

$$x + 5 = 0 \quad \text{or} \quad x - 2 = 0$$
$$x = -5 \quad \text{or} \quad x = 2.$$

The solution set is $\{-5, 2\}$. (Don't forget that all real numbers are complex numbers. That is, -5 can be written as $-5 + 0i$ and 2 as $2 + 0i$.) ■

To summarize our work with quadratic equations in Chapter 10 and this section, use the following approach to solving a quadratic equation.

1. If the equation is in a form where the property "if $x^2 = a$, then $x = \pm\sqrt{a}$" applies, use it. (See Examples 1 and 2.)

2. If the quadratic expression is obviously factorable using integers, factor it, and apply the property "if $ab = 0$, then $a = 0$ or $b = 0$." (See Example 5.)

3. If numbers 1 and 2 don't apply use either the quadratic formula or the process of completing the square. (See Examples 3 and 4.)

Problem Set 11.2

Solve each of the following quadratic equations and check your solutions.

1. $x^2 = -64$
2. $x^2 = -49$
3. $(x - 2)^2 = -1$
4. $(x + 3)^2 = -16$
5. $(x + 5)^2 = -13$
6. $(x - 7)^2 = -21$
7. $(x - 3)^2 = -18$
8. $(x + 4)^2 = -28$
9. $(5x - 1)^2 = 9$
10. $(7x + 3)^2 = 1$
11. $a^2 - 3a - 4 = 0$
12. $a^2 + 2a - 35 = 0$
13. $t^2 + 6t = -12$
14. $t^2 - 4t = -9$
15. $n^2 - 6n + 13 = 0$
16. $n^2 - 4n + 5 = 0$
17. $x^2 - 4x + 20 = 0$
18. $x^2 + 2x + 5 = 0$
19. $3x^2 - 2x + 1 = 0$
20. $2x^2 + x + 1 = 0$
21. $2x^2 - 3x - 5 = 0$
22. $3x^2 - 5x - 2 = 0$
23. $y^2 - 2y = -19$
24. $y^2 + 8y = -24$
25. $x^2 - 4x + 7 = 0$
26. $x^2 - 2x + 3 = 0$
27. $4x^2 - x + 2 = 0$
28. $5x^2 + 2x + 1 = 0$
29. $6x^2 + 2x + 1 = 0$
30. $7x^2 + 3x + 3 = 0$

11.3 Fractional Exponents

The beginning of Chapter 9 defined and discussed the concept of square root. For your convenience and for the sake of continuity, let's repeat a small portion of that material at this time.

To **square a number** means to raise it to the second power—that is, to use the number as a factor twice.

$$3^2 = 3 \cdot 3 = 9;$$
$$7^2 = 7 \cdot 7 = 49;$$
$$(-3)^2 = (-3)(-3) = 9;$$
$$(-7)^2 = (-7)(-7) = 49.$$

A **square root of a number** is one of its two equal factors. Thus, 3 is a square root of 9 because $3 \cdot 3 = 9$. Likewise, -3 is also a square root of 9 because $(-3)(-3) = 9$. The concept of square root is defined as follows.

Definition 11.3

a is a square root of b if $a^2 = b$.

The following generalizations are a direct consequence of Definition 11.3.

1. Every positive real number has two square roots; one is positive and the other is negative. They are opposites of each other.

2. Negative real numbers have no real number square roots. (This follows from Definition 11.3 since any nonzero real number is positive when squared.)

3. The square root of 0 is 0.

The symbol $\sqrt{}$, called **a radical sign**, is used to indicate the nonnegative square root. Consider the following examples.

$$\sqrt{49} = 7 \qquad (\sqrt{49} \text{ indicates the nonnegative or principal square root of 49.})$$
$$-\sqrt{49} = -7 \qquad (-\sqrt{49} \text{ indicates the negative square root of 49.})$$
$$\sqrt{0} = 0 \qquad \text{(Zero has only one square root.)}$$
$$\sqrt{-4} \qquad \text{is not a real number.}$$
$$-\sqrt{-4} \qquad \text{is not a real number.}$$

To **cube a number** means to raise it to the third power, that is, to use the number as a factor three times.

$$2^3 = 2 \cdot 2 \cdot 2 = 8;$$
$$4^3 = 4 \cdot 4 \cdot 4 = 64;$$
$$\left(\frac{2}{3}\right)^3 = \frac{2}{3} \cdot \frac{2}{3} \cdot \frac{2}{3} = \frac{8}{27};$$
$$(-2)^3 = (-2)(-2)(-2) = -8.$$

A **cube root of a number** is one of its three equal factors. Thus, -2 is a cube root of -8 because $(-2)(-2)(-2) = -8$. In general, the concept of cube root can be defined as follows.

Definition 11.4

a is a cube root of b if $a^3 = b$.

The following generalizations are a direct consequence of Definition 11.4.

1. Every positive real number has one positive real number cube root.
2. Every negative real number has one negative real number cube root.
3. The cube root of 0 is 0.

(Technically, every nonzero real number has three cube roots, but only one of them is a real number. The other two cube roots are complex numbers.)

The symbol $\sqrt[3]{}$, is used to designate the real number cube root. Thus, we can write

$$\sqrt[3]{8} = 2; \qquad \sqrt[3]{\frac{1}{27}} = \frac{1}{3}; \qquad \sqrt[3]{-8} = -2; \qquad \sqrt[3]{-\frac{1}{27}} = -\frac{1}{3}.$$

The concept of *root* can be extended to fourth roots, fifth roots, sixth roots, and in general, nth roots. The following generalizations can be made.

If n is an even positive integer, then the following statements are true.

1. Every positive real number has exactly two real nth roots, one positive and one negative. For example, the real fourth roots of 16 are 2 and -2. The symbol $\sqrt[4]{}$ is used to designate the positive root. Thus, we write $\sqrt[4]{16} = 2$.
2. Negative real numbers do not have real nth roots. For example, there are no real fourth roots of -16.

If n is an odd positive integer greater than one, then the following statements are true.

1. Every real number has exactly one real nth root and this root is designated by the symbol $\sqrt[n]{}$
2. The real nth root of a positive number is positive. For example, the fifth root of 32 is 2 and we write $\sqrt[5]{32} = 2$.
3. The nth root of a negative number is negative. For example, the fifth root of -32 is -2 and we write $\sqrt[5]{-32} = -2$.

■ The Merging of Exponents and Roots

In Section 5.6 we used the basic properties of positive integral exponents to motivate a definition for the use of zero and negative integers as exponents. Now let's use the properties of integral exponents to motivate definitions for the use of all rational numbers as exponents. This material is commonly referred to as "fractional exponents."

Consider the following comparisons.

From the meaning of root we know that

If $(b^n)^m = b^{mn}$ is to hold when n equals a rational number of the form $\frac{1}{p}$, where p is a positive integer greater than one, then

$$(\sqrt{5})^2 = 5;$$
$$(\sqrt[3]{8})^3 = 8;$$
$$(\sqrt[4]{21})^4 = 21.$$

$$\left(5^{\frac{1}{2}}\right)^2 = 5^{2\left(\frac{1}{2}\right)} = 5^1 = 5;$$
$$\left(8^{\frac{1}{3}}\right)^3 = 8^{3\left(\frac{1}{3}\right)} = 8^1 = 8;$$
$$\left(21^{\frac{1}{4}}\right)^4 = 21^{4\left(\frac{1}{4}\right)} = 21^1 = 21.$$

The following definition is motivated by such examples.

Definition 11.5

If b is a real number, n a positive integer greater than one, and $\sqrt[n]{b}$ exists, then

$$b^{\frac{1}{n}} = \sqrt[n]{b}.$$

Definition 11.5 states that $b^{\frac{1}{n}}$ means the nth root of b. The following examples illustrate this definition.

$$25^{\frac{1}{2}} = \sqrt{25} = 5; \qquad 16^{\frac{1}{4}} = \sqrt[4]{16} = 2;$$

$$8^{\frac{1}{3}} = \sqrt[3]{8} = 2; \qquad \left(\frac{36}{49}\right)^{\frac{1}{2}} = \sqrt{\frac{36}{49}} = \frac{6}{7};$$

$$(-27)^{\frac{1}{3}} = \sqrt[3]{-27} = -3; \qquad (-32)^{\frac{1}{5}} = \sqrt[5]{-32} = -2.$$

Now the following definition provides the basis for the use of *all* rational numbers as exponents.

> ## Definition 11.6
>
> If $\dfrac{m}{n}$ is a rational number, where n is a positive integer greater than one, and b is a real number such that $\sqrt[n]{b}$ exists, then
>
> $$b^{\frac{m}{n}} = \sqrt[n]{b^m} = (\sqrt[n]{b})^m.$$

Whether we use the form $\sqrt[n]{b^m}$ or $(\sqrt[n]{b})^m$ for computational purposes depends somewhat on the magnitude of the problem. Let's use both forms on two problems to illustrate this point.

$$8^{\frac{2}{3}} = \sqrt[3]{8^2} \quad \text{or} \quad 8^{\frac{2}{3}} = (\sqrt[3]{8})^2$$
$$= \sqrt[3]{64} \qquad\qquad = (2)^2$$
$$= 4. \qquad\qquad = 4.$$

$$27^{\frac{2}{3}} = \sqrt[3]{27^2} \quad \text{or} \quad 27^{\frac{2}{3}} = (\sqrt[3]{27})^2$$
$$= \sqrt[3]{729} \qquad\qquad = 3^2$$
$$= 9. \qquad\qquad = 9.$$

To compute $8^{\frac{2}{3}}$, either form seems to work about as well as the other one. However, to compute $27^{\frac{2}{3}}$, it should be obvious that $(\sqrt[3]{27})^2$ is much easier to handle than $\sqrt[3]{27^2}$.

Remember that in Section 5.6 we used the definition $b^{-n} = \dfrac{1}{b^n}$ as a basis for our work with negative integral exponents. This definition, extended to all rational numbers, and Definition 11.6 are used in the following examples.

$$25^{\frac{3}{2}} = (\sqrt{25})^3 = 5^3 = 125;$$
$$16^{\frac{3}{4}} = (\sqrt[4]{16})^3 = 2^3 = 8;$$
$$36^{-\frac{1}{2}} = \frac{1}{36^{\frac{1}{2}}} = \frac{1}{\sqrt{36}} = \frac{1}{6};$$
$$(-8)^{-\frac{1}{3}} = \frac{1}{(-8)^{\frac{1}{3}}} = \frac{1}{\sqrt[3]{-8}} = \frac{1}{-2} = -\frac{1}{2};$$
$$-8^{\frac{2}{3}} = -(\sqrt[3]{8})^2 = -(2)^2 = -4.$$

The basic properties of exponents discussed in Chapter 5 are true for all rational numbers. They provide the basis for simplifying algebraic expressions containing rational exponents, as the following examples illustrate. Our objective is to simplify and express the final result using only positive exponents.

1. $x^{\frac{1}{2}} \cdot x^{\frac{1}{3}} = x^{\frac{1}{2}+\frac{1}{3}} = x^{\frac{3}{6}+\frac{2}{6}} = x^{\frac{5}{6}}.$

2. $\left(2a^{\frac{3}{4}}\right)\left(3a^{\frac{1}{3}}\right) = 2 \cdot 3 \cdot a^{\frac{3}{4}} \cdot a^{\frac{1}{3}} = 6a^{\frac{3}{4}+\frac{1}{3}} = 6a^{\frac{9}{12}+\frac{4}{12}} = 6a^{\frac{13}{12}}.$

3. $\left(3n^{\frac{1}{2}}\right)\left(5n^{-\frac{1}{6}}\right) = 3 \cdot 5 \cdot n^{\frac{1}{2}} \cdot n^{-\frac{1}{6}}$

$$= 15n^{\frac{1}{2}+\left(-\frac{1}{6}\right)}$$

$$= 15n^{\frac{3}{6}-\frac{1}{6}}$$

$$= 15n^{\frac{2}{6}}$$

$$= 15n^{\frac{1}{3}}.$$

4. $\left(2x^{\frac{1}{2}}y^{\frac{1}{3}}\right)^2 = (2)^2\left(x^{\frac{1}{2}}\right)^2\left(y^{\frac{1}{3}}\right)^2$ $\qquad\qquad (b^n)^m = b^{mn}$

$$= 4xy^{\frac{2}{3}}.$$

5. $\dfrac{x^{\frac{1}{3}}}{x^{\frac{1}{2}}} = x^{\frac{1}{3}-\frac{1}{2}} = x^{\frac{2}{6}-\frac{3}{6}} = x^{-\frac{1}{6}} = \dfrac{1}{x^{\frac{1}{6}}}.$ $\qquad \dfrac{b^n}{b^m} = b^{n-m}$

6. $\dfrac{12x^{\frac{3}{4}}}{2x^{-\frac{1}{4}}} = \dfrac{12}{2}x^{\frac{3}{4}-\left(-\frac{1}{4}\right)} = 6x^{\frac{4}{4}} = 6x.$

7. $\left(\dfrac{2x^{\frac{1}{2}}}{3y^{\frac{2}{3}}}\right)^2 = \dfrac{(2)^2\left(x^{\frac{1}{2}}\right)^2}{(3)^2\left(y^{\frac{2}{3}}\right)^2} = \dfrac{4x}{9y^{\frac{4}{3}}}.$

8. $3^{\frac{1}{2}} \cdot 3^{\frac{1}{3}} = 3^{\frac{1}{2}+\frac{1}{3}} = 3^{\frac{3}{6}+\frac{2}{6}} = 3^{\frac{5}{6}}.$

9. $\dfrac{2^{\frac{3}{4}}}{2^{\frac{1}{4}}} = 2^{\frac{3}{4}-\frac{1}{4}} = 2^{\frac{2}{4}} = 2^{\frac{1}{2}}.$

Problem Set 11.3

Evaluate each of the following numerical expressions.

1. $\sqrt{81}$

2. $\sqrt{\dfrac{49}{4}}$

3. $-\sqrt{100}$

4. $-\sqrt{121}$

5. $\sqrt[3]{125}$

6. $\sqrt[3]{\dfrac{27}{8}}$

7. $\sqrt[3]{-64}$

8. $\sqrt[3]{-\dfrac{8}{27}}$

9. $\dfrac{\sqrt[3]{64}}{\sqrt{49}}$

10. $\dfrac{\sqrt{144}}{\sqrt[3]{8}}$

11. $\sqrt[4]{81}$

12. $\sqrt[4]{\dfrac{16}{81}}$

13. $\sqrt[5]{-243}$ **14.** $\sqrt[5]{1}$ **15.** $64^{\frac{1}{2}}$ **16.** $64^{\frac{1}{3}}$

17. $64^{\frac{2}{3}}$ **18.** $(-27)^{\frac{1}{3}}$ **19.** $(-64)^{\frac{2}{3}}$ **20.** $16^{\frac{3}{2}}$

21. $4^{\frac{5}{2}}$ **22.** $16^{\frac{1}{4}}$ **23.** $32^{-\frac{1}{5}}$ **24.** $-16^{\frac{1}{2}}$

25. $-27^{\frac{1}{3}}$ **26.** $8^{-\frac{2}{3}}$ **27.** $16^{-\frac{3}{4}}$ **28.** $\left(\dfrac{1}{2}\right)^{-2}$

29. $\left(\dfrac{2}{3}\right)^{-3}$ **30.** $\left(-\dfrac{1}{8}\right)^{-\frac{1}{3}}$ **31.** $\left(\dfrac{16}{64}\right)^{-\frac{1}{2}}$ **32.** $81^{\frac{3}{4}}$

33. $125^{\frac{4}{3}}$ **34.** $-16^{\frac{5}{2}}$ **35.** $-16^{\frac{5}{4}}$ **36.** $\left(\dfrac{1}{27}\right)^{-\frac{2}{3}}$

37. $\left(\dfrac{1}{32}\right)^{\frac{3}{5}}$ **38.** $(-8)^{\frac{4}{3}}$ **39.** $2^{\frac{1}{3}} \cdot 2^{\frac{2}{3}}$ **40.** $2^{\frac{3}{4}} \cdot 2^{\frac{5}{4}}$

41. $3^{\frac{4}{3}} \cdot 3^{\frac{5}{3}}$ **42.** $2^{\frac{3}{2}} \cdot 2^{-\frac{1}{2}}$ **43.** $\dfrac{2^{\frac{1}{2}}}{2^{\frac{1}{2}}}$ **44.** $\dfrac{3^{\frac{1}{3}}}{3^{-\frac{2}{3}}}$

45. $\dfrac{3^{-\frac{2}{3}}}{3^{\frac{1}{3}}}$ **46.** $\dfrac{2^{\frac{3}{4}}}{2^{-\frac{1}{4}}}$ **47.** $\dfrac{2^{\frac{9}{4}}}{2^{\frac{1}{4}}}$ **48.** $\dfrac{3^{-\frac{1}{2}}}{3^{-\frac{1}{2}}}$

49. $\dfrac{7^{\frac{4}{3}}}{7^{-\frac{2}{3}}}$ **50.** $\dfrac{5^{\frac{3}{5}}}{5^{-\frac{7}{5}}}$

Simplify each of the following and express final results using positive exponents only. For example,

$$\left(2x^{\frac{1}{3}}\right)\left(3x^{\frac{1}{2}}\right) = 6x^{\frac{5}{6}}.$$

51. $x^{\frac{1}{2}} \cdot x^{\frac{1}{4}}$ **52.** $x^{\frac{1}{3}} \cdot x^{\frac{1}{4}}$ **53.** $a^{\frac{2}{3}} \cdot a^{\frac{3}{4}}$

54. $a^{\frac{1}{3}} \cdot a^{-\frac{1}{4}}$ **55.** $\left(3x^{\frac{1}{4}}\right)\left(5x^{\frac{1}{3}}\right)$ **56.** $\left(2x^{\frac{1}{2}}\right)\left(5x^{\frac{1}{3}}\right)$

57. $\left(4x^{\frac{2}{3}}\right)\left(6x^{\frac{1}{4}}\right)$ **58.** $\left(3x^{\frac{2}{5}}\right)\left(6x^{\frac{1}{4}}\right)$ **59.** $\left(2y^{\frac{2}{3}}\right)\left(y^{-\frac{1}{4}}\right)$

60. $\left(y^{-\frac{1}{3}}\right)\left(4y^{\frac{2}{5}}\right)$ **61.** $\left(5n^{\frac{3}{4}}\right)\left(2n^{-\frac{1}{2}}\right)$ **62.** $\left(7n^{-\frac{1}{3}}\right)\left(8n^{\frac{5}{6}}\right)$

63. $\left(2x^{\frac{1}{3}}\right)\left(x^{-\frac{1}{2}}\right)$ **64.** $\left(x^{\frac{2}{5}}\right)\left(3x^{-\frac{1}{2}}\right)$ **65.** $\left(5x^{\frac{1}{2}}y\right)^{2}$

66. $\left(2x^{\frac{1}{3}}y^{2}\right)^{3}$ **67.** $\left(4x^{\frac{1}{4}}y^{\frac{1}{2}}\right)^{3}$ **68.** $(9x^{2}y^{4})^{\frac{1}{2}}$

69. $(8x^{6}y^{3})^{\frac{1}{3}}$ **70.** $\dfrac{18x^{\frac{1}{2}}}{9x^{\frac{1}{3}}}$ **71.** $\dfrac{24x^{\frac{3}{5}}}{6x^{\frac{1}{3}}}$

72. $\dfrac{56a^{\frac{1}{6}}}{7a^{\frac{1}{4}}}$ **73.** $\dfrac{48b^{\frac{1}{3}}}{12b^{\frac{3}{4}}}$ **74.** $\dfrac{16n^{\frac{1}{3}}}{8n^{-\frac{2}{3}}}$

75. $\dfrac{27n^{-\frac{1}{3}}}{9n^{-\frac{1}{3}}}$

76. $\dfrac{5x^{\frac{2}{5}}}{3x^{-\frac{1}{3}}}$

77. $\left(\dfrac{3x^{\frac{1}{3}}}{2x^{\frac{1}{2}}}\right)^{2}$

78. $\left(\dfrac{2a^{\frac{1}{2}}}{3a^{\frac{1}{4}}}\right)^{3}$

79. $\left(\dfrac{5x^{\frac{1}{2}}}{6y^{\frac{1}{3}}}\right)^{3}$

80. $\left(\dfrac{4y^{\frac{2}{5}}}{3x^{\frac{1}{3}}}\right)^{2}$

11.4 Relations and Functions

In the last two sections of this chapter we shall work with a concept that plays an important role throughout much of mathematics, namely, the concept of a **function**. Functions are used to unify mathematics and also to serve as a meaningful way of applying mathematics to many real world problems. They provide us with a means of studying quantities that vary with one another, that is, when a change in one quantity causes a corresponding change in another. We shall consider the general concept of a function in this section and then deal with two special kinds of functions in the last section.

Mathematically, a function is a special kind of **relation**. Thus, we will begin with a simple definition for a relation.

Definition 11.7

A **relation** is a set of ordered pairs of real numbers.

The ordered pairs of a relation may be generated by various means, such as a graph or chart. However, the most common way of generating ordered pairs is by the use of equations. Thus, we say that equations are used to describe relations. Each of the following equations describes a relation between the variables x and y, and we have listed some of the ordered pairs of that relation.

1. $x^2 + y^2 = 4$: $(1, \sqrt{3}), (1, -\sqrt{3}), (0, 2), (0, -2), \ldots$

2. $x = y^2$: $(0, 0), (4, 2), (4, -2), (9, 3), (9, -3), \ldots$

3. $y - 2x = 3$: $(0, 3), (1, 5), (-1, 1), (2, 7), (3, 9), \ldots$

4. $y = \dfrac{1}{x - 2}$: $\left(0, -\dfrac{1}{2}\right), (1, -1), (3, 1), \left(-2, -\dfrac{1}{4}\right), \left(5, \dfrac{1}{3}\right), \ldots$

5. $y = x^2$: $(0, 0), (1, 1), (-1, 1), (2, 4), (-2, 4), \ldots$

Now direct your attention to the last three equations above and the ordered pairs associated with each equation. Note that for each equation no two ordered pairs have the same first coordinate. Such a set of ordered pairs is said to be a **function**.

> **Definition 11.8**
>
> A **function** is a relation where no two ordered pairs have the same first coordinate.

The **domain** of a function is the set of all first coordinates and the **range** is the set of all second coordinates. Usually the domain and range of a function can be determined from the equation used to describe the function. For example, the domain of the function described by $y = 2x + 3$ is the set of all real numbers since any real number can be substituted for x. The range is also all real numbers since y will take on all real number values. The domain of the function described by $y = x^2$ is the set of all real numbers, but the range is only the nonnegative real numbers because squaring a number always produces a nonnegative result.

The domain of a function is frequently of more concern than the range. We should be aware of any necessary restrictions on x. Consider the following examples.

EXAMPLE 1 Specify the domain for each of the following.

(a) $y = \dfrac{1}{x - 2}$ (b) $y = \dfrac{1}{x^2 - 9}$ (c) $y = \sqrt{x - 1}$

Solution

(a) We can replace x with any real number except 2, since 2 makes the denominator 0. Thus, the domain is all real numbers except 2.

(b) We want to eliminate any value of x that will make the denominator 0. Thus, we solve the equation $x^2 - 9 = 0$.

$$x^2 - 9 = 0$$
$$(x + 3)(x - 3) = 0$$
$$x + 3 = 0 \quad \text{or} \quad x - 3 = 0$$
$$x = -3 \quad \text{or} \quad x = 3.$$

The domain is the set of all real numbers except -3 and 3.

(c) Because of the radical sign, the radicand, $x - 1$, must be nonnegative.

$$x - 1 \geq 0$$
$$x \geq 1.$$

The domain is the set of all real numbers greater than or equal to 1. ■

■ Functional Notation

Thus far we have been using the regular notation for writing equations to describe functions; that is, equations such as $y = x^2$ and $y = \dfrac{1}{x - 2}$, where y is expressed in terms of x, have been used to specify certain functions. There is a special **functional notation** that is very convenient to use when working with the function concept.

The notation $f(x)$ is read "f of x" and is defined to be the value of the function f at x. (Do not interpret $f(x)$ to mean f times x!) Instead of writing $y = x^2$ we can write $f(x) = x^2$. Therefore, $f(2)$ means the value of the function f at 2, which is $2^2 = 4$. So we write $f(2) = 4$. It makes a convenient way of expressing various values of the function. Let's illustrate that idea with another example.

EXAMPLE 2 If $f(x) = x^2 - 6$, find $f(0)$, $f(1)$, $f(2)$, $f(3)$, $f(-1)$, and $f(h)$.

Solution
$$f(x) = x^2 - 6.$$
$$f(0) = 0^2 - 6 = -6.$$
$$f(1) = 1^2 - 6 = -5.$$
$$f(2) = 2^2 - 6 = -2.$$
$$f(3) = 3^2 - 6 = 3.$$
$$f(-1) = (-1)^2 - 6 = -5.$$
$$f(h) = h^2 - 6.$$
■

When we are working with more than one function in the same problem, we use different letters to designate the different functions, as the next example illustrates.

EXAMPLE 3 If $f(x) = 2x + 5$ and $g(x) = x^2 - 2x + 1$, find $f(2)$, $f(-3)$, $g(-1)$, and $g(4)$.

Solution

$$f(x) = 2x + 5. \qquad\qquad g(x) = x^2 - 2x + 1.$$
$$f(2) = 2(2) + 5 = 9. \qquad\qquad g(-1) = (-1)^2 - 2(-1) + 1 = 4.$$
$$f(-3) = 2(-3) + 5 = -1. \qquad\qquad g(4) = 4^2 - 2(4) + 1 = 9.$$
■

Problem Set 11.4

Determine the domain and range for each of the following equations.

1. $y = x + 2$ 2. $y = 3x - 7$ 3. $y = x^2 + 2$ 4. $y = x^2 - 1$

5. $y = x^3$ 6. $y = x^4$ 7. $y = \sqrt{x}$ 8. $y = \sqrt{x - 3}$

Determine the domain for each of the following functions.

9. $y = x^2 + 2x - 1$ 10. $y = x^2 - 3x - 1$ 11. $y = \sqrt{x + 4}$

12. $y = \sqrt{2x - 3}$ 13. $f(x) = 2x + 7$ 14. $f(x) = x^3 + 2$

15. $f(x) = \dfrac{2}{x - 3}$ 16. $f(x) = \dfrac{3x}{2x - 1}$ 17. $y = \dfrac{3x}{x^2 - 4}$

18. $y = \dfrac{5}{x^2 - 25}$

19. $f(x) = \dfrac{-2}{x^2 + 4}$

20. $f(x) = \dfrac{3x}{x^2 + 9}$

21. $f(x) = \dfrac{4}{(x + 2)(x - 3)}$

22. $f(x) = \dfrac{5}{(2x - 3)(x - 1)}$

23. $f(x) = \dfrac{1}{x^2 - 5x + 6}$

24. $f(x) = \dfrac{2}{x^2 + 7x + 12}$

25. $y = \dfrac{-2}{x^2 + 4x}$

26. $y = \dfrac{-5}{x^2 - 7x}$

27. If $f(x) = 3x + 4$, find $f(0)$, $f(1)$, $f(-1)$, and $f(6)$.

28. If $f(x) = -2x + 5$, find $f(2)$, $f(-2)$, $f(-3)$, and $f(5)$.

29. If $f(x) = -5x - 1$, find $f(3)$, $f(-4)$, $f(-5)$, and $f(t)$.

30. If $f(x) = 7x - 3$, find $f(-1)$, $f(0)$, $f(4)$, and $f(a)$.

31. If $g(x) = \dfrac{2}{3}x + \dfrac{3}{4}$, find $g(3)$, $g\left(\dfrac{1}{2}\right)$, $g\left(-\dfrac{1}{3}\right)$, and $g(-2)$.

32. If $g(x) = -\dfrac{1}{2}x + \dfrac{5}{6}$, find $g(1)$, $g(-1)$, $g\left(\dfrac{2}{3}\right)$, and $g\left(-\dfrac{1}{3}\right)$.

33. If $f(x) = x^2 - 4$, find $f(2)$, $f(-2)$, $f(7)$, and $f(0)$.

34. If $f(x) = 2x^2 + x - 1$, find $f(2)$, $f(-3)$, $f(4)$, and $f(-1)$.

35. If $f(x) = -x^2 + 1$, find $f(-1)$, $f(2)$, $f(-2)$, and $f(-3)$.

36. If $f(x) = -2x^2 - 3x - 1$, find $f(1)$, $f(0)$, $f(-1)$, and $f(-2)$.

37. If $f(x) = \sqrt{x}$, find $f(4)$, $f(25)$, $f(12)$, and $f(18)$.

38. If $f(x) = \sqrt{x - 1}$, find $f(10)$, $f(17)$, $f(25)$, and $f(45)$.

39. If $f(x) = 4x + 3$ and $g(x) = x^2 - 2x$, find $f(5)$, $f(-6)$, $g(-1)$, and $g(4)$.

40. If $f(x) = -2x - 7$ and $g(x) = 2x^2 + 1$, find $f(-2)$, $f(4)$, $g(-2)$, and $g(4)$.

41. If $f(x) = 3x^2 - x + 4$ and $g(x) = -3x + 5$, find $f(-1)$, $f(4)$, $g(-1)$, and $g(4)$.

42. If $f(x) = -2x^2 + 3x + 1$ and $g(x) = -6x$, find $f(3)$, $f(-5)$, $g(2)$, and $g(-4)$.

11.5 Special Functions

As you continue to study mathematics you will frequently encounter the concept of a function. Functions are a part of almost all branches of mathematics. Basically, the study of different functions centers around (1) the equations that describe them, (2) the graphs they produce, and (3) the real world problems they help to solve. In this section we shall consider three special types of functions: constant functions, linear functions, and quadratic functions.

Any function that can be written in the form

$$f(x) = c,$$

where c is a real number, is called a **constant function**. The following are examples of constant functions.

$$f(x) = 2; \qquad f(x) = -\frac{1}{3}; \qquad f(x) = 0.$$

Graphing constant functions is quite easy because the graph of every constant function is a horizontal line.

EXAMPLE 1 Graph $f(x) = 2$.

Solution   The function $f(x) = 2$ means that the value of 2 is assigned to every value of x. Thus,

$$f(-3) = 2, \qquad f(0) = 2, \qquad f(2) = 2, \qquad \text{and so on.}$$

Therefore, the points $(-3, 2)$, $(0, 2)$, $(2, 2)$, and so on, are all points of the graph, which is the horizontal line in Figure 11.1.   ■

Figure 11.1

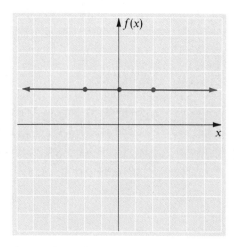

Note that in Figure 11.1 we labeled the vertical axis $f(x)$. It could also be labeled y since $f(x) = 2$ and $y = 2$ mean the same thing. We will continue to use $f(x)$ merely to help you adjust to the functional notation.

■ Linear Function

Any function that can be written in the form

$$f(x) = ax + b,$$

where a and b are real numbers, $a \neq 0$, is called a **linear function**. The following are examples of linear functions.

$$f(x) = 2x + 5;$$
$$f(x) = -3x + 6;$$
$$f(x) = -\frac{1}{2}x + \frac{3}{4}.$$

Graphing linear functions is also quite easy since the graph of every linear function is a straight line.

EXAMPLE 2 Graph $f(x) = 2x + 1$.

Solution
$$f(0) = 2(0) + 1 = 1$$
$$f(2) = 2(2) + 1 = 5$$
$$f(-1) = 2(-1) + 1 = -1.$$

The points $(0, 1), (2, 5)$, and $(-1, -1)$ are on the graph and the line can be drawn as in Figure 11.2. ■

Figure 11.2

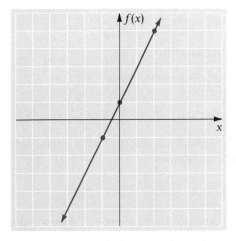

In Section 8.4, $y = mx + b$ was referred to as the slope-intercept form of the equation of a straight line. In this section we have used the equation $f(x) = ax + b$ to describe linear functions. In both equations, the coefficient of x is the slope of the line and the constant term is the y-intercept. This fact can be used to help graph linear functions.

EXAMPLE 3 Graph $f(x) = -3x - 2$.

Solution The y-intercept is -2; therefore, the point $(0, -2)$ is on the graph. Since the slope is $-3\left(-3 = \dfrac{-3}{1}\right)$, we can locate another point by moving 3 units down and 1 unit to the right of $(0, -2)$. This point is $(1, -5)$. The line can be drawn as in Figure 11.3. ■

Figure 11.3

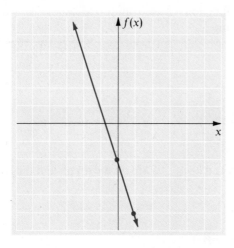

■ **Quadratic Functions**

Any function that can be written in the form

$$f(x) = ax^2 + bx + c,$$

where a, b, and c are real numbers, $a \neq 0$, is called a **quadratic function**. The following are examples of quadratic functions.

$$f(x) = 2x^2; \qquad f(x) = x^2 + 5x - 7; \qquad f(x) = -3x^2 - 2.$$

Graphing all of the variations of quadratic functions requires considerably more effort than did graphing either the constant or linear functions. We shall not consider all such variations in this text but will use a few examples to illustrate some of the basic characteristics of quadratic functions. Let's begin with the following example.

EXAMPLE 4 Graph $f(x) = x^2$.

Solution We can generate some points of the graph by finding some functional values.

$$f(0) = 0^2 = 0 \longrightarrow (0, 0).$$
$$f(1) = 1^2 = 1 \longrightarrow (1, 1).$$
$$f(2) = 2^2 = 4 \longrightarrow (2, 4).$$

$$f(-1) = (-1)^2 = 1 \longrightarrow (-1, 1).$$
$$f(-2) = (-2)^2 = 4 \longrightarrow (-2, 4).$$

Plotting these points and joining them with a smooth curve produces the graph, called a **parabola**, in Figure 11.4. ■

Figure 11.4

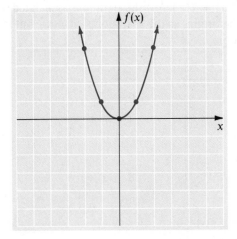

The graph of every quadratic function is a parabola. As we work with parabolas, the vocabulary indicated in Figure 11.5 will be helpful.

Figure 11.5

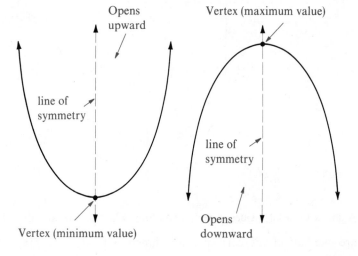

The graphing of parabolas can be conveniently explained by studying various forms of equations used to describe quadratic functions. We shall only begin this process in this text by studying equations of the form $f(x) = ax^2$, where $a \neq 0$. Consider the following examples.

EXAMPLE 5 Graph $f(x) = 2x^2$.

Solution Let's compare the graph of $f(x) = 2x^2$ to the graph of $f(x) = x^2$ by setting up a table of values. The functional values of $f(x) = 2x^2$ are twice the corresponding functional values of $f(x) = x^2$. Thus, the parabola associated with $f(x) = 2x^2$ has the *same vertex*, the origin, as the graph of $f(x) = x^2$, but is "narrower," as indicated in Figure 11.6.

x	$f(x) = x^2$	$f(x) = 2x^2$
0	0	0
1	1	2
2	4	8
−1	1	2
−2	4	8

Figure 11.6

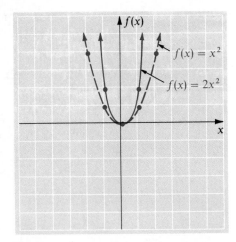

EXAMPLE 6 Graph $f(x) = \dfrac{1}{2}x^2$.

Solution Again a table of values is very revealing. The functional values of $f(x) = \dfrac{1}{2}x^2$ are one-half of the corresponding functional values of $f(x) = x^2$. Thus, the parabola associated with $f(x) = \dfrac{1}{2}x^2$ has the same vertex as the graph of $f(x) = x^2$, but is "wider," as indicated in Figure 11.7.

x	$f(x) = x^2$	$f(x) = \dfrac{1}{2}x^2$
0	0	0
1	1	$\dfrac{1}{2}$
2	4	2
-1	1	$\dfrac{1}{2}$
-2	4	2

Figure 11.7

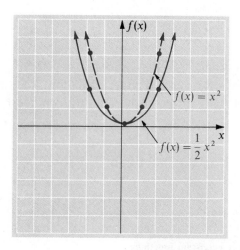

EXAMPLE 7 Graph $f(x) = -x^2$.

Solution Consider the following table of values. Functional values of $f(x) = -x^2$ are opposites of corresponding functional values of $f(x) = x^2$. Therefore, the graph of $f(x) = -x^2$ is a reflection across the x-axis of the graph of $f(x) = x^2$, as shown in Figure 11.8.

x	$f(x) = x^2$	$f(x) = -x^2$
0	0	0
1	1	-1
2	4	-4
-1	1	-1
-2	4	-4

Figure 11.8

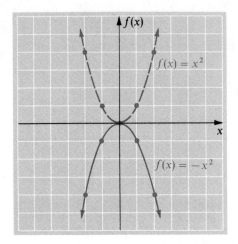

In general, the graph of a quadratic function of the form $f(x) = ax^2$ has its vertex at the origin and opens upward if a is positive and downward if a is negative. The parabola is "narrower" than the parabola associated with $f(x) = x^2$ if $|a| > 1$ and "wider" if $|a| < 1$.

Problem Set 11.5

Graph each of the following constant functions.

1. $f(x) = 4$ **2.** $f(x) = 1$ **3.** $f(x) = -2$

4. $f(x) = -3$ **5.** $f(x) = \dfrac{1}{2}$ **6.** $f(x) = 0$

Graph each of the following linear functions.

7. $f(x) = 3x + 2$ **8.** $f(x) = 4x + 5$ **9.** $f(x) = 2x - 5$

10. $f(x) = 3x - 7$ **11.** $f(x) = -4x + 1$ **12.** $f(x) = -2x + 3$

13. $f(x) = -5x - 1$ **14.** $f(x) = -x - 2$ **15.** $f(x) = 3x$

16. $f(x) = x$ **17.** $f(x) = -x$ **18.** $f(x) = -4x$

Graph each of the following quadratic functions.

19. $f(x) = 3x^2$ **20.** $f(x) = \dfrac{1}{2}x^2$ **21.** $f(x) = \dfrac{1}{4}x^2$

22. $f(x) = -2x^2$ **23.** $f(x) = -3x^2$ **24.** $f(x) = -\dfrac{1}{2}x^2$

25. $f(x) = -\dfrac{1}{4}x^2$

Chapter 11 Summary

(11.1) A **complex number** is any number that can be expressed in the form $a + bi$, where a and b are real numbers and $i = \sqrt{-1}$.

The property, $\sqrt{ab} = \sqrt{a}\sqrt{b}$, holds if only one of a or b is negative and therefore is used as a basis for simplifying radicals containing negative radicands.

Addition and **subtraction** of complex numbers are described as follows.

$$(a + bi) + (c + di) = (a + c) + (b + d)i;$$
$$(a + bi) - (c + di) = (a - c) + (b - d)i.$$

We find the **product** of two complex numbers in the same way as we find the product of two binomials.

(11.2) Every quadratic equation of the form $ax^2 + bx + c = 0$, where $a, b,$ and c are real numbers and $a \neq 0$, has a solution (solutions) in the set of complex numbers.

(11.3) The following definitions merge the concepts of *root* and *fractional exponents.*

1. a is a square root of b if $a^2 = b$.
2. a is a cube root of b if $a^3 = b$.
3. a is an nth root of b if $a^n = b$.
4. $b^{\frac{1}{n}}$ means $\sqrt[n]{b}$.
5. $b^{\frac{m}{n}}$ means $\sqrt[n]{b^m}$, which equals $(\sqrt[n]{b})^m$.

(11.4) A **relation** is a set of ordered pairs. A **function** is a relation where no two ordered pairs have the same first coordinate.

The notation $f(x)$ is read "f of x" and is defined to be the value of the function f at x. Therefore, if $f(x) = 2x + 3$, then $f(1) = 2(1) + 3 = 5$.

(11.5) Any function that can be written in the form $f(x) = c$, where c is a real number, is called a **constant function**. The graph of every constant function is a **horizontal line**.

Any function that can be written in the form $f(x) = ax + b$, where a and b are real numbers and $a \neq 0$, is called a **linear function**. The graph of every linear function is a **straight line**.

Any function that can be written in the form $f(x) = ax^2 + bx + c$, where a, b, and c are real numbers and $a \neq 0$, is a **quadratic function**. The graph of every quadratic function is a **parabola**. The graph of every quadratic function of the form $f(x) = ax^2$ is a parabola with its vertex at the origin and symmetrical about the y-axis.

Chapter 11 Review Problem Set

Write each of the following in terms of i and simplify as much as possible.

1. $\sqrt{-64}$ **2.** $\sqrt{-29}$ **3.** $\sqrt{-54}$

4. $\sqrt{-\dfrac{9}{4}}$ **5.** $\sqrt{-108}$ **6.** $\sqrt{-96}$

Perform the following indicated operations on complex numbers.

7. $(5 - 7i) + (-4 + 9i)$ **8.** $(-3 + 2i) + (-4 - 7i)$

9. $(6 - 9i) - (4 - 5i)$ **10.** $(-5 + 3i) - (-8 + 7i)$

11. $(7 - 2i) - (6 - 4i) + (-2 + i)$ **12.** $(-4 + i) - (-4 - i) - (6 - 8i)$

13. $(2 + 5i)(3 + 8i)$ **14.** $(4 - 3i)(1 - 2i)$

15. $(-1 + i)(-2 + 6i)$ **16.** $(-3 - 3i)(7 + 8i)$

17. $(2 + 9i)(2 - 9i)$ **18.** $(-3 + 7i)(-3 - 7i)$

19. $(-3 - 8i)(3 + 8i)$ **20.** $(6 + 9i)(-1 - i)$

Solve each of the following quadratic equations.

21. $(x - 6)^2 = -25$ **22.** $n^2 + 2n = -7$ **23.** $x^2 - 2x + 17 = 0$

24. $x^2 - x + 7 = 0$ **25.** $2x^2 - x + 3 = 0$ **26.** $6x^2 - 11x + 3 = 0$

27. $-x^2 + 5x - 7 = 0$ **28.** $-2x^2 - 3x - 6 = 0$ **29.** $3x^2 + x + 5 = 0$

30. $x(4x + 1) = -3$

Evaluate each of the following numerical expressions.

31. $\sqrt{\dfrac{64}{36}}$ **32.** $-\sqrt{1}$ **33.** $\sqrt[3]{\dfrac{27}{64}}$ **34.** $\sqrt{-125}$ **35.** $\sqrt[4]{\dfrac{81}{16}}$

36. $25^{\frac{3}{2}}$ **37.** $8^{\frac{5}{3}}$ **38.** $(-8)^{\frac{5}{3}}$ **39.** 4^{-2} **40.** $4^{-\frac{1}{2}}$

41. $(32)^{-\frac{2}{5}}$ **42.** $\left(\dfrac{2}{3}\right)^{-1}$ **43.** $2^{\frac{7}{4}} \cdot 2^{\frac{5}{4}}$ **44.** $3^{\frac{1}{3}} \cdot 3^{\frac{5}{3}}$ **45.** $\dfrac{3^{\frac{1}{3}}}{3^{\frac{4}{3}}}$

For Problems 46–53, simplify and express final results with positive exponents only.

46. $x^{\frac{5}{6}} \cdot x^{\frac{5}{6}}$

47. $\left(3x^{\frac{1}{4}}\right)\left(2x^{\frac{3}{5}}\right)$

48. $\left(9a^{\frac{1}{2}}\right)\left(4a^{-\frac{1}{3}}\right)$

49. $\left(3x^{\frac{1}{3}}y^{\frac{2}{3}}\right)^3$

50. $\left(25x^4y^6\right)^{\frac{1}{2}}$

51. $\dfrac{39n^{\frac{3}{5}}}{3n^{\frac{1}{4}}}$

52. $\dfrac{64n^{\frac{5}{8}}}{16n^{\frac{7}{8}}}$

53. $\left(\dfrac{6x^{\frac{2}{7}}}{3x^{-\frac{5}{7}}}\right)^3$

54. If $f(x) = x^2 - 3x - 1$, find $f(2)$, $f(-1)$, and $f(-4)$.

55. If $f(x) = -2x^2 - 7x + 2$, find $f(3)$, $f(-3)$, and $f(-4)$.

For Problems 56–60, graph each of the functions.

56. $f(x) = 3$

57. $f(x) = 2x - 3$

58. $f(x) = -3x - 1$

59. $f(x) = 4x^2$

60. $f(x) = -4x^2$

Answers to Odd-Numbered Problems and All Chapter Review Problems

Chapter 1

■ **Problem Set 1.1** (page 6)

1. 16   **3.** 35   **5.** 51   **7.** 72   **9.** 82   **11.** 55   **13.** 60   **15.** 66   **17.** 26
19. 2   **21.** 47   **23.** 21   **25.** 11   **27.** 15   **29.** 14   **31.** 79   **33.** 6   **35.** 74
37. 12   **39.** 187   **41.** 884   **43.** 9   **45.** 18   **47.** 55   **49.** 99   **51.** 72   **53.** 11
55. 48   **57.** 21   **59.** 40   **61.** 81   **63.** 100   **65.** 84   **67.** 170   **69.** 164
71. 153

■ **Problem Set 1.2** (page 12)

1. True   **3.** False   **5.** True   **7.** True   **9.** True   **11.** False   **13.** True   **15.** False
17. True   **19.** False   **21.** Prime   **23.** Prime   **25.** Composite   **27.** Prime
29. Composite   **31.** $2 \cdot 13$ **33.** $2 \cdot 2 \cdot 3 \cdot 3$ **35.** $7 \cdot 7$ **37.** $2 \cdot 2 \cdot 2 \cdot 7$ **39.** $2 \cdot 2 \cdot 2 \cdot 3 \cdot 5$
41. $3 \cdot 3 \cdot 3 \cdot 5$ **43.** 4   **45.** 8   **47.** 9   **49.** 12   **51.** 18   **53.** 12   **55.** 24

1. 392   **63.** 168   **65.** 90   **67.** All other even numbers are divisible by 2.
3. xy **75.** $2 \cdot 2 \cdot 19$ **77.** $3 \cdot 41$ **79.** $5 \cdot 23$ **81.** $3 \cdot 3 \cdot 7 \cdot 7$

Problem Set 1.3 (page 18)

3. -4 **5.** -7 **7.** 6   **9.** -6 **11.** 8   **13.** -11 **15.** -15 **17.** -7
19. -31 **21.** -19 **23.** 9   **25.** -61 **27.** -18 **29.** -92 **31.** -5 **33.** -13
35. 12   **37.** 6   **39.** -1 **41.** -45 **43.** -29 **45.** 27   **47.** -65 **49.** -29
51. -11 **53.** -1 **55.** -8 **57.** -13 **59.** -35 **61.** -15 **63.** -32 **65.** 2
67. -4 **69.** -31 **71.** -9 **73.** 18   **75.** 8   **77.** -29 **79.** -7 **81.** 15
83. 1   **85.** 36   **87.** -39 **89.** -24 **91.** 7   **93.** -1 **95.** 10   **97.** 9
99. -17 **101.** -3 **103.** -10 **105.** -3 **107.** 11   **109.** 5   **111.** -65
113. -100 **115.** -25 **117.** 130   **119.** 80   **121. (a)** 319   **(b)** 357

Problem Set 1.4 (page 24)

1. -30 **3.** -9 **5.** 7   **7.** -56 **9.** 60   **11.** -12 **13.** -126 **15.** 154
17. -9 **19.** 11   **21.** 225   **23.** -14 **25.** 0   **27.** 23   **29.** -19 **31.** 90
33. 14   **35.** undefined   **37.** -4 **39.** -972 **41.** -47 **43.** 18   **45.** 69   **47.** 4
49. 4   **51.** -6 **53.** 31   **55.** 4   **57.** 28   **59.** -7 **61.** 10   **63.** -59 **65.** 66
67. 7   **69.** 69   **71.** -7 **73.** 126   **75.** -70 **77.** 15   **79.** -10 **81.** -25
83. 77   **85.** 104   **87.** 14

Problem Set 1.5 (page 32)

1. Distributive property   **3.** Associative property for addition
5. Commutative property for multiplication   **7.** Additive inverse property
9. Identity property for addition   **11.** Associative property for multiplication   **13.** 56   **15.** 7
17. 1800   **19.** -14400 **21.** -3700 **23.** 5900   **25.** -338 **27.** -38 **29.** 7
31. $-5x$ **33.** $-3m$ **35.** $-11y$ **37.** $-3x - 2y$ **39.** $-16a - 4b$ **41.** $-7xy + 3x$
43. $10x + 5$ **45.** $6xy - 4$ **47.** $-6a - 5b$ **49.** $5ab - 11a$ **51.** $8x + 36$ **53.** $11x + 28$
55. $8x + 44$ **57.** $5a + 29$ **59.** $3m + 29$ **61.** $-8y + 6$ **63.** -5 **65.** -40 **67.** 72
69. -18 **71.** 37   **73.** -74 **75.** 180   **77.** 34   **79.** -65

Chapter 1 Review Problem Set (page 35)

1. -3 **2.** -25 **3.** -5 **4.** -15 **5.** -1 **6.** 2   **7.** -156 **8.** 252   **9.** 6
10. -13 **11.** Prime   **12.** Composite   **13.** Composite   **14.** Composite   **15.** Composite
16. $2 \cdot 2 \cdot 2 \cdot 3$ **17.** $3 \cdot 3 \cdot 7$ **18.** $3 \cdot 19$ **19.** $2 \cdot 2 \cdot 2 \cdot 2 \cdot 2 \cdot 2$ **20.** $2 \cdot 2 \cdot 3 \cdot 7$ **21.** 18
22. 12   **23.** 180   **24.** 945   **25.** 66   **26.** -7 **27.** -2 **28.** 4   **29.** -18
30. 12   **31.** -34 **32.** -27 **33.** -38 **34.** -93 **35.** 2   **36.** 3   **37.** 35
38. 27   **39.** $8x$ **40.** $-5y - 9$ **41.** $-5x + 4y$ **42.** $13a - 6b$ **43.** $-ab - 2a$
44. $-3xy - y$ **45.** $10x + 74$ **46.** $2x + 7$ **47.** $-7x - 18$ **48.** $-3x + 12$
49. $-2a + 4$ **50.** $-2a - 4$ **51.** -59 **52.** -57 **53.** 2   **54.** 1   **55.** 12
56. 13   **57.** 22   **58.** 32   **59.** -9 **60.** 37   **61.** -39 **62.** -32 **63.** 9
64. -44

Chapter 2

■ Problem Set 2.1 (page 43)

1. $\dfrac{2}{3}$ **3.** $\dfrac{2}{3}$ **5.** $\dfrac{5}{3}$ **7.** $-\dfrac{1}{6}$ **9.** $-\dfrac{3}{4}$ **11.** $\dfrac{27}{28}$ **13.** $\dfrac{6}{11}$ **15.** $\dfrac{3x}{7y}$ **17.** $\dfrac{2x}{5}$

19. $-\dfrac{5a}{13c}$ **21.** $\dfrac{8z}{7x}$ **23.** $\dfrac{5b}{7}$ **25.** $\dfrac{15}{28}$ **27.** $\dfrac{10}{21}$ **29.** $\dfrac{3}{10}$ **31.** $-\dfrac{4}{3}$ **33.** $\dfrac{7}{5}$

35. $-\dfrac{3}{10}$ **37.** $\dfrac{1}{4}$ **39.** -27 **41.** $\dfrac{35}{27}$ **43.** $\dfrac{8}{21}$ **45.** $-\dfrac{5}{6y}$ **47.** $2a$ **49.** $\dfrac{2}{5}$

51. $\dfrac{y}{2x}$ **53.** $\dfrac{20}{13}$ **55.** $-\dfrac{7}{9}$ **57.** $\dfrac{2}{9}$ **59.** $\dfrac{2}{5}$ **61.** $\dfrac{13}{28}$ **63.** $\dfrac{8}{5}$ **65.** -4 **67.** $\dfrac{36}{49}$

69. 1 **71.** $\dfrac{2}{3}$ **73.** $\dfrac{20}{9}$ **75.** (a) 8 (c) 40 (e) 5

■ Problem Set 2.2 (page 51)

1. $\dfrac{5}{7}$ **3.** $\dfrac{5}{9}$ **5.** 3 **7.** $\dfrac{2}{3}$ **9.** $-\dfrac{1}{2}$ **11.** $\dfrac{2}{3}$ **13.** $\dfrac{15}{x}$ **15.** $\dfrac{2}{y}$ **17.** $\dfrac{8}{15}$ **19.** $\dfrac{9}{16}$

21. $\dfrac{37}{30}$ **23.** $\dfrac{59}{96}$ **25.** $-\dfrac{19}{72}$ **27.** $-\dfrac{1}{24}$ **29.** $-\dfrac{1}{3}$ **31.** $-\dfrac{1}{6}$ **33.** $\dfrac{3y+4x}{xy}$

35. $\dfrac{7b-2a}{ab}$ **37.** $\dfrac{11}{2x}$ **39.** $\dfrac{4}{3x}$ **41.** $-\dfrac{2}{5x}$ **43.** $\dfrac{19}{6y}$ **45.** $\dfrac{1}{24y}$ **47.** $-\dfrac{17}{24n}$

49. $\dfrac{5y+7x}{3xy}$ **51.** $\dfrac{32y+15x}{20xy}$ **53.** $\dfrac{63y-20x}{36xy}$ **55.** $\dfrac{-6y-5x}{4xy}$ **57.** $\dfrac{3x+2}{x}$ **59.** $\dfrac{4x-3}{2x}$

61. $\dfrac{1}{4}$ **63.** $\dfrac{37}{30}$ **65.** $\dfrac{1}{3}$ **67.** $-\dfrac{12}{5}$ **69.** $-\dfrac{1}{30}$ **71.** 14 **73.** 68 **75.** $\dfrac{7}{26}$

77. $\dfrac{11}{15}x$ **79.** $\dfrac{5}{24}a$ **81.** $\dfrac{4}{3}x$ **83.** $\dfrac{13}{20}n$ **85.** $\dfrac{20}{9}n$ **87.** $-\dfrac{79}{36}n$ **89.** $\dfrac{13}{14}x+\dfrac{9}{8}y$

91. $-\dfrac{11}{45}x-\dfrac{9}{20}y$

■ Problem Set 2.3 (page 62)

1. 0.62 **3.** 1.45 **5.** 3.8 **7.** -3.3 **9.** 7.5 **11.** 7.8 **13.** -0.9 **15.** -7.8
17. 1.16 **19.** -0.272 **21.** -24.3 **23.** 44.8 **25.** 0.0156 **27.** 1.2 **29.** -7.4
31. 0.38 **33.** 7.2 **35.** -0.42 **37.** 0.76 **39.** 4.7 **41.** 4.3 **43.** -14.8 **45.** 1.3
47. $-1.2x$ **49.** $3n$ **51.** $0.5t$ **53.** $-5.8x+2.8y$ **55.** $0.1x+1.2$ **57.** $-3x-2.3$

59. $4.6x-8$ **61.** $\dfrac{11}{12}$ **63.** $\dfrac{4}{3}$ **65.** 17.3 **67.** -97.8 **69.** 2.2 **71.** 13.75 **73.** 0.6

■ Problem Set 2.4 (page 68)

1. 64 **3.** 81 **5.** -8 **7.** -9 **9.** 16 **11.** $\dfrac{16}{81}$ **13.** $-\dfrac{1}{8}$ **15.** $\dfrac{9}{4}$ **17.** 0.027

19. -1.44 **21.** -47 **23.** -33 **25.** 11 **27.** -75 **29.** -60 **31.** 31

33. -13 **35.** $9x^2$ **37.** $12xy^2$ **39.** $-18x^4y$ **41.** $15xy$ **43.** $12x^4$ **45.** $8a^5$

47. $-8x^2$ **49.** $4y^3$ **51.** $-2x^2 + 6y^2$ **53.** $-\dfrac{11}{60}n^2$ **55.** $\dfrac{3y}{5}$ **57.** $\dfrac{11}{3y}$ **59.** $\dfrac{7b^2}{17a}$

61. $-\dfrac{3ac}{4}$ **63.** $\dfrac{x^2y^2}{4}$ **65.** $\dfrac{4x}{9}$ **67.** $\dfrac{5}{12ab}$ **69.** $\dfrac{6y^2 + 5x}{xy^2}$ **71.** $\dfrac{5 - 7x^2}{x^4}$ **73.** $\dfrac{3 + 12x^2}{2x^3}$

75. $\dfrac{13}{12x^2}$ **77.** $\dfrac{11b^2 - 14a^2}{a^2b^2}$ **79.** $\dfrac{3 - 8x}{6x^3}$ **81.** $\dfrac{3y - 4x - 5}{xy}$ **83.** 79 **85.** $\dfrac{23}{36}$ **87.** $\dfrac{25}{4}$

89. -64 **91.** -25 **93.** -33 **95.** $.45$

■ Problem Set 2.5 (page 74)

Answers may vary somewhat for Problems 1–12.

1. The difference of a and b **3.** One-third of the product of B and h

5. Two times the quantity, l plus w **7.** The quotient of A divided by w

9. The quantity, a plus b, divided by 2   **11.** Two more than three times y **13.** $l + w$ **15.** ab

17. $\dfrac{d}{t}$ **19.** lwh **21.** $y - x$ **23.** $xy + 2$ **25.** $7 - y^2$ **27.** $\dfrac{x - y}{4}$ **29.** $10 - x$

31. $10(n + 2)$ **33.** $xy - 7$ **35.** $xy - 12$ **37.** $35 - n$ **39.** $n + 45$ **41.** $y + 10$

43. $2x - 3$ **45.** $10d + 25q$ **47.** $\dfrac{d}{t}$ **49.** $\dfrac{d}{p}$ **51.** $\dfrac{d}{12}$ **53.** $n + 1$ **55.** $n + 2$

57. $3y - 2$ **59.** $36y + 12f$ **61.** $\dfrac{f}{3}$ **63.** $8w$ **65.** $3l - 4$ **67.** $48f + 72$

■ Chapter 2 Review Problem Set (page 76)

1. 64 **2.** -27 **3.** -16 **4.** $\dfrac{9}{16}$ **5.** $\dfrac{49}{36}$ **6.** 0.216 **7.** 0.0144 **8.** 0.0036

9. $-\dfrac{8}{27}$ **10.** $\dfrac{1}{16}$ **11.** $\dfrac{19}{24}$ **12.** $\dfrac{39}{70}$ **13.** $\dfrac{1}{15}$ **14.** $\dfrac{14y + 9x}{2xy}$ **15.** $\dfrac{5x - 8y}{x^2y}$ **16.** $\dfrac{7y}{20}$

17. $\dfrac{4x^3}{5y^2}$ **18.** $\dfrac{2}{7}$ **19.** 1 **20.** $\dfrac{27n^2}{28}$ **21.** $\dfrac{1}{24}$ **22.** $-\dfrac{13}{8}$ **23.** $\dfrac{7}{9}$ **24.** $\dfrac{29}{12}$ **25.** $\dfrac{1}{2}$

26. 0.67 **27.** 0.49 **28.** 2.4 **29.** -0.11 **30.** 1.76 **31.** $\dfrac{5}{56}x^2 + \dfrac{7}{20}y^2$

32. $-0.58ab + 0.36bc$ **33.** $\dfrac{11x}{24}$ **34.** $2.2a + 1.7b$ **35.** $-\dfrac{1}{10}n$ **36.** $\dfrac{41}{20}n$ **37.** $\dfrac{19}{42}$

38. $-\dfrac{1}{72}$ **39.** -0.75 **40.** -0.35 **41.** $\dfrac{1}{17}$ **42.** -8 **43.** $72 - n$ **44.** $p + 10d$

45. $\dfrac{x}{60}$ **46.** $2y - 3$ **47.** $5n + 3$ **48.** $36y + 12f$ **49.** $100m$ **50.** $5n + 10d + 25q$

51. $n - 5$ **52.** $5 - n$ **53.** $10(x - 2)$ **54.** $10x - 2$ **55.** $x^2 - y^3$ **56.** $\dfrac{d}{r}$ **57.** $x^2 + 9$

58. $(x + 9)^2$ **59.** $x^3 + y^3$ **60.** $xy - 4$

Chapter 3

■ Problem Set 3.1 (page 85)

1. $\{8\}$ **3.** $\{-6\}$ **5.** $\{-9\}$ **7.** $\{-6\}$ **9.** $\{13\}$ **11.** $\{48\}$ **13.** $\{23\}$ **15.** $\{-7\}$

17. $\left\{\dfrac{17}{12}\right\}$ **19.** $\left\{-\dfrac{4}{15}\right\}$ **21.** $\{.27\}$ **23.** $\{-3.5\}$ **25.** $\{-17\}$ **27.** $\{-35\}$ **29.** $\{-8\}$

31. $\{-17\}$ **33.** $\left\{\dfrac{37}{5}\right\}$ **35.** $\{-3\}$ **37.** $\left\{\dfrac{13}{2}\right\}$ **39.** $\{144\}$ **41.** $\{24\}$ **43.** $\{-15\}$

45. $\{24\}$ **47.** $\{-35\}$ **49.** $\left\{\dfrac{3}{10}\right\}$ **51.** $\left\{-\dfrac{9}{10}\right\}$ **53.** $\left\{\dfrac{1}{2}\right\}$ **55.** $\left\{-\dfrac{1}{3}\right\}$ **57.** $\left\{\dfrac{27}{32}\right\}$

59. $\left\{-\dfrac{5}{14}\right\}$ **61.** $\left\{-\dfrac{7}{5}\right\}$ **63.** $\left\{-\dfrac{1}{12}\right\}$ **65.** $\left\{-\dfrac{3}{20}\right\}$

■ Problem Set 3.2 (page 91)

1. $\{4\}$ **3.** $\{6\}$ **5.** $\{8\}$ **7.** $\{11\}$ **9.** $\left\{\dfrac{17}{6}\right\}$ **11.** $\left\{\dfrac{19}{2}\right\}$ **13.** $\{6\}$ **15.** $\{-1\}$

17. $\{-5\}$ **19.** $\{-6\}$ **21.** $\left\{\dfrac{11}{2}\right\}$ **23.** $\{-2\}$ **25.** $\left\{\dfrac{10}{7}\right\}$ **27.** $\{18\}$ **29.** $\left\{-\dfrac{25}{4}\right\}$

31. $\{-7\}$ **33.** $\left\{-\dfrac{24}{7}\right\}$ **35.** $\left\{\dfrac{5}{2}\right\}$ **37.** $\left\{\dfrac{4}{17}\right\}$ **39.** $\left\{-\dfrac{12}{5}\right\}$ **41.** 9 **43.** 22

45. \$18 **47.** 35 years old **49.** \$6.50 **51.** 6 **53.** 5 **55.** 11 **57.** 8 **59.** 3
61. \$300 **63.** 4 meters **65.** 14 cars **67.** 5 hours

■ Problem Set 3.3 (page 97)

1. $\{5\}$ **3.** $\{-8\}$ **5.** $\left\{\dfrac{8}{5}\right\}$ **7.** $\{-11\}$ **9.** $\left\{-\dfrac{5}{2}\right\}$ **11.** $\{-9\}$ **13.** $\{2\}$

15. $\{-3\}$ **17.** $\left\{\dfrac{13}{2}\right\}$ **19.** $\left\{\dfrac{5}{3}\right\}$ **21.** $\{17\}$ **23.** $\left\{-\dfrac{13}{2}\right\}$ **25.** $\left\{\dfrac{16}{3}\right\}$ **27.** $\{2\}$

29. $\left\{-\dfrac{1}{3}\right\}$ **31.** $\left\{-\dfrac{19}{10}\right\}$ **33.** 17 **35.** 35 and 37 **37.** 36, 38, and 40 **39.** $\dfrac{3}{2}$
41. -6 **43.** $32°$ and $58°$ **45.** $50°$ and $130°$ **47.** $65°$ and $75°$ **49.** \$26 per share
51. \$9 per hour **53.** 150 males and 450 females **55.** 690 voters **57.** 6 feet
59. 29 yards by 10 yards

■ Problem Set 3.4 (page 105)

1. $\{1\}$ **3.** $\{10\}$ **5.** $\{-9\}$ **7.** $\left\{\dfrac{29}{4}\right\}$ **9.** $\left\{-\dfrac{17}{3}\right\}$ **11.** $\{10\}$ **13.** $\{44\}$ **15.** $\{26\}$

17. $\{-38\}$ **19.** $\left\{\dfrac{11}{6}\right\}$ **21.** $\{3\}$ **23.** $\{-1\}$ **25.** $\{-2\}$ **27.** $\{16\}$ **29.** $\left\{\dfrac{22}{3}\right\}$

31. $\{-2\}$ **33.** $\left\{-\dfrac{1}{6}\right\}$ **35.** $\{-57\}$ **37.** $\left\{-\dfrac{7}{5}\right\}$ **39.** $\{2\}$ **41.** $\{-3\}$ **43.** $\left\{\dfrac{27}{10}\right\}$

45. $\left\{\dfrac{3}{28}\right\}$ **47.** $\left\{\dfrac{18}{5}\right\}$ **49.** $\left\{\dfrac{24}{7}\right\}$ **51.** $\{5\}$ **53.** $\{0\}$ **55.** $\left\{-\dfrac{51}{10}\right\}$ **57.** $\{-12\}$
59. $\{15\}$ **61.** 7 and 8   **63.** 14, 15, and 16   **65.** 6 and 11   **67.** 48   **69.** 12 and 18
71. 8 feet and 12 feet   **73.** 12 nickels, 17 dimes, and 40 quarters
75. 40 nickels, 80 dimes, and 90 quarters   **77.** 8 dimes and 10 quarters
79. 75 pennies, 150 nickels, and 225 dimes   **81.** 30°   **83.** 20°, 50°, and 110°   **85.** 40°
87. (a) {all reals}   **(c)** $\{0\}$ **(e)** $\varnothing$ **(g)** $\varnothing$ **(i)** $\{0\}$

■ Problem Set 3.5 (page 114)

1. $\{9\}$ **3.** $\{10\}$ **5.** $\left\{\dfrac{15}{2}\right\}$ **7.** $\{-22\}$ **9.** $\{-4\}$ **11.** $\{34\}$ **13.** $\{6\}$ **15.** $\left\{-\dfrac{8}{5}\right\}$

17. $\{7\}$ **19.** $\{50\}$ **21.** $\{120\}$ **23.** $\left\{\dfrac{9}{7}\right\}$ **25.** 55%   **27.** 60%   **29.** $16\dfrac{2}{3}\%$

31. $37\dfrac{1}{2}\%$ **33.** 150%   **35.** 240%   **37.** 2.66   **39.** 42   **41.** 80%   **43.** 60

45. 115%   **47.** 90   **49.** 15 feet by $19\dfrac{1}{2}$ feet   **51.** 330 miles   **53.** 60 centimeters

55. 7.5 pounds   **57.** $33\dfrac{1}{3}$ pounds   **59.** 90,000   **61.** 15 inches by 10 inches   **63.** $300

65. $150,000   **67.** $\varnothing$ **69.** $\varnothing$

■ Problem Set 3.6 (page 120)

1. $\{1.11\}$ **3.** $\{6.6\}$ **5.** $\{.48\}$ **7.** $\{80\}$ **9.** $\{3\}$ **11.** $\{50\}$ **13.** $\{70\}$ **15.** $\{200\}$
17. $\{450\}$ **19.** $\{150\}$ **21.** $\{2200\}$ **23.** $50   **25.** $36   **27.** $20.80   **29.** 30%
31. $8.50   **33.** $4.65   **35.** $1000   **37.** 40%   **39.** $400 at 9% and $650 at 10%
41. $500 at 9% and $700 at 12%   **43.** $4000   **45.** $800 at 10% and $1500 at 12%
47. $3000 at 8% and $2400 at 10%   **49.** Yes, if the profit is figured as a percent of the selling price.

■ Chapter 3 Review Problem Set (page 123)

1. $\{-3\}$ **2.** $\{1\}$ **3.** $\left\{-\dfrac{3}{4}\right\}$ **4.** $\{9\}$ **5.** $\{-4\}$ **6.** $\left\{\dfrac{40}{3}\right\}$ **7.** $\left\{\dfrac{9}{4}\right\}$ **8.** $\left\{-\dfrac{15}{8}\right\}$

9. $\{-7\}$ **10.** $\left\{\dfrac{2}{41}\right\}$ **11.** $\left\{\dfrac{19}{7}\right\}$ **12.** $\left\{\dfrac{1}{2}\right\}$ **13.** $\left\{\dfrac{17}{12}\right\}$ **14.** $\{5\}$ **15.** $\{-32\}$

16. $\{-12\}$ **17.** $\{21\}$ **18.** $\{-60\}$ **19.** $\{10\}$ **20.** $\left\{-\dfrac{11}{4}\right\}$ **21.** $\left\{-\dfrac{8}{5}\right\}$ **22.** $\{800\}$

23. $\{16\}$ **24.** $\{73\}$ **25.** $\left\{\dfrac{5}{21}\right\}$ **26.** 24   **27.** 7   **28.** 33   **29.** 8   **30.** 40

31. 16 and 24   **32.** 60%   **33.** 18   **34.** 40 and 56   **35.** 8 nickels and 22 dimes
36. 8 nickels, 25 dimes, and 50 quarters   **37.** 52°   **38.** 20°   **39.** 30 gallons
40. $900 at 9% and $1200 at 11%   **41.** $40   **42.** 35%   **43.** 700 miles

Chapter 4

■ Problem Set 4.1 (page 131)

1. 7   **3.** 500   **5.** 20   **7.** 48   **9.** 9   **11.** 46 centimeters   **13.** 15 inches
15. 504 square feet   **17.** \$2   **19.** 7 inches   **21.** 150π square centimeters

23. $\dfrac{1}{4}\pi$ square yards   **25.** $S = 324\pi$ square inches and $V = 972\pi$ cubic inches

27. $V = 1152\pi$ cubic feet and $S = 416\pi$ square feet   **29.** 12 inches   **31.** 8 feet   **33.** $h = \dfrac{V}{B}$

35. $B = \dfrac{3V}{h}$ **37.** $w = \dfrac{P - 2l}{2}$ **39.** $h = \dfrac{3V}{\pi r^2}$ **41.** $C = \dfrac{5}{9}(F - 32)$ **43.** $h = \dfrac{A - 2\pi r^2}{2\pi r}$

45. $x = \dfrac{9 - 7y}{3}$ **47.** $y = \dfrac{9x - 13}{6}$ **49.** $x = \dfrac{11y - 14}{2}$ **51.** $x = \dfrac{-y - 4}{3}$ **53.** $y = \dfrac{3}{2}x$

55. $y = \dfrac{ax - c}{b}$ **57.** $x = \dfrac{2y - 22}{5}$ **59.** $y = mx + b$

■ Problem Set 4.2 (page 137)

1. $12\dfrac{1}{2}$ years   **3.** 20 years   **5.** The width is 14 inches and the length is 42 inches.

7. The width is 12 centimeters and the length is 34 centimeters.   **9.** 80 square inches
11. 24 feet, 31 feet, and 45 feet   **13.** 6 centimeters, 19 centimeters, and 21 centimeters

15. 12 centimeters   **17.** 7 centimeters   **19.** 9 hours   **21.** $2\dfrac{1}{2}$ hours   **23.** 55 miles per hour

25. 64 and 72 miles per hour   **27.** 60 miles

■ Problem Set 4.3 (page 142)

1. 12.5 milliliters   **3.** 15 centiliters   **5.** $7\dfrac{1}{2}$ quarts of the 30% solution and $2\dfrac{1}{2}$ quarts of the

50% solution   **7.** 5 gallons   **9.** 3 quarts   **11.** 12 gallons   **13.** 16.25%
15. The square is 6 inches by 6 inches and the rectangle is 9 inches long and 3 inches wide.   **17.** 40 minutes
19. Pam is 9 and Bill is 18.   **21.** Abby is 14 and her mother is 35.   **23.** 1 year old and 10 years old
25. 56 miles

■ Problem Set 4.4 (page 149)

1. True   **3.** False   **5.** False   **7.** True   **9.** True

11. $\{x \mid x > -2\}$

13. $\{x \mid x \le 3\}$

15. $\{x \mid x > 2\}$

17. $\{x \mid x \le -2\}$

19. $\{x\,|\,x < -1\}$

21. $\{x\,|\,x < 2\}$

23. $\{x\,|\,x < -20\}$ **25.** $\{x\,|\,x \geq -9\}$ **27.** $\{x\,|\,x > 9\}$ **29.** $\left\{x\,\middle|\,x < \dfrac{10}{3}\right\}$ **31.** $\{x\,|\,x < -8\}$

33. $\{n\,|\,n \geq 8\}$ **35.** $\left\{n\,\middle|\,n > -\dfrac{24}{7}\right\}$ **37.** $\{n\,|\,n > 7\}$ **39.** $\{x\,|\,x > 5\}$ **41.** $\{x\,|\,x \leq 6\}$

43. $\{x\,|\,x \leq -21\}$ **45.** $\left\{x\,\middle|\,x < \dfrac{8}{3}\right\}$ **47.** $\left\{x\,\middle|\,x < \dfrac{5}{4}\right\}$ **49.** $\{x\,|\,x < 1\}$ **51.** $\{t\,|\,t \geq 4\}$

53. $\{x\,|\,x > 14\}$ **55.** $\left\{x\,\middle|\,x > \dfrac{3}{2}\right\}$ **57.** $\left\{t\,\middle|\,t \geq \dfrac{1}{4}\right\}$ **59.** $\left\{x\,\middle|\,x < -\dfrac{9}{4}\right\}$ **61.** $\varnothing$

63. $\{x\,|\,x$ is any real number$\}$ **65.** $\varnothing$ **67.** $\{x\,|\,x$ is any real number$\}$

■ Problem Set 4.5 (page 154)

1. $\{x\,|\,x > 2\}$ **3.** $\{x\,|\,x < -1\}$ **5.** $\left\{x\,\middle|\,x > -\dfrac{10}{3}\right\}$ **7.** $\{n\,|\,n \geq -11\}$ **9.** $\{t\,|\,t \leq 11\}$

11. $\left\{x\,\middle|\,x > -\dfrac{11}{5}\right\}$ **13.** $\left\{x\,\middle|\,x < \dfrac{5}{2}\right\}$ **15.** $\{x\,|\,x \leq 8\}$ **17.** $\left\{n\,\middle|\,n > \dfrac{3}{2}\right\}$ **19.** $\{y\,|\,y > -3\}$

21. $\left\{x\,\middle|\,x < \dfrac{5}{2}\right\}$ **23.** $\{x\,|\,x < 8\}$ **25.** $\{x\,|\,x < 21\}$ **27.** $\{x\,|\,x < 6\}$ **29.** $\left\{n\,\middle|\,n > -\dfrac{17}{2}\right\}$

31. $\{n\,|\,n \leq 42\}$ **33.** $\left\{n\,\middle|\,n > -\dfrac{9}{2}\right\}$ **35.** $\left\{x\,\middle|\,x > \dfrac{4}{3}\right\}$ **37.** $\{n\,|\,n \geq 4\}$ **39.** $\{t\,|\,t > 300\}$

41. $\{x\,|\,x \leq 50\}$ **43.** $\{x\,|\,x > 0\}$ **45.** $\{x\,|\,x > 64\}$ **47.** $\left\{n\,\middle|\,n > \dfrac{33}{5}\right\}$ **49.** $\left\{x\,\middle|\,x \geq -\dfrac{16}{3}\right\}$

51. ⟶ -1 2

53. ⟶ -2 1

55. ⟶ -2 2

57. ⟶ 2

59. ⟶ -4

61. ⟶ $\varnothing$

63. ⟶ 0 2

65. ⟶ all reals

67. All numbers greater than 7 **69.** 15 inches **71.** 158 or better **73.** Better than 90
75. More than 12% **77.** 77 or less

■ Problem Set 4.6 (page 160)

1. $\{-4, 4\}$

-4 4

3. $\{x\,|\,x > -1$ and $x < 1\}$

-1 1

5. $\{x\,|\,x \leq -2$ or $x \geq 2\}$

-2 2

7. $\{-3, -1\}$

-3 -1

9. $\{-1, 3\}$

11. $\{x \mid x \geq 0 \text{ and } x \leq 4\}$

13. $\{x \mid x < -4 \text{ or } x > 2\}$

15. $\{-2, 1\}$

17. $\left\{-\dfrac{2}{5}, \dfrac{6}{5}\right\}$

19. $\{x \mid x \leq 1 \text{ or } x \geq 2\}$

21. $\left\{x \mid x > -\dfrac{5}{4} \text{ and } x < -\dfrac{1}{4}\right\}$

23. $\{-2\}$

25. $\left\{x \mid x \neq \dfrac{2}{3}\right\}$

27. $\left\{-\dfrac{16}{3}, 6\right\}$　　**29.** $\{x \mid x < -5 \text{ or } x > 4\}$

31. $\left\{x \mid x > -\dfrac{14}{3} \text{ and } x < 8\right\}$　　**33.** $\left\{-6, \dfrac{16}{3}\right\}$　　**35.** $\left\{x \mid x \geq -6 \text{ and } x \leq \dfrac{19}{2}\right\}$

37. $\left\{x \mid x \leq -\dfrac{21}{5} \text{ or } x \geq 3\right\}$　　**39.** $\{x \mid x > -6 \text{ and } x < 2\}$　　**41.** $\left\{x \mid x < -\dfrac{5}{2} \text{ or } x > \dfrac{7}{2}\right\}$　　**43.** $\{0\}$

45. $\{x \mid x \text{ is any real number}\}$　　**47.** $\varnothing$　　**49.** $\{-6\}$　　**51.** $\{x \mid 4 < x < 14\}$

53. $\{x \mid -1 \leq x \leq 4\}$　　**55.** $\{x \mid -11 < x < 7\}$　　**57.** $\{x \mid -7 < x < -1\}$　　**59.** $\{x \mid -1 < x < 6\}$

■ Chapter 4 Review Problem Set　(page 162)

1. 6　　**2.** 25　　**3.** $t = \dfrac{A - P}{Pr}$　　**4.** $x = \dfrac{13 + 3y}{2}$　　**5.** 77 square inches　　**6.** 6 centimeters

7. 15 feet　　**8.** $\{x \mid x > 4\}$　　**9.** $\{x \mid x > -4\}$　　**10.** $\{x \mid x \geq 13\}$　　**11.** $\left\{x \mid x \geq \dfrac{11}{2}\right\}$

12. $\{x \mid x > 35\}$　　**13.** $\left\{x \mid x < \dfrac{26}{5}\right\}$　　**14.** $\{n \mid n < 2\}$　　**15.** $\left\{n \mid n > \dfrac{5}{11}\right\}$　　**16.** $\{s \mid s \geq 6\}$

17. $\{t \mid t \leq 100\}$　　**18.** $\{y \mid y < 24\}$　　**19.** $\{x \mid x > 10\}$　　**20.** $\left\{n \mid n < \dfrac{2}{11}\right\}$　　**21.** $\{n \mid n > 33\}$

22. $\{n \mid n \leq 120\}$　　**23.** $\left\{n \mid n \leq -\dfrac{180}{13}\right\}$　　**24.** $\left\{x \mid x > \dfrac{9}{2}\right\}$　　**25.** $\left\{x \mid x < -\dfrac{43}{3}\right\}$

26.

27.

all reals

28.

29.

30.

31.

32.

33.

34.

35.

36. The width is 6 meters and the length is 17 meters. **37.** $1\frac{1}{2}$ hours **38.** 20 liters

39. 89 or better **40.** $34°$ and $99°$ **41.** 5 hours **42.** 18 gallons **43.** 88 or better
44. 26% **45.** 10 years ago

Chapter 5

■ Problem Set 5.1 (page 170)

1. 3 **3.** 2 **5.** 3 **7.** 2 **9.** $8x + 11$ **11.** $4y + 10$ **13.** $2x^2 - 2x - 23$
15. $17x - 19$ **17.** $6x^2 - 5x - 4$ **19.** $5x + 5$ **21.** $-2x - 5$ **23.** $-3x + 7$
25. $2x^2 + 15x - 6$ **27.** $5n^2 + 2n + 3$ **29.** $-3x^3 + 2x^2 - 11$ **31.** $9x - 2$ **33.** $2a + 15$
35. $-2x^2 + 5$ **37.** $6x^3 + 12x^2 - 5$ **39.** $4x^3 - 8x^2 + 13x + 12$ **41.** $x + 11$
43. $-3x - 14$ **45.** $-x^2 - 13x - 12$ **47.** $x^2 - 11x - 8$ **49.** $-10a - 3b$
51. $-n^2 + 2n - 17$ **53.** $8x + 1$ **55.** $-5n - 1$ **57.** $-2a + 6$ **59.** $11x + 6$
61. $6x + 7$ **63.** $-5n + 7$

■ Problem Set 5.2 (page 175)

1. $45x^2$ **3.** $21x^3$ **5.** $-6x^2y^2$ **7.** $14x^3y$ **9.** $-48a^3b^3$ **11.** $5x^4y$ **13.** $104a^3b^2c^2$

15. $30x^6$ **17.** $-56x^2y^3$ **19.** $-6a^2b^3$ **21.** $72c^3d^3$ **23.** $\frac{2}{5}x^3y^5$ **25.** $-\frac{2}{9}a^2b^5$

27. $0.28x^8$ **29.** $-6.4a^4b^2$ **31.** $4x^8$ **33.** $9a^4b^6$ **35.** $27x^6$ **37.** $-64x^{12}$
39. $81x^8y^{10}$ **41.** $16x^8y^4$ **43.** $81a^{12}b^8$ **45.** $x^{12}y^6$ **47.** $15x^2 + 10x$ **49.** $18x^3 - 6x^2$
51. $-28x^3 + 16x$ **53.** $2x^3 - 8x^2 + 12x$ **55.** $-18a^3 + 30a^2 + 42a$ **57.** $28x^3y - 7x^2y + 35xy$
59. $-9x^3y + 2x^2y + 6xy$ **61.** $13x + 22y$ **63.** $-2x - 9y$ **65.** $4x^3 - 3x^2 - 14x$
67. $-x + 14$ **69.** $-7x + 12$ **71.** $18x^5$ **73.** $-432x^5$ **75.** $25x^7y^8$ **77.** $-a^{12}b^5c^9$
79. $-16x^{11}y^{17}$ **81.** x^{4n} **83.** x^{5n+1} **85.** x^{4n-2} **87.** $6x^{3n}$ **89.** $-30x^{5n}$

■ Problem Set 5.3 (page 180)

1. $xy + 3x + 2y + 6$ **3.** $xy + x - 4y - 4$ **5.** $xy - 6x - 5y + 30$ **7.** $xy + xz + x + 2y + 2z + 2$
9. $6xy + 2x + 9y + 3$ **11.** $x^2 + 10x + 21$ **13.** $x^2 + 5x - 24$ **15.** $x^2 - 6x - 7$
17. $n^2 - 10n + 24$ **19.** $3n^2 + 19n + 6$ **21.** $15x^2 + 29x - 14$ **23.** $x^3 + 7x^2 + 21x + 27$
25. $x^3 + 3x^2 - 10x - 24$ **27.** $2x^3 - 7x^2 - 22x + 35$ **29.** $8a^3 - 14a^2 + 23a - 9$
31. $3a^3 + 2a^2 - 8a - 5$ **33.** $x^4 + 7x^3 + 17x^2 + 23x + 12$ **35.** $x^4 - 3x^3 - 34x^2 + 33x + 63$
37. $x^2 + 11x + 18$ **39.** $x^2 + 4x - 12$ **41.** $x^2 - 8x - 33$ **43.** $n^2 - 7n + 12$
45. $n^2 + 18n + 72$ **47.** $y^2 - 4y - 21$ **49.** $y^2 - 19y + 84$ **51.** $x^2 + 2x - 35$
53. $x^2 - 6x - 112$ **55.** $a^2 + a - 90$ **57.** $2a^2 + 13a + 6$ **59.** $5x^2 + 33x - 14$
61. $6x^2 - 11x - 7$ **63.** $12a^2 - 7a - 12$ **65.** $12n^2 - 28n + 15$ **67.** $14x^2 + 13x - 12$
69. $45 - 19x + 2x^2$ **71.** $-8x^2 + 22x - 15$ **73.** $-9x^2 + 9x + 4$ **75.** $72n^2 - 5n - 12$

77. $27 - 21x + 2x^2$ **79.** $20x^2 - 7x - 6$ **81.** $x^2 + 14x + 49$ **83.** $25x^2 - 4$
85. $x^2 - 2x + 1$ **87.** $9x^2 + 42x + 49$ **89.** $4x^2 - 12x + 9$ **91.** $4x^2 - 9y^2$
93. $1 - 10n + 25n^2$ **95.** $9x^2 + 24xy + 16y^2$ **97.** $9 + 24y + 16y^2$ **99.** $1 - 49n^2$
101. $16a^2 - 56ab + 49b^2$ **103.** $x^2 + 16xy + 64y^2$ **105.** $25x^2 - 121y^2$ **107.** $64x^3 - x$
109. $-32x^3 + 2xy^2$ **111.** $x^3 + 6x^2 + 12x + 8$ **113.** $x^3 - 9x^2 + 27x - 27$
115. $8n^3 + 12n^2 + 6n + 1$ **117.** $27n^3 - 54n^2 + 36n - 8$ **119.** Same answers as Problems 111–117.

■ Problem Set 5.4 (page 184)

1. x^8 **3.** $2x^2$ **5.** $-8n^4$ **7.** -8 **9.** $13xy^2$ **11.** $7ab^2$ **13.** $18xy^4$
15. $-32x^5y^2$ **17.** $-8x^5y^4$ **19.** -1 **21.** $14ab^2c^4$ **23.** $16yz^4$ **25.** $4x^2 + 6x^3$
27. $3x^3 - 8x$ **29.** $-7n^3 + 9$ **31.** $5x^4 - 8x^3 - 12x$ **33.** $4n^5 - 8n^2 + 13$ **35.** $5a^6 + 8a^2$
37. $-3xy + 5y$ **39.** $-8ab - 10a^2b^3$ **41.** $-3bc + 13b^2c^4$ **43.** $-9xy^2 + 12x^2y^3$
45. $-3x^4 - 5x^2 + 7$ **47.** $-3a^2 + 7a + 13b$ **49.** $-1 + 5xy^2 - 7xy^5$

■ Problem Set 5.5 (page 189)

1. $x + 12$ **3.** $x + 2$ **5.** $x + 8$ with a remainder of 4   **7.** $x + 4$ with a remainder of -7
9. $5n + 4$ **11.** $8y - 3$ with a remainder of 2   **13.** $4x - 7$ **15.** $3x + 2$ with a remainder of -6
17. $2x^2 + 3x + 4$ **19.** $5n^2 - 4n - 3$ **21.** $n^2 + 6n - 4$ **23.** $x^2 + 3x + 9$
25. $9x^2 + 12x + 16$ **27.** $3n - 8$ with a remainder of 17   **29.** $3t + 2$ with a remainder of 6
31. $3n^2 - n - 4$ **33.** $4x^2 - 5x + 5$ with a remainder of -3 **35.** $x + 4$ with a remainder of $5x - 1$
37. $2x - 12$ with a remainder of $49x - 5$ **39.** $x^3 - 2x^2 + 4x - 8$

■ Problem Set 5.6 (page 195)

1. $\dfrac{1}{9}$ **3.** $\dfrac{1}{64}$ **5.** $\dfrac{2}{3}$ **7.** 16 **9.** 1 **11.** $-\dfrac{27}{8}$ **13.** $\dfrac{1}{4}$ **15.** $-\dfrac{1}{9}$ **17.** $\dfrac{27}{64}$

19. $\dfrac{1}{8}$ **21.** 27 **23.** 1000 **25.** $\dfrac{1}{1000}$ or $.001$ **27.** 18 **29.** 144 **31.** x^5 **33.** $\dfrac{1}{n^2}$

35. $\dfrac{1}{a^5}$ **37.** $8x$ **39.** $\dfrac{27}{x^4}$ **41.** $-\dfrac{15}{y^3}$ **43.** 96 **45.** x^{10} **47.** $\dfrac{1}{n^4}$ **49.** $2n^2$

51. $-\dfrac{3}{x^4}$ **53.** 4 **55.** x^6 **57.** $\dfrac{1}{x^4}$ **59.** $\dfrac{1}{x^3y^4}$ **61.** $\dfrac{1}{x^6y^3}$ **63.** $\dfrac{8}{n^6}$ **65.** $\dfrac{1}{16n^6}$

67. $\dfrac{81}{a^8}$ **69.** $\dfrac{x^2}{25}$ **71.** $\dfrac{x^2y}{2}$ **73.** $\dfrac{y}{x^2}$ **75.** a^4b^8 **77.** $\dfrac{x^2}{y^6}$ **79.** x **81.** $\dfrac{1}{8x^3}$

83. $\dfrac{x^4}{4}$ **85.** $(3.21)(10^2)$ **87.** $(8)(10^3)$ **89.** $(2.46)(10^{-3})$ **91.** $(1.79)(10^{-5})$ **93.** $(8.7)(10^7)$

95. 8000 **97.** $52,100$ **99.** $11,400,000$ **101.** $.07$ **103.** $.000987$ **105.** $.00000864$
107. $.84$ **109.** 450 **111.** $4,000,000$ **113.** $.0000002$ **115.** $.3$ **117.** $.000007$

■ Chapter 5 Review Problem Set (page 198)

1. $8x^2 - 13x + 2$ **2.** $3y^2 + 11y - 9$ **3.** $3x^2 + 2x - 9$ **4.** $-8x^2 + 18$ **5.** $11x + 8$
6. $-9x^2 + 8x - 20$ **7.** $2y^2 - 54y + 18$ **8.** $-13a - 30$ **9.** $-27a - 7$ **10.** $n - 2$
11. $-5n^2 - 2n$ **12.** $17n^2 - 14n - 16$ **13.** $35x^6$ **14.** $-54x^8$ **15.** $24x^3y^5$
16. $-6a^4b^9$ **17.** $8a^6b^9$ **18.** $9x^2y^4$ **19.** $35x^2 + 15x$ **20.** $-24x^3 + 3x^2$
21. $x^2 + 17x + 72$ **22.** $3x^2 + 10x + 7$ **23.** $x^2 - 3x - 10$ **24.** $y^2 - 13y + 36$
25. $14x^2 - x - 3$ **26.** $20a^2 - 3a - 56$ **27.** $9a^2 - 30a + 25$ **28.** $2x^3 + 17x^2 + 26x - 24$
29. $30n^2 + 19n - 5$ **30.** $12n^2 + 13n - 4$ **31.** $4n^2 - 1$ **32.** $16n^2 - 25$ **33.** $4a^2 + 28a + 49$

34. $9a^2 + 30a + 25$ **35.** $x^3 - 3x^2 + 8x - 12$ **36.** $2x^3 + 7x^2 + 10x - 7$
37. $a^3 + 15a^2 + 75a + 125$ **38.** $a^3 - 18a^2 + 108a - 216$ **39.** $x^4 + x^3 + 2x^2 - 7x - 5$
40. $n^4 - 5n^3 - 11n^2 - 30n - 4$ **41.** $-12x^3y^3$ **42.** $7a^3b^4$ **43.** $3x^2y - 9x^4$
44. $10a^4b^9 - 13a^3b^7$ **45.** $14x^2 - 10x - 8$ **46.** $x + 4, R = -21$ **47.** $7x - 6$

48. $2x^2 + x + 4, R = 4$ **49.** 13 **50.** 25 **51.** $\dfrac{1}{16}$ **52.** 1 **53.** -1 **54.** 9

55. $\dfrac{16}{9}$ **56.** $\dfrac{1}{4}$ **57.** -8 **58.** $\dfrac{11}{18}$ **59.** $\dfrac{5}{4}$ **60.** $\dfrac{1}{25}$ **61.** $\dfrac{1}{x^3}$ **62.** $12x^3$ **63.** x^2

64. $\dfrac{1}{x^2}$ **65.** $8a^6$ **66.** $\dfrac{4}{n}$ **67.** $\dfrac{x^2}{y}$ **68.** $\dfrac{b^6}{a^4}$ **69.** $\dfrac{1}{2x}$ **70.** $\dfrac{1}{9n^4}$ **71.** $\dfrac{n^3}{8}$

72. $-12b$ **73.** 610 **74.** $56,000$ **75.** 0.08 **76.** 0.00092 **77.** $(9)(10^3)$ **78.** $(4.7)(10)$
79. $(4.7)(10^{-2})$ **80.** $(2.1)(10)^{-4}$ **81.** 0.48 **82.** 4.2 **83.** 2000 **84.** 0.00000002

Chapter 6

■ Problem Set 6.1 (page 206)

1. $6y$ **3.** $12xy$ **5.** $14ab^2$ **7.** $2x$ **9.** $8a^2b^2$ **11.** $4(2x + 3y)$ **13.** $7y(2x - 3)$
15. $9x(2x + 5)$ **17.** $6xy(2y - 5x)$ **19.** $12a^2b(3 - 5ab^3)$ **21.** $xy^2(16y + 25x)$
23. $8(8ab - 9cd)$ **25.** $9a^2b(b^3 - 3)$ **27.** $4x^4y(13y + 15x^2)$ **29.** $8x^2y(5y + 1)$
31. $3x(4 + 5y + 7x)$ **33.** $x(2x^2 - 3x + 4)$ **35.** $4y^2(11y^3 - 6y - 5)$ **37.** $7ab(2ab^2 + 5b - 7a^2)$
39. $(y + 1)(x + z)$ **41.** $(b - 4)(a - c)$ **43.** $(x + 3)(x + 6)$ **45.** $(x + 1)(2x - 3)$
47. $(x + y)(5 + b)$ **49.** $(x - y)(b - c)$ **51.** $(a + b)(c + 1)$ **53.** $(x + 5)(x + 12)$
55. $(x - 2)(x - 8)$ **57.** $(2x + 1)(x - 5)$ **59.** $(2n - 1)(3n - 4)$ **61.** $\{0, 8\}$ **63.** $\{-1, 0\}$

65. $\{0, 5\}$ **67.** $\left\{0, \dfrac{3}{2}\right\}$ **69.** $\left\{-\dfrac{3}{7}, 0\right\}$ **71.** $\{-5, 0\}$ **73.** $\left\{0, \dfrac{3}{2}\right\}$ **75.** $\{0, 7\}$

77. $\{0, 13\}$ **79.** $\left\{-\dfrac{5}{2}, 0\right\}$ **81.** 0 or 9 **83.** 20 units **85.** $\dfrac{4}{\pi}$

87. The square is 3 inches by 3 inches and the rectangle is 3 inches by 6 inches. **89. (a)** \$116 **(c)** \$750

91. $x = 0$ or $x = \dfrac{c}{b^2}$ **93.** $y = \dfrac{c}{1 + a - b}$

■ Problem Set 6.2 (page 212)

1. $(x - 1)(x + 1)$ **3.** $(x - 10)(x + 10)$ **5.** $(x - 2y)(x + 2y)$ **7.** $(3x - y)(3x + y)$
9. $(6a - 5b)(6a + 5b)$ **11.** $(1 - 2n)(1 + 2n)$ **13.** $5(x - 2)(x + 2)$ **15.** $8(x^2 + 4)$
17. $2(x - 3y)(x + 3y)$ **19.** $x(x - 5)(x + 5)$ **21.** Not factorable **23.** $9x(5x - 4y)$
25. $4(3 - x)(3 + x)$ **27.** $4a^2(a^2 + 4)$ **29.** $(x - 3)(x + 3)(x^2 + 9)$ **31.** $x^2(x^2 + 1)$
33. $3x(x^2 + 16)$ **35.** $5x(1 - 2x)(1 + 2x)$ **37.** $4(x - 4)(x + 4)$ **39.** $3xy(5x - 2y)(5x + 2y)$

41. $\{-3, 3\}$ **43.** $\{-2, 2\}$ **45.** $\left\{-\dfrac{4}{3}, \dfrac{4}{3}\right\}$ **47.** $\{-11, 11\}$ **49.** $\left\{-\dfrac{2}{5}, \dfrac{2}{5}\right\}$ **51.** $\{-5, 5\}$

53. $\{-4, 0, 4\}$ **55.** $\{-4, 0, 4\}$ **57.** $\left\{-\dfrac{1}{3}, \dfrac{1}{3}\right\}$ **59.** $\{-10, 0, 10\}$ **61.** $\left\{-\dfrac{9}{8}, \dfrac{9}{8}\right\}$

63. $\left\{-\dfrac{1}{2}, 0, \dfrac{1}{2}\right\}$ **65.** -7 or 7 **67.** $-4, 0,$ or 4 **69.** 3 inches and 15 inches
71. The length is 20 centimeters and the width is 8 centimeters. **73.** 4 meters and 8 meters

75. 5 centimeters　　**77.** $(x + 1)(x^2 - x + 1)$　　**79.** $(n - 3)(n^2 + 3n + 9)$
81. $(2x + 3y)(4x^2 - 6xy + 9y^2)$　　**83.** $(1 - 2x)(1 + 2x + 4x^2)$　　**85.** $(x + 2y)(x^2 - 2xy + 4y^2)$
87. $(ab - 1)(a^2b^2 + ab + 1)$　　**89.** $(2 + n)(4 - 2n + n^2)$　　**91.** $(3n - 5)(9n^2 + 15n + 25)$

■ Problem Set 6.3 (page 220)

1. $(x + 4)(x + 6)$　　**3.** $(x + 5)(x + 8)$　　**5.** $(x - 2)(x - 9)$　　**7.** $(n - 7)(n - 4)$　　**9.** $(n + 9)(n - 3)$
11. $(n - 10)(n + 4)$　　**13.** Not factorable　　**15.** $(x - 6)(x - 12)$　　**17.** $(x + 11)(x - 6)$
19. $(y - 9)(y + 8)$　　**21.** $(x + 5)(x + 16)$　　**23.** $(x + 12)(x - 6)$　　**25.** Not factorable
27. $(x - 2y)(x + 5y)$　　**29.** $(a - 8b)(a + 4b)$　　**31.** $\{-7, -3\}$　　**33.** $\{3, 6\}$　　**35.** $\{-2, 5\}$
37. $\{-9, 4\}$　　**39.** $\{-4, 10\}$　　**41.** $\{-8, 7\}$　　**43.** $\{2, 14\}$　　**45.** $\{-12, 1\}$　　**47.** $\{2, 8\}$
49. $\{-6, 4\}$　　**51.** 7 and 8 or -7 and -8　　**53.** 12 and 14
55. $-4, -3, -2,$ and -1 or 7, 8, 9, and 10　　**57.** 4 and 7 or 0 and 3
59. The length is 9 inches and the width is 6 inches.　　**61.** 9 centimeters by 6 centimeters
63. 7 rows and 12 trees per row　　**65.** 8 feet, 15 feet, and 17 feet　　**67.** 6 inches and 8 inches
69. $(x^a + 6)(x^a + 4)$　　**71.** $(x^a - 4)(x^a + 2)$　　**73.** (a) $\{-18, -12\}$　　(c) $\{16, 24\}$　　(e) $\{-24, 18\}$

■ Problem Set 6.4 (page 227)

1. $(3x + 1)(x + 2)$　　**3.** $(2x + 5)(3x + 2)$　　**5.** $(4x - 1)(x - 6)$　　**7.** $(4x - 5)(3x - 4)$
9. $(5y + 2)(y - 7)$　　**11.** $(2n - 3)(n + 8)$　　**13.** Not factorable　　**15.** $(3x + 7)(6x + 1)$
17. $(7x - 2)(x - 4)$　　**19.** $(4x + 7)(2x - 3)$　　**21.** $(3t + 2)(3t - 7)$　　**23.** $(12y - 5)(y + 7)$
25. Not factorable　　**27.** $(7x + 3)(2x + 7)$　　**29.** $(4x - 3)(5x - 4)$　　**31.** $(4n + 3)(4n - 5)$
33. $(6x - 5)(4x - 5)$　　**35.** $(2x + 9)(x + 8)$　　**37.** $(3a + 1)(7a - 2)$　　**39.** $(12a + 5)(a - 3)$
41. $(2x + 3)(2x + 3)$　　**43.** $(2x - y)(3x - y)$　　**45.** $(4x + 3y)(5x - 2y)$　　**47.** $(5x - 2)(x - 6)$

49. $(8x + 1)(x - 7)$　　**51.** $\left\{-6, -\dfrac{1}{2}\right\}$　　**53.** $\left\{-\dfrac{2}{3}, -\dfrac{1}{4}\right\}$　　**55.** $\left\{\dfrac{1}{3}, 8\right\}$　　**57.** $\left\{\dfrac{2}{5}, \dfrac{7}{3}\right\}$

59. $\left\{-7, \dfrac{5}{6}\right\}$　　**61.** $\left\{-\dfrac{3}{8}, \dfrac{3}{2}\right\}$　　**63.** $\left\{-\dfrac{2}{3}, \dfrac{4}{3}\right\}$　　**65.** $\left\{\dfrac{2}{5}, \dfrac{5}{2}\right\}$　　**67.** $\left\{-\dfrac{5}{2}, -\dfrac{2}{3}\right\}$　　**69.** $\left\{-\dfrac{5}{4}, \dfrac{1}{4}\right\}$

71. $\left\{-\dfrac{3}{7}, \dfrac{7}{5}\right\}$　　**73.** $\left\{\dfrac{5}{4}, 10\right\}$　　**75.** $\left\{-7, \dfrac{3}{7}\right\}$　　**77.** $\left\{-\dfrac{5}{12}, 4\right\}$　　**79.** $\left\{-\dfrac{7}{2}, \dfrac{4}{9}\right\}$
81. (a) $(4x + 5)(5x + 4)$　　(c) $(6x - 5)(5x + 8)$

■ Problem Set 6.5 (page 235)

1. $(x + 2)^2$　　**3.** $(x - 5)^2$　　**5.** $(3n + 2)^2$　　**7.** $(4a - 1)^2$　　**9.** $(2 + 9x)^2$　　**11.** $(4x - 3y)^2$
13. $(2x + 1)(x + 8)$　　**15.** $2x(x - 6)(x + 6)$　　**17.** $(n - 12)(n + 5)$　　**19.** Not factorable
21. $8(x^2 + 9)$　　**23.** $(3x + 5)^2$　　**25.** $5(x + 2)(3x + 7)$　　**27.** $(4x - 3)(6x + 5)$　　**29.** $(x + 5)(y - 8)$
31. $(5x - y)(4x + 7y)$　　**33.** $3(2x - 3)(4x + 9)$　　**35.** $6(2x^2 + x + 5)$　　**37.** $5(x - 2)(x + 2)(x^2 + 4)$

39. $(x + 6y)^2$　　**41.** $\{0, 5\}$　　**43.** $\{-3, 12\}$　　**45.** $\{-2, 0, 2\}$　　**47.** $\left\{-\dfrac{2}{3}, \dfrac{11}{2}\right\}$　　**49.** $\left\{\dfrac{1}{3}, \dfrac{3}{4}\right\}$

51. $\{-3, -1\}$　　**53.** $\{0, 6\}$　　**55.** $\{-4, 0, 6\}$　　**57.** $\left\{\dfrac{2}{5}, \dfrac{6}{5}\right\}$　　**59.** $\{12, 16\}$　　**61.** $\left\{-\dfrac{10}{3}, 1\right\}$

63. $\{0, 6\}$　　**65.** $\left\{\dfrac{4}{3}\right\}$　　**67.** $\{-5, 0\}$　　**69.** $\left\{-\dfrac{4}{3}, \dfrac{5}{8}\right\}$　　**71.** $\dfrac{5}{4}$ and 12 or -5 and -3

73. -1 and 1 or $-\dfrac{1}{2}$ and 2　　**75.** 4 and 9 or $-\dfrac{24}{5}$ and $-\dfrac{43}{5}$　　**77.** 6 rows and 9 chairs per row

79. One square is 6 feet by 6 feet and the other one is 18 feet by 18 feet. **81.** 11 centimeters long and 5 centimeters wide **83.** The side is 17 inches long and the altitude to that side is 6 inches long.

85. $1\frac{1}{2}$ inches **87.** 6 inches and 12 inches

■ Chapter 6 Review Problem Set (page 238)

1. $(x - 2)(x - 7)$ **2.** $3x(x + 7)$ **3.** $(3x + 2)(3x - 2)$ **4.** $(2x - 1)(2x + 5)$ **5.** $(5x - 6)^2$
6. $n(n + 5)(n + 8)$ **7.** $(y + 12)(y - 1)$ **8.** $3xy(y + 2x)$ **9.** $(x + 1)(x - 1)(x^2 + 1)$
10. $(6n + 5)(3n - 1)$ **11.** Not factorable **12.** $(4x - 7)(x + 1)$ **13.** $3(n + 6)(n - 5)$
14. $x(x + y)(x - y)$ **15.** $(2x - y)(x + 2y)$ **16.** $2(n - 4)(2n + 5)$ **17.** $(x + y)(5 + a)$
18. $(7t - 4)(3t + 1)$ **19.** $2x(x + 1)(x - 1)$ **20.** $3x(x + 6)(x - 6)$ **21.** $(4x + 5)^2$
22. $(y - 3)(x - 2)$ **23.** $(5x + y)(3x - 2y)$ **24.** $n^2(2n - 1)(3n - 1)$ **25.** $\{-6, 2\}$ **26.** $\{0, 11\}$
27. $\left\{-4, \frac{5}{2}\right\}$ **28.** $\left\{-\frac{8}{3}, \frac{1}{3}\right\}$ **29.** $\{-2, 2\}$ **30.** $\left\{-\frac{5}{4}\right\}$ **31.** $\{-1, 0, 1\}$ **32.** $\left\{-\frac{2}{7}, -\frac{9}{4}\right\}$
33. $\{-7, 4\}$ **34.** $\{-5, 5\}$ **35.** $\left\{-6, \frac{3}{5}\right\}$ **36.** $\left\{-\frac{7}{2}, 1\right\}$ **37.** $\{-2, 0, 2\}$ **38.** $\{8, 12\}$
39. $\left\{-5, \frac{3}{4}\right\}$ **40.** $\{-2, 3\}$ **41.** $\left\{-\frac{7}{3}, \frac{5}{2}\right\}$ **42.** $\{-9, 6\}$ **43.** $\left\{-5, \frac{3}{2}\right\}$ **44.** $\left\{\frac{4}{3}, \frac{5}{2}\right\}$
45. $-\frac{8}{3}$ and $-\frac{19}{3}$ or 4 and 7 **46.** The length is 8 centimeters and the width is 2 centimeters.
47. A 2-by-2-inch square and a 10-by-10-inch square **48.** 8 by 15 by 17
49. $-\frac{13}{6}$ and -12 or 2 and 13 **50.** 7, 9, and 11 **51.** 4 shelves **52.** A 5-by-5-yard square
and a 5-by-40-yard rectangle **53.** -18 and -17 or 17 and 18 **54.** 6 units
55. 2 meters and 7 meters **56.** -1 and 1 or 9 and 11 **57.** 2 centimeters **58.** 15 feet

Chapter 7

■ Problem Set 7.1 (page 245)

1. $\frac{3x}{7y}$ **3.** $\frac{3y}{8}$ **5.** $-\frac{3xy}{5}$ **7.** $\frac{3y}{4x^2}$ **9.** $-\frac{2b^2}{9}$ **11.** $\frac{4xy}{9z}$ **13.** $\frac{y}{x - 2}$ **15.** $\frac{2x + 3y}{3}$
17. $\frac{x + 2}{x - 7}$ **19.** -1 **21.** -3 **23.** $-4x$ **25.** $\frac{x + 1}{3x}$ **27.** $\frac{x + y}{x}$ **29.** $\frac{x(2x - 5y)}{2(x + 4y)}$
31. $\frac{n}{n + 1}$ **33.** $\frac{2n - 1}{n - 3}$ **35.** $\frac{2x + 7}{3x + 4}$ **37.** $\frac{3}{4(x - 1)}$ **39.** $\frac{x + 9}{x + 3}$ **41.** $\frac{2a - 1}{3a - 1}$
43. $\frac{x + 3y}{2x + y}$ **45.** $-\frac{x - 3}{x}$ **47.** $\frac{n + 7}{8}$ **49.** $\frac{2n - 3}{n + 1}$ **51.** $\frac{y - 12}{y - 14}$ **53.** $\frac{1 + x}{x}$
55. $\frac{2 + x}{4 + 5x}$ **57.** $-\frac{x + 9}{x + 6}$ **59.** $-\frac{1}{2}$ **61.** $\frac{y - 3}{y + 5}$ **63.** $\frac{x + 1}{x + 5}$ **65.** $\frac{1}{x^6}$ **67.** $\frac{1}{x^3 y^2}$
69. $-\frac{4}{a^3}$

■ Problem Set 7.2 (page 249)

1. $\dfrac{1}{6}$ **3.** $-\dfrac{9}{14}$ **5.** $-\dfrac{17}{19}$ **7.** $\dfrac{2x^2}{7y}$ **9.** $-\dfrac{3n^2}{10}$ **11.** $2a$ **13.** $\dfrac{10b^3}{27}$ **15.** $\dfrac{3x^2y}{2}$

17. $-\dfrac{4}{5b^3}$ **19.** $\dfrac{s}{17}$ **21.** $\dfrac{x-y}{x}$ **23.** $\dfrac{1}{5}$ **25.** $\dfrac{3a}{14}$ **27.** $\dfrac{5(x+6)}{x+9}$ **29.** $\dfrac{3y(2x-y)}{2(x+y)}$

31. $\dfrac{a}{(5a+2)(3a+1)}$ **33.** $\dfrac{2x(x+4)}{5y(x+8)}$ **35.** $\dfrac{5}{x+y}$ **37.** $\dfrac{4t+1}{3(4t+3)}$ **39.** 4 **41.** $\dfrac{y^2}{3}$

43. $\dfrac{x-1}{y^2(1-y)(x-y)}$ **45.** $\dfrac{x+6}{x}$

■ Problem Set 7.3 (page 255)

1. $\dfrac{17}{x}$ **3.** $\dfrac{2}{3x}$ **5.** $\dfrac{4}{n}$ **7.** $-\dfrac{1}{x^2}$ **9.** $\dfrac{x+4}{x}$ **11.** $-\dfrac{3}{x-1}$ **13.** 1 **15.** $\dfrac{5t+2}{4}$

17. $\dfrac{3a+8}{3}$ **19.** $\dfrac{5n+4}{4}$ **21.** $-n-1$ **23.** $\dfrac{-3x-5}{7x}$ **25.** $\dfrac{9}{4}$ **27.** 3 **29.** $-\dfrac{1}{7x}$

31. $a-2$ **33.** $\dfrac{3}{x-6}$ **35.** $\dfrac{13x}{8}$ **37.** $-\dfrac{3n}{4}$ **39.** $\dfrac{11y}{12}$ **41.** $\dfrac{47x}{21}$ **43.** $\dfrac{14x}{15}$ **45.** $\dfrac{13n}{24}$

47. $\dfrac{7x-14}{10}$ **49.** $\dfrac{4x}{9}$ **51.** $\dfrac{18n+11}{12}$ **53.** $\dfrac{9n-14}{18}$ **55.** $\dfrac{7x}{24}$ **57.** $\dfrac{-23x-18}{60}$

59. $\dfrac{19}{24x}$ **61.** $\dfrac{1}{18y}$ **63.** $\dfrac{20x-33}{48x^2}$ **65.** $\dfrac{25}{12x}$ **67.** $\dfrac{10x-35}{x(x-5)}$ **69.** $\dfrac{-n+3}{n(n-1)}$

71. $\dfrac{-2n+16}{n(n+4)}$ **73.** $\dfrac{6}{x(2x+1)}$ **75.** $\dfrac{10x+12}{(x+4)(x-3)}$ **77.** $\dfrac{-6x+21}{(x-2)(x+1)}$ **79.** $\dfrac{x+7}{(2x-1)(3x+1)}$

81. $\dfrac{9}{x-1}$ **83.** 1 **85.** $\dfrac{5x+4}{(x+3)(x-3)}$

■ Problem Set 7.4 (page 263)

1. $\dfrac{3x-8}{x(x-4)}$ **3.** $\dfrac{-5x-3}{x(x+2)}$ **5.** $\dfrac{8n-50}{n(n-6)}$ **7.** $-\dfrac{4}{n+1}$ **9.** $\dfrac{5x-7}{2x(x-1)}$ **11.** $\dfrac{5x-17}{(x+4)(x-4)}$

13. $\dfrac{4}{x+1}$ **15.** $\dfrac{11a-6}{a(a-2)(a+2)}$ **17.** $\dfrac{12}{x(x-6)(x+6)}$ **19.** $\dfrac{n+8}{3(n+4)(n-4)}$ **21.** $\dfrac{19x}{6(3x+2)}$

23. $\dfrac{-2x+17}{15(x+1)}$ **25.** $\dfrac{5x+6}{(x+3)(x+4)(x-3)}$ **27.** $\dfrac{x^2-10x-20}{(x+2)(x+4)(x-5)}$ **29.** $\dfrac{a-b}{ab}$

31. $\dfrac{8x-14}{(x-5)(x+5)}$ **33.** $\dfrac{15x+4}{x(x-2)(x+2)}$ **35.** $\dfrac{4x-4}{(x+5)(x+2)}$ **37.** $\dfrac{-2x+6}{(3x-5)(x+4)}$ **39.** $\dfrac{3}{x+4}$

41. $-\dfrac{1}{2}$ **43.** $\dfrac{10}{3}$ **45.** $\dfrac{28}{27}$ **47.** $\dfrac{y}{3x}$ **49.** $\dfrac{2y+3x}{5y-x}$ **51.** $\dfrac{x^2-4y}{7xy-3x^2}$ **53.** $\dfrac{6+2x}{3+4x}$

55. $\dfrac{9-24x}{10x+42}$ **57.** $\dfrac{x^2+2x}{6x+4}$ **59.** $\dfrac{-2x+3}{4x-1}$ **61.** $\dfrac{m}{40}$ **63.** $\dfrac{k}{r}$ **65.** $\dfrac{d}{l}$ **67.** $\dfrac{34}{n}$

69. $\dfrac{47}{l}$ **71.** $\dfrac{96}{b}$ **73.** $\dfrac{-3n^2+2n+8}{n+4}$ **75.** $\dfrac{5x^2+6x+2}{3x+1}$

■ Problem Set 7.5 (page 270)

1. $\{12\}$ **3.** $\left\{-\dfrac{2}{21}\right\}$ **5.** $\{4\}$ **7.** $\{-1\}$ **9.** $\{3\}$ **11.** $\{-38\}$ **13.** $\left\{-\dfrac{13}{8}\right\}$

15. $\{2\}$ **17.** $\{6\}$ **19.** $\left\{\dfrac{5}{18}\right\}$ **21.** $\left\{-\dfrac{7}{10}\right\}$ **23.** $\left\{-\dfrac{1}{3}\right\}$ **25.** $\{8\}$ **27.** $\{37\}$

29. $\varnothing$ **31.** $\{39\}$ **33.** $\left\{-\dfrac{5}{4}\right\}$ **35.** $\{-6\}$ **37.** $\left\{-\dfrac{2}{3}\right\}$ **39.** $\left\{\dfrac{12}{7}\right\}$ **41.** $\dfrac{40}{48}$

43. 10 **45.** 7 and 58 **47.** $\dfrac{14}{10}$ **49.** 15 miles per hour

51. 50 miles per hour for Dave and 54 miles per hour for Kent **53.** $\varnothing$ **55.** $\{0\}$

■ Problem Set 7.6 (page 278)

1. $\{-2\}$ **3.** $\{5\}$ **5.** $\left\{\dfrac{9}{2}\right\}$ **7.** $\left\{-\dfrac{49}{10}\right\}$ **9.** $\left\{\dfrac{2}{3}\right\}$ **11.** $\left\{\dfrac{4}{3}\right\}$ **13.** $\left\{\dfrac{3}{2}\right\}$ **15.** $\left\{\dfrac{11}{3}\right\}$

17. $\left\{\dfrac{13}{4}\right\}$ **19.** $\left\{-1, -\dfrac{5}{8}\right\}$ **21.** $\left\{\dfrac{1}{4}, 4\right\}$ **23.** $\left\{-\dfrac{5}{2}, 6\right\}$ **25.** $\{5\}$ **27.** $\{-21\}$

29. $\{2\}$ **31.** $\{-8, 1\}$ **33.** $\dfrac{1}{2}$ or 4 **35.** $-\dfrac{2}{5}$ or $\dfrac{5}{2}$ **37.** 17 miles per hour for Tom and

20 miles per hour for Celia **39.** 16 miles per hour for the trip out and 12 miles per hour for the return trip
41. 30 minutes **43.** 60 minutes for Mike and 120 minutes for Barry **45.** 15 hours **47.** 4 minutes
49. 40 words per minute for Paul and 60 words per minute for Amelia **51.** $\varnothing$
53. All reals except -4 and 4

■ Chapter 7 Review Problem Set (page 281)

1. $\dfrac{7x}{9y^2}$ **2.** $\dfrac{x}{x+3}$ **3.** $\dfrac{3n+5}{n+2}$ **4.** $\dfrac{4a+3}{5a-2}$ **5.** $\dfrac{3x}{8}$ **6.** $x(x-3)$ **7.** $\dfrac{n-7}{n^2}$

8. $\dfrac{2a+1}{a+6}$ **9.** $\dfrac{22x-19}{20}$ **10.** $\dfrac{43x-3}{12x^2}$ **11.** $\dfrac{10n-7}{n(n-1)}$ **12.** $\dfrac{-a+8}{(a-4)(a-2)}$ **13.** $\dfrac{5x+9}{4x(x-3)}$

14. $\dfrac{5x-4}{(x+5)(x-5)(x+2)}$ **15.** $\dfrac{6x-37}{(x-7)(x+3)}$ **16.** $\left\{\dfrac{20}{17}\right\}$ **17.** $\left\{-\dfrac{61}{60}\right\}$ **18.** $\{9\}$ **19.** $\left\{\dfrac{28}{3}\right\}$

20. $\left\{\dfrac{3}{4}\right\}$ **21.** $\{1\}$ **22.** $\{-7\}$ **23.** $\left\{\dfrac{1}{7}\right\}$ **24.** $\left\{\dfrac{1}{2}, 2\right\}$ **25.** $\{-5, 10\}$ **26.** $\left\{-\dfrac{1}{5}\right\}$

27. $\{-1\}$ **28.** 7 and 68 **29.** Becky $2\dfrac{2}{3}$ hours, Nancy 8 hours **30.** 1 or 2 **31.** $\dfrac{36}{72}$

32. Todd's rate is 15 miles per hour and Lanette's rate is 22 miles per hour. **33.** 8 miles per hour
34. 60 minutes **35.** Corinne's rate is 56 words per minute and Sue's rate is 50 words per minute.

Chapter 8

■ Problem Set 8.1 (page 291)

1. $y = \dfrac{13-3x}{7}$ **3.** $x = 3y + 9$ **5.** $y = \dfrac{x+14}{5}$ **7.** $x = \dfrac{y-7}{3}$ **9.** $y = \dfrac{2x-5}{3}$

11.

13.

15.

17.

19.

21.

23.

25.

27.

29.

31.

33.

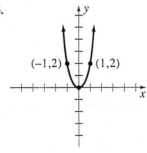

■ Problem Set 8.2 (page 296)

1.

3.

5.

7.

9.

11.

13.

15.

17.

19.

21.

23.

25.

27.

29.

31.

33.

35.

37.

39.

■ Problem Set 8.3 (page 301)

1. $\dfrac{3}{4}$ **3.** $\dfrac{7}{5}$ **5.** $-\dfrac{6}{5}$ **7.** $-\dfrac{10}{3}$ **9.** $\dfrac{3}{4}$ **11.** 0   **13.** $-\dfrac{3}{2}$ **15.** undefined   **17.** 1

19. $\dfrac{b-d}{a-c}$ or $\dfrac{d-b}{c-a}$ **21.** 4   **23.** -6 **25–31.** answers will vary.   **33.** negative

35. positive   **37.** zero   **39.** negative   **41.** $-\dfrac{3}{2}$ **43.** $\dfrac{5}{4}$ **45.** $-\dfrac{1}{5}$ **47.** 2   **49.** 0

51. $\dfrac{2}{5}$ **53.** $\dfrac{6}{5}$ **55.** -3 **57.** 4   **59.** $\dfrac{2}{3}$

■ Problem Set 8.4 (page 308)

1. $2x - 3y = -5$ **3.** $x - 2y = 7$ **5.** $x + 3y = 20$ **7.** $y = -7$ **9.** $4x + 9y = 0$
11. $3x - y = -16$ **13.** $7x - 5y = -1$ **15.** $3x + 2y = 5$ **17.** $x - y = 1$ **19.** $5x - 3y = 0$
21. $4x + 7y = 28$ **23.** $3x - 5y = -10$ **25.** $2x - y = 1$ **27.** $x + 6y = -24$

29. $2x + 2y = 5$ **31.** $10x + 18y = -9$ **33.** $m = -2, b = -5$ **35.** $m = \dfrac{3}{5}, b = -3$

37. $m = \dfrac{4}{9}, b = 2$ **39.** $m = \dfrac{3}{4}, b = -4$ **41.** $m = -\dfrac{2}{11}, b = -1$ **43.** $m = -\dfrac{9}{7}, b = 0$

45. perpendicular **47.** intersecting lines that are not perpendicular **49.** parallel
51. $2x - 3y = -1$

■ Problem Set 8.5 (page 314)

1. No **3.** Yes **5.** Yes **7.** Yes **9.** No **11.** $\{(2, -1)\}$ **13.** $\{(2, 1)\}$ **15.** $\varnothing$
17. $\{(0, 0)\}$ **19.** $\{(1, -1)\}$ **21.** Infinitely many **23.** $\{(1, 3)\}$ **25.** $\{(3, -2)\}$ **27.** $\{(2, 4)\}$
29. $\{(-2, -3)\}$

■ Problem Set 8.6 (page 321)

1. $\{(6, 8)\}$ **3.** $\{(-5, -4)\}$ **5.** $\{(-6, 12)\}$ **7.** $\{(5, -2)\}$ **9.** $\left\{\left(\dfrac{11}{4}, \dfrac{9}{8}\right)\right\}$ **11.** $\left\{\left(-\dfrac{2}{3}, \dfrac{2}{3}\right)\right\}$

13. $\{(-4, 5)\}$ **15.** $\{(4, 1)\}$ **17.** $\left\{\left(\dfrac{3}{2}, -3\right)\right\}$ **19.** $\left\{\left(-\dfrac{18}{71}, \dfrac{5}{71}\right)\right\}$ **21.** $\{(250, 500)\}$

23. $\{(100, 200)\}$ **25.** 9 and 21 **27.** 8 and 15 **29.** 6 and 12 **31.** \$.15 per lemon and \$.30
per apple **33.** 7 dimes and 3 quarters **35.** 14 books at \$12 each and 21 books at \$14 each
37. 4 gallons of 10% and 6 gallons of 15% **39.** \$500 at 10% and \$800 at 12% **41.** $\{(12, 24)\}$
43. $\varnothing$

■ Problem Set 8.7 (page 330)

1. $\{(5, 9)\}$ **3.** $\{-4, 10)\}$ **5.** $\left\{\left(\dfrac{3}{2}, 4\right)\right\}$ **7.** $\left\{\left(-\dfrac{1}{3}, \dfrac{4}{3}\right)\right\}$ **9.** $\{(18, 24)\}$ **11.** $\{(-10, -15)\}$

13. $\{(43, 11)\}$ **15.** $\{(-2. 6)\}$ **17.** $\{(-6, -13)\}$ **19.** $\left\{\left(\dfrac{9}{8}, \dfrac{7}{12}\right)\right\}$ **21.** $\left\{\left(\dfrac{6}{31}, \dfrac{9}{31}\right)\right\}$

23. $\{(100, 400)\}$ **25.** $\{(3, 10\}$ **27.** $\{(-2, -6)\}$ **29.** $\left\{\left(-\dfrac{6}{7}, \dfrac{6}{7}\right)\right\}$ **31.** $\{(10, 12)\}$

33. $\left\{\left(\dfrac{3}{5}, \dfrac{12}{5}\right)\right\}$ **35.** $\left\{\left(-\dfrac{5}{2}, 6\right)\right\}$ **37.** $\left\{\left(\dfrac{1}{2}, 4\right)\right\}$ **39.** $\{(-4, 9)\}$ **41.** $\left\{\left(5, \dfrac{5}{2}\right)\right\}$

43. $\{(12, 4)\}$ **45.** $\left\{\left(-\dfrac{1}{11}, -\dfrac{10}{11}\right)\right\}$ **47.** 12 and 34 **49.** 35 double rooms and 15 single rooms

51. 45 **53.** 18 dimes and 41 quarters **55.** 93 **57.** \$150 at 8% and \$400 at 9%

59. 7.5 liters of 30% and 2.5 liters of 70% **61.** $\left\{\left(-1, -\dfrac{7}{6}\right)\right\}$ **63.** Infinitely many

■ Chapter 8 Review Problem Set (page 334)

1.

2.

3.

4.

5.

6.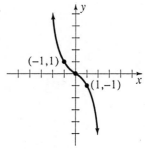

7. $-\dfrac{9}{5}$ **8.** $\dfrac{5}{6}$ **9.** $5x + 7y = -11$ **10.** $8x - 3y = 1$ **11,** $2x - 9y = 9$ **12.** $x = 2$

13. $\{(3, -2)\}$ **14.** $\{(7, 13)\}$ **15.** $\{(16, -5)\}$ **16.** $\left\{\left(\dfrac{41}{23}, \dfrac{19}{23}\right)\right\}$ **17.** $\{(10, 25)\}$

18. $\{(-6, -8)\}$ **19.** $\{(400, 600)\}$ **20.** $\varnothing$ **21.** $\left\{\left(\dfrac{5}{16}, -\dfrac{17}{16}\right)\right\}$ **22.** $t = 4$ and $u = 8$

23. $t = 8$ and $u = 4$ **24.** $t = 3$ and $u = 7$ **25.** $\{(-9, 6)\}$ **26.** 38 and 75   **27.** \$250 at 9% and \$300 at 11%   **28.** 18 nickels and 25 dimes   **29.** length of 19 inches and width of 6 inches   **30.** 49   **31.** 36   **32.** 32° and 58°   **33.** 50° and 130°   **34.** \$1.15 for a cheeseburger and \$0.75 for a milkshake   **35.** \$0.89 for prune juice and \$0.59 for tomato juice

Chapter 9

■ Problem Set 9.1 (page 342)

1. 7   **3.** −8   **5.** 11   **7.** 60   **9.** −40   **11.** 80   **13.** 4.359   **15.** 7.071
17. 8.660   **19.** 9.695   **21.** 14   **23.** 73   **25.** 29   **27.** 86   **29.** 6   **31.** 66
33. 34   **35.** 97   **37.** 81   **39.** 58   **41.** $21\sqrt{2}$ **43.** $8\sqrt{7}$ **45.** $-9\sqrt{3}$ **47.** $6\sqrt{5}$
49. $-\sqrt{2} + 2\sqrt{3}$ **51.** $-9\sqrt{7} - 2\sqrt{10}$ **53.** 17.3   **55.** 13.4   **57.** −1.4   **59.** 26.5
61. 4.1   **63.** 2.1   **65.** −29.9   **67.** $\dfrac{2}{3}$ **69.** .2   **71.** .03

■ Problem Set 9.2 (page 347)

1. $2\sqrt{6}$ **3.** $3\sqrt{2}$ **5.** $3\sqrt{3}$ **7.** $2\sqrt{10}$ **9.** $2\sqrt{7}$ **11.** $4\sqrt{5}$ **13.** $3\sqrt{13}$ **15.** $24\sqrt{2}$

17. $15\sqrt{3}$ **19.** $-10\sqrt{5}$ **21.** $-32\sqrt{6}$ **23.** $3\sqrt{2}$ **25.** $\frac{3}{2}\sqrt{3}$ **27.** $-2\sqrt{5}$

29. $-5\sqrt{6}$ **31.** $xy\sqrt{y}$ **33.** $x\sqrt{2y}$ **35.** $2x\sqrt{2}$ **37.** $3a\sqrt{3ab}$ **39.** $12b^2\sqrt{ab}$

41. $3x^2y\sqrt{7}$ **43.** $12x\sqrt{3}$ **45.** $-36x^3\sqrt{2x}$ **47.** $\frac{2}{3}\sqrt{6xy}$ **49.** $\frac{3}{2}x^2y^3\sqrt{13x}$ **51.** $-\frac{26}{3}a^4$

53. $33\sqrt{2}$ **55.** $3\sqrt{5}$ **57.** $27\sqrt{2}$ **59.** $-2\sqrt{7}$ **61.** $9\sqrt{3}$ **63.** $2\sqrt{5}$

65. $-15\sqrt{2}-55\sqrt{5}$ **67. (a)** 11.3 **(c)** 12.2 **(e)** 11.6

■ Problem Set 9.3 (page 353)

1. $\frac{4}{5}$ **3.** -3 **5.** $\frac{1}{8}$ **7.** $\frac{13}{12}$ **9.** $-\frac{5}{16}$ **11.** $\frac{\sqrt{19}}{5}$ **13.** $\frac{2\sqrt{2}}{7}$ **15.** $\frac{2\sqrt{7}}{3}$

17. $\frac{\sqrt{3}}{3}$ **19.** $\frac{\sqrt{6}}{2}$ **21.** $\frac{\sqrt{10}}{4}$ **23.** $\sqrt{7}$ **25.** 3 **27.** $\frac{\sqrt{10}}{6}$ **29.** $\frac{2\sqrt{3}}{9}$ **31.** $\frac{\sqrt{6}}{12}$

33. $\frac{2\sqrt{15}}{5}$ **35.** $\frac{4\sqrt{6}}{9}$ **37.** $\frac{\sqrt{21}}{8}$ **39.** $\frac{\sqrt{37}}{3}$ **41.** $\frac{3\sqrt{x}}{x}$ **43.** $\frac{5\sqrt{2x}}{2x}$ **45.** $\frac{\sqrt{3x}}{x}$

47. $\frac{2\sqrt{3}}{x}$ **49.** $\frac{\sqrt{10xy}}{5y}$ **51.** $\frac{\sqrt{15xy}}{9y}$ **53.** $\frac{x\sqrt{xy}}{2y}$ **55.** $\frac{3\sqrt{x}}{x^2}$ **57.** $\frac{4\sqrt{x}}{x^4}$ **59.** $\frac{3\sqrt{xy}}{2y^2}$

61. $\frac{22\sqrt{3}}{3}$ **63.** $\frac{19\sqrt{10}}{5}$ **65.** $-3\sqrt{5}$ **67.** $-\frac{14\sqrt{6}}{3}$ **69.** $-6\sqrt{3}$ **71.** $-7\sqrt{15}$

73. (a) .707 **(c)** .707

■ Problem Set 9.4 (page 358)

1. $\sqrt{35}$ **3.** $4\sqrt{3}$ **5.** $5\sqrt{2}$ **7.** 9 **9.** $4\sqrt{6}$ **11.** $15\sqrt{21}$ **13.** $-6\sqrt{14}$ **15.** 72

17. $40\sqrt{6}$ **19.** $24\sqrt{5}$ **21.** $\sqrt{6}+\sqrt{10}$ **23.** $2\sqrt{3}-5\sqrt{6}$ **25.** $3\sqrt{2}+\sqrt{21}$

27. $6\sqrt{2}-4\sqrt{6}$ **29.** $4\sqrt{6}-8\sqrt{15}$ **31.** $56+15\sqrt{2}$ **33.** $-9-2\sqrt{6}$

35. $3\sqrt{2}+2\sqrt{6}+4\sqrt{3}+6$ **37.** 15 **39.** 15 **41.** 51 **43.** $x\sqrt{y}$ **45.** $3\sqrt{2xy}$

47. $12a\sqrt{b}$ **49.** $x\sqrt{6}-2\sqrt{3xy}$ **51.** $x+2\sqrt{x}-15$ **53.** $x-49$ **55.** $\frac{-3\sqrt{2}+12}{14}$

57. $4\sqrt{6}+8$ **59.** $\sqrt{5}-\sqrt{3}$ **61.** $\frac{-20-30\sqrt{3}}{23}$ **63.** $\frac{4\sqrt{x}+8}{x-4}$ **65.** $\frac{x-3\sqrt{x}}{x-9}$

67. $\frac{a+7\sqrt{a}+10}{a-25}$ **69.** $\frac{6+2\sqrt{2}+3\sqrt{3}+\sqrt{6}}{7}$

■ Problem Set 9.5 (page 362)

1. $\{49\}$ **3.** $\{18\}$ **5.** $\varnothing$ **7.** $\left\{\frac{9}{4}\right\}$ **9.** $\left\{\frac{4}{9}\right\}$ **11.** $\{14\}$ **13.** $\varnothing$ **15.** $\left\{\frac{7}{3}\right\}$

17. $\{36\}$ **19.** $\{6\}$ **21.** $\{2\}$ **23.** $\left\{\frac{21}{4}\right\}$ **25.** $\{-3,-2\}$ **27.** $\{6\}$ **29.** $\{2\}$

31. $\{12\}$ **33.** $\{25\}$ **35.** $\{3\}$ **37.** $\left\{-\frac{1}{2}\right\}$ **39.** $\{2\}$ **41.** $\{8\}$ **43.** $\{5\}$

■ Chapter 9 Review Problem Set (page 364)

1. 8 **2.** -7 **3.** 40 **4.** $\dfrac{9}{5}$ **5.** $-\dfrac{2}{3}$ **6.** $\dfrac{7}{6}$ **7.** $2\sqrt{5}$ **8.** $4\sqrt{2}$ **9.** $10\sqrt{2}$

10. $4\sqrt{5}$ **11.** $-6\sqrt{6}$ **12.** $\dfrac{2\sqrt{3}}{7}$ **13.** $\dfrac{6\sqrt{7}}{7}$ **14.** $\dfrac{\sqrt{14}}{4}$ **15.** $\dfrac{\sqrt{3}}{3}$ **16.** $\dfrac{3\sqrt{10}}{5}$

17. 2 **18.** $\dfrac{5\sqrt{6}}{6}$ **19.** $\dfrac{-\sqrt{6}}{3}$ **20.** $\dfrac{2\sqrt{2}}{3}$ **21.** 5.196 **22.** 1.155 **23.** 17.320

24. -6.928 **25.** $2ab\sqrt{3b}$ **26.** $5y^2\sqrt{2x}$ **27.** $3x^3\sqrt{2x}$ **28.** $-8t^2\sqrt{2t}$ **29.** $4y\sqrt{3x}$

30. $\dfrac{3xy^2\sqrt{6xy}}{2}$ **31.** $\dfrac{\sqrt{10xy}}{5y}$ **32.** $\dfrac{3\sqrt{2xy}}{2y}$ **33.** $\dfrac{2\sqrt{x}}{x}$ **34.** $\dfrac{x\sqrt{2x}}{3}$ **35.** $\dfrac{3\sqrt{xy}}{4y^2}$

36. $\dfrac{-2\sqrt{x}}{5}$ **37.** $6\sqrt{2}$ **38.** $18\sqrt{2}$ **39.** -40 **40.** $24\sqrt{21}$ **41.** $4 + 2\sqrt{10}$

42. $6\sqrt{10} - 12\sqrt{15}$ **43.** $3 + \sqrt{21} + \sqrt{15} + \sqrt{35}$ **44.** $-24 - 7\sqrt{6}$ **45.** $4 + 5\sqrt{42}$

46. $-18 - \sqrt{5}$ **47.** $\dfrac{5(\sqrt{7} + \sqrt{5})}{2}$ **48.** $3\sqrt{2} + 2\sqrt{3}$ **49.** $\dfrac{3\sqrt{2} + \sqrt{6}}{6}$ **50.** $\dfrac{3\sqrt{42} - 4\sqrt{15}}{23}$

51. $18\sqrt{2}$ **52.** $-7\sqrt{2x}$ **53.** $-12\sqrt{3}$ **54.** $\dfrac{16\sqrt{10}}{5}$ **55.** $\dfrac{52\sqrt{5}}{5}$ **56.** $\dfrac{-17\sqrt{6}}{3}$

57. $\{6\}$ **58.** $\{4\}$ **59.** $\{0, 9\}$ **60.** $\{-5, -4\}$ **61.** $\{3\}$ **62.** $\{1\}$

Chapter 10

■ Problem Set 10.1 (page 372)

1. $\{-15, 0\}$ **3.** $\{0, 12\}$ **5.** $\{0, 5\}$ **7.** $\{1, 8\}$ **9.** $\{-2, 7\}$ **11.** $\{-6, 1\}$

13. $\left\{-\dfrac{5}{3}, \dfrac{1}{2}\right\}$ **15.** $\left\{\dfrac{2}{5}, \dfrac{5}{6}\right\}$ **17.** $\left\{\dfrac{1}{2}\right\}$ **19.** $\{-8, 8\}$ **21.** $\left\{-\dfrac{5}{3}, \dfrac{5}{3}\right\}$ **23.** $\{-4, 4\}$

25. $\{-\sqrt{14}, \sqrt{14}\}$ **27.** $\varnothing$ **29.** $\{-4\sqrt{2}, 4\sqrt{2}\}$ **31.** $\{-3\sqrt{2}, 3\sqrt{2}\}$ **33.** $\left\{-\dfrac{3\sqrt{2}}{2}, \dfrac{3\sqrt{2}}{2}\right\}$

35. $\left\{-\dfrac{5\sqrt{2}}{4}, \dfrac{5\sqrt{2}}{4}\right\}$ **37.** $\{-1, 3\}$ **39.** $\{-8, 2\}$ **41.** $\left\{-\dfrac{5}{3}, 3\right\}$ **43.** $\{-6 - \sqrt{5}, -6 + \sqrt{5}\}$

45. $\{1 - 2\sqrt{2}, 1 + 2\sqrt{2}\}$ **47.** $\left\{\dfrac{-3 - 2\sqrt{5}}{2}, \dfrac{-3 + 2\sqrt{5}}{2}\right\}$ **49.** $\varnothing$ **51.** $\left\{\dfrac{5 - 2\sqrt{10}}{3}, \dfrac{5 + 2\sqrt{10}}{3}\right\}$

53. $\left\{-\dfrac{3}{7}, \dfrac{5}{7}\right\}$ **55.** $\{-8 - 5\sqrt{2}, -8 + 5\sqrt{2}\}$ **57.** $5\sqrt{2}$ inches **59.** $2\sqrt{7}$ meters

61. $2\sqrt{11}$ feet **63.** $a = 4$ inches and $b = 4\sqrt{3}$ inches **65.** $c = 12$ feet and $b = 6\sqrt{3}$ feet

67. $a = 4\sqrt{3}$ meters and $c = 8\sqrt{3}$ meters **69.** $a = 10$ inches and $c = 10\sqrt{2}$ inches

71. $a = b = \dfrac{9\sqrt{2}}{2}$ meters **73. (a)** negative **(c)** positive **(e)** positive **(g)** negative

■ Problem Set 10.2 (page 378)

1. $\{-4 - \sqrt{17}, -4 + \sqrt{17}\}$ **3.** $\{-5 - \sqrt{23}, -5 + \sqrt{23}\}$ **5.** $\{2 - 2\sqrt{2}, 2 + 2\sqrt{2}\}$ **7.** $\varnothing$

9. $\{-1 - 3\sqrt{2}, -1 + 3\sqrt{2}\}$ **11.** $\left\{\dfrac{-1 - \sqrt{13}}{2}, \dfrac{-1 + \sqrt{13}}{2}\right\}$ **13.** $\left\{\dfrac{5 - \sqrt{33}}{2}, \dfrac{5 + \sqrt{33}}{2}\right\}$

15. $\left\{\dfrac{-4 - \sqrt{22}}{2}, \dfrac{-4 + \sqrt{22}}{2}\right\}$ **17.** $\left\{\dfrac{-6 - \sqrt{42}}{3}, \dfrac{-6 + \sqrt{42}}{3}\right\}$ **19.** $\left\{\dfrac{2 - \sqrt{2}}{2}, \dfrac{2 + \sqrt{2}}{2}\right\}$ **21.** $\varnothing$

23. $\left\{\dfrac{9 - \sqrt{65}}{2}, \dfrac{9 + \sqrt{65}}{2}\right\}$ **25.** $\left\{\dfrac{-3 - \sqrt{17}}{4}, \dfrac{-3 + \sqrt{17}}{4}\right\}$ **27.** $\left\{\dfrac{-1 - \sqrt{7}}{3}, \dfrac{-1 + \sqrt{7}}{3}\right\}$

29. $\{-14, 12\}$ **31.** $\{-11, 15\}$ **33.** $\{-6, 2\}$ **35.** $\{-9, -3\}$ **37.** $\{-5, 8\}$ **39.** $\left\{\dfrac{1}{2}, 4\right\}$

41. $\left\{-\dfrac{5}{2}, \dfrac{3}{2}\right\}$ **43.** $\left\{\dfrac{-b + \sqrt{b^2 - 4ac}}{2a}, \dfrac{-b - \sqrt{b^2 - 4ac}}{2a}\right\}$

■ Problem Set 10.3 (page 383)

1. $\{-1, 6\}$ **3.** $\{-9, 4\}$ **5.** $\{1 - \sqrt{6}, 1 + \sqrt{6}\}$ **7.** $\left\{\dfrac{5 - \sqrt{33}}{2}, \dfrac{5 + \sqrt{33}}{2}\right\}$ **9.** $\varnothing$

11. $\{-2 - \sqrt{2}, -2 + \sqrt{2}\}$ **13.** $\{0, 6\}$ **15.** $\left\{0, \dfrac{7}{2}\right\}$ **17.** $\{16, 18\}$ **19.** $\{-10, 8\}$

21. $\{-2\}$ **23.** $\left\{-\dfrac{2}{3}, \dfrac{1}{2}\right\}$ **25.** $\left\{-1, \dfrac{2}{5}\right\}$ **27.** $\left\{-\dfrac{5}{4}, -\dfrac{1}{3}\right\}$ **29.** $\left\{\dfrac{-5 - \sqrt{73}}{4}, \dfrac{-5 + \sqrt{73}}{4}\right\}$

31. $\left\{\dfrac{-2 - \sqrt{7}}{3}, \dfrac{-2 + \sqrt{7}}{3}\right\}$ **33.** $\left\{-\dfrac{3}{4}\right\}$ **35.** $\left\{\dfrac{-2 - \sqrt{5}}{2}, \dfrac{-2 + \sqrt{5}}{2}\right\}$

37. $\left\{\dfrac{-9 - \sqrt{57}}{12}, \dfrac{-9 + \sqrt{57}}{12}\right\}$ **39.** $\left\{\dfrac{1 - \sqrt{33}}{4}, \dfrac{1 + \sqrt{33}}{4}\right\}$ **41.** $\varnothing$

43. $\left\{\dfrac{-5 - \sqrt{137}}{14}, \dfrac{-5 + \sqrt{137}}{14}\right\}$ **45.** $\left\{\dfrac{1 - \sqrt{85}}{6}, \dfrac{1 + \sqrt{85}}{6}\right\}$ **47.** $\{-14, -9\}$

■ Problem Set 10.4 (page 387)

1. $\{-9, 5\}$ **3.** $\left\{-\dfrac{13}{5}, \dfrac{1}{5}\right\}$ **5.** $\{-1, 2\}$ **7.** $\left\{0, \dfrac{8}{3}\right\}$ **9.** $\left\{\dfrac{1}{3}\right\}$ **11.** $\left\{0, \dfrac{2\sqrt{2}}{5}\right\}$

13. $\{7 - 2\sqrt{17}, 7 + 2\sqrt{17}\}$ **15.** $\left\{-1, \dfrac{7}{5}\right\}$ **17.** $\left\{-\dfrac{5}{3}, -\dfrac{1}{5}\right\}$ **19.** $\{\sqrt{2} - 3, \sqrt{2} + 3\}$

21. $\{-12, 7\}$ **23.** $\left\{\dfrac{3 - \sqrt{33}}{4}, \dfrac{3 + \sqrt{33}}{4}\right\}$ **25.** $\{-1, 4\}$ **27.** $\varnothing$ **29.** $\{16, 30\}$

31. $\left\{\dfrac{-1 - \sqrt{13}}{2}, \dfrac{-1 + \sqrt{13}}{2}\right\}$ **33.** $\left\{\dfrac{3}{4}, \dfrac{4}{3}\right\}$ **35.** $\{-13, 1\}$ **37.** $\{11, 17\}$

39. $\left\{\dfrac{3 - \sqrt{3}}{2}, \dfrac{3 + \sqrt{3}}{2}\right\}$ **41.** $\left\{-1, -\dfrac{2}{3}\right\}$ **43.** $\left\{\dfrac{-5 - \sqrt{193}}{6}, \dfrac{-5 + \sqrt{193}}{6}\right\}$ **45.** $\{-5, 3\}$

■ Problem Set 10.5 (page 393)

1. 17 and 18 **3.** 19 and 25 **5.** $3 + \sqrt{5}$ and $3 - \sqrt{5}$ **7.** $\sqrt{2}$ or $\dfrac{\sqrt{2}}{2}$ **9.** 12, 14, and 16

11. -8 or 8 **13.** 8 meters by 12 meters **15.** 15 centimeters by 25 centimeters
17. 27 feet by 78 feet **19.** 15 rows and 20 seats per row **21.** 7 feet by 9 feet

23. 6 inches and 8 inches   **25.** $1\frac{1}{2}$ inches   **27.** 30 students   **29.** 24 inches and 32 inches

31. 18 miles per hour

■ Chapter 10 Review Problem Set (page 395)

1. $\{-6, -1\}$ **2.** $\{-4 - \sqrt{13}, -4 + \sqrt{13}\}$ **3.** $\left\{\frac{2}{7}, \frac{1}{3}\right\}$ **4.** $\{0, 17\}$ **5.** $\{-4, 1\}$

6. $\{11, 15\}$ **7.** $\left\{\frac{-7 - \sqrt{61}}{6}, \frac{-7 + \sqrt{61}}{6}\right\}$ **8.** $\left\{\frac{1}{2}\right\}$ **9.** $\varnothing$ **10.** $\{-5, -1\}$

11. $\{2 - \sqrt{2}, 2 + \sqrt{2}\}$ **12.** $\{-2 - 3\sqrt{2}, -2 + 3\sqrt{2}\}$ **13.** $\{-3\sqrt{5}, 3\sqrt{5}\}$ **14.** $\{-3, 9\}$

15. $\{0, 1\}$ **16.** $\{2 - \sqrt{13}, 2 + \sqrt{13}\}$ **17.** $\{20, 24\}$ **18.** $\{2 - 2\sqrt{2}, 2 + 2\sqrt{2}\}$ **19.** $\left\{-\frac{8}{5}, 1\right\}$

20. $\left\{\frac{-1 - \sqrt{73}}{12}, \frac{-1 + \sqrt{73}}{12}\right\}$ **21.** $\left\{\frac{1}{2}, 4\right\}$ **22.** $\left\{-\frac{4}{3}, -1\right\}$ **23.** 9 inches by 12 inches

24. 18 and 19   **25.** 7, 9, and 11   **26.** $\sqrt{5}$ meters and $3\sqrt{5}$ meters   **27.** 4 yards and 6 yards

28. 40 shares at $20 per share   **29.** 10 meters   **30.** Jay's rate was 45 miles per hour and Jean's rate was

48 miles per hour; or Jay's rate was $7\frac{1}{2}$ miles per hour and Jean's rate was $10\frac{1}{2}$ miles per hour.

31. $6\sqrt{2}$ inches   **32.** $\frac{8\sqrt{3}}{3}$ centimeters

Chapter 11

■ Problem Set 11.1 (page 403)

1. $8i$ **3.** $\frac{5}{3}i$ **5.** $i\sqrt{11}$ **7.** $5i\sqrt{2}$ **9.** $4i\sqrt{3}$ **11.** $3i\sqrt{6}$ **13.** $8 + 17i$

15. $10 - 10i$ **17.** $4 + 2i$ **19.** $-2 - 6i$ **21.** $-5 + 3i$ **23.** $-12 - 16i$ **25.** $-10 - 4i$

27. $-1 + 12i$ **29.** $-20 - 8i$ **31.** $-56 + 0i$ **33.** $-6 + 12i$ **35.** $-24 + 20i$

37. $-2 + 23i$ **39.** $59 - 17i$ **41.** $-21 - 12i$ **43.** $-26 + 15i$ **45.** $-9 + 40i$

47. $61 + 0i$ **49.** $5 + 0i$

■ Problem Set 11.2 (page 407)

1. $\{-8i, 8i\}$ **3.** $\{2 - i, 2 + i\}$ **5.** $\{-5 - i\sqrt{13}, -5 + i\sqrt{13}\}$ **7.** $\{3 - 3i\sqrt{2}, 3 + 3i\sqrt{2}\}$

9. $\left\{-\frac{2}{5}, \frac{4}{5}\right\}$ **11.** $\{-1, 4\}$ **13.** $\{-3 - i\sqrt{3}, -3 + i\sqrt{3}\}$ **15.** $\{3 - 2i, 3 + 2i\}$

17. $\{2 - 4i, 2 + 4i\}$ **19.** $\left\{\frac{1 - i\sqrt{2}}{3}, \frac{1 + i\sqrt{2}}{3}\right\}$ **21.** $\left\{-1, \frac{5}{2}\right\}$ **23.** $\{1 - 3i\sqrt{2}, 1 + 3i\sqrt{2}\}$

25. $\{2 - i\sqrt{3}, 2 + i\sqrt{3}\}$ **27.** $\left\{\frac{1 - i\sqrt{31}}{8}, \frac{1 + i\sqrt{31}}{8}\right\}$ **29.** $\left\{\frac{-1 - i\sqrt{5}}{6}, \frac{-1 + i\sqrt{5}}{6}\right\}$

■ Problem Set 11.3 (page 412)

1. 9   **3.** -10 **5.** 5   **7.** -4 **9.** $\dfrac{4}{7}$ **11.** 3   **13.** -3 **15.** 8   **17.** 16

19. 16   **21.** 32   **23.** $\dfrac{1}{2}$ **25.** -3 **27.** $\dfrac{1}{8}$ **29.** $\dfrac{27}{8}$ **31.** 2   **33.** 625   **35.** -32

37. $\dfrac{1}{8}$ **39.** 2   **41.** 27   **43.** 1   **45.** $\dfrac{1}{3}$ **47.** 4   **49.** 49   **51.** $x^{\frac{3}{4}}$ **53.** $a^{\frac{17}{12}}$

55. $15x^{\frac{7}{12}}$ **57.** $24x^{\frac{11}{12}}$ **59.** $2y^{\frac{5}{12}}$ **61.** $10n^{\frac{1}{4}}$ **63.** $\dfrac{2}{x^{\frac{1}{6}}}$ **65.** $25xy^2$ **67.** $64x^{\frac{3}{4}}y^{\frac{3}{2}}$

69. $2x^2y$ **71.** $4x^{\frac{4}{15}}$ **73.** $\dfrac{4}{b^{\frac{5}{12}}}$ **75.** 3   **77.** $\dfrac{9}{4x^{\frac{1}{3}}}$ **79.** $\dfrac{125x^{\frac{3}{2}}}{216y}$

■ Problem Set 11.4 (page 416)

1. Domain: all reals   **3.** Domain: all reals   **5.** Domain: all reals
Range: all reals   Range: all reals greater than or equal to 2   Range: all reals
7. Domain: nonnegative reals   **9.** all reals   **11.** all reals greater than or equal to -4
Range: nonnegative reals
13. all reals   **15.** all reals except 3   **17.** all reals except 2 and -2 **19.** all reals
21. all reals except -2 and 3   **23.** all reals except 2 and 3   **25.** all reals except -4 and 0

27. 4; 7; 1; 22   **29.** -16; 19; 24; $-5t - 1$ **31.** $\dfrac{11}{4}; \dfrac{13}{12}; \dfrac{19}{36}; -\dfrac{7}{12}$ **33.** 0; 0; 45; -4

35. 0; -3; -3; -8 **37.** 2; 5; $2\sqrt{3}$; $3\sqrt{2}$ **39.** 23; -21; 3; 8   **41.** 8; 48; 8; -7

■ Problem Set 11.5 (page 424)

13.

15.

17.

19.

21.

23.

25.

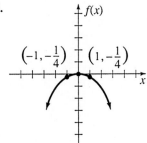

■ **Chapter 11 Review Problem Set (page 426)**

1. $8i$ **2.** $i\sqrt{29}$ **3.** $3i\sqrt{6}$ **4.** $\dfrac{3}{2}i$ **5.** $6i\sqrt{3}$ **6.** $4i\sqrt{6}$ **7.** $1 + 2i$ **8.** $-7 - 5i$

9. $2 - 4i$ **10.** $3 - 4i$ **11.** $-1 + 3i$ **12.** $-6 + 10i$ **13.** $-34 + 31i$ **14.** $-2 - 11i$

15. $-4 - 8i$ **16.** $3 - 45i$ **17.** 85 **18.** 58 **19.** $55 - 48i$ **20.** $3 - 15i$

21. $\{6 - 5i, 6 + 5i\}$ **22.** $\{-1 - i\sqrt{6}, -1 + i\sqrt{6}\}$ **23.** $\{1 - 4i, 1 + 4i\}$

24. $\left\{\dfrac{1 - 3i\sqrt{3}}{2}, \dfrac{1 + 3i\sqrt{3}}{2}\right\}$ **25.** $\left\{\dfrac{1 - i\sqrt{23}}{4}, \dfrac{1 + i\sqrt{23}}{4}\right\}$ **26.** $\left\{\dfrac{1}{3}, \dfrac{3}{2}\right\}$ **27.** $\left\{\dfrac{5 - i\sqrt{3}}{2}, \dfrac{5 + i\sqrt{3}}{2}\right\}$

28. $\left\{\dfrac{-3 - i\sqrt{29}}{4}, \dfrac{-3 + i\sqrt{29}}{4}\right\}$ **29.** $\left\{\dfrac{-1 - i\sqrt{59}}{6}, \dfrac{-1 + i\sqrt{59}}{6}\right\}$ **30.** $\left\{\dfrac{-1 - i\sqrt{47}}{8}, \dfrac{-1 + i\sqrt{47}}{8}\right\}$

31. $\dfrac{4}{3}$ **32.** -1 **33.** $\dfrac{3}{4}$ **34.** -5 **35.** $\dfrac{3}{2}$ **36.** 125 **37.** 32 -   **38.** -32 **39.** $\dfrac{1}{16}$

40. $\dfrac{1}{2}$　　**41.** $\dfrac{1}{4}$　　**42.** $\dfrac{3}{2}$　　**43.** 8　　**44.** 9　　**45.** $\dfrac{1}{3}$　　**46.** $x^{\frac{5}{3}}$　　**47.** $6x^{\frac{17}{20}}$　　**48.** $36a^{\frac{1}{6}}$

49. $27xy^2$　　**50.** $5x^2y^3$　　**51.** $13n^{\frac{7}{20}}$　　**52.** $\dfrac{4}{n^{\frac{1}{4}}}$　　**53.** $8x^3$

54. $f(2) = -3, f(-1) = 3, f(-4) = 27$　　**55.** $f(3) = -37, f(-3) = 5, f(-4) = -2$

56.

57.

58.

59.

60.

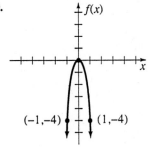

Index